Abraham Lincoln's Cyphering Book and Ten other Extraordinary Cyphering Books

Nerida F. Ellerton • M. A. (Ken) Clements

Abraham Lincoln's Cyphering Book and Ten other Extraordinary Cyphering Books

Foreword by Fred Rickey

To Bill and Faith
Thanks so much for enriching our time in the United States.
Nerida and Ken
October, 2014

Springer

Nerida F. Ellerton • M. A. (Ken) Clements
Department of Mathematics
Illinois State University
Normal, IL, USA

Chapter 6, which deals with Abraham Lincoln's cyphering book, was written by Nerida F. Ellerton, Valeria Aguirre Holguín (New Mexico State University), and M. A. (Ken) Clements.

ISBN 978-3-319-02501-8 ISBN 978-3-319-02502-5 (eBook)
DOI 10.1007/978-3-319-02502-5
Springer Cham Heidelberg New York Dordrecht London

Library of Congress Control Number: 2014930303

Printed on acid-free paper

Springer is part of Springer Science+Business Media (www.springer.com)

Contents

Overall Book Abstract, and Individual Abstracts for the 12 Chapters

Overall Abstract

This book analyzes 11 extraordinary cyphering books prepared by students in Great Britain or in North America between 1630 and 1836. Altogether, there are 150 figures, most of them full-color images of pages, or sections of pages, in the cyphering books. Changes in school mathematics curricula and in the teaching and learning of mathematics in Europe and North America during the period from 1630 to 1880 are discussed in ways consistent with the authors' main thesis, originally presented in *Rewriting the History of School Mathematics in North America 1607–1861* (Springer, 2012), that school mathematics was controlled by expectations associated with the cyphering tradition, especially the business-oriented *abbaco* curriculum. The 11 extraordinary cyphering books which are analyzed were prepared by:

1. An anonymous person in a New England settlement in the 1660s;
2. Thomas Prust and John Collings (England and North America, 1702–1721);
3. Martha and Elisabeth Ryan (North Carolina, 1776–1781);
4. Mary Walters (Maryland, 1820);
5. Abraham Lincoln (Indiana, 1819–1826);
6. Thomas Dixson (England, 1630);
7. George Bickham (London, c. 1740);
8. An anonymous British midshipman (c. 1791);
9. Charles Page (Royal Mathematical School, Christ's Hospital, London, 1825); and
10. William Beattie (who prepared two of the cyphering books—one around 1810 in Scotland, and the second, in 1835 and 1836, possibly as a special gift for his friend, King William IV, of England).

Chapter 6, which was co-authored by Nerida Ellerton, Valeria Aguirre Holguín and Ken Clements, offers the first major analysis of the 22 extant pages of Abraham Lincoln's cyphering book—which is the oldest surviving Lincoln manuscript. The analysis reveals that between 1819 and 1826 the future President, with much determination, prepared his cyphering book. What made his cyphering book extraordinary was the fact that he prepared it at all, given the circumstances which made it difficult for him to study.

In the final chapter, six major lessons for 21st-century school mathematics are drawn from the analyses set out in the preceding 11 chapters. The six lessons relate to relationships between the intended, implemented and attained curricula associated with the cyphering era, to the great importance attached to the cyphering books by those who prepared them, to how ethnomathematical circumstances influenced the intended, implemented and attained curricula, and to differences between the cyphering traditions in Great Britain and North America.

Individual Chapter Abstracts

Chapter 1: Cyphering Books and the Cyphering Tradition in North America and Great Britain, 1630–1880

Abstract: The cyphering tradition controlled school mathematics in Western Europe and North America during the seventeenth and eighteenth centuries and the first half of the nineteenth century. In this book the authors analyze 11 extraordinary cyphering books—7 arithmetic cyphering books, 3 navigation cyphering books, and 1 mensuration/surveying cyphering book. Of these, 6 were prepared entirely in Great Britain, 4 entirely in North America, and 1 partly in Great Britain and partly in North America. After defining key terms—such as "cyphering book," "cyphering tradition," "intended curriculum," "implemented curriculum," and "attained curriculum"—this first chapter summarizes characteristics of the 11 extraordinary cyphering books that will be considered. Finally, six research issues—which will frame subsequent analyses—are stated.

Chapter 2: Primitive Beginnings, Circa 1667

Abstract: Although the name of the person who prepared this seven-page manuscript, and of the place where it was prepared, are unknown, the last owner indicated that the manuscript was purchased from a very long-established family in Maine, New England. Importantly, the paper is clearly watermarked, and the watermark indicates the paper was made in 1666 or 1667. We have assumed that the paper was brought from Europe to New England during the second half of the 1660s, and that the writing in the manuscript was done in New England. If that was indeed the case then this is easily the oldest of known extant North American cyphering books and, in that sense, is extraordinary. A second notable feature of the manuscript is that the mathematics and spelling are at a very low level. Two other interesting features of the manuscript are: (a) the person who prepared it seemed to want to be able to claim that he or she had cyphered to the rule of three because, all of a sudden, that rule is mentioned on the last page; and (b) one of the topics mentioned was "halving," which was *not* a normal part of the *abbaco* curriculum sequence for arithmetic, but *was* part of the Sacrobosco curriculum sequence for arithmetic.

Chapter 3: "Thomas Prust his Booke Amen 1702"

Abstract: The person who prepared most of the entries for the cyphering book studied in this chapter was Thomas Prust, who was born in England in 1690 and migrated to New England in 1720. There is a handwritten note in the cyphering book indicating that in 1720 Thomas passed it on to James Collings. Thomas became a shopkeeper and the training he received when he was preparing his beautiful, largely *abbaco*-inspired, cyphering book would have been beneficial to him during his time as a shopkeeper. He used the galley algorithm when performing division calculations, and was determined to master the rule of three. Many of the problems in the cyphering book were linked to real-life situations. The record indicates that James Collings became a sea captain, and he probably took this cyphering book with him on his voyages. Although we do not know how many later students were guided by this extraordinary manuscript, of one thing we can be certain: this was originally Thomas Prust's "booke," and over 300 years later most of it still remains intact.

Chapter 4: Daughters of the Revolution: Martha and Elisabeth Ryan's Cyphering Book, Circa 1780

Abstract: There are three independently remarkable characteristics of this manuscript, and there is a fourth feature which is also of special interest. The first extraordinary characteristic is the patriotic fervor of the two sisters who prepared it. Illustrations of ships and patriotic inscriptions reflecting the Revolutionary times—such as "Liberty," "George Washington," and "Poor freedom is better than rich slavery"—are to be found on many of the pages. The second extraordinary aspect of this manuscript is that although it was prepared by two teenage sisters around 1780 the arithmetic went well beyond the rule of three to topics like alligation, fellowship, equation of payments, and single and double false position—it was very unusual for females at that time to study such topics. The third, and most obvious, remarkable aspect of the manuscript is the extremely high quality of the penmanship, calligraphy and artwork. Martha and Elisabeth Ryan lived in North Carolina, and this manuscript is an example of an early "southern" cyphering book—of which relatively few of those prepared by girls have survived.

Chapter 5: With a Tinge of Green: Mary Walters' Cyphering Book, 1820

Abstract: In 1820 Mary Walters was based in Baltimore, Maryland. She had emigrated from Ireland to North America and her cyphering book, prepared in 1820, provides evidence that in Maryland some Irish-heritage school teachers were responsible for engaging students in a form of mathematics education that had more than a tinge of green. Certainly, like all who followed the *abbaco* arithmetic curriculum, Mary was expected to learn how to add, subtract, multiply and divide counting numbers, and to apply elementary measurement principles in a range of everyday situations. When dealing with tasks involving the measurement of quantities, she needed to be able to change from one set of units to another. She also needed to learn to apply the direct rule of three in a wide range of real-life scenarios in which ratio and proportional thinking were needed. But the entries in Mary's cyphering book also made it clear that her teacher(s) wanted her to inject herself into problem-based scenarios by which she would learn to apply elementary arithmetic principles to tasks concerned with human-interest, life-and-death situations which often generated emotions like jealousy, love, and desire. In particular, she was expected to appreciate the power of stories, poetry and song to convey interesting problem scenarios.

Chapter 6: He would be Good: Abraham Lincoln's Early Mathematics, 1819–1826

Abstract: This chapter, jointly written by Nerida Ellerton, Valeria Aguirre Holguín, and Ken Clements, offers a comprehensive analysis of the earliest extant handwritten manuscript of Abraham Lincoln—22 pages (11 leaves), from a cyphering book that Lincoln prepared between 1819 and 1826. The pages are examined from historical, mathematical and educational perspectives, and an order in which they were written is conjectured, with justifications provided. This is the first time all 22 pages have been examined, it having become received tradition that there were only 20 surviving pages. The authors argue that if the corpus of all extant U.S. cyphering books were to be studied, then the future President's cyphering book would not be particularly outstanding from calligraphic, penmanship, or abstract mathematical points of view. Analysis reveals, nevertheless, that Lincoln succeeded in his quest to prepare an attractive cyphering book in which almost all entries were arithmetically correct—despite his having to cope with rough frontier circumstances.

Chapter 7: Sacrobosco's Heritage: Thomas Dixson's Cyphering Book, 1630–1632

Abstract: The 302-page handwritten manuscript by Thomas Dixson which is the focus of attention in this chapter is 46 years older than the oldest of the British cyphering books mentioned by John Denniss (2012). The analysis presented here suggests that the curriculum sequence put forward by Johannes de Sacrobosco early in the thirteenth century, for the study of arithmetic, may have been more influential in seventeenth-century England than Leonardo of Pisa's *abbaco* curriculum sequence. Some attention in this chapter is given to the so-called "galley" algorithm for division, which Dixson used. In keeping with the standard Sacrobosco curriculum sequence, Thomas Dixson included, early in his arithmetic studies, sections on arithmetical and geometrical progressions, and on common fractions—those topics would have come much later if the *abbaco* curriculum had been followed.

Chapter 8: Writing as if Arithmeticke: George Bickham's Cyphering Book, Circa 1740

Abstract: Of the 11 extraordinary cyphering books considered in this volume, George Bickham's was the only one that was specifically prepared for sale to others. We do not know how many copies of this book were printed, but we do know that the copy which we now hold is one of very few which remain. Bickham was a writing master who was regarded, in his time, as one of world's best penmen. As was usually the case with writing masters, he paid much attention to the kind of writing and calligraphy needed to prepare a strikingly beautiful cyphering book, and this book provides testimony to that fact. However, there are hardly any model examples, or exercises for the student to do. The emphasis seemed to be on form rather than mathematical substance, and the language of commercial arithmetic is strongly emphasized. As would have been expected, topics are presented according to the normal *abbaco* sequence.

Chapter 9: Bound for Botany Bay? Circa 1791

Abstract: The 240-page navigation cyphering book featured in this chapter includes 80 artistic creations, many of them exquisite water-color miniatures. These miniatures are remarkable in the sense that they depict buildings, towers, hills, streams, etc., on which are labeled heights, widths, distances, angles, etc., relevant to mathematical tasks which are given in the text. Some of the labels are so small that it is difficult to read them with the naked eye. An important part of this chapter relates the beautiful artwork in the manuscript to the 18th- and 19th-century push to include serious drawing education in curricula designed for middle-class students and artisans. Another intriguing aspect of the chapter is that it draws attention to mathematical compromises, by mathematicians, with respect to trigonometric functions and logarithms. The title of the chapter, "Bound for Botany Bay? Circa 1791" was chosen because of the possibility that the diagrams and other artwork in the chapter depict actual scenes that the person who created the navigation cyphering book saw from a ship (or ships) that was (or were) part of the First Fleet from England to Botany Bay (New Holland) in 1787 and 1788.

Chapter 10: From the Royal Mathematical School: Charles Page, 1825

Abstract: Charles Page was born in 1810, and between 1818 and 1825 he was a student at the Royal Mathematical School within Christ's Hospital, London. Over his final two years as a "mathemat" (i.e., student in the Royal Mathematical School) he prepared the 671-page navigation cyphering book which is the subject of analysis in this chapter. Throughout the manuscript the standard of penmanship, calligraphy, artwork, and mathematics is at a very high level—indeed, much of the mathematics in this cyphering book is at a higher level than the mathematics in any of the other extraordinary cyphering books considered in this volume. Careful examination of the entries in Charles Page's manuscript suggests, however, that many of them were taken down from dictation. In the section on quadratic equations there are unchecked errors, and we were left with the impression that although the appearance of pages was invariably impressive, the quality of Charles' learning about mathematics and about navigation may have been less than impressive. Details in this chapter provide a glimpse of the expectations on mathemats in the 1820s.

Chapter 11: Fit for a King? William Beattie, Circa 1810 and 1835–1836

Abstract: William Beattie prepared both of the manuscripts that are analyzed in this chapter. The first has 466 pages and is concerned mainly with mensuration. Many of the notes and examples in that manuscript are from a 1748 London edition of William Hawney's classic text, *The Compleat*

Measurer. The second manuscript has 309 pages and concerns navigation. It is based almost entirely on a 1796 edition of John Hamilton Moore's *The New Practical Navigator*. It is conjectured that although the title page of the measurement manuscript is dated 1835, in fact the manuscript was prepared by Beattie around 1810, when he was a student at Clarencefield Academy in Dumfriesshire, Scotland. Beattie became a medical doctor, and served as personal physician and then private secretary to the man who would become King William IV—and would be remembered as the "Sailor King." It is conjectured that the navigation cyphering book, dated 1836 on its title page, was specially prepared by Beattie for his ailing King (who died in 1837). Although the penmanship and calligraphy in both manuscripts are of a high standard, Beattie seemed to value the mensuration manuscript more highly than the navigation manuscript.

Chapter 12: Lessons from Extraordinary Cyphering Books

Abstract: Six lessons about school mathematics, and its history, are drawn from the analyses of the 11 extraordinary cyphering books. These lessons relate to: (a) the intended curriculum that the *abbaco* and Sacrobosco traditions, and the British and North American textbook authors, defined for school mathematics during the period 1600 through 1880; (b) the implemented curriculum, as evidenced by entries in cyphering books; (c) the lack of hard data with respect to the attained curriculum—that is to say, with respect to what was actually learned by students during the cyphering era; (d) the remarkable value that many students placed upon their cyphering books; (e) the ethnomathematical circumstances which surrounded the preparation of the extraordinary cyphering books; and (f) the educational significance of qualitative differences between British and North American cyphering books.

Preface

In Chapter 4 of our *Rewriting the History of School Mathematics in North America 1607–1861: The Central Importance of the Cyphering Tradition* (Ellerton & Clements, 2012), also published by Springer, we summarized our journey which enabled us to recognize the large influence on school mathematics, over many centuries, of the cyphering tradition. Since writing that book we have continued our investigations into the cyphering tradition, and have now analyzed about 1000 handwritten cyphering books—most of them prepared in the United States or in Great Britain between 1630 and 1880.

The cyphering tradition *controlled* the intended and implemented curricula of pre-college mathematics education in Great Britain and North America between 1600 and 1850. And, although the influence of that tradition declined sharply in the second half of the nineteenth century, it lingers almost ghost-like, affecting school mathematics programs in every country in the world. Much has been written about historical circumstances which influenced presidents, prime ministers, great literary figures, large corporations, and about the rise and fall of civilizations. However, very few books have been written on the history of school mathematics—something which has affected, and continues to affect, vast numbers of people, everywhere.

"But," one might respond, "not everyone is greatly interested in the history of electricity, or of the automobile, or of the internet, and yet electricity, cars, and the internet, are all vitally important in today's society. One simply does not have enough time to take a scholarly interest in everything that is important. So why should one have an interest in the history of school mathematics?" If you are a mathematics education researcher, or education historian, then you might have wanted to add that scholars must be selective in what they choose to research, and that so far as school mathematics is concerned the most important task is to improve *present* practices. To such an argument we would respond by pointing out that most people have an interest in their family roots—they want to know from whom, and from where, they have come. From that perspective, it would appear to be reasonable to expect mathematics teachers and mathematics education researchers, and those concerned with school curricular issues, to have some interest in the roots of today's school mathematics curricula and practices.

This book does more than provide a complement to our *Rewriting the History of School Mathematics in North America 1607–1861*. Here, we extend our analysis to the history of the cyphering tradition in the United Kingdom, covering the period from about 1630 to 1850. As well, the book adds flesh to our earlier analysis of the cyphering tradition in North America. The methodology employed is the historical case study. Eleven cyphering books were chosen as cases to be studied, and each was, in its own peculiar way, "extraordinary." The analyses presented are more qualitative than quantitative.

One of the chapters provides the most detailed analysis of Abraham Lincoln's cyphering book ever presented—that chapter is the first to examine all 22 surviving pages of what is easily the oldest surviving handwritten manuscript prepared by the future president. Another chapter analyzes the hauntingly beautiful composite manuscript of Martha and Elisabeth Ryan, who, around 1780, studied arithmetic to a level well beyond what was expected of girls in North America. Another manuscript was probably prepared somewhere in New England in the late 1660s—if it was, then it is easily the oldest extant North American cyphering book. One chapter analyzes the incredible 671-page navigation cyphering book prepared by Charles Page, a boy attending the Royal Mathematical School at Christ's Hospital, London, in the 1820s. Another chapter presents an analysis of a navigation cyphering book that may have been prepared as a gift for the King of England. Space does not permit each manuscript to be mentioned here, but each is, in its own way, remarkable.

Throughout the book a critical stance has been adopted. Some of the analyses reveal that students and their teachers were more interested in preparing attractive books than in developing an understanding of the mathematics that should have been the main focus of their attention. The final chapter begins with a discussion of whether it makes sense to claim that "lessons for today" can sensibly be identified from history. Then six lessons that we believe emerge from our analyses in this book are, in fact, identified.

Those who read this book will, we trust, gain a deeper knowledge of the history of school mathematics. They should also gain new perspectives on today's problems, and challenges, with respect to school mathematics. Perhaps, too, they will be inspired by the "stories" emerging from our analyses of the extraordinary cyphering books. We were.

This book analyzes cyphering books prepared at various times in either North America or Great Britain between 1630 and 1836. During that period there were name changes with respect to both nations in which the books were prepared. The term "United States of America," or its abbreviated form, the "United States," will be used to denote the nation comprising those states which, after 1776, comprised the United States of America. The term "Great Britain" will be used to describe the nation comprising England, Scotland and Wales, which was formed by the Act of Union passed by the Scottish Parliament and by Westminster in 1707 establishing the Kingdom of Great Britain. After the Irish Parliament voted to join the Union in 1801 the then Kingdom of Great Britain became the United Kingdom of Great Britain and Ireland.

In this book many of the figures are photographic images of handwritten pages from cyphering books. Often it is difficult to decipher the writing—sometimes (as with the surviving pages of Abraham Lincoln's cyphering book), the pages were very worn, and the writing faded. Also, the ways letters were formed have changed—so that an "*s*" will often appear to a modern reader like an "*f*." The writing is sometimes very small, and on numerous occasions we used enlarged images for analysis.

A careful reader of this volume might notice that, from time to time, there is repetition of points made earlier in the book. Individual chapters of this book will be available for purchase in electronic form, and therefore it makes sense to repeat material important for a chapter, even if that material has been discussed earlier in the book. We have attempted to limit such repetition to cases where what is being repeated represents essential knowledge if the present chapter is to be understood as a stand-alone document. Another feature, in line with opportunities provided by technology, is that there is a reference list at the end of *each* chapter, *and* a *composite* reference list after all 12 chapters have been presented.

Toward the end of some of the chapters we have acknowledged those who have assisted our research in major ways with respect to those chapters. Although we will not repeat those acknowledgements here, we do wish to thank all who have so generously extended to us their time and expertise as we progressed along "the way." That general acknowledgement takes into account numerous persons working in archives in various parts of the United States of America and the United Kingdom.

Of course, we would offer a special "thank you" to Fred Rickey, who has written the Foreword to this book. Fred, a fellow researcher in the history of mathematics education, is someone whose work we greatly admire. Fred, and two of his colleagues, are currently in the process of writing a book on the cyphering manuscripts prepared by George Washington in the 1740s. We await the outcome of this research with much anticipation. We also wish to thank Valeria Aguirre Holguín, of New Mexico State University, for her contributions to the chapter on Abraham Lincoln's cyphering book, and George Seelinger and other colleagues within the Department of Mathematics at Illinois State University.

Nerida F. Ellerton
M. A. (Ken) Clements
Department of Mathematics, Illinois State University
January 2014

Foreword

Igitur, eme, lege, fruere.[1]

In their previous book, *Rewriting the History of School Mathematics in North America 1607-1861: The Central Role of Cyphering Books* (2012), Nerida Ellerton and Ken Clements convincingly established that the long-forgotten and centuries-old tradition of students preparing their own cyphering books was central to their study of mathematics. Surprisingly few of these cyphering books have survived even though they were highly valued by their creators. Even so, our authors have amassed a personal collection of some 350 cyphering books and examined another 650 in various libraries. Out of this abundance of resources, eleven "exceptional" cyphering books have been selected for detailed examination in this volume. These eleven were selected for a variety of reasons: their age, the lovely handwriting, the beauty of the diagrams, their geographic origins, their mathematical content, and the contexts in which they were written.

The mathematics in these cyphering books is the mathematics of the practitioner, not the theoretician. This is not pure mathematics, but, to use the archaic term, mixed mathematics. Besides the arithmetic of the integers, there is business arithmetic, practical geometry, logarithms and trigonometry, fortification, navigation, and surveying. Of course, none of these cyphering books deals with all of these topics but each has a common arithmetical core which derives from the medieval abbacists. This work describes the provenance and mathematical content of the eleven cyphering books, setting each in its historical context.

The mathematics in these eleven wonderful cyphering books is completely algorithmic. Rules are stated which indicate how to solve a class of problems, and the student is expected to memorize and follow them. There is almost no explanation of why these algorithms work the way they are expected to.

Many obsolete mathematical terms will be encountered—alligation, compound addition, evolution, exchange, fellowship, the galley method, reduction, superficies—and they are carefully explained as they arise, thereby giving insight into how mathematics was taught and learned in Great Britain and in North America between 1630 and 1836. The most interesting of these terms are the various forms of the rule of three, an algorithm for dealing with proportions. There is little elementary symbolic algebra in these cyphering books—the rule of three provides a way of avoiding it. Not all students progressed to the rule of three in their study of mathematics, but those who did were truly accomplished. They were regarded as well prepared to run a business or to oversee a plantation. The more advanced were prepared to pilot a ship, to become professional surveyors, or to become schoolteachers who would use their own cyphering books as texts. Of course, not all teachers were well prepared.

At a meeting of the History and Pedagogy of Mathematics group at West Point in April of 2013, Valeria Aguirre Holguín, a graduate student at New Mexico State University, gave a presentation entitled "New Information on the Provenance of Abraham Lincoln's Arithmetic Copybook." It was at this time that the three authors began joint work on the Lincoln chapter. Amazingly, in June of 2009, Ellerton and Clements had identified a leaf of Lincoln's cyphering book in the Houghton Library at Harvard University, a result announced in a June 2013 press release.

[1] "Therefore, buy, read, enjoy [this book]," Copernicus, *De revolutionibus orbium coelestium* (1543).

This brought the number of identified leaves to 11 (that is to say, 22 pages); hopefully someone, someday, will identify more leaves from Lincoln's cyphering book. The 11 extant leaves are carefully examined in a chapter that will undoubtedly become the most studied chapter in the book.

Louis C. Karpinski lists hundreds of elementary mathematics books in his *Bibliography of Mathematical Works Printed in America Through 1850.* But textbooks such as these were expensive and not easy to acquire, so students continued to prepare their own cyphering books until the time of the Civil War when educational philosophy changed. But those that have survived bear witness to the pride that the authors took in preparing their own personal cyphering books.

As with any seminal work, this one will engender numerous future research questions. Paramount among these is one only partially answered here—a question which still perplexes researchers today: Based on their written work, how do we know how much mathematics students have comprehended and retained for their future work? It will be interesting to match the signatures in the corpus of cyphering books with the way those individuals used mathematics later in their lives. One sign of a wonderful book is the further research it encourages.

West Point, New York V. Frederick Rickey

Chapter 1

Cyphering Books and the Cyphering Tradition in North America and Great Britain, 1630–1880

Abstract: The cyphering tradition controlled school mathematics in Western Europe and North America during the seventeenth and eighteenth centuries and the first half of the nineteenth century. In this book the authors analyze 11 extraordinary cyphering books—7 arithmetic cyphering books, 3 navigation cyphering books, and 1 mensuration/surveying cyphering book. Of these, 6 were prepared entirely in Great Britain, 4 entirely in North America, and 1 partly in Great Britain and partly in North America. After defining key terms—such as "cyphering book," "cyphering tradition," "intended curriculum," "implemented curriculum," and "attained curriculum"—this first chapter summarizes characteristics of the 11 extraordinary cyphering books that will be considered. Finally, six research issues—which will frame subsequent analyses—are stated.

Key Words: *Abbaco* curriculum; Abraham Lincoln; arithmetic education; attained curriculum; ciphering book; ciphering tradition; copybook; cyphering book; history of education; history of school mathematics; implemented curriculum; intended curriculum; navigation cyphering book; Royal Mathematical School at Christ's Hospital; Noah Webster.

What is a Cyphering Book, and what was the Cyphering Tradition?

What is a Cyphering Book?

In this book the term "cyphering book" will be used to refer to a handwritten manuscript, usually prepared at some time during the period 1600–1880, which focused on mathematical content and had all of the following properties:

1. Either the manuscript was written by a student who, through the act of preparing it, was expected to learn and be able to apply whatever content was under consideration; or, it was prepared by a teacher who wished to use it as a model which could be followed by students preparing their own cyphering books.
2. Usually, all entries in the manuscript appeared in ink—as handwritten notes, or problem solutions, or as illustrations. Headings and sub-headings were presented in decorative, calligraphic style. Occasionally, water-color illustrations were prepared.
3. The manuscript was dedicated to setting out rules, cases, model examples and exercises associated with a sequence of mathematical topics. Although most cyphering books were specifically concerned with arithmetic, especially commercial arithmetic, some were dedicated to algebra, or geometry, or trigonometry, or to mathematics associated with mensuration, navigation, surveying, fortification, etc.
4. The topics covered were sequenced so that they became progressively more difficult. The content also reflected the expectation that, normally, no child less than 10 years of age would be assigned the task of preparing a cyphering book (Ellerton & Clements, 2012).

Nerida F. Ellerton and M. A. (Ken) Clements, *Abraham Lincoln's Cyphering Book and Ten other Extraordinary Cyphering Books*,
DOI 10.1007/978-3-319-02502-5_1, © Springer International Publishing Switzerland 2014

In their earlier book, Ellerton and Clements (2012) analyzed 212 North American cyphering books prepared between 1701 and 1861. They now have 350 North American cyphering books in their collection, and have studied another 650 North American cyphering books—all prepared between 1700 and 1861—held in archives at the Phillips Library (Salem, Massachusetts), Harvard University, Yale University, the University of North Carolina (at Chapel Hill), the University of Michigan, the College of William and Mary, and the Rockefeller Library (both in Williamsburg, Virginia), the Library of Congress, and the New York Public Library. They also own 150 cyphering books from other nations, including 100 from Great Britain—the third largest collection of British cyphering books known to exist (Denniss, 2012).

As a result of efforts to standardize American-English spelling, led by Noah Webster during the period 1780 through 1830 (Ellerton & Clements, 2008), the term "ciphering book" (with an "*i*") gradually replaced "cyphering book" (with a "*y*") in the United States of America. Although the term "copy book" (or "copybook") was sometimes used to describe a cyphering book, it was a more general term, and was applied to any book in which handwritten texts had been "copied." It has most commonly been used in relation to handwritten manuscripts in which there was an attempt to use fine penmanship and/or high calligraphy, especially when the copying was modeled on a well-regarded system of penmanship (see, e.g., Bickham, c. 1740).

The *Abbaco* Curriculum

A cyphering tradition which found its way into Europe from Arabic nations after about 1200 more or less controlled, for over 600 years, how European and North American school students who were at least 10 years of age learned mathematics (Ellerton & Clements, 2012; Van Egmond, 1980). From about the thirteenth century onwards, sharp rises in the numbers of international trading and banking companies in Southern European city-states prompted the formation of vernacular schools across Europe in which commercial mathematics, accounting and writing were taught by "reckoning masters" to sons of the merchant class.

During the period 1200–1850 almost all Western European and North American students aged 10 years, or more, who learned mathematics at school or with a private tutor were expected to prepare cyphering books, and the intended curriculum derived from what has been called the *abbaco* tradition. That curriculum defined a well-sequenced set of topics and was an integral part of the approach to business arithmetic adopted by European nations. For many years it was accepted by scholars that the tradition was introduced into Europe through Leonardo of Pisa's (Fibonacci's) *Liber Abbaci*—which was written around 1200 CE (see, e.g., Yeldham, 1926)—but in recent years that tradition has been challenged. There is evidence that even before 1200 there were features of the tradition already to be found in parts of Southern Europe (Høyrup, 2005).

According to Jens Høyrup (2008), *abbacus* schools, which operated in Italy from the thirteenth century, were

> … primarily frequented by merchant and artisan youth for … two years (around the age of 11), who were taught the mathematics needed for commercial life: calculation with the Hindu-Arabic numerals; the rule of three; how to deal with the complicated metrological and monetary systems; alloying; partnership; simple and composite discount; the use of "single false position"; and area computation. Smaller towns might employ a master; in towns like Florence and Venice, private *abbacus* schools could flourish. In both situations *abbacus* masters had to compete, either for communal positions or for the enrolment of students. (pp. 4–5)

In schools, the cyphering tradition came to be associated with a typical organizational structure, which included three recitation components:

- First, the teacher prescribed what an individual student would do. Sometimes the teacher would write the headings and the problems to be solved into a student's cyphering book (Dickens, 1850; Walkingame, 1802). Each student would then prepare for the next recitation session.
- Second, the student would work individually on finding solutions to set exercises, often on a slate or on scraps of paper. These would be shown to the teacher during another recitation session. If approved by the teacher, the student would then be told to complete his or her cyphering book entries which included introductory statements for the given topic, rules, cases, model examples, and exercises. Sometimes the teacher provided a student with an older cyphering book, or textbook, to show what was required. Writing would be done with a quill and home-made ink (Ellerton & Clements, 2012).
- Finally, the student would show the teacher his or her completed cyphering book entries. It was expected that headings would be in fine calligraphy, and general penmanship would be of a good quality. A cyphering book was to be a guidebook for life—a text that could be consulted if and when the need arose—and it was important for all entries to be correct.

In each of these recitation components, the teacher would talk to individual students, asking verbal questions. At the lowest level, this verbal aspect of recitation may have involved the teacher expecting a student to recite rules which should have been memorized. At the highest level, teachers would ask probing questions that tested a student's understanding (Adams, 1848; Babcock, 1829; Cobb, 1835; Colburn, 1824; Emerson, 1838; Sterry & Sterry, 1795). According to Lyman Cobb (1835), a diligent teacher would talk with each learner, in recitation sessions, as often as once a day, probing the student's understanding of the principles upon which the rules were founded. Cobb (1835) added:

> The teacher should not permit him [the learner] to commence a new sum, or engage in a new rule, until he is fully and thoroughly acquainted with the principles of the rule in which he has been working. Young scholars are generally anxious to make rapid progress in passing through the Arithmetick. This propensity, however laudable, should not be indulged at the expense of a partial knowledge of the subject. The teacher should endeavour to convince the scholar that, in order to make his progress advantageous, he must perfectly understand each rule and its principles. (p. 2)

Thus, the cyphering tradition incorporated a form of pedagogy which combined individual, yet supervised, learning, with follow-up evaluative sessions, usually through one-on-one recitations.

Figure 1.1 shows reproductions of two etchings by Abraham Bosse, a French-Huguenot artist who, in the 1630s, depicted the kind of classrooms that existed in middle-class Western European schools at that time. In the early 17th century, the Huguenots, the Dutch, and groups within quite a few other Western European nations (e.g., Italy), valued arithmetic, and saw it as a vital part of the education of all children.

In both of the etchings reproduced in Figure 1.1, a teacher and a student can be seen engaged in the process of recitation—the teacher was checking an individual student's work and, perhaps, asking questions or setting new work for the student to do. Some students were waiting to see the teacher, and others were working, individually, preparing their copybooks for when it would be their turn for recitation. In the North American colonies, most school rooms were less elaborate than those shown in Figure 1.1, and often teachers were not qualified to teach arithmetic beyond numeration and the four operations with whole numbers. These early North American students rarely owned a commercially-distributed arithmetic textbook (Ellerton & Clements, 2012).

Figure 1.1. Le Maître d'Ecole (c. 1635) and *La Maîtresse d'Ecole* (c. 1638) by Abraham Bosse, (1611–1678). Held by the Metropolitan Museum of Art, New York (public domain images).

In Jan Steen's (1670) depiction of a village school in France, reproduced in Figure 1.2, there was little to be learnt from the teacher who was sharpening a quill and seemed to be unaware of the chaos surrounding him. A schoolmistress (the "usher") was checking a child's work—but others fought, sung, made fun of the teachers, and even slept. On the right a child was handing spectacles to an owl—a symbol Steen sometimes used to denote "foolishness."

Figure 1.2. Jan Steen's (c. 1670), "The Village School" in France. The original painting is held by the National Gallery of Scotland, in Edinburgh, Scotland (public domain image).

The cyphering tradition was duly translated to North America from Europe, and it quickly gained control over pre-college intended mathematics curricular sequences. Monroe (1917), drawing on research by William Eliot Griffis (1909) and William Heard Kilpatrick (1912), pointed out that cyphering was always included in the curriculum of the Dutch-language schools established in New Amsterdam—which would become New York—between the 1630s and the mid-1660s.

The approach to mathematics education inherent in the cyphering tradition represented a radical departure from the humanist, classical medieval forms of mathematics often found in European universities between 1200 and 1700. At the universities, mathematics was presented to students in Latin. At its worst, the study of mathematics at a university at that time did not go much further than a statement and illustration of rules for determining moveable feast days in the church calendar; at its best, however, it featured rigorous proofs and exercises (Duhem, 1985).

Eleven Extraordinary Cyphering Books

This volume analyzes 11 extraordinary cyphering books in varying degrees of detail. Each of the 11 manuscripts was initially prepared at some time between 1630 and 1835, in either Great Britain or North America. It will be useful here to indicate why each can be regarded as "extraordinary." A summary of the 11 manuscripts is given in Table 1.1.

Table 1.1
Summary of the 11 Extraordinary Cyphering Books

Chapter in this Volume	When and where the Manuscript was Prepared; Number of Pages	Author(s) of the Manuscript	Present Location of the Manuscript	Type of Cyphering Book	Why is this Manuscript "Extraordinary"?
2	c. 1667; Maine, New England (?); 7 pages.	Unknown	Ellerton/ Clements collection, Illinois	Arithmetic	This is probably the oldest extant North American cyphering book. The level of mathematics is extremely low, but the last page mentions the rule of three.
3	1702–1703; Great Britain, then 1720–1721, Portsmouth, New Hampshire; 88 pages.	Thomas Prust, and John Collings	Ellerton/ Clements collection	Arithmetic	Thomas Prust began preparing this book in England, but brought it to Portsmouth, NH, in 1720, and handed it on to his step-son, John Collings. The manuscript features beautiful calligraphy.
4	1776–1777 and 1780–1781; Bertie County, North Carolina; 151 pages.	Martha and Elisabeth Ryan	Wilson Library, University of North Carolina	Arithmetic	This features amazing penmanship, calligraphy, and artwork. The content goes well beyond the rule of three. It is held in the Southern Historical Collection of the Wilson Library.
5	1820–1821; Maryland (USA); 80 pages.	Mary Walters	Ellerton/ Clements collection	Arithmetic	This features a problem-based approach to arithmetic via Irish ballads, poetry, and song.
6	1819(?) –1826; Spencer County, Indiana (USA); 22 pages.	Abraham Lincoln	11 leaves are in 11 different locations (all in USA)	Arithmetic	This is the oldest extant manuscript of the future President. The content goes beyond the rule of three. A page-by-page analysis of content is presented.
7	1630–1632; Great Britain; 302 pages.	Thomas Dixson	Butler Library, Columbia University	Arithmetic	This is a very early British cyphering book. Penmanship and calligraphy are good, and much attention is paid to "galley" division.
8	c. 1740; London; 24 pages.	George Bickham (the Elder)	Ellerton/ Clements collection	Arithmetic	This cyphering book was prepared by one of the world's best-known exponents of fine penmanship and calligraphy.
9	c. 1791; Great Britain; 240 pages.	Unknown	Ellerton/ Clements collection	Navigation	This manuscript was probably prepared by a midshipman during one or more voyages. It features exquisite miniature water-color illustrations.
10	1820–1825; London; 671 pages.	Charles Page	Ellerton/ Clements collection	Navigation	This was prepared at the Royal Mathematical School, Christ's Hospital, London
11	c. 1810; Scotland; 466 pages.	William Beattie	Ellerton/ Clements collection	Mensuration /Surveying	This was probably prepared at Clarencefield Academy, in Dumfriesshire, Scotland
11	c. 1835; London; 309 pages..	William Beattie	Ellerton/ Clements collection	Navigation	This may have been prepared especially for the "Sailor King," King William IV of England. William Beattie was the King's private secretary.

Just two of the eleven extraordinary cyphering books analyzed in this book were prepared by females—which is consistent with recent literature indicating that in North America in the eighteenth and nineteenth centuries between 15 percent and 20 percent of cyphering books were prepared by females (Ellerton & Clements, 2012). Although eleven extraordinary cyphering books have been summarized in Table 1.1, two of those were prepared by the same person—William Beattie. Beattie prepared the mensuration/surveying manuscript sometime around 1810, when he was still at school in Scotland, and he prepared the navigation manuscript about 25 years later, when he was a close confidant of King William IV, of England.

Authors of seven of the eleven cyphering books were British. One of these books—that by Thomas Prust—was begun in England in 1702 but was subsequently brought to North America, where it was passed on to a second owner (John Collings), who made additional entries. Three of the other cyphering books—those by Martha and Elisabeth Ryan, Mary Walters, and Abraham Lincoln—were prepared in North America, and the fourth was probably prepared in Maine in New England, on paper made in Europe but brought to the New World in 1666 or 1667.

Cyphering Books, and Intended, Implemented, and Attained Curricula

Throughout this book it will be argued that the preparation of cyphering books was a very important component of what students actually did at school during the cyphering era. Handwritten entries typically included introductory notes on topics, rules, cases, model examples, and completed exercises. Sometimes, the notes were copied from a textbook or from an older cyphering book, and sometimes they were dictated by a teacher (Ellerton & Clements, 2012). For most cyphering books we do not know whether students merely copied material, or whether they copied and memorized it, or whether they attempted to comprehend what they wrote so that they could apply the knowledge in appropriate situations. That is to say, the cyphering books are largely silent on how well students learned what they transcribed.

Entries in cyphering books testify unmistakably to what Westbury (1980) and Travers and Westbury (1989) have called the *implemented* curriculum—that is say the actual mathematics that students studied. By contrast, textbooks, and documents prepared by education institutions like schools and colleges summarizing what students would be expected to learn, indicated the *intended* curriculum—which may or may not have matched the implemented curriculum. The *attained* curriculum is a term denoting what students actually learned. Cyphering books showed what students had studied but did not, and could not, show what they had learned.

During recitations, good teachers attempted to identify, through questioning, what students did and did not understand (McMurry & McMurry, 1903). A cyphering book said a lot about the kinds of mathematics students had studied but very little about how well they had learned it. Yet, if one carefully and systematically reads the entries in a cyphering book one can often catch glimpses of the degrees of understanding of the person who prepared the book. Thus, for example, in the cyphering book prepared by Abraham Lincoln (see Chapter 6), a careful reader can sense that Abraham had a driving determination to know what he was doing. He was not happy merely to copy. By contrast, with George Bickham's (c. 1740) cyphering book (Chapter 8), a reader is likely to feel that the author seemed to care only about the quality of the penmanship and calligraphy. In each of the chapters, comment will be made on the extent to which the person who prepared a cyphering book seemed to want to know, to understand, what he or she was writing about.

Six Prominent Issues

Ellerton and Clements's (2012) original book on the rise and fall of the cyphering tradition in North American between 1607 and 1861 was mainly concerned with establishing, beyond reasonable doubt, that the *abbaco* curriculum, as it found expression in a cyphering tradition inherited from Western Europe, effectively controlled the teaching and learning of school mathematics in the New World. Analyses in that book took advantage of the largest-ever data set of *primary* sources (comprising mainly, but not exclusively, textbooks and cyphering books) ever assembled in relation to the early history of school mathematics in North America.

In that earlier book, Ellerton and Clements (2012) presented a broad-brush analysis of the cyphering tradition as it operated in North America between 1607 and 1861. They reached the conclusion that it was through the cyphering tradition that some citizens became numerate, as they learned that problems demanding quantitative solutions came in categories, and that each category could be subdivided so that particular rules and cases would apply. This realization led Ellerton and Clements to conceptualize what they called the IRCEE ("Introduction, Rule, Case, Example, Exercise") genre. The "promiscuous problem" sets found in many of the cyphering books in their principal data set were intended to force students to ask themselves—"To which type of main problem category does this problem relate?" And, "If I want to solve this problem, which rule and which case will be most appropriate?"

The cyphering tradition provided the glue which held together, but at the same time constrained, the development of pre-college mathematics education curricula and teaching and learning in North America between 1607 and 1861. It was the force of that tradition which explains why the same arithmetic topics were studied, in more or less the same sequence, right across North America—in Boston, in New York, across Canada, and in schools in small rural communities. Ellerton and Clements (2012) concluded that the cyphering tradition "provided teachers with a subconscious base for the intended, implemented and attained aspects of school mathematics," and in particular provided "the backdrop against which the PCA (Problem, Calculation, Answer) genre, that still bedevils thinking and practice in school mathematics in the United States, arose" (p. 151).

This present volume provides qualitative, fine-grain analyses of early school mathematics in Great Britain and North America. The basic framework for analysis will be a concept of curriculum as having intended, implemented, and attained components (Westbury, 1980). It will be assumed that details of intended curricula are most readily available through contemporary textbooks, which in themselves are important primary sources. It will also be assumed that the best available evidence for implemented curricula is provided by cyphering books. This present volume is mainly concerned with analyzing implemented curricula and, because of that, the commercially-published textbooks are regarded as complementary rather than fundamental to learning. It has been useful, nevertheless, to be in a position to compare entries in cyphering books with sections in textbooks.

Denniss (2012), Michael (1993), and Stedall (2012) have provided evidence that between 1750 and 1860 most entries in British cyphering books were copied from textbooks, especially from those authored by the prolific Francis Walkingame. One only has to read the prefaces in the numerous editions of Walkingame's arithmetics (see, for example, Walkingame, 1802) to be convinced that that author assumed that students would use a textbook as the basis for making entries into cyphering books. It was likely that this textbook-driven approach to cyphering would be favored by some teachers in the newly-formed United States of America—especially when arithmetic textbooks written by North American authors became more readily available.

In Benjamin Workman's (1788) adaptation of John Gough's *A Treatise of Arithmetic in Theory and Practice,* there was an attempt to fit Gough's text, originally written for students in Ireland, to students in North America. Workman emphasized that he had taken great pains to correct all errors in Gough's original text, and that he had also noticed that there was a huge number of errors in the arithmetic by Thomas Dilworth, which was often used in North America. Workman (1788) then commented on the difficulties a teacher of arithmetic in North America who used a textbook as the basis for setting exercises for students to do in their in their cyphering books might have:

> When his pupils find answers different from those in the book, he obliges them to revise the work; for knowing the book to be right, he insists the mistake lies with them; but if he knows the book to abound with errors, he is obliged to work the questions himself before he can certainly tell them their work is wrong. (p. v)

Workman wrote at a time when North American teachers like Nicolas Pike (1788) recognized that preparing an arithmetic textbook for students in the schools of the recently-established United States of America could be a profitable venture. In the United States, many teachers had taught the subject without having access to published texts (Ellerton & Clements, 2012), the only texts available having been handwritten cyphering books—often prepared by the teachers themselves.

Ellerton and Clements hold a large collection of North American school mathematics textbooks dating from 1729 through 1950. In that collection there is at least one edition of most of the mathematics books listed by Karpinski (1980) as having been published in North America up to 1850. They also hold a major collection of 17th-, 18th-, and 19th-century school mathematics textbooks published in Great Britain. Those textbooks, combined with their set of 350 North American cyphering books and 100 British cyphering books, provided them with a solid primary data set which was readily available to them as they conducted the qualitative analyses reported in the following chapters.

The Six Issues

At the outset, we (Ellerton and Clements) identified the following six issues for detailed study. Although our analyses of individual cyphering books were not constrained by the need to take account of these issues, they were, nonetheless, never far away from our thinking.

1. For a particular cyphering book, what was the *intended* mathematics curriculum?
2. For a particular cyphering book, what was the *implemented* mathematics curriculum, and to what extent did that differ from the intended curriculum?
3. Was it possible to get glimpses of the *attained* curriculum merely by examining a cyphering book?
4. For a particular cyphering book, to what extent was there a sense of individual valuing of the mathematics, of ownership, of a "this-is-my-book," mentality?
5. What were the contextual and ethnomathematical circumstances which surrounded the preparation of the cyphering book, and did those circumstances have an impact on the student who prepared it?
6. To what extent did the cyphering tradition in Great Britain differ from the cyphering tradition in North America?

Following the chapter-by-chapter analyses, these six issues will be explicitly addressed in the final chapter (Chapter 12) of this book.

A Window into the History of School Education

Within the academic mathematical community there has long been a strong sense of the need to pay due attention to the history of the development of mathematical ideas. However, among mathematics educators there does not seem to have been the same kind of determination to investigate, critically, the history of school mathematics. In recent years that deficiency within the international mathematics education community has begun to be addressed—for example, in September 2013 the first author (Ellerton) presented a paper at the Third International Conference on the History of Mathematics Education (3ICHME) held at Uppsala University, Sweden. About 40 people attended the Conference, but interestingly only two other scholars from the United States of America were present. The time is ripe for much greater attention to be paid to the history of school mathematics in all parts of the world, including Europe and North America. From that perspective, this book provides a window into the cyphering tradition which was so important in the early history of school mathematics in Western nations, including that history within the United States of America, Canada, and the United Kingdom.

In the chapters which follow, features of extraordinary cyphering books are identified and used to draw attention to central and, if you like, ordinary, features of the cyphering tradition. Each chapter will be well illustrated, showing pages from the cyphering books under consideration. The aim, throughout, has been to make readers aware of key features of the *abbaco* curriculum and, more generally, of the cyphering tradition, which controlled school mathematics in Western nations for many centuries. The use of the word "controlled" in the last sentence could raise academic eyebrows, but that is the word we feel is most appropriate.

By reproducing many pages and sections of pages from the 11 cyphering books, and by providing accompanying analyses, we hope to make readers fully aware of mathematical, ethnomathematical, pedagogical and individual factors which influenced school mathematics during the cyphering era. Although the related concepts of intended, implemented and attained curricula (Travers & Westbury, 1989; Westbury, 1980) provide the framework for analysis in this book, it is interesting to note that during the cyphering era the word "curriculum" was unknown.

The analyses conducted in the preparation of this book were informed by many primary sources from both North America and the United Kingdom. Six major issues were identified. Thus, for example, it is concluded that commercially-published textbooks played a much more important role in the British version of the cyphering tradition than in the North American version. However, within-nation differences were probably greater than between-nation differences. Thus, for example, in Chapter 6, which deals with Abraham Lincoln's cyphering book, it is argued that the task of preparing a cyphering book in rural Indiana in the 1820s was an altogether different thing from preparing a cyphering book in an exclusive private secondary school in, say, Massachusetts.

Since the publication of Ellerton and Clements's (2012) book on the cyphering tradition in North America at least three books which include analyses of cyphering books have appeared in the United Kingdom (Denniss, 2012; Stedall, 2012; Wardhaugh, 2012). Many other similar books are needed, written by scholars from a range of nations, which take account of the cyphering era in the history of school mathematics.

The scholarship which generated this book lies at the intersection of mathematics, mathematics education, and the history of education. It is important that books such as this one be read by scholars in all three of those domains. Mathematicians need to learn to distinguish between the history of mathematics, on the one hand, and the history of mathematics education, on the other, and to reflect on if and when such a distinction is helpful. Mathematics educators need to pay much more attention than they have to the history of their field of study. And, scholars with a general interest in the history of school education will profit from becoming better acquainted with aspects of the teaching and learning of a subject which has long been a major component of the school curriculum. Too often, textbooks on the history of education have paid scant attention to the history of school mathematics, despite the fact that there can be no doubt that during the seventeenth, eighteenth and nineteenth centuries the cyphering tradition was, increasingly, a vitally important aspect of general school education, not only in Western European nations but also in North America.

References

Adams, D. (1848). *Adams's new arithmetic: Arithmetic in which the principles of operating by numbers are analytically explained and synthetically explained, thus combining the advantages to be derived both from the inductive and synthetic mode of instructing.* New York, NY: Collins & Brother.

Babcock, T. H. (1829). *The practical arithmetic; in which the principles of operating by numbers are analytically explained and synthetically applied ... Adapted to the use of schools and academies in the United States.* New York, NY: G & C & H. Carvill.

Bickham, G. (c. 1740). *Youth's instructor in the art of numbers, a new cyphering book in which is shewn variety of penmanship, command of hand: Engrav'd for the use of schools.* London, United Kingdom: William & Cluer Dicey.

Cobb, L. (1835). *Cobb's ciphering book, No. 1, containing all the sums and questions for theoretical and practical exercises in Cobb's Explanatory Arithmetic No 1.* Elmira, NY: Birdsall & Huntley.

Colburn, W. (1824). *Arithmetic; being a sequel to first lessons in arithmetic.* Boston, MA: Cummings, Hilliard, & Co

Denniss, J. (2012). *Figuring it out: Children's arithmetical manuscripts 1680–1880.* Oxford, United Kingdom: Huxley Scientific Press.

Dickens, C. (1850). *David Copperfield.* London, United Kingdom: Bradbury & Evans.

Duhem, P. (1985). *Medieval cosmology, theories of infinity, place, time, void, and the plurality of worlds.* Chicago, IL: Chicago University Press.

Ellerton, N. F., & Clements, M. A. (2008). An opportunity lost in the history of school mathematics: Noah Webster and Nicolas Pike. In O. Figueras, J. L. Cortina, S. Alatorre, & A. Mepúlveda (Eds.), *Proceedings of the Joint Meeting of PME 32 and PME-NA XXX* (Vol. 2, pp. 447–454). Morelia, Mexico: Cinvestav-UMSWH.

Ellerton, N. F., & Clements, M. A. (2012). *Rewriting the history of school mathematics in North America 1607–1861.* New York, NY: Springer.

Emerson, F. (1838). *Key to the North American arithmetic, Part second and third. For the use of teachers.* Boston, MA: G. W. Palmer & Co.

Griffis, W. E. (1909). *The story of New Netherland.* Boston, MA: The Riverside Press.

Høyrup, J. (2005). Leonardo Fibonacci and *abacco* culture: A proposal to invert the roles. *Revue d'Histoire des Mathématiques, 11*, 23–56.

Høyrup, J. (2008). The tortuous ways toward a new understanding of algebra in the Italian *Abbacus* School (14th–16th centuries). In O. Figueras, J. L. Cortina, A. Alatorre, T. Rojano & S. Sepulveda (Eds.), *Proceedings of the Joint Meeting of PME 32 and PME-NA XXX* (Vol. 1, pp. 1–20). Morelia, Mexico. International Group for the Psychology of Mathematics.

Karpinski, L. C. (1980). *Bibliography of mathematical works printed in America through 1850* (2nd ed.). Ann Arbor, MI: University of Michigan Press.

Kilpatrick, W. H. (1912). *The Dutch schools of New Netherland and colonial New York.* Washington, DC: US Bureau of Education.

McMurry, C. A., & McMurry, F. M. (1903). *The method of recitation.* New York, NY: The Macmillan Company.

Michael, I. (1993). The textbook as a commodity: Walkingame's *The Tutor's Assistant. Paradigm, 12*, 2–10.

Monroe, W. S. (1917). *Development of arithmetic as a school subject.* Washington, DC: Government Printing Office.

Pike, N. (1788). *A new and complete system of arithmetic, composed for the use of citizens of the United States.* Newbury-Port, MA: John Mycall.

Stedall, J. (2012). *The history of mathematics: A very short introduction.* Oxford, United Kingdom: Oxford University Press.

Sterry, C., & Sterry, J. (1795). *A complete exercise book, in arithmetic: Designed for the use of schools in the United States*. Norwich, CT: John Sterry & Co.

Travers, K. J., & Westbury, I. (Eds.). (1989). *The IEA Study of Mathematics I: Analysis of mathematics curricula.* Oxford, United Kingdom. Pergamon.

Van Egmond, W. (1980). *Practical mathematics in the Italian Renaissance: A catalog of Italian abbacus manuscripts and printed books to 1600.* Firenze, Italy: Istituto E Museo di Storia Della Scienza.

Walkingame, F. (1802). *The tutor's assistant, being a compendium of arithmetic, and a complete question-book*. London, United Kingdom: C. Whittingham.

Wardhaugh, B. (2012). *Poor Robin's prophecies: A curious almanac, and the everyday mathematics of Georgian Britain.* Oxford, United Kingdom: Oxford University Press.

Westbury, I. (1980). Change and stability in the curriculum: An overview of the questions. In H. G. Steiner (Ed.), *Comparative studies of mathematics curricula: Change and stability 1960–1980* (pp. 12–36). Bielefeld, Germany: Institut für Didaktik der Mathematik-Universität Bielefeld.

Workman, B. (1788). *A treatise on arithmetic in theory and practice.* Philadelphia, PA: J. McCulloch.

Yeldham, F. A. (1926). *The story of reckoning in the Middle Ages.* London, United Kingdom: Harrap.

Chapter 2

Primitive Beginnings, Circa 1667

Abstract: Although the name of the person who prepared this seven-page manuscript, and of the place where it was prepared, are unknown, the last owner indicated that the manuscript was purchased from a very long-established family in Maine, New England. Importantly, the paper is clearly watermarked, and the watermark indicates the paper was made in 1666 or 1667. We have assumed that the paper was brought from Europe to New England during the second half of the 1660s, and that the writing in the manuscript was done in New England. If that was indeed the case then this is easily the oldest of known extant North American cyphering books and, in that sense, is extraordinary. A second notable feature of the manuscript is that the mathematics and spelling are at a very low level. Two other interesting features of the manuscript are: (a) the person who prepared it seemed to want to be able to claim that he or she had cyphered to the rule of three because, all of a sudden, that rule is mentioned on the last page; and (b) one of the topics mentioned was "halving," which was *not* a normal part of the *abbaco* curriculum sequence for arithmetic, but *was* part of the Sacrobosco curriculum sequence for arithmetic.

Key Words: *Abbaco* curriculum sequence; arithmetic; ciphering book; ciphering tradition; cyphering book; cyphering tradition; rule of three; Sacrobosco.

This chapter will describe, and assess the historical significance of, a seven-page manuscript which might be the oldest extant cyphering book prepared in North America. The document became part of our collection of North American and British cyphering books in June 2013 when we purchased it from an online auction. The manuscript had been advertised in the following way:

> This is a curious and quite possibly ultra-rare early (c. 17th century) American mathematics lesson book, handwritten on watermarked, hand-laid paper!
>
> Measuring 8" by 6," and spanning 16 pages, 7 of which contain manuscript handwriting. Language is Olde English, with handwriting characteristics undoubtedly mid 1600s–early 1700s. Recovered from a Camden, Maine, auction; assumed American through association, as other documents were all American in the sale. No further identifying characteristics other than watermark consisting of a crest with a crown, and a central cross, as shown. Paper with some staining, creasing and folding, held by original strings loosely. A few interesting doodles accompany the mathematical rules, as shown as well. Quite possibly the earliest cypher style manuscript we have seen.

The advertisement was accompanied by photographs of a few pages of the manuscript and, after looking carefully at those photographs, our intuitions told us that, yes, this did seem to be a manuscript that was prepared some time during the period from the mid-1600s to the early 1700s.

After purchasing the manuscript, we were surprised how little handwriting there was on many pages. Taking the lead from the advertisement, we examined the watermarks in the paper, and two clear watermarks enabled us to identify, through the Gravell Watermark Archive, that the paper could be dated to 1666 (referenced as Watermark ARMS.047.1). Unfortunately, the name of the person who prepared the manuscript, and the place where it was prepared, were not given and,

Nerida F. Ellerton and M. A. (Ken) Clements, *Abraham Lincoln's Cyphering Book and Ten other Extraordinary Cyphering Books*, DOI 10.1007/978-3-319-02502-5_2, © Springer International Publishing Switzerland 2014

almost certainly, will never be known. However, taking note of the seller's statement that it was purchased at an auction in Camden, Maine, and that it was among other very early American documents, we shall assume that the manuscript was prepared in New England in the late 1660s. We shall also assume that the paper on which the handwriting occurred had been brought to New England from Europe, either by an early settler from Europe or by a commercial exporter. Admittedly, those are large assumptions. However, we recognize that the manuscript would still be interesting if the handwriting was done, totally, in Great Britain before the manuscript was, somehow, and at some unknown time, brought to North America.

If this manuscript was prepared in North America in the 1660s then it is easily the oldest North American cyphering book that we have seen or heard of. And that would justify the use of the word "extraordinary" to describe it. But we have a second reason for regarding the manuscript as "extraordinary." Of all the cyphering books that we have ever seen, this is, mathematically speaking, the *worst*—although, perhaps, "worst" might not be the most appropriate word to describe it. Why we say that will become clear as we examine its pages—which we shall now do.

The Text of the Manuscript

The sections on the pages are numbered, with the first section being numbered 15 and the last 23. Section 21 is missing. Sections 19 and 20 occupy the same page in the handwritten manuscript, and the same is true of Sections 23 and 24. No page has a large amount of handwriting, and several pages have ruled lines (in ink), to guide the writing.

The pages are reproduced in Figures 2.1 through 2.7.

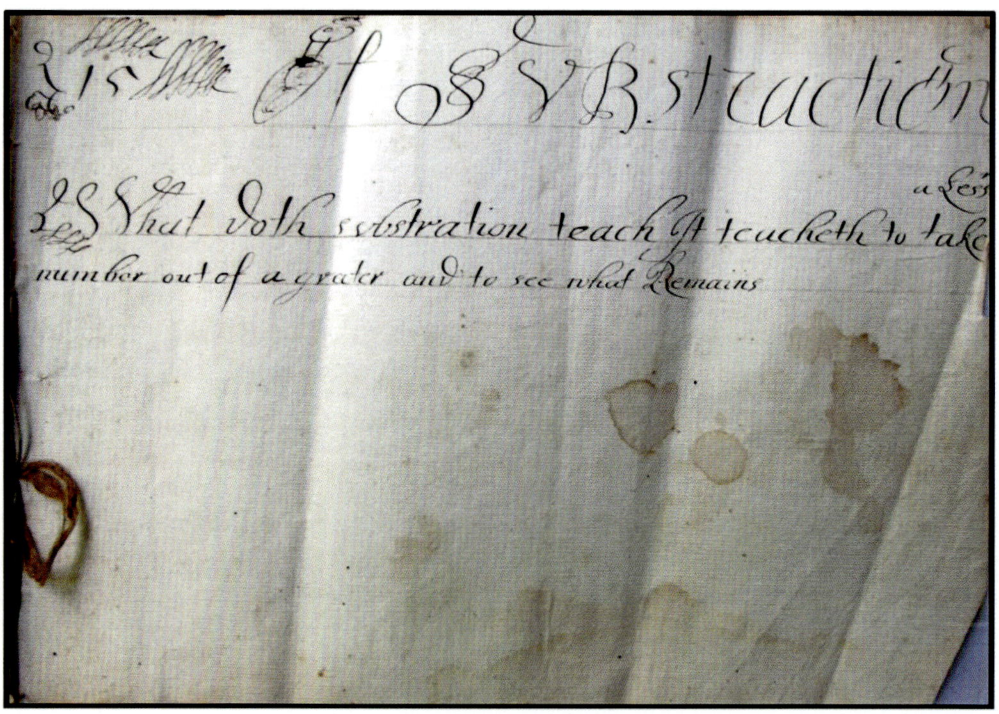

15 Of Substraction

What doth substraction teach It teacheth to take a Less number out of a greater and to see what Remains

Figure 2.1. "Substraction" c. 1666 (Section 15).

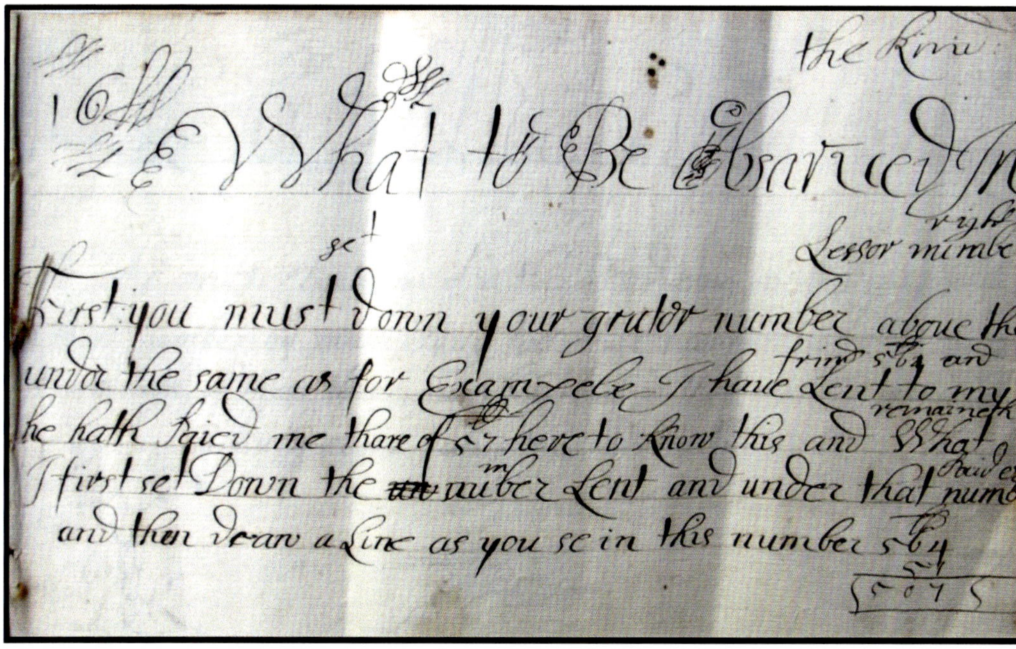

16 What to Be Carried In

Lessor number right

First you must set down your greater number above the under the same as for Example I have Lent to my friend 564 and he hath paid me thare of 57 here to know this and what remaineth I first set Down the number Lent and under that number paid and then draw a Line as you se in this number 564
57
507

Figure 2.2. "What to be Carried in" (Section 16).

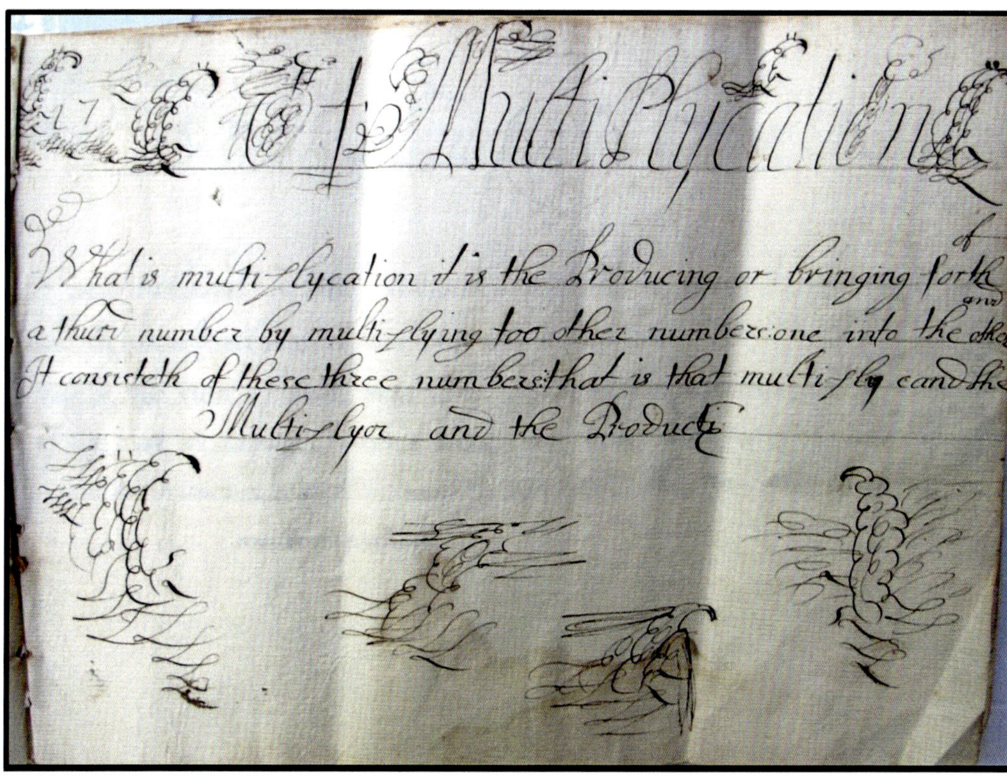

17 Of Multiplycation

What is multiplycation it is the Producing or bringing forth of a third number by multiplying too other numbers one into the other and It consisteth of these three numbers that is that multiplycand the Multiplyor and the Product

Figure 2.3. "Muliplycation" (Section 17).

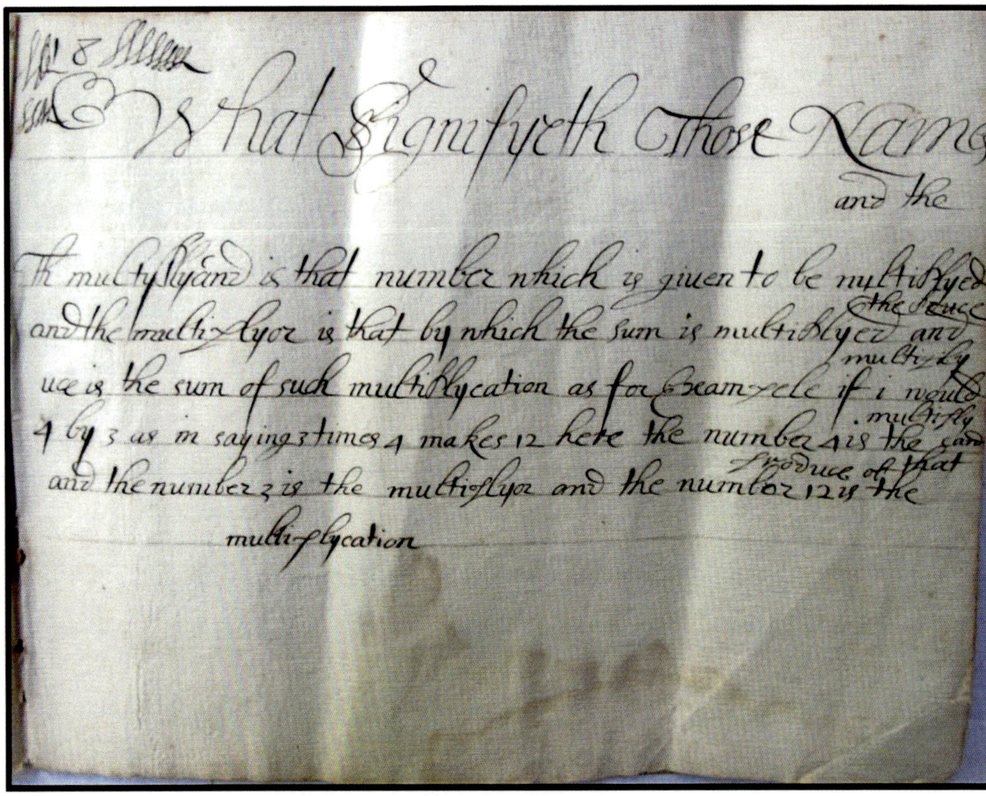
No 18 What Signifyeth Those Names and the
The multyplyand is that number which is given to be multiplyed and the multiplyer is that by which the sum is multiplyed and the produce is the sum of such multiplycation as for Examples if i would multiply 4 by 3 as in saying 3 times 4 makes 12 here the number 4 is the multiplyand and the number 3 is the multiplyer and the number 12 is the produce of that multiplycation

Figure 2.4. "What Signifyeth Those Names" (Section 18).

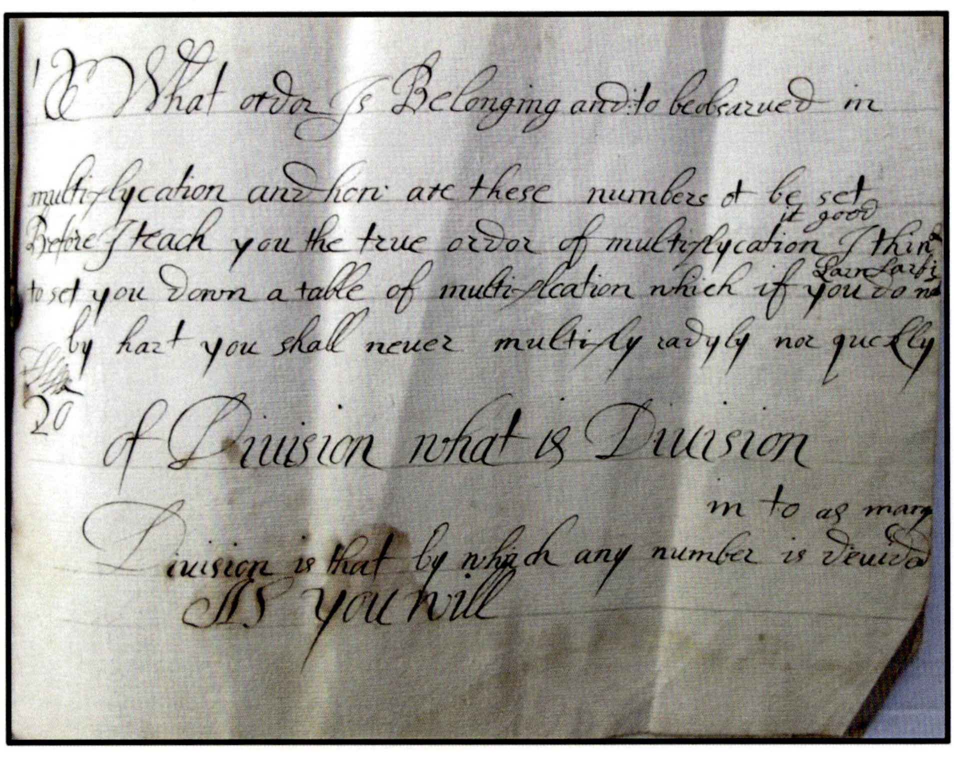
19 What order Is Belonging and to be observed in multiplycation and how are these numbers to be set
Before I teach you the true order of multiplycation I think it good to set you down a table of multiplycation which if you do not learn by hart you shall never multiply radyly nor quickly
20 of Division what is Division
Division is that by which any number is devided in to as many As you will

Figure 2.5. "What order is Belonging and to be obsearved in multiplycation and how are these numbers to be set," and "of Division what is Division" (Sections 19 and 20).

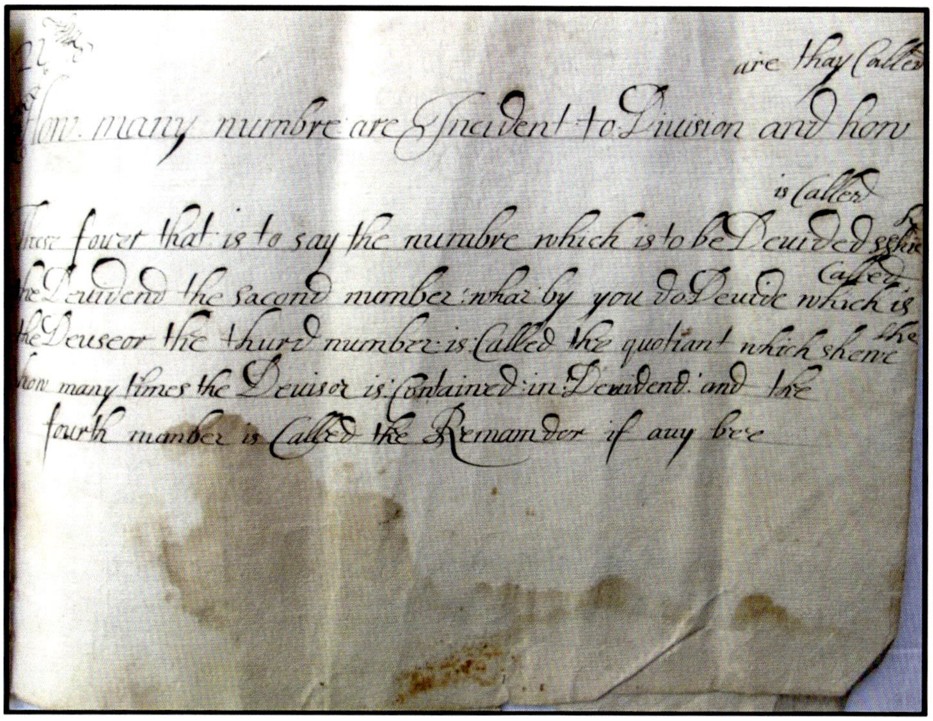

22 are thay Calle[d]
How many numbre are Incident to Division and how
is Called
Thee fower that is to say the numbre which is to be Deuided
the Deuidend the sacond number whar by you do Deuide which is Called
the Deuisor the third number is Called the quotiant which shewe the
how many times the Deuisor is Contained in Deuidend and the
fourth number is Called the Remainder if any bee

Figure 2.6. "How many numbre are incident to Division and how are they called" (Section 22).

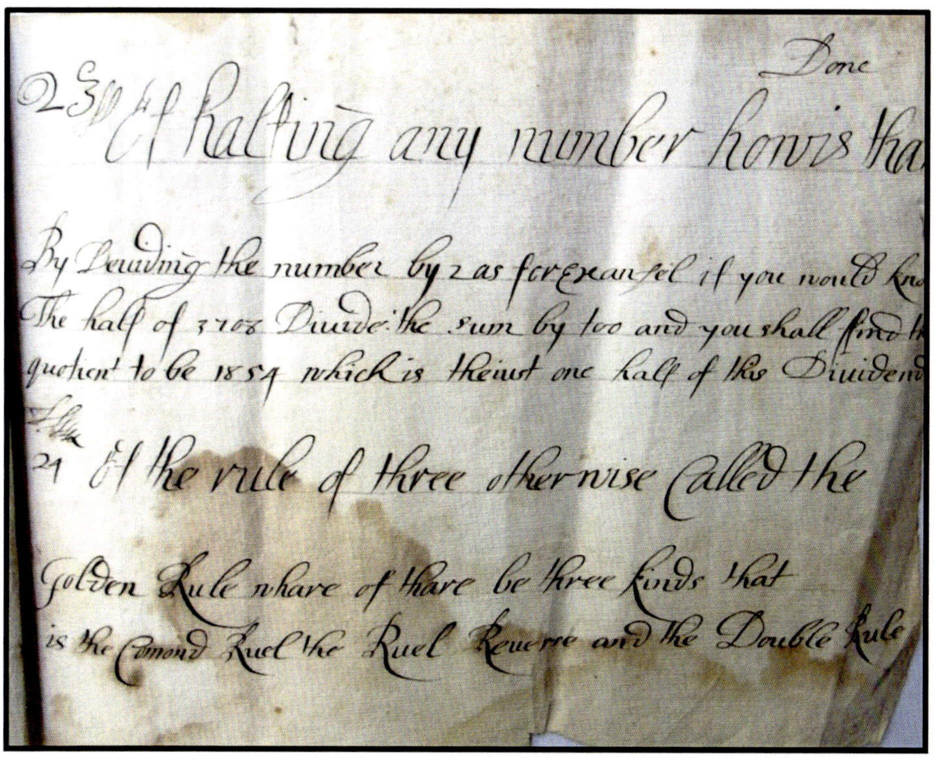

23 Of halfing any number how is tha[t] Done
By Deuiding the number by 2 as for example if you would kn[o]
The half of 3708 Diuide the sum by too and you shall find th[e]
quotient to be 1854 which is the iust one half of the Diuiden[d]
24 Of the rule of three otherwise Called the
Golden Rule whare of thare be three kinds that
is the Comond Ruel the Ruel Reuerse and the Double Rule

Figure 2.7. "Of halfing any number how that is Done," and "Of the rule of three otherwise called the Golden Rule" (Sections 23 and 24).

For readers who found it difficult to read the text in any or all of Figures 2.1 through 2.7, the following provides completely unedited versions of what appears on the pages. Occasionally we could not decipher a particular word, and in those cases we have followed the word with (?). Since no question mark was used in the manuscript text, no question mark is included in these transcripts.

Section 15: This was headed "Substraction, a Lesson." The text asked "What doth substraction teach," and answered with "It teacheth to take number out of a grator and to see what remains."

Section 16: The heading was "What to be Carried in." The text stated: "First you must down your grator number above the lesser number under the same as for example I have lent to my friend 564 and he hath saied me thare of 57 here to know this and what remaineth. I first set down the number lent and under that number paid, and then draw a line as you se in this number

$$\begin{array}{r} 564 \\ \underline{57} \\ \underline{507} \end{array}$$

Section 17: The heading was "Multiplycation." The text asked: "What is multiplycation" and the answer given was: "It is the producing or bringing forth of a third number by multiplying too other numbers one into the other. It consisteth of these three numbers: that is that multiplycand, the Multiplyor and the Product." There were numerous unrelated doodles on this page.

Section 18: The heading was "What Signifyeth Those Names." The text responded to this question with: "The multiplycand is that number which is given to be multiplyed and the multiplyor is that by which the sum is multiplyed and the produce is the sum of such multiplycation as for exampele if i would multiply 4 by 3 as in saying 3 times 4 makes 12 here the number 4 is the multiplycand and the number 3 is the multiplyor and the number 12 is the produce of that multiplycation.

Section 19: The heading was "What order is Belonging and to be obsearved in multiplycation and how are these numbers to be set." Then followed: "Before I teach you the true order of multiplycation I think it good to set down a table of multiplcation which if you do not Larn [learn] by hart you shall never multiply radyly [readily] nor quckly."

Section 20 (which is on the same page as *Section* 19, and immediately below the notes for *Section* 19). Following the heading, "Of Division what is Division," the response was "Division is that by which any number is deuided in to as many as you will."

Section 22: The heading was "How many numbers are Incident to Division and how are they called." Then followed: "These fouer [Therefore?] that is to say the numbre which is to be deuided is called the Deuidend, the sacond number whar by you do Deuide which is called the Deuseor the thurd number is called the quotient which shewe the how many times the Deuisor is Contained in Deuidend and fourth number is called the Remamder if any bee."

Section 23: The heading was "Of halfing any number how is that Done." Then followed—"By Deuiding the number by 2 as for exempel if you would know the half of 3708 Divide the sum by too and you shall find the quotient to be 1854 which is the inst [instantly?] one half of this dividend." Then followed: *Section 24:* "Of the rule of three otherwise called the Golden Rule whare of thare be three kinds that is the Comond Ruel and Ruel Reverse and Double Rule." [Note that in this last statement we have "Rule" three times and "Ruel" twice.]

Brief Commentary on the Text

Although Cohen (1982) claimed that in the 18th century local schools in the North American European settlements did not teach arithmetic as part of their regular course of instruction, Monroe (1917) named quite a few 17th-century local schools that offered cyphering, and it is likely that this practice would have continued into the 18th century. Arithmetic skills, and the associated penmanship skills, were obviously useful in colonial society. Money (1993) reported that in 1700 there were at least 17 private tutors within the European settlements in North America who advertised that they were willing to receive students for cyphering, and that between 1720 and 1740 this number grew to 49. So, it is quite feasible that the manuscript under examination in this chapter was prepared in one of the very early New England schools.

The first of the seven pages shows Section 15. It is likely that the notes were copied from, or perhaps dictated from, an existing manuscript—perhaps from another cyphering book—and that the number "15" on the first page corresponded to the number of a statement or section on one of the pages in that other document. That said, if 14, or so, earlier sections did exist, at some time, there can be little doubt that Sections 1 through 14 would have been concerned with "numeration," and with "addition," for within cyphering books those topics almost always preceded "subtraction."

In the heading "Substraction, a Lesson" in Section 15, the word "substraction" rather than "subtraction" was used. Although one can find "substraction"—instead of subtraction—in some cyphering books, the correct word, in English, was then (as it is still) "subtraction." It seems that sometimes there was confusion regarding the spelling because the corresponding French word was "soustration," which had an "r" after the first "t."

After the heading on Section 15, the text asked "What doth substraction teach." The answer given was: "It teacheth to take a number out of a grator and to see what remains." Such a definition would not be acceptable in the 21st century because for centuries it has been conceptually possible to subtract a larger number from a smaller number. But around 1666 that possibility was not contemplated.

It is difficult to make sense of the notes on Section 16 regarding how to subtract. However, the use of words like "Carried" and "lent" suggest that the person who prepared the manuscript was trying to find the words to describe a standard subtraction algorithm. From that point of view it is interesting that the one example given, 564 – 57, could involve "carrying." It seems that even at this early stage a language involving lending and carrying had been developed in an attempt to facilitate the teaching and learning of subtraction.

Sections 17, 18 and 19 were all concerned with elementary multiplication. The emphasis on appropriate language (multiplicand, multiplier, and product) points to an attempt to analyze the process of multiplication. Although no multiplication table is given, the comment that one would need to learn a multiplication table by heart if one wanted to be able to multiply "readily" and "quickly," would be heard over the centuries—whenever and wherever children were being asked to learn to multiply.

The distinction, in Section 22, between four components of division (i.e., dividend, divisor, quotient, and remainder) was another example of an apparent emphasis on appropriate terminology. The actual process of division was not described, and one wonders whether the person who prepared the cyphering book ever learned how to carry out formal divisions.

In Section 23, the issue of "halfing" is raised, and an actual example (half of 3708) is presented, and the answer (1854) given. However, other than to say that one needed to divide by 2, there was no discussion of how the answer would be obtained. The emphasis on finding one-half of a number was part of Sacrobosco's curriculum—he called the process of finding one-half, "mediation"

(Yeldham, 1926, p. 77). Although that process was never an important component of the *abbaco* curriculum defined by Leonardo of Pisa (Van Egmond, 1976), throughout the period 1200–1800 doubling and halving were dealt with in a minority of arithmetic textbooks across Europe (Jackson, 1906). Jackson (1906) stated that most contemporary commentators regarded doubling as simply another example of multiplication, and therefore did not see any need for it to be introduced as an independent topic. Indeed, one commentator wrote: "I do not understand what influences those stupid ones, since to double is to multiply by 2 and to halve is to divide by 2" (quoted in Jackson, 1906, p. 76). However, although, from a purely logical point of view the commentator made a valid point, from a psychological perspective it is well known that for children there is something special about doubling and halving.

The statement in Section 24 concerned the "golden rule," or the "rule of three." As the statement indicated, there was a "common" (or direct) rule of three, a "reverse" rule of three (the more common word in this context was "inverse"), and a double rule of three. No actual statements of these various rules of three were written down in this cyphering book, and one wonders whether the person who prepared the cyphering book ever actually heard, or learned to apply, any of the rules. For centuries it was a mark of prestige to have studied "up to the rule of three," and one suspects that the person who prepared this cyphering book wanted to be able to say that he (or she) had done that.

The Historical Significance of the Manuscript

Our decision to include this seven-page manuscript as one of 11 "extraordinary" cyphering books for analysis in this book was made on the basis that, despite its brevity, it qualifies as extraordinary for two reasons. First, it is very early—and if indeed it was prepared in Maine in the late 1660s, then it is easily the oldest of the known extant North American cyphering books. And, second, the quality of the mathematical discussion presented in the book is so very low—easily the lowest we have seen in any cyphering book. That is saying something, considering we (Ellerton and Clements) have carefully examined about 1000 cyphering books.

Let us assume, for the moment, that this cyphering book is the oldest extant cyphering book prepared in North America, and that the book was actually prepared somewhere in New England. Because of the British background of most settlers in New England, one would expect that it would not have taken long for the cyphering approach to arithmetic education, so widely used "at home," to be transplanted in the early New England settlements. However, such an approach was less likely to be as successful in the New World as it had been in Great Britain. In early New England none of the main requirements for success—necessary *mathematical knowledge* among teachers, sufficient *teaching power*, appropriate *student readiness* for learning arithmetic, and adequate *resources*—was satisfied (Ellerton & Clements, 2012). Furthermore, the early settlers in New England had not expected to have to endure the harsh winters which befell them. For them it could have been helpful to find a potentially useful activity—like cyphering—to occupy the minds of boisterous youngsters who found themselves more or less compelled to remain in small homes with not much to do.

Early North American settlements and states were not ready for a problem-solving mathematics regime like that followed by Thomas Dixson (see Chapter 7). Most of the settlers had had very little formal training on anything to do with mathematics and there were relatively few, if any, persons with sufficiently strong mathematical backgrounds to be in a position to teach mathematics effectively. Of the few who might have known something about mathematics, the demands of daily living in a new and difficult world could have precluded their having the time to take seriously the challenge of teaching formal mathematics to their children (Cohen, 2003). This state of affairs would have had an ongoing effect whereby because children did not learn much mathematics beyond mere counting, the next generation, the children's children, would not have had access to persons who could teach them mathematics beyond counting. The only mathematics textbooks were those brought over from the "mother country," and these were both difficult to get and expensive, in the New World. Little wonder, then, that very few of the children growing up in New England learned more than the absolute rudiments of elementary mathematics—beyond being able to read and write the Hindu-Arabic numerals which appeared on most hornbooks.

That would explain why the quality of mathematics in the cyphering book examined in this chapter was so low. Putting the matter succinctly, it is likely that neither the student nor his or her teacher (if a teacher was available) would have known much about the four operations on numbers, especially multiplication and division. But even in these early times, cyphering "up to the rule of three" was thought to be something desirable to do, and so this cyphering book closed with the student being able to say to himself (or herself) that he (or she) had reached the rule of three.

Undoubtedly, the approach to mathematics in the seven pages emphasized form rather than substance. With division, for example, the emphasis was on knowing the meanings of terminology like "dividend," "divisor," "quotient" and "remainder," but not on learning how to find the actual values of the quotient and the remainder.

It would be unfair to be completely negative with respect to the cyphering book under examination. Whoever prepared the pages of this book wrote in pages which originated in Europe. In the New World paper was expensive, and even the task of getting suitable quills with which to write, and of making suitable ink, was difficult and time-consuming. The pages were watermarked, and likely to have been expensive. The decision within a New England community in the 1660s to use such pages for notes on arithmetic pointed to a community which valued education.

The person who prepared the seven pages had actually learned to write, which of itself was no mean feat within an emerging culture in which opportunity for significant written forms of education was often lacking. Being able to write words like "multiplycand," and "halfing," and "deusior," however imperfect those words might have been, represented an achievement. To cypher "up to the rule of three" was the aim and, most likely, the person who prepared this book would have claimed that he (or she) had done that. In its own way, this cyphering book might represent the beginnings of formal mathematics education in the New World.

There is an old saying: "Well begun half done." It is doubtful whether the quality of the cyphering book examined in this chapter justified the expression "well begun." But at least there had been a beginning. The very existence of this manuscript testified to a determination among early colonists that their children should learn to read, to write, and to cypher. The New World was one of new horizons, and it seems that at least some parents wanted their children to experience forms of education that went beyond the child-minding, dame-school approach, with its hornbook culture. From that viewpoint, this small 1660s manuscript represented an important step forward.

References

Cohen, P. C. (1982). *A calculating people: The spread of numeracy in early America.* Chicago, IL: University of Chicago Press.

Cohen, P. C. (2003). Numeracy in 19th-century America. In G. M. A. Stanic & J. Kilpatrick (Eds.), *A history of school mathematics* (Vol. 1, pp. 43–76). Reston, VA: National Council of Teachers of Mathematics.

Ellerton, N. F., & Clements, M. A. (2012). *Rewriting the history of school mathematics in North America 1607–1861*. New York, NY: Springer.

Jackson, L. L. (1906). *The educational significance of sixteenth century arithmetic.* New York, NY: Teachers College Press.

Money, J. (1993). Teaching in the market place, or "*Caesar adsum jam forte: Pompey aderat*": The retaining of knowledge in provincial England during the 18th century. In J. Brewer & R. Porter (Eds.), *Consumption and the world of goods* (pp. 335–377). London, United Kingdom: Routledge.

Monroe, W. S. (1917). *Development of arithmetic as a school subject.* Washington, DC: Government Printing Office.

Van Egmond, W. (1976). *The commercial revolution and the beginnings of Western mathematics in Renaissance Florence, 1300–1500.* PhD dissertation, Indiana University.

Yeldham, F. A. (1926). *The story of reckoning in the Middle Ages.* London, United Kingdom: George A. Harrap.

Chapter 3
"Thomas Prust his Booke Amen 1702"

Abstract: The person who prepared most of the entries for the cyphering book studied in this chapter was Thomas Prust, who was born in England in 1690 and migrated to New England in 1720. There is a handwritten note in the cyphering book indicating that in 1720 Thomas passed it on to James Collings. Thomas became a shopkeeper and the training he received when he was preparing his beautiful, largely *abbaco*-inspired, cyphering book would have been beneficial to him during his time as a shopkeeper. He used the galley algorithm when performing division calculations, and was determined to master the rule of three. Many of the problems in the cyphering book were linked to real-life situations. The record indicates that James Collings became a sea captain, and he probably took this cyphering book with him on his voyages. Although we do not know how many later students were guided by this extraordinary manuscript, of one thing we can be certain: this was originally Thomas Prust's "booke," and over 300 years later most of it still remains intact.

Key Words: *Abbaco* curriculum; arithmetic; calligraphy; casting out nines; ciphering book; cyphering book; galley method for division; problem-based learning; rule of three; Sacrobosco.

Thomas Prust, who sometimes wrote his family name as Priest, signed his name on many of the folio-sized pages which remain from the cyphering book that he prepared during 1702 and 1703. But not all of the 88 pages in the cyphering book were prepared by Thomas, for about 10 were prepared by John Collings, Prust's step-son. Prust passed his cyphering book on to Collings in 1720, and on three occasions Collings indicated, in writing, that the book was now his. Unfortunately, Prust never indicated *where* he was when he was preparing his book. Genealogical records indicate that Collings was in Portsmouth, New Hampshire when, in the early 1720s, he began adding new pages to the cyphering book that he received from his step-father.

Genealogical records also suggest that the Thomas Prust who prepared this book was born in England in 1690 and migrated to New England in 1720. Later in 1720 he married Sarah Collins from Portsmouth, New Hampshire. There is a handwritten note in the cyphering book indicating that the book was passed on to James Collings by his "father-in-law" and earlier in the book James Collings declared, on November 20th, 1720, that the book was now his property. It seems that Thomas Prust was James Collings' father-in-law (which, at the time, sometimes meant what "step-father" means today), and that Thomas passed on his highly-valued cyphering book to the 13 year-old James, who was his new wife's son by a previous marriage.

If that conjecture is correct, then Thomas prepared his part of the cyphering book as a boy growing up in a well-to-do family in England. However, the book was passed on and used in North America, where it has remained for almost 300 years. It is reasonable to regard this as an American cyphering book—despite the fact that most of it was prepared by Thomas Prust when he was a boy in England.

It was the artist, Erik Erikson, who recognized the quality of the calligraphy in Thomas Prust's manuscript when he purchased it at an antiquarian book show in Boston 25 years ago. Erikson began collecting cyphering books in the 1970s as he found them to be a source of inspiration for some of his art. We acquired the manuscript from Erikson in 2006, and it has now become one of the

Nerida F. Ellerton and M. A. (Ken) Clements, *Abraham Lincoln's Cyphering Book and Ten other Extraordinary Cyphering Books*, DOI 10.1007/978-3-319-02502-5_3, © Springer International Publishing Switzerland 2014

cornerstones and inspirations of our collection—not only in terms of its age, but also in light of its fine calligraphy and the authenticity of the tradition it represents.

High Calligraphy in Thomas Prust's 1702–1703 Manuscript

Some of the calligraphy and penmanship in the early sections of Thomas Prust's cyphering is spectacular—as will become evident in some of the illustrations in the figures in this chapter which are reproductions of pages or parts of pages from what remains of Thomas Prust's manuscript. However, in the later pages the entries are more mundane—perhaps the most beautifully prepared pages were removed, but more likely Thomas lost much of his early resolve to make his cyphering book as attractive as he could possibly make it.

The first illustration (Figure 3.1) shows the cover page, on which Thomas inscribed his name in high calligraphy. Figure 3.2 shows the first page of mathematics in the book. Again, pride of ownership is evident in the remarkable calligraphic display for "Addition," and in the lovely designs that appeared on what was obviously a carefully planned page.

The original forms of Figures 3.1 and 3.2 are works of art in their own right. The final appearance of both pages was obviously carefully planned before the calligraphy was put in place. Thomas appeared to be announcing: "Reader, please note, this is to be my book for life." In keeping with a resolve, of that kind, throughout the book Thomas usually made sure that his entries were arithmetically correct. Toward the end of the cyphering book, however, the quality of calligraphy, penmanship, and arithmetical accuracy declined.

If one can overlook the "old worlde English," the first sentence on the page shown in Figure 3.2 has flair. In its unedited form it stated: "Addition is an arithmeticall gathering of two or more numbers into one total summe in which the rule we must observe to begin at the right hand and invite the particular summes of the several rows under every one in its own place." Then followed four different addition tasks, each of which involved large numbers. It is not known whether the notes, and the examples, were taken from a textbook, or from another cyphering book, or were dictated by a teacher.

The next page of the cyphering book began with a word problem that was clearly linked to the concerns of the mercantile world. An unedited version of this problem is:

> A merchant hath 4 ships safely arrived from this yeare 1702 in which hee had severall goods visit in the good ship called the *Endever* which came from Lisborn 8 chestes of suger and 25 tapnals of figs and in the good ship called *Exchang* which came from Lisborn with 52 chest of shuger and 125 tapnals of figs and in the ship called the *Prosperous* which came from Lisbon with 564 bars of iron and 30 bags of wooll and in the *Green Dragon* 431 bars of iron and 22 bags of wooll. Now the question is to know how many chest of suger, tapnals of figs, bars of iron, and bags of wooll this merchant hath come home severally in all these 4 shipps.

Figure 3.1. The cover page of Thomas Prust's cyphering book.

Figure 3.2. Addition—The first page in Thomas Prust's cyphering book with mathematics.

That was the first of many real-life word problems in the cyphering book. Then followed two problems involving the addition of large sums of money (expressed in pounds, shillings and pence). With respect to the above word problem, it should be noted that in the 17th and 18th centuries, merchants consistently shipped goods to the American colonies from Lisbon (Portugal) (Lounsbury, 1930; Lydon, 1965; McCusker & Menard, 1985).

On page 5 the topic "Substraction" was introduced, once again by a heading in high calligraphy (see Figure 3.3). As noted in the last chapter, although one can find "substraction"—instead of subtraction—in some cyphering books, the correct word, in English, was then (as it still is) "subtraction." The words that Thomas used were: "Substraction teacheth to abate, drawe or take a lesser number from a greater or an equall from an equall where by we find the rest remainder excess or diffrance."

Substraction teacheth to abate drawe
or take alesser Number from a greater or an
equall from an equall where by we find
ye rest remainder exces or diffrance

Figure 3.3. Thomas Prust's introduction to "substraction."

Following this introduction, subtraction tasks arising from pairs of very large numbers, or quantities involving money or weights and measures, were shown, and "proofs" that the answers obtained were correct were shown by using the inverse operation of addition. Figure 3.4 shows two word problems that Thomas solved by using "substraction"—both were set in the context of a merchant buying and selling goods. But the most impressive part of this page was the calligraphic "Thomas Prust his booke Amen 1702" at the bottom of the page.

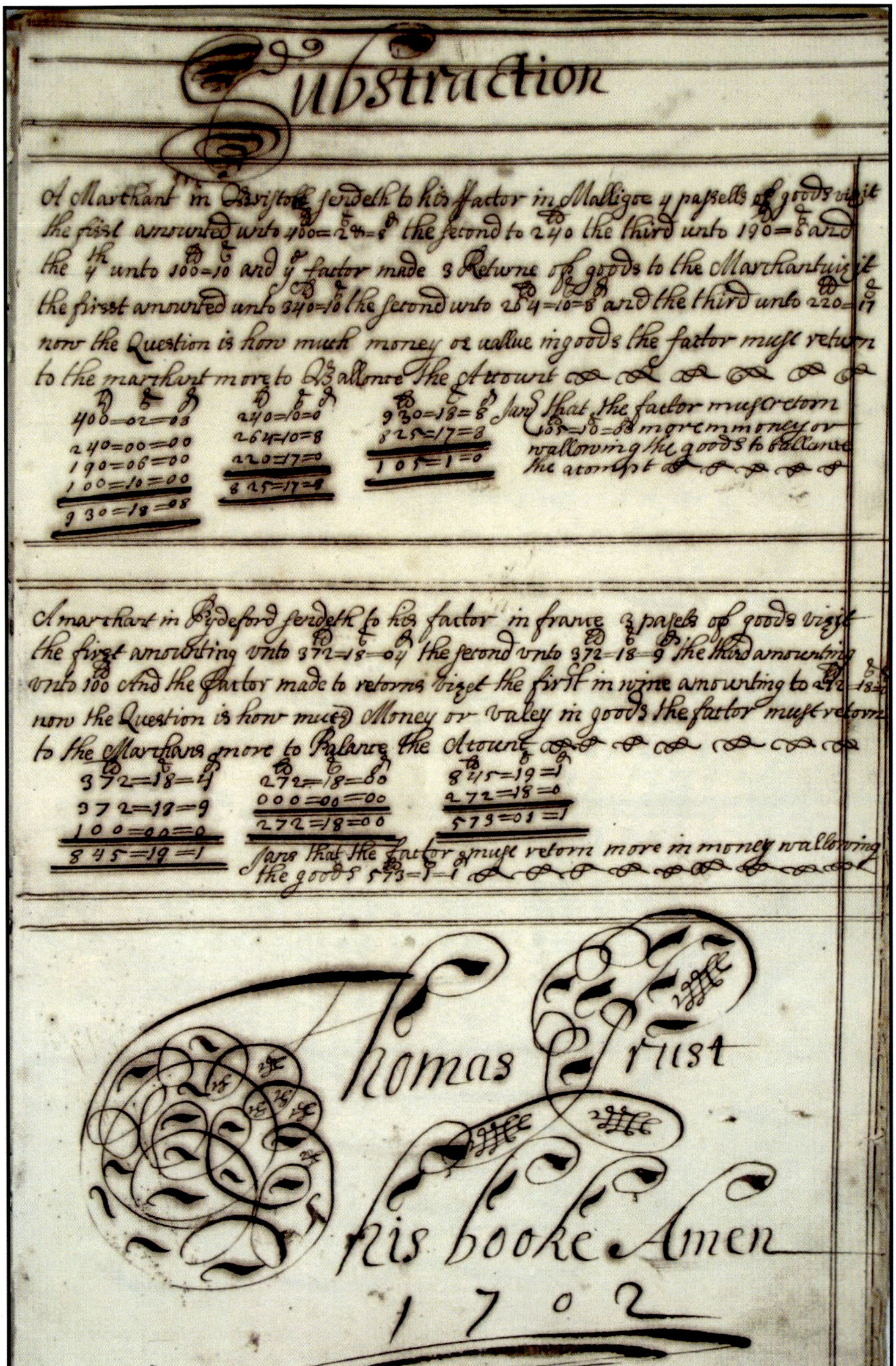

Substraction

A Marchant in Bristoll sendeth to his factor in Malligoe 4 passells of goods vizit the first amounted unto 400=2=8 the second to 240 the third unto 190=6 and the 4th unto 100=10 and ye factor made 3 Returne of goods to the Marchant vizit the first amounted unto 240=10 the second unto 264=10=8 and the third unto 220=17 now the Question is how much money or vallue in goods the factor must return to the marchant more to Ballance the Account

400=02=08
240=00=00
190=06=00
100=10=00
930=18=08

240=10=0
264=10=8
220=17=0
825=17=8

930=18=8
825=17=8
105=1=0

Ans that the factor must retorn 105=1=0 more in money or vallowing the goods to ballance the acount

A marchant in Bydeford sendeth to his factor in france 3 passels of goods vizit the first amounting unto 372=18=04 the second unto 372=18=9 the third amounting unto 100 And the factor made to retorns vizet the first in wine amounting to 272=18 now the Question is how much Money or valey in goods the factor must retorn to the Marchant more to Balance the Acount

372=18=4
372=18=9
100=00=0
845=19=1

272=18=00
000=00=00
272=18=00

845=19=1
272=18=0
573=01=1

Ans that the factor must retorn more in money vallowing the goods 573=1=1

Thomas Prust his booke Amen 1702

Figure 3.4. "Substraction," and "Thomas Prust his booke Amen 1702."

Figure 3.5 is from Page 8 of the Prust manuscript. After announcing the topic of multiplication in attractive calligraphic form, definitions of relevant terminology, not unlike definitions of similar terminology given in the circa 1667 cyphering book studied in Chapter 2, were given: "Multiplication compendiously performes the worke of many additions for which rule belongs three principale members: first, the multiplicand or numbers to be multiplyed together; secondly, the multiplyer or number by which the first is multiplyed; thirdly the product or number produced by multiplying of the two former ..." The most spectacular part, however, was the example given. The multiplicand was 76987693276987698769876 and the multiplier was 492123456789. A long multiplication is shown, with a 35-digit answer which could not fit across the page—so Thomas moved onto an oblique line on the right side of the page.

Of course, the issue of whether the calculations were all correct was a consideration—but Thomas showed a check which suggested that no mistake had been made. The check was done by a process known as "casting out nines."

If one adds up the digits in the 23-digit multiplicand, one gets 160, and if that is divided by 9 the remainder is 7. If one adds up the 12 digits in the multiplier, one gets 60, and if that is divided by 9 the remainder is 6. They are the digits shown on the top and left side of the cross in Figure 3.5. Then, if these two are multiplied, and the product divided by 9, the remainder is 6, which is the digit shown on the right of the cross. What remains to complete the check is add up the 35 digits in the overall product, divide the number obtained by 9, and the remainder should equal 6 (the number on the right). If the 35 digits are summed then the answer is 177, and when that is divided by 9 the remainder is 6, which is inserted at the bottom of the cross. Since the same digit appears at the bottom of the cross and at the right of the cross, the check is complete and it can be assumed that the calculations were correct.

However, in fact, there were several errors in the calculations. For example, the fourth digit from the top in the oblique part of the product should be 8, and not 9 (that can easily be checked by adding the final four columns). That means that there was at least one other calculational error. Thus, the casting-out-nines check indicated a correct product had been obtained when it had not. Nevertheless, the casting-out-nines check can be useful, and was used by other persons who prepared cyphering books that are considered in this book.

Nearly all of the word problems in Thomas Prust's manuscript were related to everyday events. In a section on "multiplycation practice," for example, the following delightfully-worded problem was solved (original spelling retained):

> A tayler mared his daughter to a sayler and gave her 100 needels and to every needle he houngs 20 threeds and to every end of every threed houngs 12 purses and in every purs he puts 12 penses now the question is, what portion of money the tayler gave with his daughter in marage to the sayler?

Figure 3.6 shows Thomas Prust's statement of the problem and his solution. The shape with numbers enclosed on the right of Figure 3.5 shows a check of the multiplication calculations using what was known as the "galley" method for division. That method will be discussed in some detail in Chapter 7, when Thomas Dixson's cyphering book will be the main subject of attention. Thomas Prust's introduction to division is shown in Figure 3.7. The actual division task shown was 789123456789123456789 divided by 123456789.

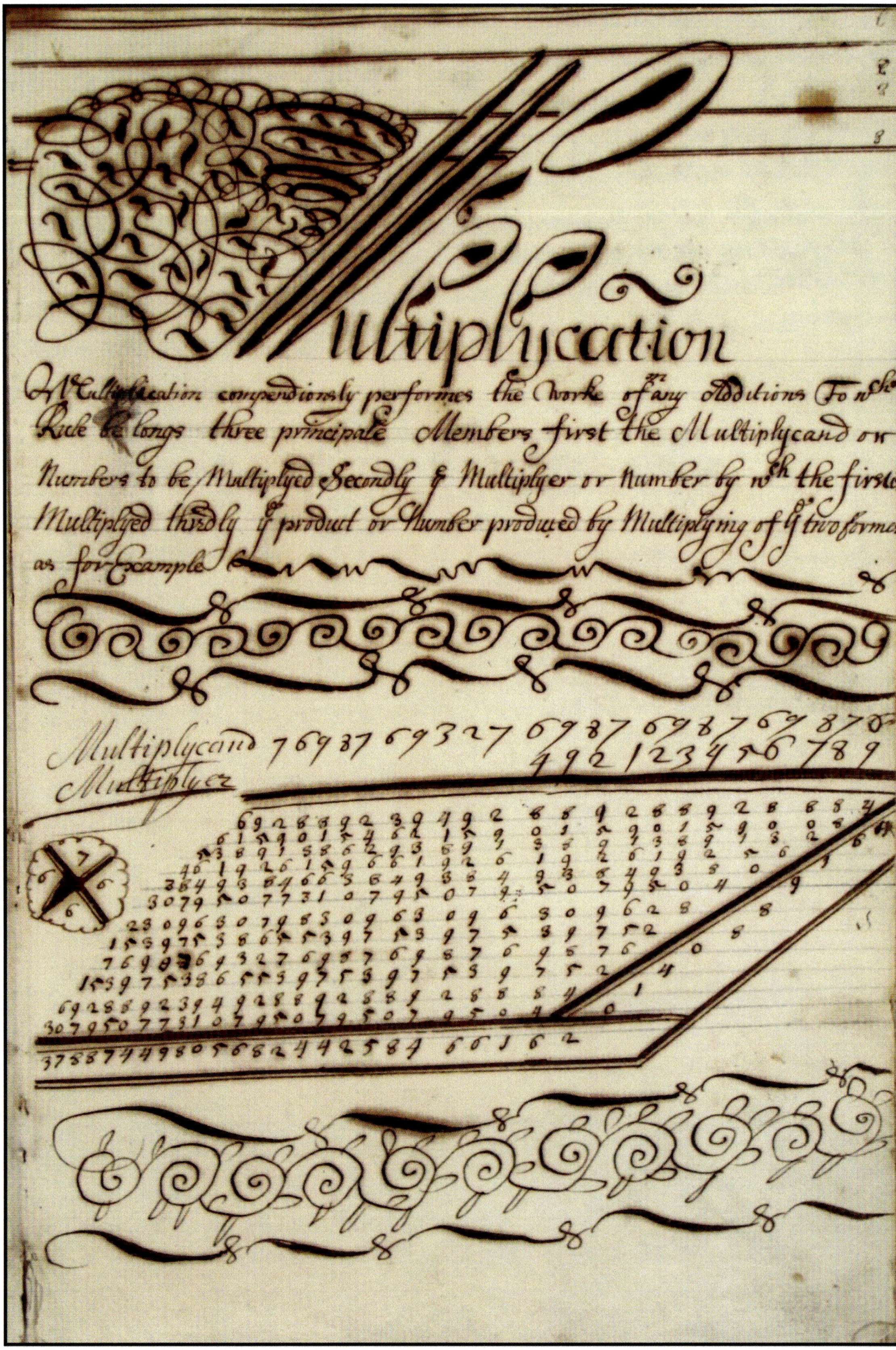

Figure 3.5. Multiplying a 23-digit number by a 12-digit number, to get a 35-digit number.

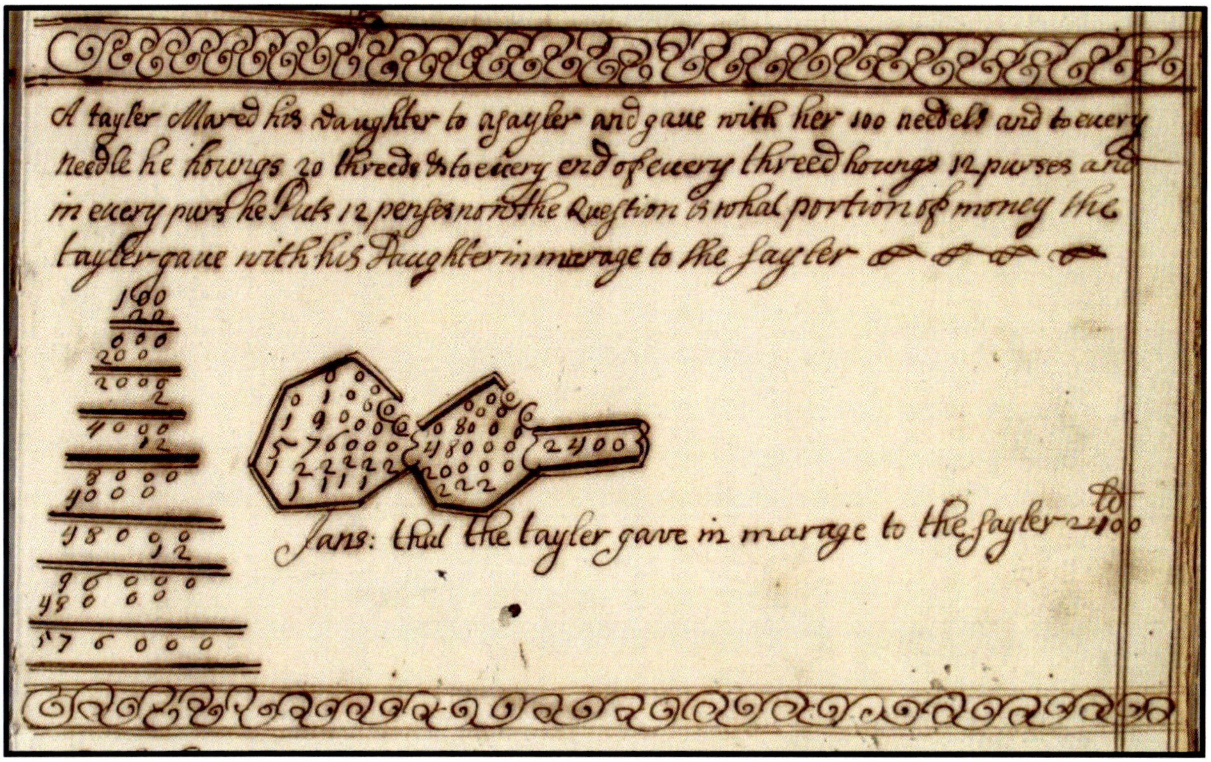

Figure 3.6. "A taylor mared his daughter to a sayler": From Thomas Prust's (1702) cyphering book.

Following the sections on the four operations on whole numbers, Thomas Prust moved on to topics in which he applied the algorithms he had developed and illustrated. The first of these was mediation (or partitioning), and then came reduction, the various rules of three (direct, inverse and double), fellowship, practice and common fractions.

"Mediation" was probably the only topic in the book for which no applications were given. Mediation was concerned with partitioning a larger number into equal smaller numbers—that is to say, dividing a whole number by *n,* where *n* could be any whole number. In Thomas Prust's cyphering book, the value of *n* ranged from 2 through 10. Mediation was a topic preferred by Sacrobosco (also known as John of Holywood), a thirteenth-century writer (Yelland, 1926). Its inclusion within Thomas Prust's book pointed to the book's English origins. In Chapters 6 and 7 of this book, the ideas of Johannes de Sacrobosco will be further discussed—here it suffices to note that up to about 1700 his writings were particularly influential in Great Britain, especially in relation to the kind of arithmetic needed to enhance mercantile efficiency.

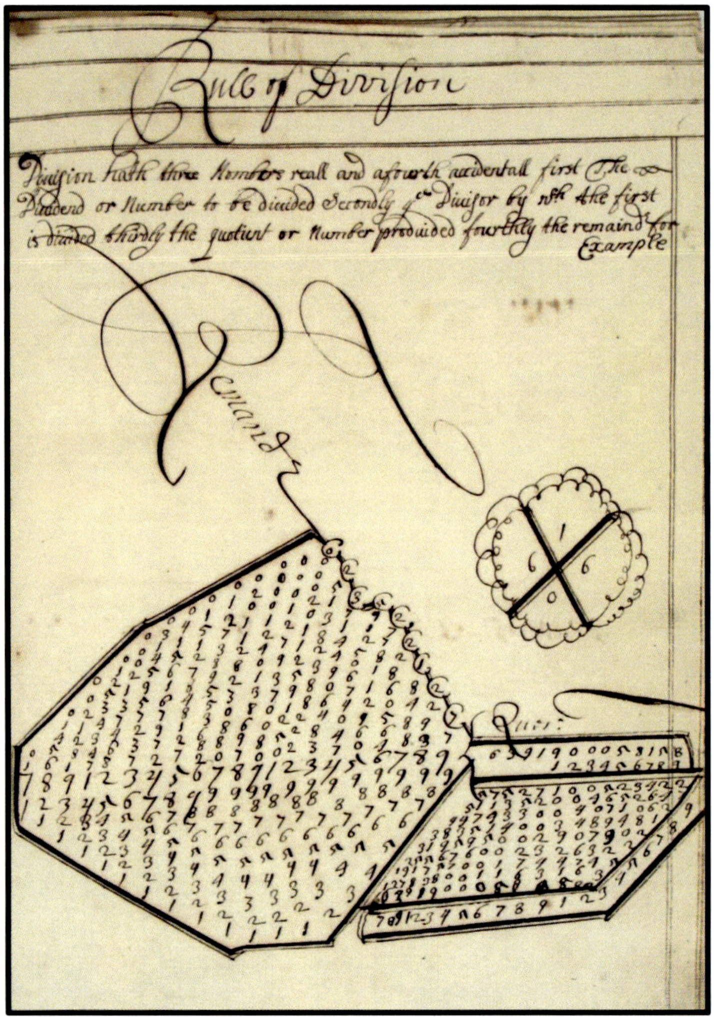

Figure 3.7. The rule of division, and an example of division using the galley algorithm.

The topic presented after mediation was reduction. This required a knowledge of units for measuring quantities, and on this Thomas Prust prepared the page reproduced as Figure 3.8. "Reduction" tasks were of two kinds. The first was to find the number of smaller units in a larger unit (like, for example, "how many inches are there in $2\frac{1}{4}$ miles?" The second was the reverse of this—find the number of larger units in a given number of smaller units.

At the top left of Figure 3.8 it was indicated that there were 20 shillings in a pound, and also 2 angles, or 3 nobles, or 240 pence in a pound. There were 2 nobles or 13 shillings or 160 pence in a marke—but note that also there were 12 pence in a shilling, 13 times 12 did not equal 160. Much of the information shown in Figure 3.8 is now totally obsolete, but in the early 1700s it needed to be known, both in Great British and in the New England colonies in North America.

Although the correspondences in Figure 3.8 were carefully presented, the level of penmanship had slipped from what it had been earlier in the cyphering book, and Thomas never got around to including a calligraphic heading on the page. But, in the bottom right corner he still felt the need to indicate, by writing his name, that this was "his booke."

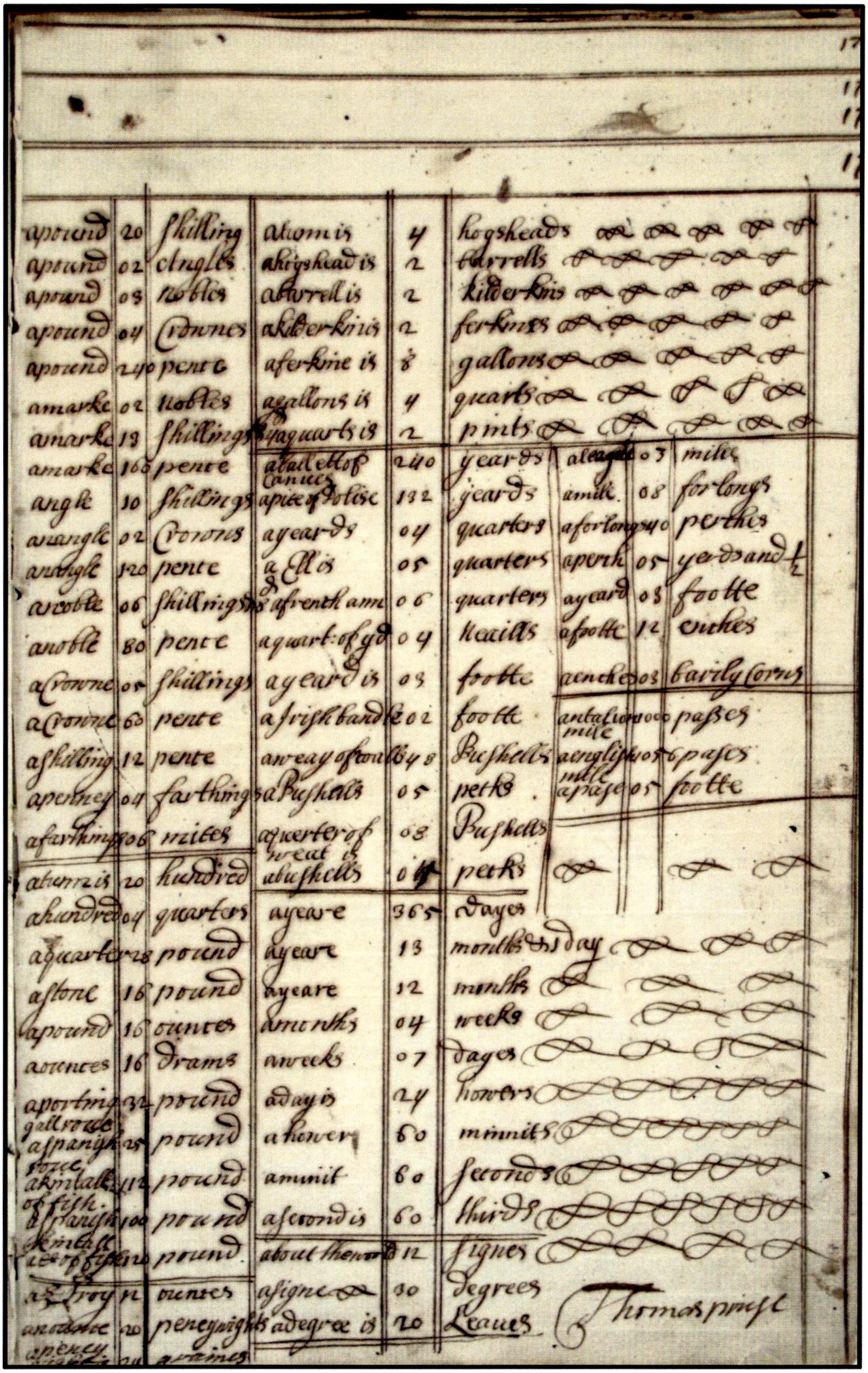

Figure 3.8. Thomas Prust's page showing unit equivalents for various quantities.

Thomas Prust dedicated many of the pages in his cyphering book to answering questions on "reduction." The first of these pages is shown in Figure 3.9.

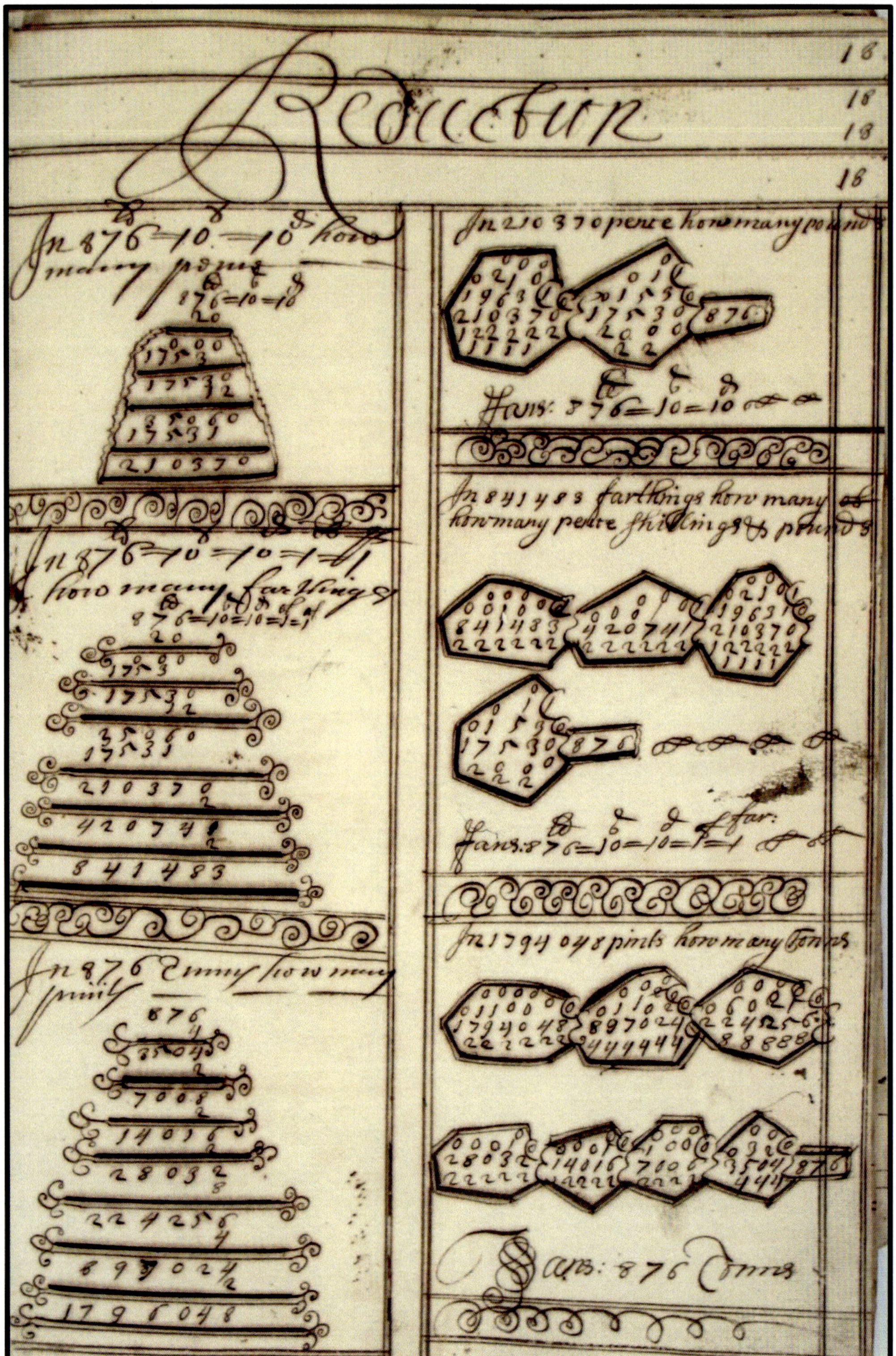

Figure 3.9. Reduction tasks—in two directions.

The first problem (top left) asked: "In 876 pounds 10 shillings and 10 pence how many pennies?" To answer this, Thomas multiplied the 876 by 20 (adding 10 as he proceeded), to obtain 17530 (shillings), and then he multiplied this by 12 (adding 10 as he proceeded). The (correct) answer that he obtained was 210370. Then (top right) he considered the reverse problem, "In 210370 pence, how many pounds?" Using the galley algorithm he first divided 210370 by 12, and then he divided the answer by 20, getting 876 pounds, 10 shillings and 10 pence. This kind of work not only required concentration but also a level of understanding associated with dealing appropriately with the meanings of remainders.

A similar structural approach to that shown with the tasks in Figure 3.9 is evident with the reduction tasks shown in Figure 3.10. The major task (left side) was: "Demand how many signes and degrees, leagues, miles, furlongs, perches, yards, feet, inches and barlicorns of those will round the compass of this whole world." Starting with 12 signes—there are 12 signs of the zodiac, each consisting of 30 degrees—Thomas recorded a sequence of multiplications that led him to the result 4105728000 barlicorns. Then, on the left side Thomas showed a series of galley divisions that enabled him check his calculations on the left side. Yes, 4105728000 barlicorns were 12 signes.

One cannot let this example pass without commenting on the vast array of unit equivalents that those learning arithmetic around 1700 were expected to know. The situation was not totally negative from an educational point of view, because the students were expected, as the examples shown in Figures 3.9 and 3.10 reveal, to be able to demonstrate arithmetically the equivalents, and in both directions. Many commentators (e.g., Cohen, 1982; Martin, 1897) have drawn attention to the conglomeration of units. In North America, ultimately, part of the problem would be relieved when, soon after the United States became an independent nation, it introduced a fully decimalized national currency—the United States of America was the first nation in the world to do that. Furthermore, in the early 1790s it almost became the first nation to introduce a fully decimalized system of weights and measures (Linklater, 2002)—but, ultimately, that initiative was rejected by the politicians and business leaders of the day, and today the United States is one of only two nations in the world not to use the metric system on a daily basis.

In the standard *abbaco* curriculum sequence, the topic "reduction" was followed by the "rule of three direct," which was definitely the most famous rule in all of arithmetic during the period from 1200 through 1850. Known as the "golden rule" (or, simply, the "rule of three"), it defined the standard approach to the crucially important study of proportionality throughout the period. It was sometimes introduced through the following poem (variants of which can be found in many cyphering books):

> The golden rule hath places three,
> The first and third must so agree,
> That of one kind they may remain,
> If to the truth thou will attain.
> Then third by second multiply,
> Divide by the first ingeniously,
> Then will your quotient shew the same,
> Which you in proper place must frame.
>
> (Quoted in Ellerton & Clements, 2012, p. 55)

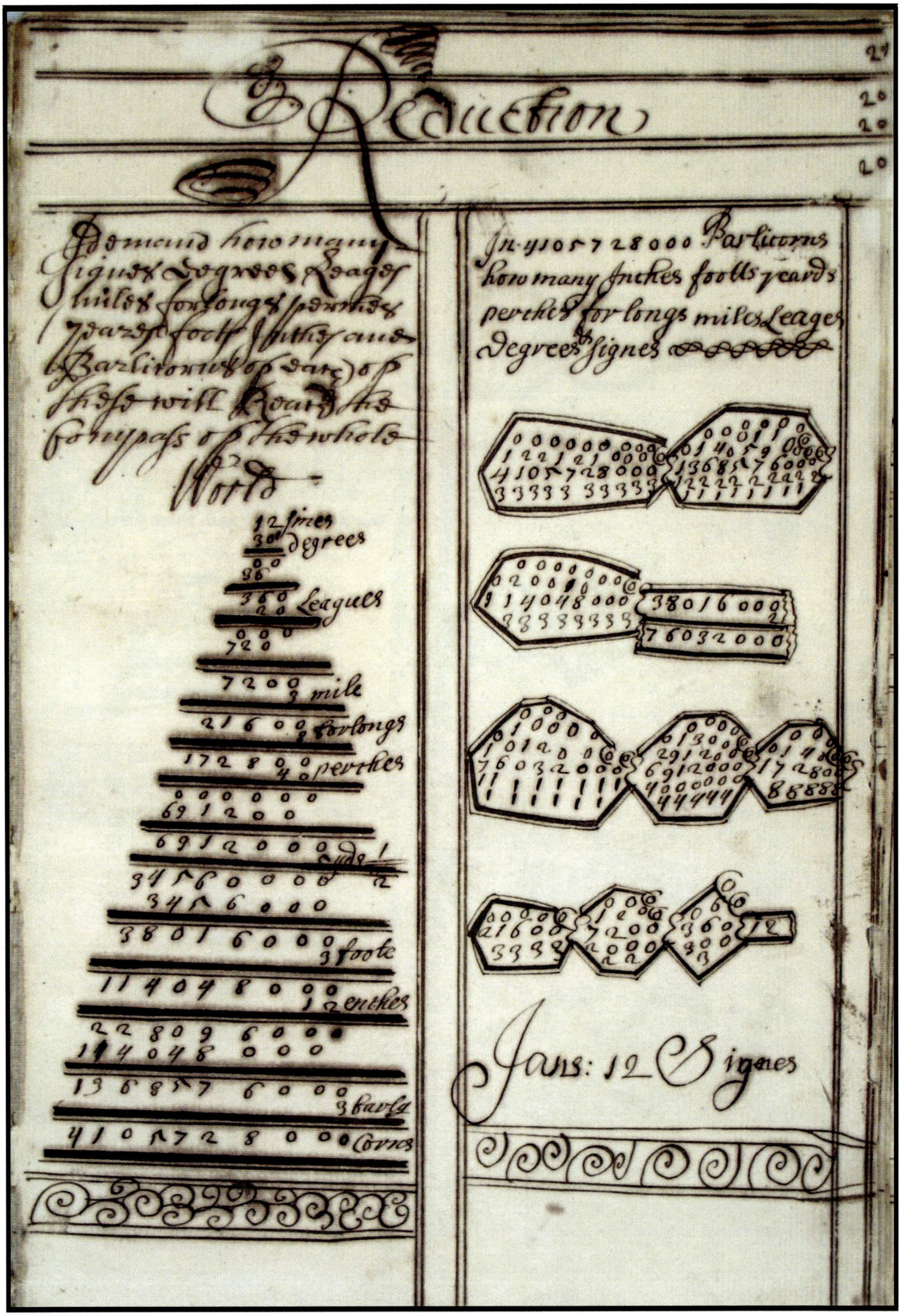

Figure 3.10. Reduction tasks involving many kinds of distance units.

Thus for example, in a preliminary exercise, Thomas Prust looked at the question "If in 319 weeks I spend 53 pounds 6 shillings, how long would a marke last?" From Figure 3.8 it can be seen that 1 marke was equivalent to 13 shillings and 4 pence. So, the question could be rephrased:

- If in 319 weeks I spend 53 pounds 6 shillings, how long would 13 shillings and 4 pence last?

Or, better still,

- If I spend 53 pounds 6 shillings in 319 weeks, then 13 shillings and 4 pence would last how many weeks?

So, in terms of the above rule-of-three poem, the first thing to do was to make the first and the third terms "agree"—that is to say, 53 pounds 6 shillings needed to be converted to 12792 pence. Once that was done then the question could have been summarized, by saying the first term was 12792, the second was 319, the third was 160, and the fourth could be obtained by multiplying 319 by 160 and then dividing that product by 12792.

Those calculations should have generated an answer of slightly less than 4 (weeks). But, Thomas Prust got exactly 4 weeks for his answer (Figure 3.11 shows his response to the task). Which raises the question: Why was there any discrepancy?

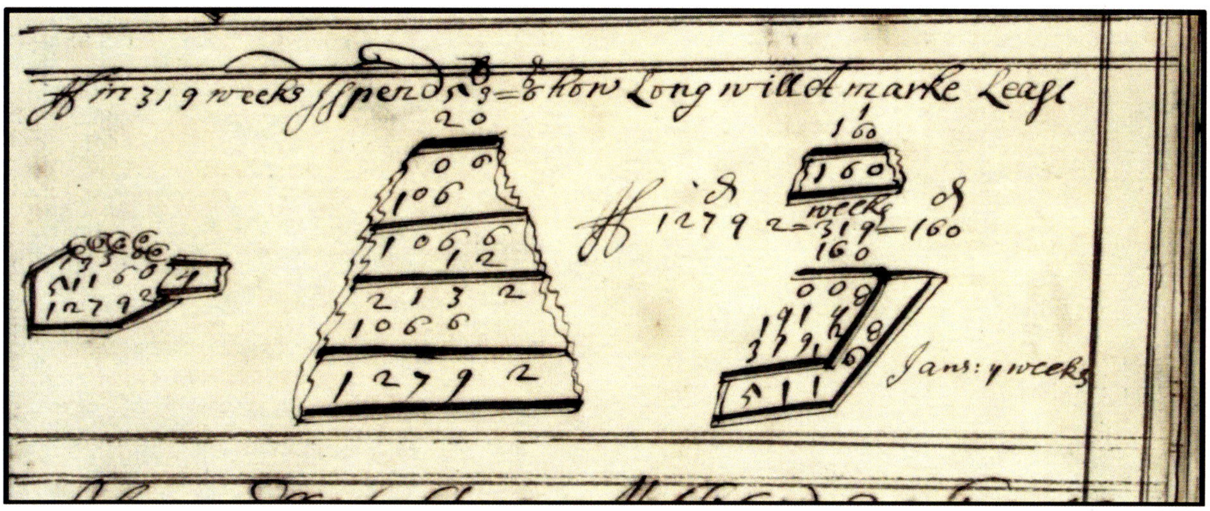

Figure 3.11. Thomas Prust's response to the task: "If in 319 weeks I spend 53 pounds 6 shillings, how long would a marke last?"

The reason for the discrepancy was that, when using the direct rule of three, Thomas arrived at a point where he needed to divide 51040 by 12792—which is equal to a number slightly less than 4. Thomas tried to use the galley method for division (see his working to the left of Figure 3.11), but without a knowledge of decimals he was unable to do what was needed. The fact that he gave an answer of 4 suggests that he had good estimation skills.

After a few preliminary rule-of-three exercises, Thomas tackled the following problem (shown in unedited form in Figure 3.12). We have not been able to decipher the question completely, but the working which follows it gives a good idea of what the problem was all about. It seems that three groups were to share the profits of a fishing investment: one was to receive one-fifth of the profits, another one-fourth, and a third group what remained of the profits.

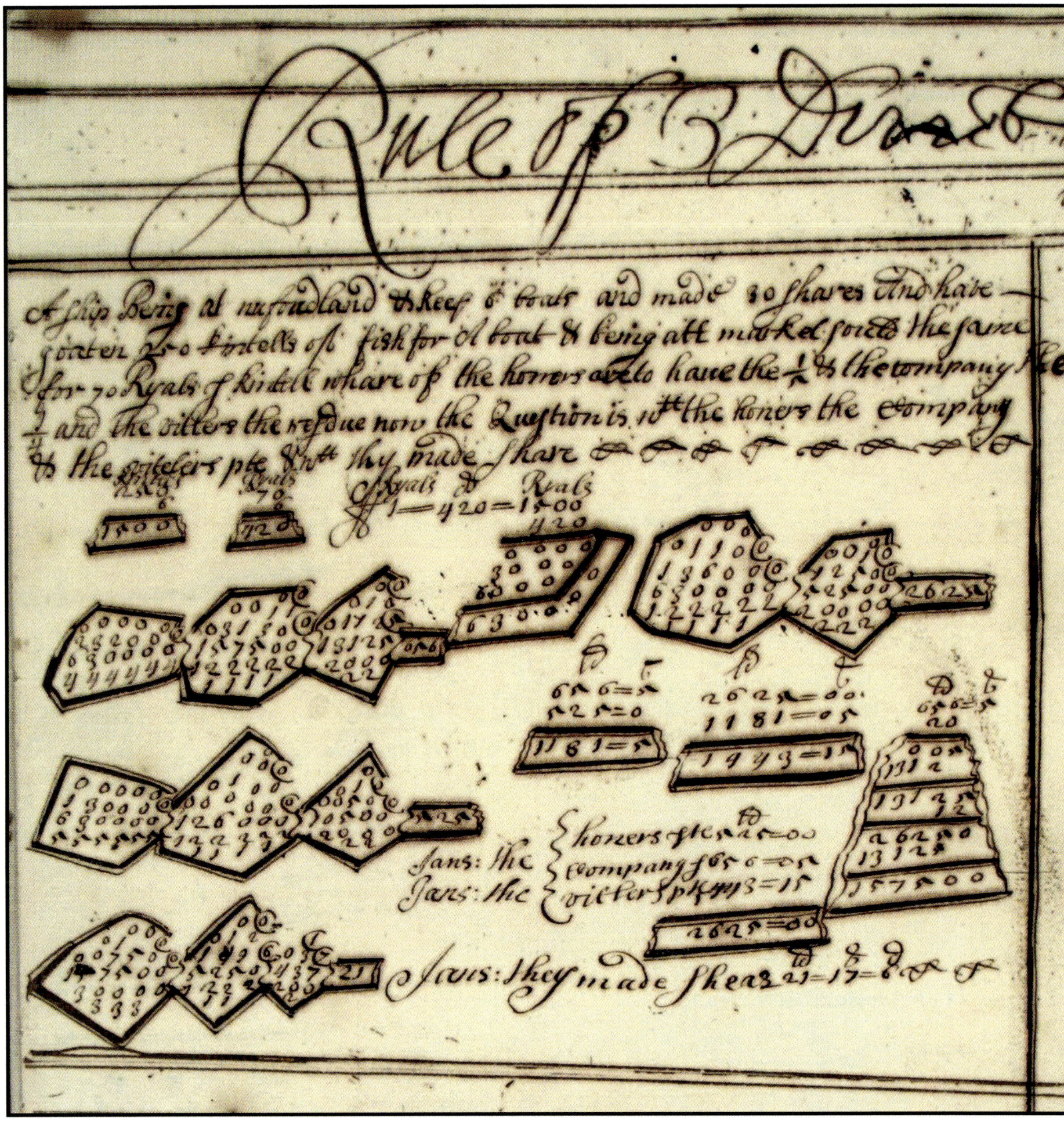

Figure 3.12. Thomas Prust's use of the rule of three when solving a real-life problem.

This was a genuine word problem, one which required the development of a strategy in order to solve it. Non-trivial unit conversions were needed, and it can be seen that at least four different divisions were needed.

Thomas Prust on Beyond-the-Rule-of-Three Tasks

Throughout the period 1200–1850 to have cyphered to the "rule of 3 direct" was regarded as a sign that a person was an efficient reckoner. Anyone who studied *beyond* the rule of three direct, in topics like the "rule of 3 inverse," or the "double rule of three," or any of simple interest, compound interest, discount, loss and gain, barter, alligation, fellowship, false position, common fractions, decimal fractions, involution and evolution, mensuration, arithmetical and geometrical progressions, and permutations and combinations, was to be regarded as an advanced reckoner, ready for employment as an applied problem solver. Thomas Prust cyphered beyond the rule of three, making entries in his book on the rule of three "reverse" (i.e., inverse), the rule of five numbers (i.e., the double rule of three), fellowship, and common fractions (in that order). The level of calligraphy and penmanship evident in the second half of Thomas Prust's cyphering book is lower than in the first half.

Thomas Prust and the Rule of Three Reverse

Thomas started this "advanced" section of his cyphering book with a statement about the meaning of "reverse" proportion. Here is an unedited version of what he wrote:

> In setting foorth this rule you are to observe the same deriction that is the rule of three dericte. But in the operation of this rule it is contrary, Therefore you must hear: Multiply the second and first number together and that produce devide by the last number and the answare comes foorth in the questions as for example ...

Undoubtedly, this wording of the rule would appear to a modern reader as clumsy and even unclear. Thomas tried to clarify matters with his first example, which in unedited form stated: "If when wheat is at 4 shillings and 2 pence per bushel then the penny white lofe doth weight 9 ounces neat, how much must the penny white lofe weight when weat is at 3 shillings per bushel." Thomas began his solution to this problem by multiplying 3 by 12 to get 36 and by converting 4 shillings and 2 pence to 50 pence. He then multiplied 50 by 9 to get 450, which, using galley division, he divided, by 36 to get 12 and 18 remainder. He expressed his answer as 12 ounzes $\frac{18}{36}$. Thomas's actual handwritten solution is shown, following the statement of the rule for "reverse," in Figure 3.13.

Using modern algebraic notation the "penny white lofe" problem could be summarized as: "If $50 \rightarrow 9$ then what is the value of x if $36 \rightarrow x$?" Notice that if the price of wheat decreases then the size of the loaf will increase—that is to say, inverse variation is involved. According to the rule that Thomas Prust stated, to get the answer, in ounces, one needed to multiply 50 by 9 and then divide by 36. And, that would, in fact, generate the correct answer. The important educational issue, of course, is what Thomas was thinking when he solved this problem. Was he merely following the rule? Or did he have a genuine understanding of the inverse proportion that was built into the problem situation? One can never know.

In some of the other cyphering books on which we focus in this volume students progressed from the rule of three direct to the rule of three inverse (see, e.g., Chapter 6, with Abraham Lincoln's cyphering book). The same educational issue, relating to the student's degree of understanding, has never gone away. Nevertheless, it is interesting to note that all of the exercises on the rule of three "reverse" that Thomas Prust solved were embedded in real-life contexts. Thomas Prust even included an18-line poem concerned with "a gentle woman" who "did her selfe betake of needle worke a carpet rare to make." The poem had built into it a simple reverse (or inverse) proportion problem, which Thomas duly solved. But to what extent did he understand, relationally, what he was doing?

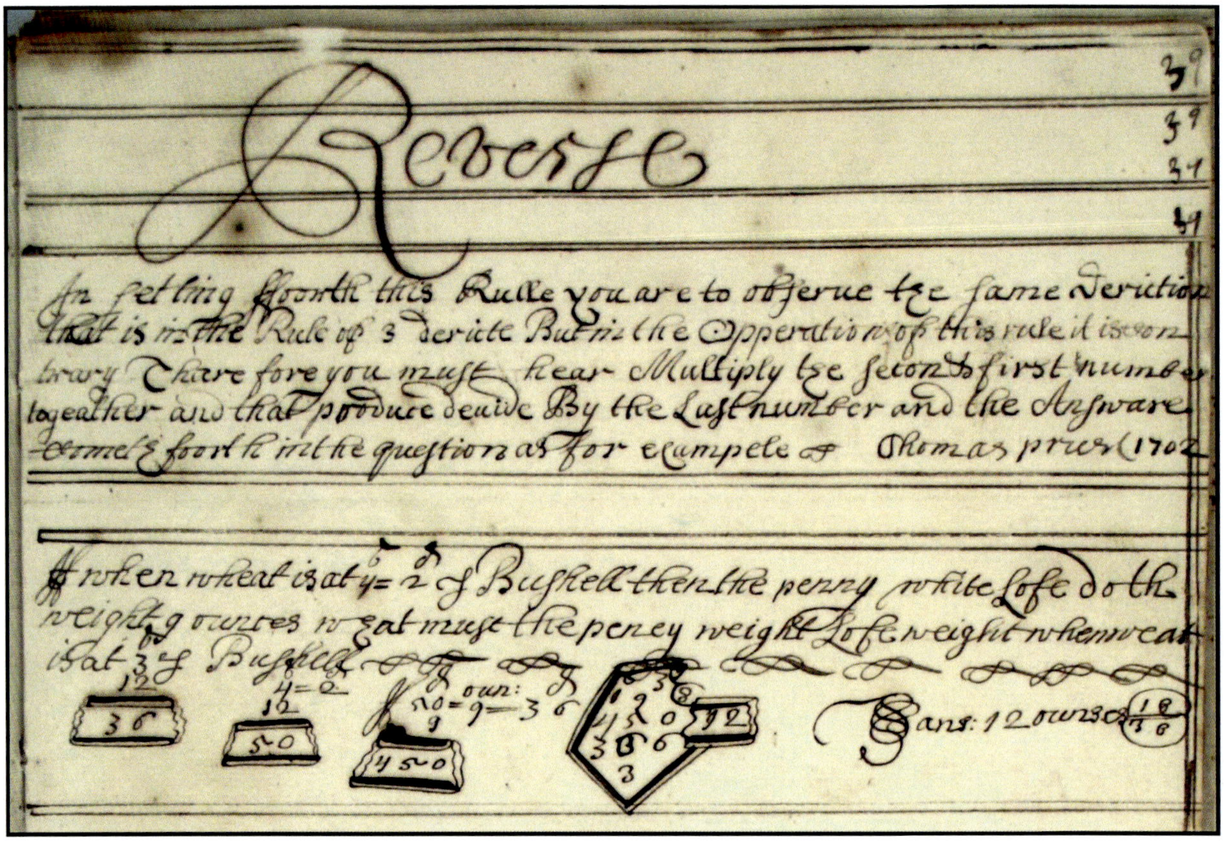

Figure 3.13. The rule of three "reverse," and the penny white lofe problem.

Thomas Prust and the Rule of Five Numbers

Immediately after his work on the rule of three "reverse," Thomas moved on to a consideration of what he termed the "rule of five numbers." The rule to which he referred was often known, even around 1700, as the "double rule of three." Thomas started with a statement about what he was calling the rule of five numbers. In unedited form, this statement was:

> If you do worke any question of the compound rule of 5 which consisteth of 5 places then you must set them down in such a order as that the first and fouerth must be of one denomination and the second and last number must be of on denomination and the third or midell must be of a denomination of sellfe such as the question requires and then your question or answare will be the same denominator your midell number of having sett them out in such order then you must work then orderly in this manner you must multiply the third and last number to gather-or-one into the other and that set out for your deuident and the multiply the first and second one into the other so that poduce make your deuiser and then deuide the one by the other and then each quistion [y]ealdeth your answare as for exampell 1702 Thomas Prust 1702.

The original text is shown in Figure 3.14.

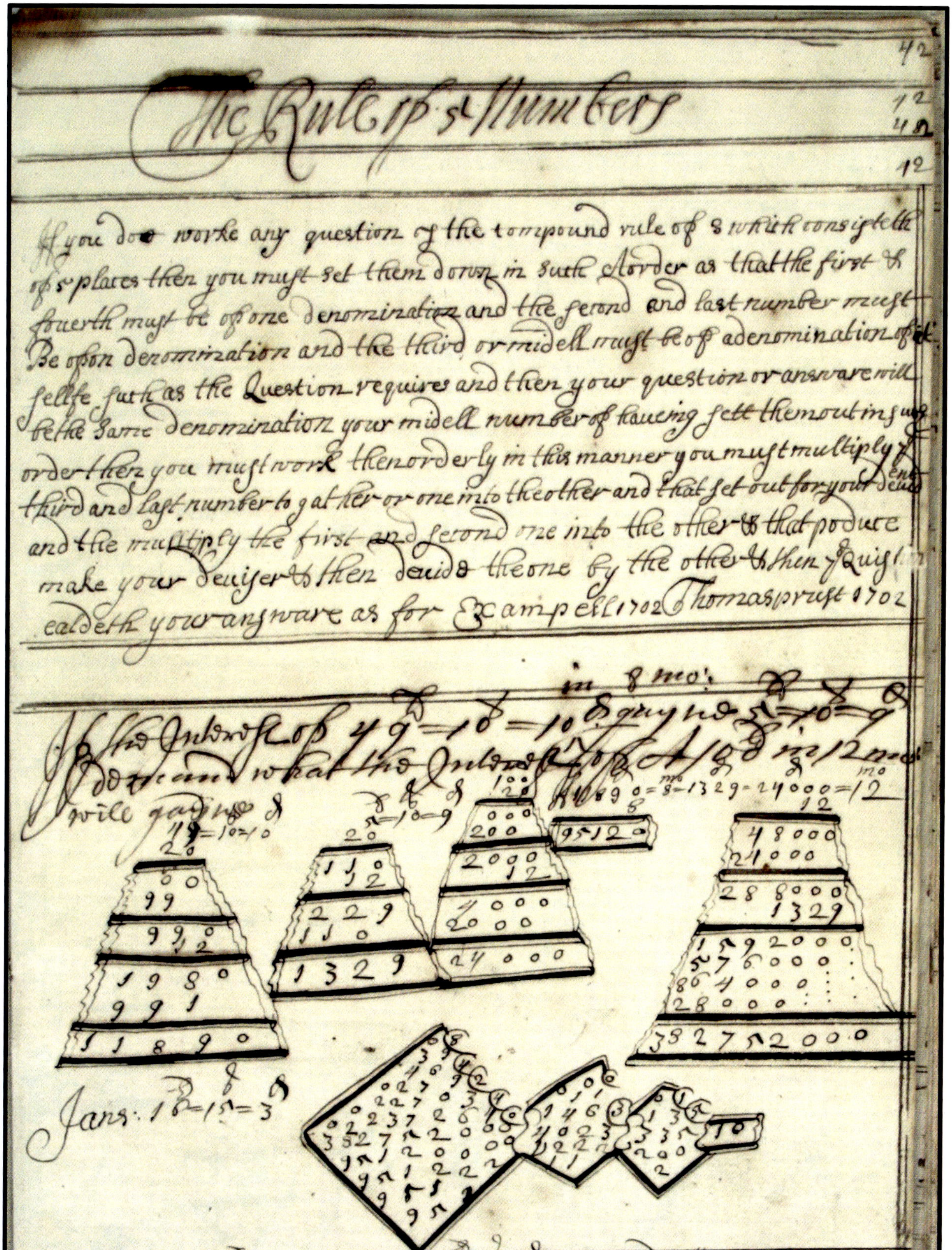

Figure 3.14. Thomas Prust's solution to a double-rule-of-three task.

The first problem on the rule of five numbers addressed by Thomas was the following:

> If the interest of 49 pounds 10 shillings and 10 pence in 8 months is gained 5 pounds 10 shillings and 9 pence, demand what the interest on 100 pounds in 12 months will gayne.

First, Thomas converted the 49 pounds 10 shillings and 10 pence, the 5 pounds 10 shillings and 9 pence, and the 100 pounds to 11890, 1529 and 24000 pence, respectively. He then multiplied the 11890 by 8 to get 95120 and the 24000 by 12 to get 28800. He then multiplied the 288000 by 1329 to get 382752000 and, using the galley algorithm, divided this number by 95120 to obtain 16 pounds 15 shillings and 3 pence. Figure 3.14 shows his calculations:

Again, one wonders how much of this Thomas understood.

Fellowship Task, and PCA Genre

Later in his cyphering book Thomas dealt with tasks involving fellowship without time (sometimes called single fellowship) and fellowship with time (sometimes called double fellowship). Figure 3.15 shows (unedited) his solution to the following "fellowship-with-time" problem (also unedited):

> 4 men compained to trade for gayne. *A* putt in 398 pounds 17 shillings and 3 pence for 8 months, *B* putt in 130 pounds 11 shillings and 10 pence for 9 months, *C* put in 174 pounds 10 shillings and 3 pence for 10 months, and *D* put in 189 pounds 10 shillings and 2 pence for 12 months, and they gained 590 pounds 12 shillings and 9 pence. I demand what is each mans pte of the gayne.

The solution shown in Figure 3.15 provides an excellent illustration of what Ellerton and Clements (2012) referred to as the Problem-Calculation-Answer (PCA) genre. The problem was stated, and then a whole lot of calculations were shown, with hardly any explanation given of the reasoning employed along the way.

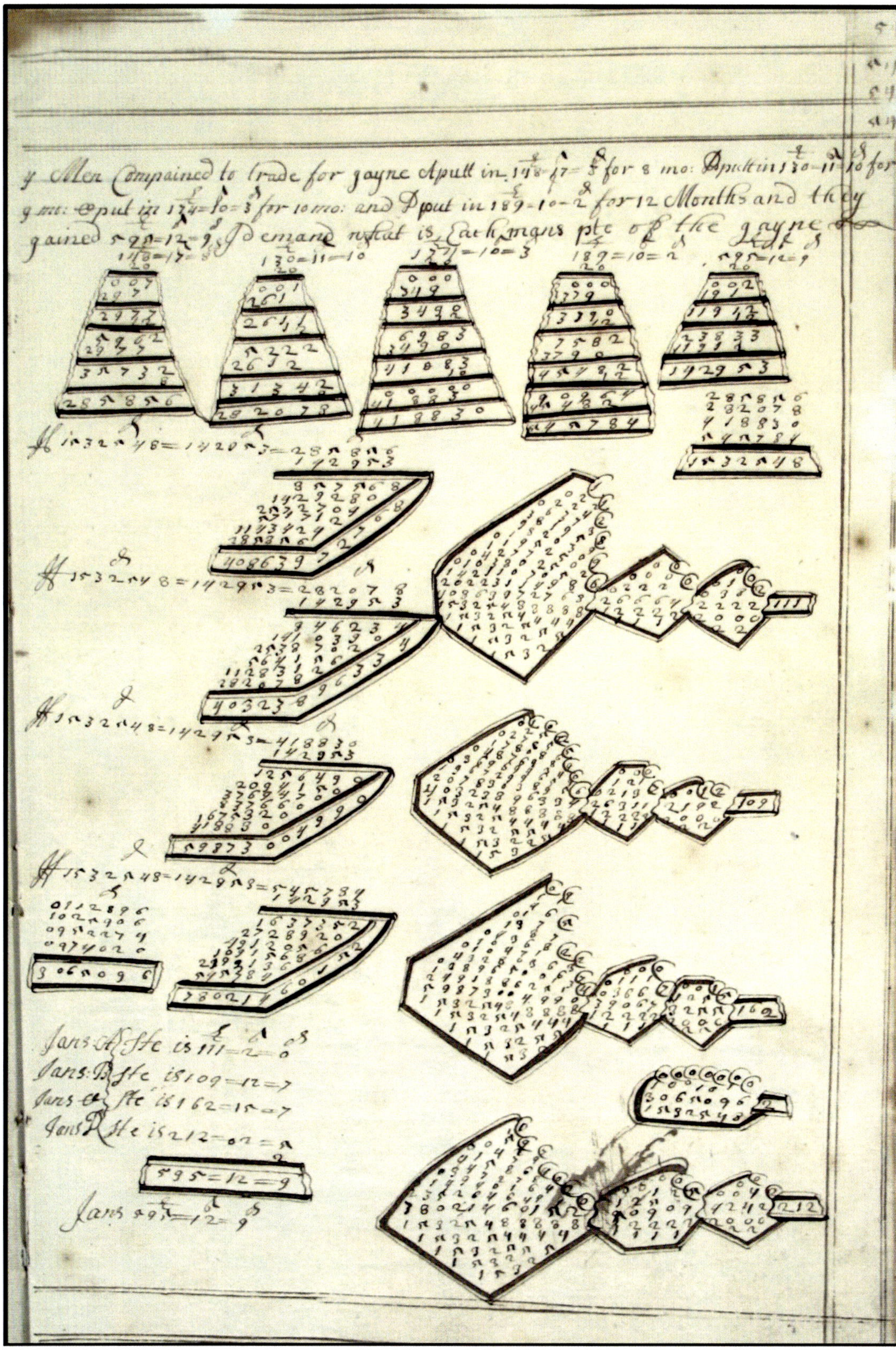

Figure 3.15. Thomas Prust's use of PCA genre in his solution to a fellowship-with-time problem.

Thomas Prust and John Collings, on Common Fractions

Thomas Prust began his work on fractions with the following (unedited) introductory statement:

> To vallue fractions bringe your question into the lowest denominator then multiply it by the numerator and divid by the denominator and if thare bee any remender you must multiply the remender upon pounds by 20 and the remender upond shillings by 12 and upon the pence by 4 and still divide by the ould denominator.

At this stage it seems that Thomas was assuming that the context of his work on fractions had to be money, expressed in pounds, shillings and pence. After the above statement, he then tackled tasks like "Vallew $\frac{1}{4} of \frac{7}{8} of \frac{5}{6}$ of 1000 pounds sterling."

After answering questions in which fraction concepts were applied in money contexts, Thomas then noted that the context for fractions could apply to many other measurement-related situations. He then proceeded to consider operations on fractions. Under the curious heading "The First Sort of Reduction of Fractions," he announced the approach he would take with fractions in the following (unedited) passage:

> The are 5 sorts of reductions of fractions very usefull for the more speedy and exact answaring of some questions and thare for nesary to bee known and the first sort showeth how to torn severall fractions of one intyre unite in to one denominatour which to doe multiply all your denominatour togeather or one into the other and product that comes of itt set out for a new denominator to all your severall ffractions. Secondly to find out a new denominater to each take the former nominator of each, severally multyply it by or into the denominator. Or after another expecting his one product is a new nominator the former denominator as for Example 1.

The spelling of words in this, and in many other, passages in Thomas Prust's cyphering book is inconsistent, suggesting that Thomas had the notes dictated to him by a teacher.

Following the above passage, Thomas worked on six questions which were intended to illustrate the class of questions involving "reduction of fractions.' Figure 3.16 is a reproduction of the entire page. The first question was "Reduce $\frac{2}{3}$ and $\frac{4}{5}$ in to one denominator." Analysis of the questions that were later answered on the same page indicate that the term "reduction of fractions" really meant, in modern language, "add these fractions together." The examples which followed on the page adopted an algorithm for which a correct response would involve multiplying the two denominators together to get the denominator in the required answer, and then "cross-multiplying each denominator with the numerator in the other fraction, and then adding" to get the required numerator. Thus, "Reduce $\frac{2}{3}$ and $\frac{4}{5}$ in to one denominator" yielded $\frac{(2\times 5)+(3\times 4)}{3\times 5}$, or $\frac{22}{15}$. The procedure was to be set out in a very definite way, and no indication was given about why it worked. This "rules-without-reason" approach had bedevilled the teaching and learning of fractions for centuries, and would continue to do so well beyond the times when Thomas Prust and John Collings prepared their pages on fractions in their cyphering book.

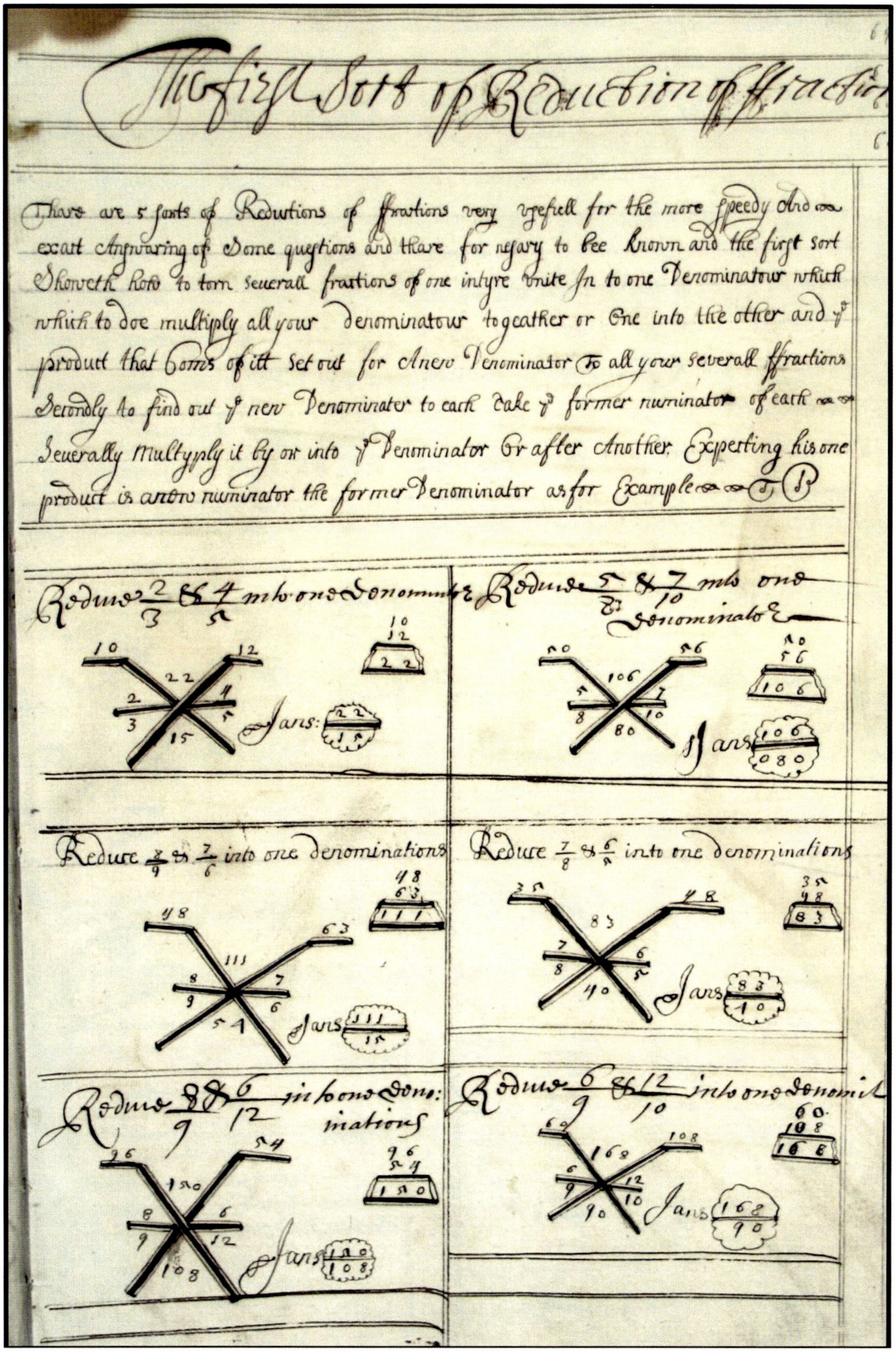

Figure 3.16. The first sort of "reduction of fractions" (in Thomas Prust's book).

The calligraphic heading on the page, as well as four of the six questions, would seem to have been written in a more mature hand than the other two questions, suggesting that Thomas's teacher was preparing the headings and writing down some of the questions—which were duly answered by Thomas. The setting out is in the characteristic "Introduction, Rule, Case, Example, Exercise" (IRCEE) cyphering-book genre identified by Ellerton and Clements (2012). But Thomas did find time to depart from the main script when he recorded:

> Many men have many minds, many birds have many kinds;
> Many mountains one the land, many rotiks one the sands;
> Many starts there is in the sky, many ships goeth one the ... [word is missing];
> Say Thomas Prust his booke amen, this he wrote to try my pen.

But, this was probably Thomas's last display of pride of ownership with respect to his "booke," because if one looks on the reverse side of the same page on which this poem appeared one finds different handwriting used in solving examples on fractions. This new handwriting was probably John Collings', and the cyphering book was now John Collings'. This new handwriting would appear on many, but not all, of the subsequent pages. Figure 3.17 shows a page on which John Collings (probably) used the rule of three to solve two problems involving fractional quantities. If the handwriting and setting out in Figures 3.16 and 3.17 are compared, differences in handwriting and setting out will be readily observed.

Concluding Comments on Thomas Prust's Cyphering Book

Whatever one might think of the idea of structuring curriculum theory in terms of intended, implemented and attained components (see, e.g., Westbury, 1980), in the context of school mathematics in Great Britain and North America in the early 1700s it would be difficult to sustain an argument that the written content and attempts at solving problems which appeared in students' cyphering books did not testify to what students actually did when they studied mathematics—that is to say, to the implemented aspect of curriculum. This chapter has attempted to give readers an idea of the mathematics that Thomas Prust dealt with as he prepared the pages of his cyphering book in 1702 and 1703. It also testified to Thomas's early determination to make his cyphering book something of which he could be proud.

The chapter has not said much about the intended curriculum, but there can be little doubt that whoever was directing Thomas's arithmetical studies wanted him to proceed through a standard *abbaco* arithmetic sequence—including the four operations on whole numbers, reduction, the rule of three and its extensions, fellowship, and common fractions.

So far as the attained curriculum was concerned, the analysis presented in this chapter has not made clear what Thomas Prust actually learned when he was preparing his cyphering book. Did he merely copy from another person's cyphering book, or a textbook, or neither? Did he take down notes given by a teacher? Did he *understand* the mathematics that he was studying—and what do we mean by "understanding," anyway, in the context of Thomas's efforts when preparing the pages of his cyphering book? We can never know the answers to those questions with any degree of certainly, but for the record we want to say that we think that often the notes were dictated (for the spelling is very inconsistent). We also think that on most occasions Thomas actually solved the problems by himself (and probably had his answers checked by a teacher).

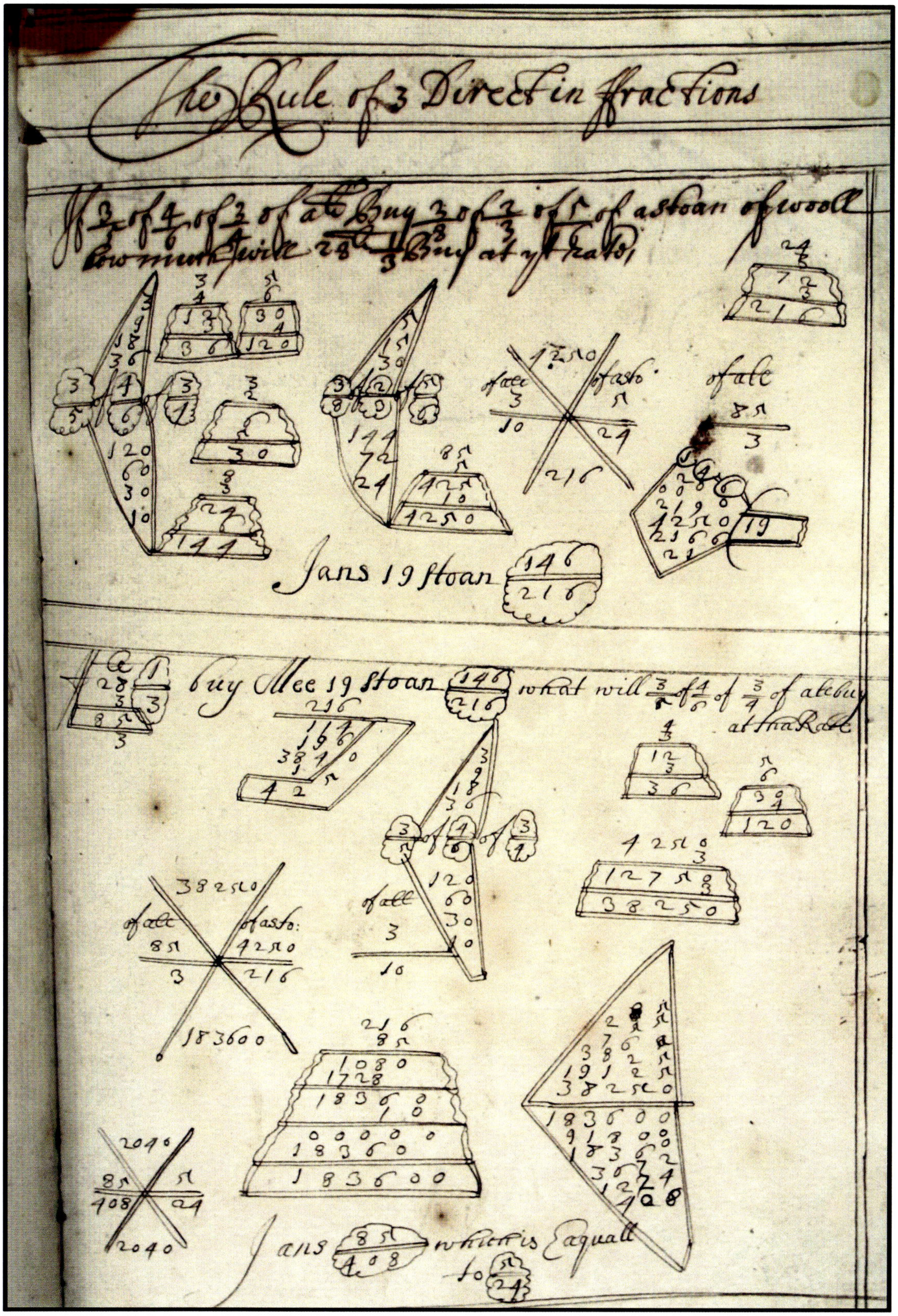

Figure 3.17. John Collings' handwriting (toward the end of Thomas Prust's book).

Genealogical records suggest that Thomas Prust died, aged 50, in Portsmouth, New Hampshire, in 1740. The record states that he had been a shopkeeper. We think that the training he received when he was preparing his beautiful cyphering book would certainly have been beneficial to him during his time as a shopkeeper. His step-son, James Collings had the cyphering book passed on to him around 1720, and the record indicates that Collings would become a sea captain. Although we do not know how many later students were guided by this extraordinary cyphering book, we do know that this was originally Thomas Prust's "booke," and that more than 300 years later most of it still remains intact.

References

Cohen, P. C. (1982). *A calculating people: The spread of numeracy in early America.* Chicago, IL: University of Chicago Press.

Ellerton, N. F., & Clements, M. A. (2012). *Rewriting the history of school mathematics in North America 1607–1861.* New York, NY: Springer.

Linklater, A. (2002). *Measuring America: How the United States was shaped by the greatest land sale in history.* New York, NY: Plume.

Lounsbury, R. G. (1930). Yankee trade at Newfoundland. *The New England Quarterly, 3*(4), 607–626.

Lydon, J. G. (1965). Fish and flour for gold: Southern Europe and the colonial America balance of payments. *The Business History Review, 39*(2), 171–183.

Martin, G. H. (1897). *The evolution of the Massachusetts public school system: A historical sketch.* New York, NY: D. Appleton & Co.

McCusker, J. J., & Menard, R. R. (1985). *The economy of British America, 1607–1789.* Chapel Hill, NC: University of North Carolina Press.

Westbury, I. (1980). Change and stability in the curriculum: An overview of the questions. In H. G. Steiner (Ed.), *Comparative studies of mathematics curricula: Change and stability 1960–1980* (pp. 12–36). Bielefeld, Germany: Institut für Didaktik der Mathematik-Universität Bielefeld.

Yeldham, F. A. (1926). *The story of reckoning in the Middle Ages.* London, United Kingdom: George A. Harrap.

Chapter 4

Daughters of the Revolution: Martha and Elisabeth Ryan's Cyphering Book, Circa 1780

Abstract: There are three independently remarkable characteristics of this manuscript, and there is a fourth feature which is also of special interest. The first extraordinary characteristic is the patriotic fervor of the two sisters who prepared it. Illustrations of ships and patriotic inscriptions reflecting the Revolutionary times—such as "Liberty," "George Washington," and "Poor freedom is better than rich slavery"—are to be found on many of the pages. The second extraordinary aspect of this manuscript is that although it was prepared by two teenage sisters around 1780 the arithmetic went well beyond the rule of three to topics like alligation, fellowship, equation of payments, and single and double false position—it was very unusual for females at that time to study such topics. The third, and most obvious, remarkable aspect of the manuscript is the extremely high quality of the penmanship, calligraphy and artwork. Martha and Elisabeth Ryan lived in North Carolina, and this manuscript is an example of an early "southern" cyphering book—of which relatively few of those prepared by girls have survived.

Keywords: American Revolution; calligraphy; cyphering book; girls and mathematics; history of school mathematics; history of education in North Carolina; penmanship.

Who Were Martha and Elisabeth Ryan?

The first 52 of the 151 pages in this composite manuscript were prepared during 1776 and 1777, and during January 1781, by Martha Ryan (who was born in 1765). The last 99 pages were prepared by Martha's sister, Elisabeth Ryan (born in October 1767) during 1780 and the first half of 1781. Genealogical and other research (see, e.g., Hathaway, 1900; Newby, 1958, p. 39; Smallwood, 1964) confirmed that the two girls were sisters—both daughters of James and Ann Ryan, of Bertie County, North Carolina.

The following summary details with respect to Martha and Elisabeth were gleaned from a range of genealogical and other websites, as well as from relevant local records and books. Every attempt has been made to check for consistency, but some details have been difficult to confirm.

- Martha was born in 1765. The strongest evidence for that is on page 10 of the composite cyphering book, where Martha stated that she was born in 1765. There is a genealogical record which suggests that Martha married on January 19, 1790, but that record could be related to another Martha in another Ryan family in North Carolina. Although some family-tree data suggest that Martha died in 1803, there are other "Martha" Ryans who died in 1803 in North Carolina, so the date of death of the Martha who prepared the cyphering book has not been established. As will be discussed later in this chapter, it is possible that she died early in 1781.
- On page 137 of the composite cyphering book Elisabeth stated that her birthday was on October 14, 1767. She died on August 22, 1815. Although Elisabeth always spelt her first name as "Elisabeth," most websites in which her name appears use "Elizabeth."

DOI 10.1007/978-3-319-02502-5_4,

- Elisabeth's father was James Ryan (1735–1777), and her mother was Ann Ryan (1737–1794). James owned a plantation in Bertie County, North Carolina, and when he died in 1777 the plantation was passed on to wife, Ann. In a 1784–1787 North Carolina census it was recorded that Ann was head of a household which included several slaves and possibly some indentured servants.
- In 1797, Elisabeth became Thomas Stevenson's second wife (National Park Service, 1998). Thomas Stevenson (1762–1801) was a plantation owner in Perquimans County, North Carolina. In 1800, Thomas and Elisabeth had a daughter, Margaret Ryan Stevenson (1800–1826).
- Margaret Ryan Stevenson would marry twice. In 1815 she married Lemuel Creasy Jr, but he died in 1816 before they had any children. In 1817 Margaret married Jonathan Hill Jacocks, and they had three children.
- After Margaret Ryan Stevenson's death in 1826, Jonathan Hill Jacocks married Grizelle P. Copeland, and it is likely that it was a great grandson from that union, Jesse C. Jacocks (National Park Service, 1998), who, in 1952, donated Martha and Elisabeth's composite cyphering book to the Philanthropic Society of the University of North Carolina. Today the manuscript is permanently housed in the Southern Historical Collection of the Wilson Library at the University of North Carolina, Chapel Hill.

It is almost certain that Martha and Elisabeth received a privileged form of education. Their paternal grandfather was Thomas Ryan (17??–1753), and in his will he indicated that his children's education should be paid for from his estate—suggesting that in the Ryan family the education of children was given high priority (Grimes, 1910). There were many hints in Martha's and Elisabeth's composite cyphering book that they were assisted by resident or visiting tutors. Many of the pages were initialed with a sideways "H"—or perhaps it was a "4"—that appeared at the bottom right corner. On page 7 in Martha's section of the cyphering book a certain Charles Dumplin appended his signature at the bottom right corner, below the words "errors excepted." On page 124 of the cyphering book—in Elisabeth's section—there is a handwritten statement "December 5th, 1780 wrote by Asa Lawrence."

Martha began cyphering sometime in 1776, when she was 10 or 11 years old. At that time, or perhaps at some later time, material covers for the cyphering book were stretched over card to form hard covers. On the insides of the covers were pasted paper on which Martha had previously written, in her best handwriting, statements like "Avoid all appearances," "Honour father and mother," and "A good girl will mind her b...." (the rest of the sentence is no longer present). She did not forget to write: "This is Martha Ryan's book."

The first, fully-dated entry in her section of what would become a composite cyphering book was on page 37. The date indicated was March 1, 1777—at which time Martha was studying the "rule of three direct." Then there was an almost four-year gap before Martha returned to preparing pages that would be included in the composite cyphering book. On page 39 the date January 4, 1781 was shown, and the topic was "the rule of three reverse, indirect proportion."

As stated previously, in 1777 Martha's father, James Ryan died. In that same year her maternal grandmother died. Two immediate-family deaths like that, coming so close to each other, would probably have been difficult for any 11 or 12-year-old. Perhaps, in the short term, there was no money freely available to continue to employ a tutor. It is possible that her tutor ("H") served in the War, and she had no tutor, or she had another tutor and prepared a different cyphering book. It should also be remembered that in 1777 the Revolutionary War was not going well for the 13 United

Colonies (McMillin, 2011a, 2011b), and that would have been a source of great anxiety for patriotic young Martha.

Martha resumed cyphering with "H," in January 1781. She dated her entry on the "double rule of three reverse" as "January 6, 1781" (p. 47), and later the same month she dated her final entry on mathematics—which was concerned with the topic, "double fellowship," as "January 30, 1781" (p. 53). Martha did leave a farewell poem, on a separate page, dated January 31, 1781.

Elisabeth began making cyphering-book entries in 1780, at least eight months before Martha resumed making cyphering-book entries. Immediately after the pages by Martha on "double fellowship" in the composite cyphering book comes Elisabeth's first page. It was concerned with the topic "addition of whole numbers" (p. 54). Between April and June 1780, Elisabeth proceeded to make beautifully-crafted entries. The dimensions of the pages in Elisabeth's section of the composite cyphering book were slightly different from those of the pages prepared by Martha (Doar, 2006), but "H" also seems to have been Elisabeth's tutor. He often initialed the pages, and the style of penmanship and calligraphy evident in Martha's cyphering-book entries in 1777 is also to be found in Elisabeth's cyphering-book entries. Many of the pages were similar to those made by Martha—they dealt with the same arithmetical topics, and the penmanship and calligraphy were similar in appearance. The first of the dated pages in Elisabeth's section of the manuscript was on multiplication, when the date April 22, 1780 was shown. Then, on May 20, 1780 and June 24, 1780, there were pages on division.

Why were the separate quires of paper used by Martha and Elisabeth Ryan combined so that they would appear to form one cyphering book? There are no data available which obviously point to an answer to that question. There are simply no records with respect to Martha after her final entry in the cyphering book on January 31, 1781. It could have been the case that in January 1781 Martha and Elisabeth were in the same room, working next to each other on individual leaves separated from their respective quires for ease of handling. Once the left half of the folded leaf had been completed, one can envisage how the leaf was carefully replaced in its correct position within the quire. Later, the quires would be combined to form the composite cyphering book. Martha may have departed from the family after January 1781. She may have died.

On July 20, 1780, Elisabeth made an entry on the rule of three direct, and the next day the entry was for "the rule of three reverse, indirect proportion" (p. 92). Then followed "the double rule of three reverse," on July 27, 1780 (p. 98), and "tare and tret," on December 5, 1780 (p. 121). The next date shown was January 31, 1781 (p. 132)—a date which corresponded to Martha's last entry. Then pages were dated on February 2, 1781 (p. 135), and February 9, 1781 (p. 150)—on the second of those dates "double position" was the subject of attention, and the date was given immediately after a quotation: "Poor freedom is better than rich slavery." The final page (p. 152) was on "reduction of vulgar fractions," and was dated March 21, 1781.

In the analysis which follows, complete pages, or sections of pages, from the composite cyphering book, will be reproduced, and comments will be made. The figures, and the comments, will testify to the excellence in penmanship, calligraphy and artwork, to the patriotic fervor of Martha and Elisabeth, and to the fact that both girls cyphered well beyond the rule of three.

It will become evident that the content of the implemented curriculum for the girls was completely in line with the intended curriculum defined by the *abbaco* tradition. Elementary numeration and the four operations on whole numbers were followed by reduction, the various versions of the rule of three, and then more advanced topics like simple interest, compound interest, discount, profit and loss, alligation, partnership, equation of payments, and single and double false position. Decimal fractions were not studied. The arithmetical emphasis in the cyphering book was on business concepts and associated calculational procedures. IRCEE ("Introduction-Rule-Case-Example-Exercise") and PCA (Problem-Calculations-Answer") genres (Ellerton & Clements, 2012) were evident throughout.

Although the entries in Elisabeth's section appear to have been confined to 1780 and 1781 (when she was 13 and 14 years of age), her section occupied 99 pages, which was 46 pages more than the 52 pages in Martha's section. A careful examination of pages in the composite cyphering book would suggest that tutor "H" helped make many of the entries, and that that was especially the case in Elisabeth's section.

In her section, Elisabeth dealt with a range of topics that Martha did not address—those extra topics included practice, tare and tret, simple and compound interest, barter, discount, profit and loss, equation of payments, alligation medial, alligation alternate, single position, double position, and vulgar fractions. Most of topics covered by both Martha and Elisabeth came early in the standard *abbaco* sequence for elementary arithmetic—up to, and slightly beyond the rule of three. However, those topics which Elisabeth covered but Martha did not, with the possible exception of practice, would be regarded as middle-level in the standard *abbaco* sequence. Both Martha and Elisabeth, and especially Elisabeth, cyphered well beyond the rule of three, and that was highly unusual for a girl in eighteenth-century North America.

Both Martha and Elisabeth used their cyphering book as a platform to express, in no uncertain terms, their high level of patriotism. Not a few of the illustrations and calligraphy throughout the composite cyphering book call for the adjective "extraordinary," especially when it is considered that in 1776 and 1777 Martha was 11 to 12 years old, and even in January 1781 Martha and Elisabeth were probably only 15 and 13 years old, respectively.

Cultural Expectations Regarding Girls and Mathematics

The above factual information provides some background to the privileged circumstances in which Martha and Elisabeth grew up. However, it says little about the difficult Revolutionary times that dominated the thinking of young Martha and Elisabeth as they were preparing their sections of what would become a composite cyphering book. It also offers no commentary on why Martha and Elisabeth were asked to prepare cyphering books involving a level of arithmetical thinking far beyond what was expected of almost all other North American girls in the eighteenth century. Even the extent to which Martha and Elisabeth had some say in what they ought to study is unknown.

Perhaps Martha's and Elisabeth's parents, James and Ann Ryan, believed that it would be useful for their daughters to learn enough arithmetic to become able to keep accounts intelligently for a plantation. Martha and Elisabeth had two brothers, Thomas and Edward, but nothing is known of their education. Aside from societal expectation, there were good reasons why all girls should have cyphered, including girls living in privileged circumstances like Martha and Elisabeth. It made good sense for the two girls to become well placed to manage accounts, and to estimate likely costs for the food, clothing, and equipment associated with the day-to-day running of a plantation. Certainly, too, well-to-do families in the eighteenth century were often heavily involved in investment activities, and a working knowledge of the commercially-oriented *abbaco* sequence would have been useful. Trading partnerships were being formed, properties purchased, homesteads built, produce and equipment bought and sold, and therefore some knowledge of how to calculate simple and compound interest, and discount, would have been useful. Bartering was commonplace, and therefore arithmetic principles associated with bartering would not have gone astray. And, with different states placing different values on the pounds, shillings and pence in local usage, achieving competence in making calculations for currency-exchange transactions would have been of considerable benefit to a family engaged in trading activities.

Ellerton and Clements (2012) pointed to how girls who lived in the coastal Massachusetts trading center of Salem in the eighteenth and early nineteenth centuries were engaged in trading activities on the wharves at Salem, as goods were moved between that North American port and distant trading centers in North America, in Europe, in Africa, in the Caribbean, and in the East Indies. In Salem, it was accepted that girls could and should do arithmetic. However, in many other parts of North America the common attitude was that girls did not really need to study arithmetic—rather, they should learn to sew and to prepare samplers—arithmetic should be left to their brothers. It was commonly held that if girls were to study any arithmetic, at all, then it should be confined to elementary rules and computations (Middlekauf, 1963).

Cohen (1982) stated that "the Harvard collection of cyphering books contains six by colonial girls, but none went beyond problems in simple interest" (p. 251*n*). Monaghan (2007) maintained that female students in the colonial and early Federal periods who cyphered did not proceed beyond the rule of three. Yet, in the 1790s, all 365 girls who attended the Moravian Seminary for Young Ladies in Bethlehem, Northampton County, Pennsylvania, prepared cyphering books (Moravian Publication Office, 1876) and, if the contents of a cyphering book begun in December 1790 by Susan Shimer can be taken as typical, the arithmetic which the girls studied was not always elementary. All of the pages of Susan Shimer's cyphering book can be seen on the web-page of the Bethlehem Digital History Project (2013).

Further research is needed to see if the enlightened attitude displayed by James and Ann Ryan toward the education of their daughters, Martha and Elisabeth, was more common than has been portrayed in the literature. Studies on attitudes toward the education of girls of eighteenth- and early nineteenth-century plantation owners might be of special interest.

Patriotism Among Women in North-Carolina at the Time of the Revolution

Martha's and Elisabeth's cyphering book is full of patriotic fervor. The two pages reproduced in Figure 4.1 attest to that fervor, and offer a glimpse of the amazing calligraphy, and artwork in the manuscript. Martha announced, at the outset, that this was "Martha Ryan's book." Although this book would be concerned primarily with arithmetic, the front double page was decorated with drawings of six warships (the Sloop *Property*, the Ship *Continental Congress*, the Schooner *Freedom*, the Sloop *Liberty*, the Ship *American*, and the Ship *Union*). Two American flags were also displayed, and the choice of "LIBERTY or DEATH" was prominently announced. As will be seen, such displays of patriotism would be found, often, in both Martha's and Elisabeth's sections in the composite cyphering book.

The strength of the words "Liberty" and "Death" might come as a surprise, but there can be little doubt that in the 1770s anti-British sentiments ran high among females in North Carolina. In 1774 the so-called "Edenton Tea Party" gained much publicity (Butler, 1976). Fifty-one North Carolina women, led by Penelope Barker, signed a statement in which they promised not only to give up drinking tea but also to refuse to purchase any British-produced products until Great Britain repealed legislation which took advantage of North American colonists by taxing colonists even though they did not have representation in the British parliament (Morley, 2002). This event gained considerable notoriety, even in England, where cartoons were published satirizing those women who participated in the Edenton Tea Party (see Figure 4.2).

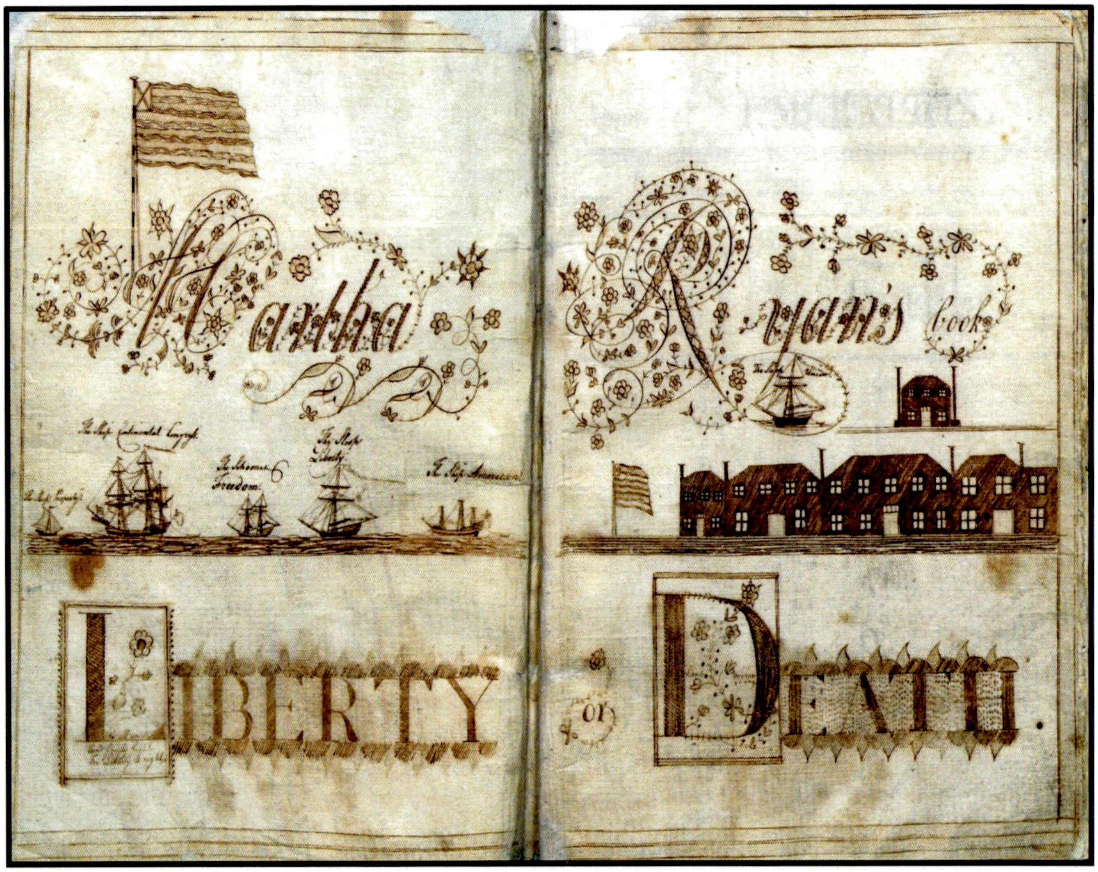

Figure 4.1. Pages 1 and 2 of Martha Ryan's (circa 1780) cyphering book (Southern Historical Collection, Wilson Library, University of North Carolina at Chapel Hill).

Figure 4.2. A March 1775 British cartoon satirizing "a society of patriotic ladies" at the Edenton Tea Party (Image courtesy of the North Carolina Office of Archives and History, Raleigh, NC).

Edenton was a town located in Chowan County, which adjoined Bertie County where Martha and Elisabeth Ryan prepared their composite cyphering book.

During the Revolutionary War, which was in full swing at the very time Martha and Elisabeth were preparing all of the entries in their manuscript, patriotic American women are said to have "led boycotts against British goods, raised funds for the Continental Army, nursed wounded soldiers, spied on the enemy, and wrote propaganda to advance the patriotic cause" (Berkin, 2005, p. xii).

Each page in Martha's and Elisabeth's cyphering book was a work of art, obviously taking many hours to complete. Just why the girls chose to make such strong, unambiguous patriotic statements in a cyphering book is not clear—but, the long-established tradition insisted that cyphering books should feature high calligraphy, fine penmanship, and impressive artwork, and that was what Martha and Elisabeth set out to achieve. Their parents owned a large plantation, and it is likely that the girls had the time as well as the material and human support needed to produce an outstanding cyphering book.

There are many unexplained aspects of the story behind Martha's and Elisabeth's composite cyphering book. Some of these will be outlined in the pages which follow.

Entries in Martha's and Elisabeth's Composite Cyphering Book

In much of the remainder of this chapter a page from the composite cyphering book will be reproduced as a figure and then, on the opposite page, commentary will be provided on entries on that page. Because the Ryan sisters followed the traditional *abbaco* arithmetic sequence, this arrangement of the chapter should give the reader a useful overview of *abbaco* traditions.

Although Martha and Elisabeth followed the traditional *abbaco* sequence for topics, it is a matter of interest that it has not been possible to link what they wrote when making their cyphering book entries for a topic with the words used, or exercises set, for that topic in any widely-used arithmetic textbook. Often the wording is sufficiently good that one suspects that a textbook has been consulted, and in the future a link between what the sisters wrote and a particular textbook may be discovered. However, as Ellerton and Clements (2012) have argued, arithmetic textbooks were not widely used as the basis for creating cyphering-book entries in North America. Rather than copying notes from a textbook, it seems to have been the case that the sentences were created by Martha's and Elisabeth's private tutor, whom we have denoted by "H." As the chapter proceeds it will be argued that not only was "H" responsible for the wording of entries, but he (or she) probably wrote many of the entries—the notes, the model examples, and perhaps even the answers to exercises—into the composite cyphering book. That statement is based on the maturity of handwriting, and on the fact that for many pages in Martha's section of the composite cyphering book Martha did write answers to exercises—and the maturity of her handwriting in those exercises was very much less than the maturity of the person who wrote the introductions to topics, the statements of rules and cases, and the model examples. That same mature handwriting is to be seen throughout both Martha's and Elisabeth's sections.

When Martha began making her cyphering book entries in 1776 she was 10 or 11 years of age. She was living in a household with servants, and the family plantation, with the aid of slave labor, was doing well. She would not have been expected to help with household chores. Her help was not essential to the smooth-running of the plantation, and she would have been able to devote as much time as she felt necessary to her cyphering. Her situation might have changed, though, in 1777, when her father died—although we do not know whether, then, she had to do more of the household chores. Certainly, there is no record of her continuing with her cyphering during the period 1778–1780. Dated pages in Elisabeth's cyphering book indicate that Elisabeth cyphered during summer months in 1780 and 1781.

Martha's and Elisabeth's Cyphering Book Entries

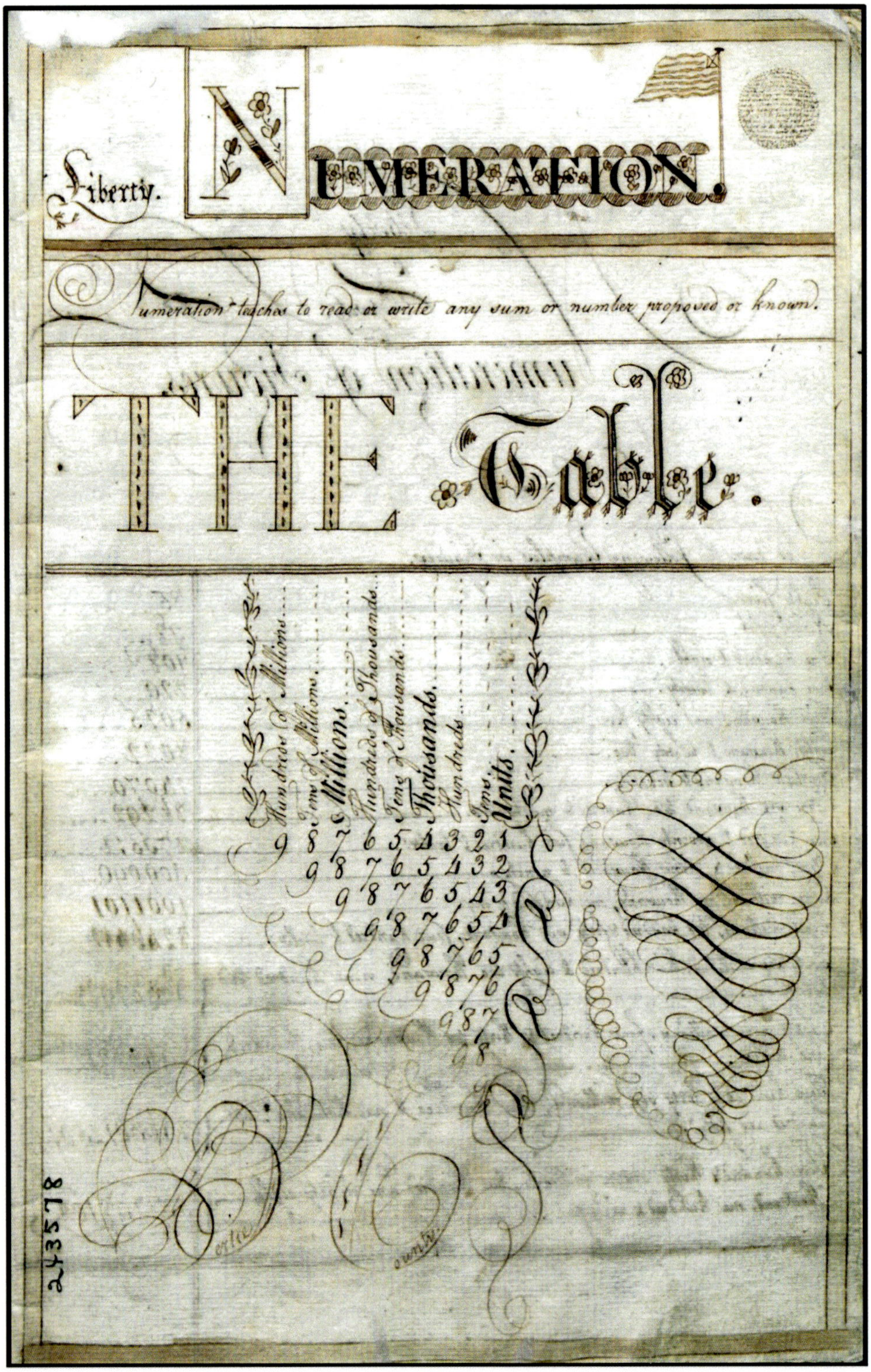

Figure 4.3. The first page on arithmetic in Martha's cyphering book (p. 3) was concerned with numeration, with numbers going to hundreds of millions (Southern Historical Collection, Wilson Library, University of North Carolina at Chapel Hill).

Martha, on Numeration

Figure 4.3 shows the first page on arithmetic in Martha's section of the cyphering book. Numeration was, traditionally, the first page in an elementary cyphering book. It was assumed that students who were beginning to cypher had already learned to count, and had some of idea of numbers from 1 to about 1000. That had been learned in dame schools or—what was more likely with Martha and Elisabeth—from private tutors or governesses. So, the first page in a cyphering book would announce that now the learner was going on to deal with bigger, harder numbers. This first page would have been prepared in 1776, when Martha was 10 or 11 years old.

A careful examination of Figure 4.3 will reveal the word "Liberty" was placed to the left of the heading "Numeration," with a depiction of the American flag to the right. The first sentence read, "Numeration teacheth to read or write any number proposed or known." A check was made of popular English-language arithmetics available in the 1770s and, in addition, a Google search was performed in an effort to identify a source for that description of numeration. But, no exact match for that wording was found, so it is unlikely that this page was copied from a commercial arithmetic. It may have been copied from a cyphering book owned by Martha's private tutor.

The high quality of the handwriting in the numeration table is such that one might expect that it was done by an adult. But there are many examples of beautiful penmanship from young children who were creating cyphering books, and so one cannot dismiss the possibility that it was done by Martha. A similar comment would apply to the calligraphic heading and to the decorative scroll below the numeration table. It is difficult for the naked eye to identify the words "Bertie County" which were written within the decorative scrolls. As stated earlier, both Martha and Elisabeth resided in Bertie County, North Carolina. Somewhat amazingly, the writing, calligraphy, and artwork on the pages in Martha's and Elisabeth's cyphering book would have been performe with a quill, and with homemade ink.

On the next page in Martha's cyphering book there was another table showing "numeration of figures." In this table the words for the numerals 85, 98, 108, 220, 5085, 8022, 14070, 86292, 270512, 909090, 1001101, 8246412, 46382921, 89432212, 246301250, 537298164 had been given and Martha wrote the numerals corresponding to those words. Certainly, some of the numbers were large for a young girl, and no commas were used in the numerals for even the large numbers.

In Elisabeth's section of the composite cyphering book there was no section on numeration. It would appear to have been the case that Elisabeth's tutor, or Elisabeth herself, decided not to prepare some of the pages that Martha had already prepared. The two sections are not identical—Elisabeth's is almost twice the length of Martha's, and on the whole Martha's manuscript is more elementary than Elisabeth's. That said, there were many topics in Martha's section which also appeared in Elisabeth's, and sometimes the wording and the examples in the two sections were the same, or almost the same.

Writing masters like George Bickham (c. 1740) and Edward Cocker (1664) insisted that all children studying arithmetic should learn to write numerals beautifully. They also argued that it was more difficult to show fine penmanship with numerals than with ordinary letters. This state of mind was present in many teachers of arithmetic in both North America and Great Britain. The argument was that young school graduates seeking employment would be greatly advantaged if they could present a prospective employer with a neatly and fully developed cyphering book, with beautifully-written numerals. Until the 1830s, most teachers in the United States of America and in Great Britain who offered cyphering classes also doubled as writing masters (Thornton, 1996). Martha's and Elisabeth's tutor, "H," probably regarded himself primarily as a "writing master."

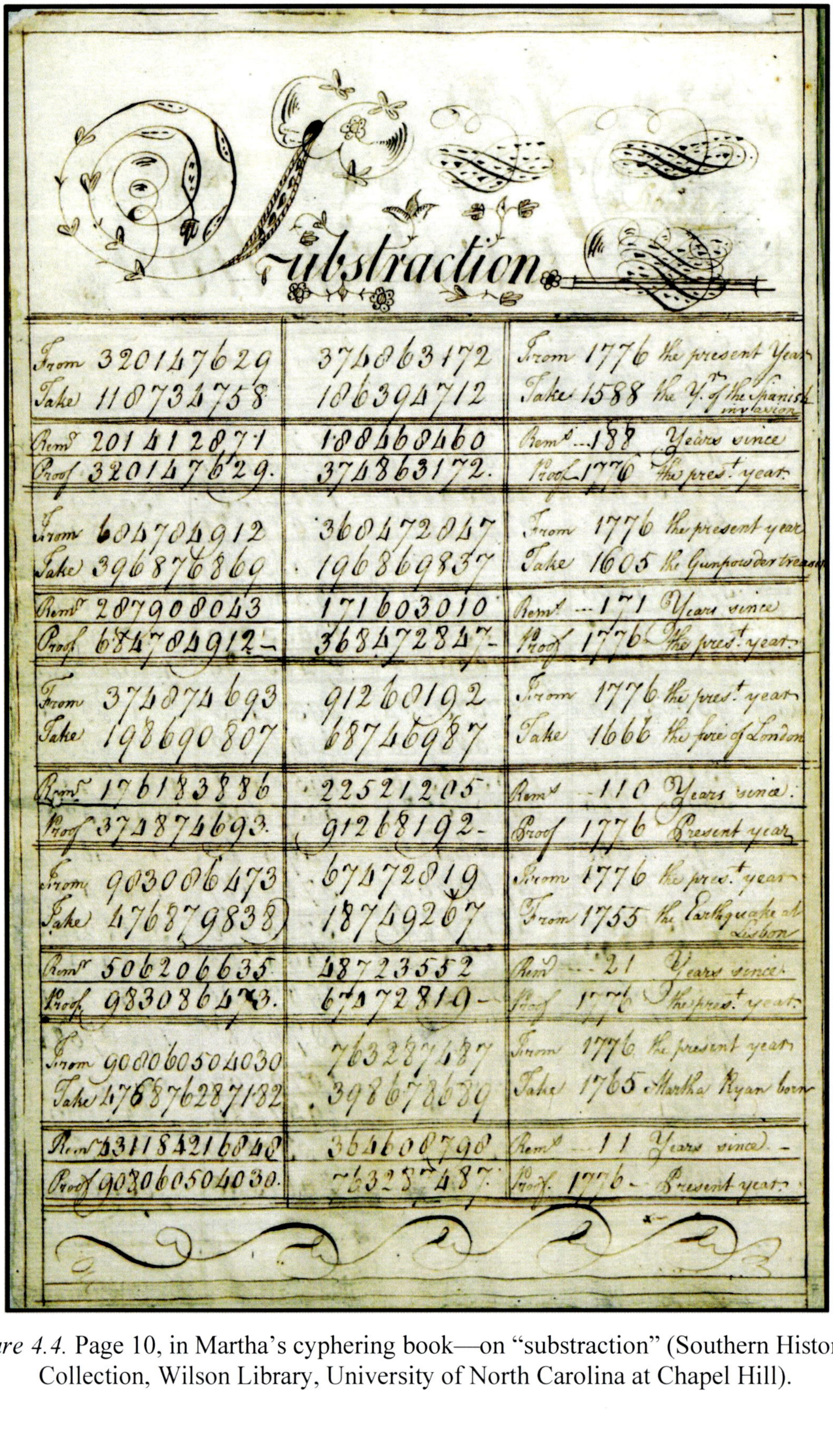

Figure 4.4. Page 10, in Martha's cyphering book—on "substraction" (Southern Historical Collection, Wilson Library, University of North Carolina at Chapel Hill).

Pages on Subtraction in the Composite Cyphering Book

Of the eleven authors of extraordinary *arithmetic* cyphering books examined for this volume, five—the unknown circa 1667 author (Chapter 2), Thomas Prust (Chapter 3), Martha and Elisabeth Ryan (this chapter) and George Bickham (Chapter 8)—used the word "substraction" rather than "subtraction," despite the latter being preferred in mathematical circles. The "Substraction" page shown in Figure 4.4 is from Martha's section of the composite cyphering book. It was the second of four pages on subtraction. In her section, Elisabeth had only two pages on "substraction."

From a mathematical perspective, the most interesting feature of the working on the page is the way the word "proof" was used for what was really nothing more than a check. After a subtraction, a "corresponding" addition was carried out and, if no error had been made, the original number would be the answer. Although that use of the word "proof" would raise eyebrows among 21st-century mathematicians, it was standard in school mathematics contexts across Europe and North America in the 17th-,18th-, and early 19th-centuries.

From a mathematics-education perspective, the most interesting aspect of Figure 4.4 was to be seen in the right hand column. It appears to have been the case that Martha prepared this page when she was 11, for in one of the exercises she wrote: "From the present year, 1776, take Martha Ryan born 1765; Rem 11 years since." That is the only clue to Martha's year of birth that we have found. Other calculations in the right-side column concerned the number of years that elapsed between a series of events and the present year of 1776—the events were: the Spanish invasion of 1588; the Gunpowder Plot, 1605; the fire of London, 1666; and the earthquake of Lisbon, 1755. This example of making school arithmetic "relevant" to the learner suggests that it would be an over-generalization to claim that during the cyphering tradition students merely copied material out of textbooks, for the purpose of memorizing it.

Altogether, there were almost three pages in the cyphering book dedicated to subtraction of money—two pages by Martha and almost one page by Elisabeth. The sums of money involved were given in pounds, shillings and pence, and Elisabeth did not do the same subtractions as Martha. However, once again the entries showed a practical emphasis, with some of the problems having the structure shown in the following example (taking from Martha's section, p. 12):

	£ s d
Owing	£ 681 . 16 . 8
Paid	187 . 5 . –
	88 . 17 . 6
	100 . – . –
	91. 18 . 6
Paid in all	£ 468 . 1. –
Remaining Dues	£ 213 .15 . 8
Proof	£ 681 . 16 . 8

Some of the exercises involved calculations with farthings (e.g., in Elisabeth's section, p. 61, there was the task "From £ 600-12-11$\frac{1}{2}$ take £ 503-15-6$\frac{3}{4}$." With such problems the common fraction notation was used to denote farthings. Yet, neither Martha nor Elisabeth had ever formally studied fractions at the time when they did these exercises. Such subtractions involved changing bases (from 4 for the farthings, to 12 for the pence, and to 20 for the shillings).

Thomas Jefferson's Comments on English Currency and his Proposed Decimal Currency

At around the same time that Martha and Elisabeth were wrestling with the arithmetic of Sterling currency, with its pounds, shillings, pence and farthings (not to mention guineas, crowns, etc.), a youthful, but already very well-known politician, based in Virginia, not too far away from Bertie County, was wrestling with the idea of doing away with British currency in the fledgling United States of America. That politician was the brilliant Thomas Jefferson, who adopted an educational stance in his handwritten *Notes on Coinage*, written before he went to France in 1785. After taking up his position in France, Jefferson (1785) caused to have published his *Notes on the Establishment of a Money Unit and of a Coinage for the United States*.

Concerning addition of money, Jefferson (1785) described the feelings of an American school boy, puzzled with "adding the farthings, taking out the fours and carrying them on; adding the pence, taking out the twelves and carrying them on; adding the shillings, taking out the twenties and carrying them on. ... But when he came to the pounds, where he had only tens to carry forward, it was easy and free from error" (p. 240). He compared traditional "English" calculations with money with corresponding calculations if a decimalized system of money were to be adopted.

Jefferson considered the operations involved in subtracting *£4 12 8* $\frac{3}{4}$ from *£8 13 11*$\frac{1}{2}$ and commented on how much easier it would be to subtract 20.61 dollars (the amount equivalent to *£4-12-8*$\frac{3}{4}$) from 38.65 dollars (the amount equivalent to *£8-13-11*$\frac{1}{2}$). As a further example, he considered the problem of dividing 8 pounds, 13 shillings, 11 pence and 2 farthings by 8. With the English system, he showed how the 8 pounds would be multiplied by 20 to get 160 shillings; then 173 shillings would be multiplied by 12 to get 2076 pence; and then 2087 pence would be multiplied by 4, and 2 added, to get 8350 farthings. Then the 8350 would be divided by 8, 4, 12 and 20 to arrive at 1 pound, 1 shilling, 8 pence and 3 farthings. He contrasted this with the kind of reasoning that might take place with his proposed decimal system, 38 dollars 6 silvers and 5 coppers (i.e., \$38.65), could quickly be divided by 8 to obtain "4 dollars, 8 silvers and 3 coppers." He commented:

> A bare inspection of the above operations will evince the labor which is occasioned by subdividing the Unit into 20ths, 240ths, 960ths, as the English do, and as we have done. And the ease of subdivision in a decimal ratio. An Englishman to pay 8 pounds 13 shillings 11 pence and 2 farthings must find, by calculation, what combination of the coins of his country will pay this sum; but an American will have the same sum to pay thus expressed \$38.65, will know by instruction only, that 3 golden pieces 8 units or dollars, 6 tenths, and 5 coppers, pay it precisely.

Jefferson went on to say that "the most *easy ratio* of multiplication and division is that by ten. Everyone knows the facility of Decimal Arithmetic. ... And even mathematical heads feel the relief of an easier, substituted for a more difficult process. Foreigners, too, who trade and travel among us, will find a great facility in understanding our coins and accounts from this ratio of subdivision. ..."

Jefferson was far from being correct when he stated that everyone knew the facility of decimal arithmetic. In fact, around 1780 relatively few North Americans knew much about decimal fractions. Certainly, several of the textbooks on arithmetic that were being used in North America had a chapter, or chapters, on decimal arithmetic (e.g., Dilworth, 1773)—although, as in the case of George Fisher's (1734) *Arithmetick*, such treatment of decimal fractions was confined to the final chapter or to an appendix.

Analyses of cyphering books establish beyond reasonable doubt that around 1780 decimal fractions were rarely dealt with in the implemented curriculum. Of the 350 manuscripts in the Ellerton/Clements collection of North American cyphering books there are 20 which were prepared in 1780 or before that and, of those, only two have at least one page dedicated to decimal fractions. One of these was by Hulett Cornwell, who prepared his manuscript in New Hemstead, in New York City in 1766; and the other was by Cornelius Houghtaling, who prepared his manuscript in New Paltz, New York, in 1775. The extent of the treatment of decimal fractions in both Cornwell's and Houghtaling's cyphering books is slight. Almost certainly, Jefferson who had had a privileged formal education, studying with a private tutor and then at the College of William and Mary in Williamsburg, Virginia, would have had an opportunity to study decimal fractions at some stage, but he may not have been aware of the arithmetic curricula being implemented in schools attended by children from less privileged circumstances.

In fact, around 1780 there were few, if any, good reasons why any schoolmaster should want to teach decimal arithmetic to children attending schools in North America. Since most people responsible for day-to-day monetary transactions—in shops, in business exchanges, in family agreements, etc.—had never learned about decimal fractions, hardly any everyday calculations were based on decimal fractions. Thus, even if teachers were aware of the existence of such numbers, there was no pressing reason to include the study of them in an arithmetic curriculum. What Thomas Jefferson (1785) did, in his *Notes on the Establishment of a Money Unit and of a Coinage for the United States,* was to point to the educational advantages that could accrue if the vast number of units for money, and for weights and measures, were to be replaced by a decimalized system in which calculations could be done with decimal fractions.

Around 1780, Martha and Elisabeth Ryan probably did not know how to calculate with decimals, for throughout their cyphering book they offered no evidence of ever having studied decimal fractions. And, as will be shown in Chapter 6, even in the mid-1820s a young Abraham Lincoln did not know much about decimals, either. Just like Martha and Elisabeth Ryan, 40 years earlier, at school Abraham did not get the opportunity to study decimal fractions formally.

The United States of America would become the first nation in the world to introduce a fully decimalized currency. Jefferson wanted more than that, though, and he argued for the introduction of a decimal system of weights and measures. In the early 1790s he almost persuaded Congress to adopt such a system. The story is superbly told by Andro Linklater (2002)—ultimately, financial issues associated with the sale of land stopped Jefferson's plan from going ahead—and today the United States is one of only two nations not to have adopted the metric system.

Elisabeth devoted about one-fourth of a page to subtraction of troy-weight quantities, the system used for measuring weights of precious metals and gemstones. There were 12 troy-weight ounces in 1 troy-weight pound, as compared with 16 in an avoirdupois pound. The troy ounce could be subdivided into 24 pennyweights and since 20 grains made one pennyweight, there were 480 grains in a troy ounce. Elisabeth subtracted 303 lbs 8 oz 15 pennyweight and 21 grains from 508 lbs 3 oz 18 pennyweight and 5 grains, to get 204 lbs 7 oz 2 pennyweight and 8 grains. Although one might wonder about the relevance of such a task for a girl aged about 14 years, one should remember that Elisabeth was the daughter of a wealthy Southern plantation owner, and as such would, as she grew older, be expected not only to wear jewelry herself but also to understand how gold and diamonds, etc., were measured. Because of that, asking Elisabeth to do exercises related to troy weight might be thought of as engaging her in a culturally-relevant task.

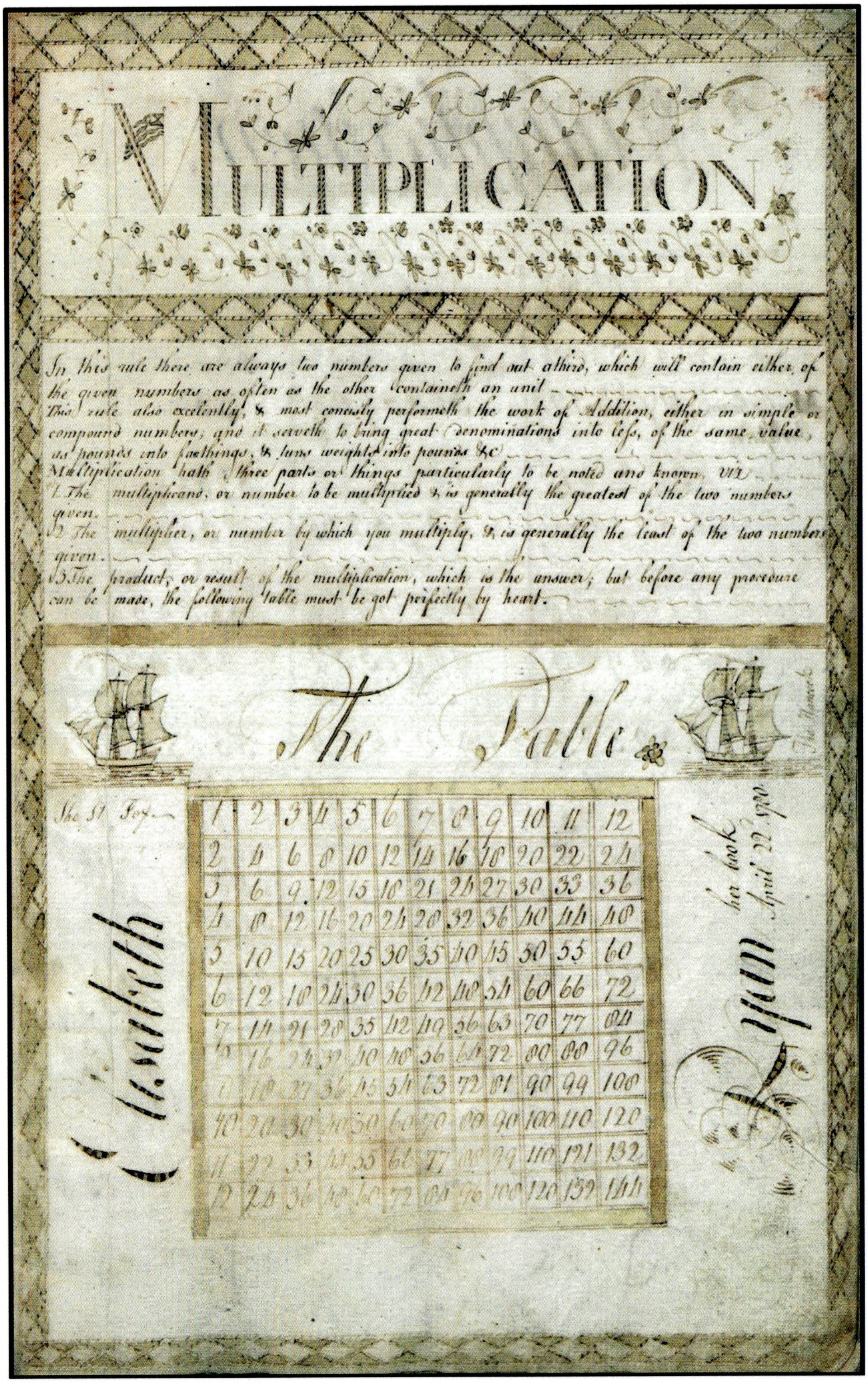

Figure 4.5. One of the 10 pages on multiplication in Elisabeth's section of the cyphering book (Southern Historical Collection, Wilson Library, University of North Carolina at Chapel Hill).

Pages on Multiplication in the Composite Cyphering Book

Figure 4.5 shows the first of 10 pages (pp. 62–71) that Elisabeth Ryan devoted to multiplication. Earlier in the composite cyphering book Martha Ryan had also devoted 10 pages to multiplication (pp. 13–22). The page shown in Figure 4.5 had the same words, and the same table as those on the first page of Martha's notes on multiplication. However, the other nine pages in each of the two sets of notes, although sometimes similar, were different. The last two pages in Elisabeth's notes on multiplication were concerned with cross multiplication, a topic which was not covered in Martha's notes.

With typical patriotic fervor, Elisabeth included two drawings of warships on her first page on multiplication. In addition, there was an American flag hidden within the first letter of the upper-case calligraphic heading "Multiplication." Elisabeth's name was written in beautiful calligraphic style, as was the statement that this was "her book." The entry was dated April 22, 1780. On the pages in which she practiced multiplication, Elisabeth emphasized the meanings of terms like "multiplicand," "multiplier," and "product." Both girls also showed "proofs" (i.e., checks) indicating that their long-multiplication calculations were correct. For each multiplication, this check was carried out by performing a second multiplication, only with the order of the multiplicand and the multiplier reversed. Thus, for example, Elisabeth multiplied 642528 by 43829 and got 28161359712. She then wrote "Proof," and multiplied 43829 by 642528 and got the same result.

Elisabeth included four pages (pp. 66–69) on "multiplication of money," and Martha included six pages (pp. 17–22) on the same theme. It was in this section that Elisabeth included some notes on elementary common fractions. She wrote:

> If there be any fractional part annexed to the given number of yards, pounds, gallons, or anything else, use $\frac{1}{4}$ or $\frac{1}{2}$ or $\frac{3}{4}$, if $\frac{1}{4}$ take the fourth part of 1 yard, &c, if $\frac{1}{2}$ take half, if $\frac{3}{4}$ first take half the price of 1 yard, &c, and then half of that and add both together will give the value of $\frac{3}{4}$. (From p. 69 of Elisabeth Ryan's section of the manuscript)

This method was then applied to finding "24 lbs & $\frac{3}{4}$ of tea at £ 25-12." In Elisabeth's solution for this problem she multiplied £ 25-12 first by 4, and then by 6, to get £ 614-8. She then added the cost for $\frac{3}{4}$ lb, which was £ 19-4, to get £ 633-12. In the language of 21st-century mathematics she found $24\frac{3}{4} \times £\ 25\text{-}12$ by calculating $6 \times (4 \times £\ 25\text{-}12) + 3\ (£\ 25\text{-}12 \div 4)$, and those calculations assumed the truth of both the associative law for multiplication as well as the distributive law. Martha learned to use a similar technique (p. 22).

Thus, although Martha did not formally study common (i.e., vulgar) fractions, and Elisabeth would devote only one page to that topic in a much later section of the cyphering book (p. 152), both were taught to apply fractions concepts in practical situations.

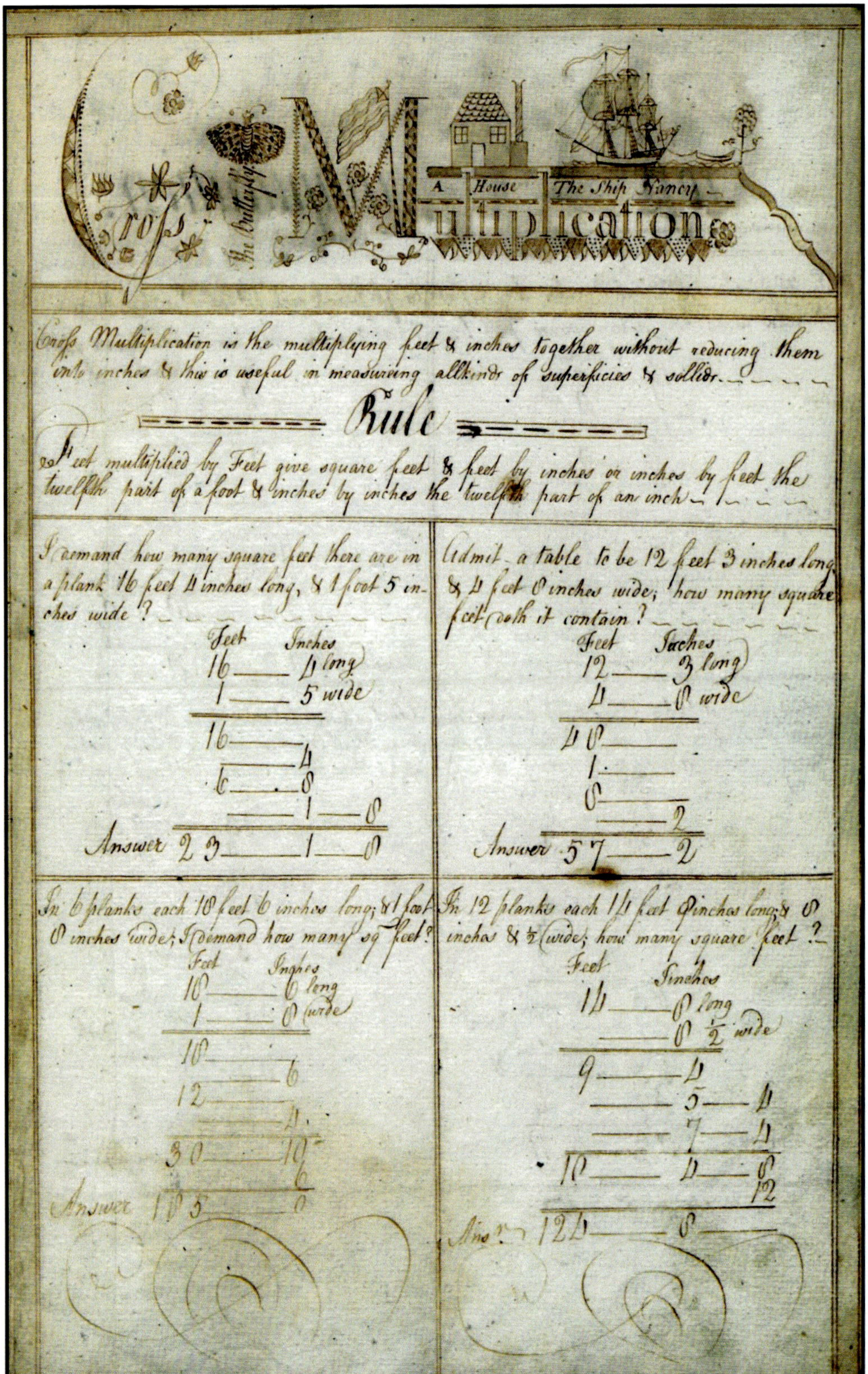

Figure 4.6. The first of two pages on cross-multiplication in Elisabeth Ryan's section (Southern Historical Collection, Wilson Library, University of North Carolina at Chapel Hill).

Cross-Multiplication

Following a very elaborate heading, which showed, among other things, illustrations of a butterfly, a warship and an American flag, Elisabeth's section included two pages of notes (pp. 71–72) on "cross-multiplication" (see Figure 4.6). The introduction stated that cross-multiplication was "the multiplying feet & inches together without reducing them into inches and this is useful in measuring all kinds of superficies and sollids (sic.)." Then a rule was given:

> Feet multiplied by feet give square feet and feet by inches or inches by feet the twelfth part of a foot & inches by inches the twelfth part of an inch.

The following example, from Elisabeth's cyphering book, revealed how the system worked:

> In 6 planks, each 18 feet 6 inches long; & 1 foot 8 inches wide; I demand, how many square feet?

A model solution was then given:

	Feet	Inches
	18	6 long
	1	8 wide
	18	
		6
	12	
		4
	30	10
		6
Answer	185	0

This "answer" was not fully explained, but the working was meant to imply that the six planks had, altogether, 185 square feet. For the 21st-century student, even a student of mathematics, the working shown would be puzzling. However, a quick check, using modern calculational methods, will show that $6 \times (18\frac{1}{2} \times 1\frac{2}{3})$ equals 185, so that, yes, the method shown in Elisabeth's notes did generate a correct answer (which was 185 square feet).

Florian Cajori (1890), a historian of mathematics education in the United States of America, severely criticized the cross-multiplication approach. He wrote:

> "Feet multiplied by feet give feet"; "Feet multiplied by inches give inches," etc. These rules, taken literally, are absurd. We can no more multiply feet by feet than we can multiply umbrellas by umbrellas. These rules are in opposition to the fundamental ideas of multiplication in arithmetic. A concrete number cannot be multiplied by a concrete number. It seems strange that so gross an error should not have been corrected ... ; but still more strange is the fact that nearly all arithmetics down to the present day should have persisted in making the mistake. (p. 16)

However, the method *is* mathematically defensible and was used widely by tradesmen in the eighteenth and nineteenth centuries (Clements & Ellerton, 2012). That said, it is nevertheless doubtful whether Elisabeth could have justified her notes on this theme.

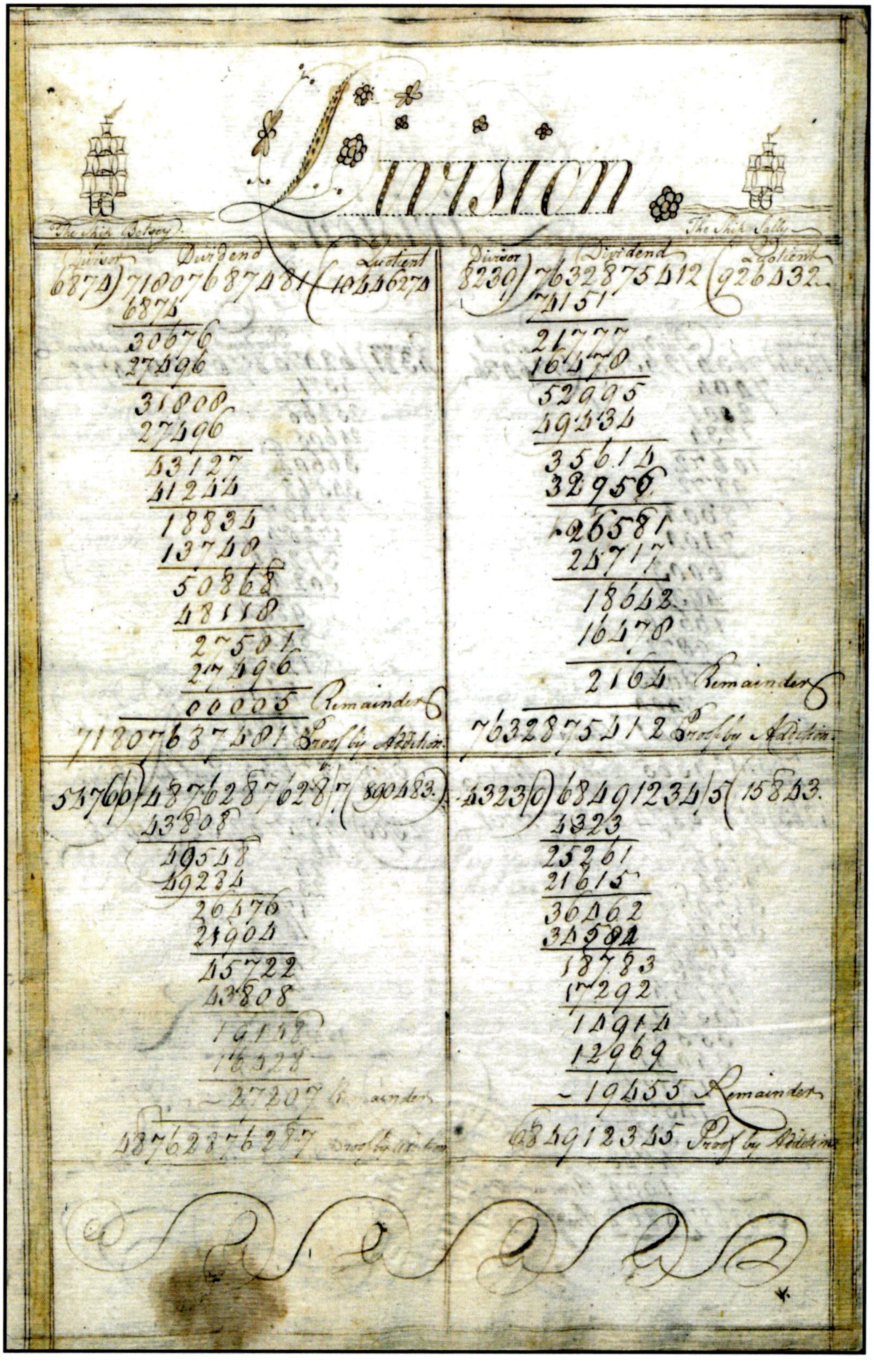

Figure 4.7. The third of five pages of notes on division, by Martha Ryan (Southern Historical Collection, Wilson Library, University of North Carolina at Chapel Hill).

The Division Operation

Figure 4.7 is from page 25 of Martha's section of the composite cyphering book, on which the calligraphic heading was embellished with sketches of the Ship *Betsey* and the Ship *Sally*. Martha dedicated five pages to division (pp. 23–27), and Elisabeth, four pages (pp. 72–75). No page in Martha's section on division is the same as a page in Elisabeth's section on division.

The examples shown in Figure 4.7 show the long division process—there is no hint of the galley method having been used. Terminology was emphasized, with definitions given of terms like "divisor," "dividend," "quotient" and "remainder." There is also a reference to a "proof by addition." Martha appears to have carried out the divisions accurately and neatly.

In the first of Martha's pages on division (p. 23) the divisors were 2, 3, 4, 5, 6, 7, 8, 9, 10, 11 and 12. Checking was carried out by multiplying the quotient by the divisor and seeing if the result was equal to the dividend. In the second page on division, the divisors were all 2- or 3-digit numbers. It was indicated that proof was by "addition," but it was not made clear what that meant.

Martha's final page on division (p. 27) was concerned with word problems in which money was involved and the operation of division was required. The problems on that page were of a practical nature.

Problem 1. Divide *£12-17-6* equally among 4 men.

Problem 2. If 10 barrels of corn cost *£6-13-4,* what is that per barrel?

Problem 3. Divide 1 shilling equally among 5 men.

Problem 4. Divide *£196-12-9* equally among 33 men

Problem 5. If 64 lbs of bitter cost *£4-5-4,* what is that per pound?

Problem 6. If 132 yards of ribband cost *£11,* what is that per yard?

Problem 7. If 12 yards of cloth cost *£10-16,* what is that per yard?

Problem 8. If 11 barrels of pork cost *£38-10,* what is that per barrel?

Problem 9. Divide *£36-18-4* equally among 16 men.

Problem 10. Divide *£133-6-8* equally among 55 men.

Problem 11. If 100 yards of cloth cost *£33-6-8,* what is that per yard?

Problem 12. If 144 yards of lawn cost *£129-12,* what is that per yard?

It is easy to imagine that, in an era when all household calculations had to be done manually, on paper, having on hand someone able to formulate, and solve such problems could have greatly assisted the budgeting and financial management of the operations of a plantation. Often everyday financial management of plantation operations fell to one of the plantation owner's educated children, or to his wife.

With the above problems, the normal method was to factor the divisor into two factors, and then divide, successively by those two factors. Thus, for example, with Problem 4, Martha divided *£196-12-9* by 3, to get *£65-10-11*. She then divided *£65-10-11* by 11, to get *£5-19-2,* with 1 remainder. Martha gave her answer as *£5-19-2*$\frac{1}{11}$.

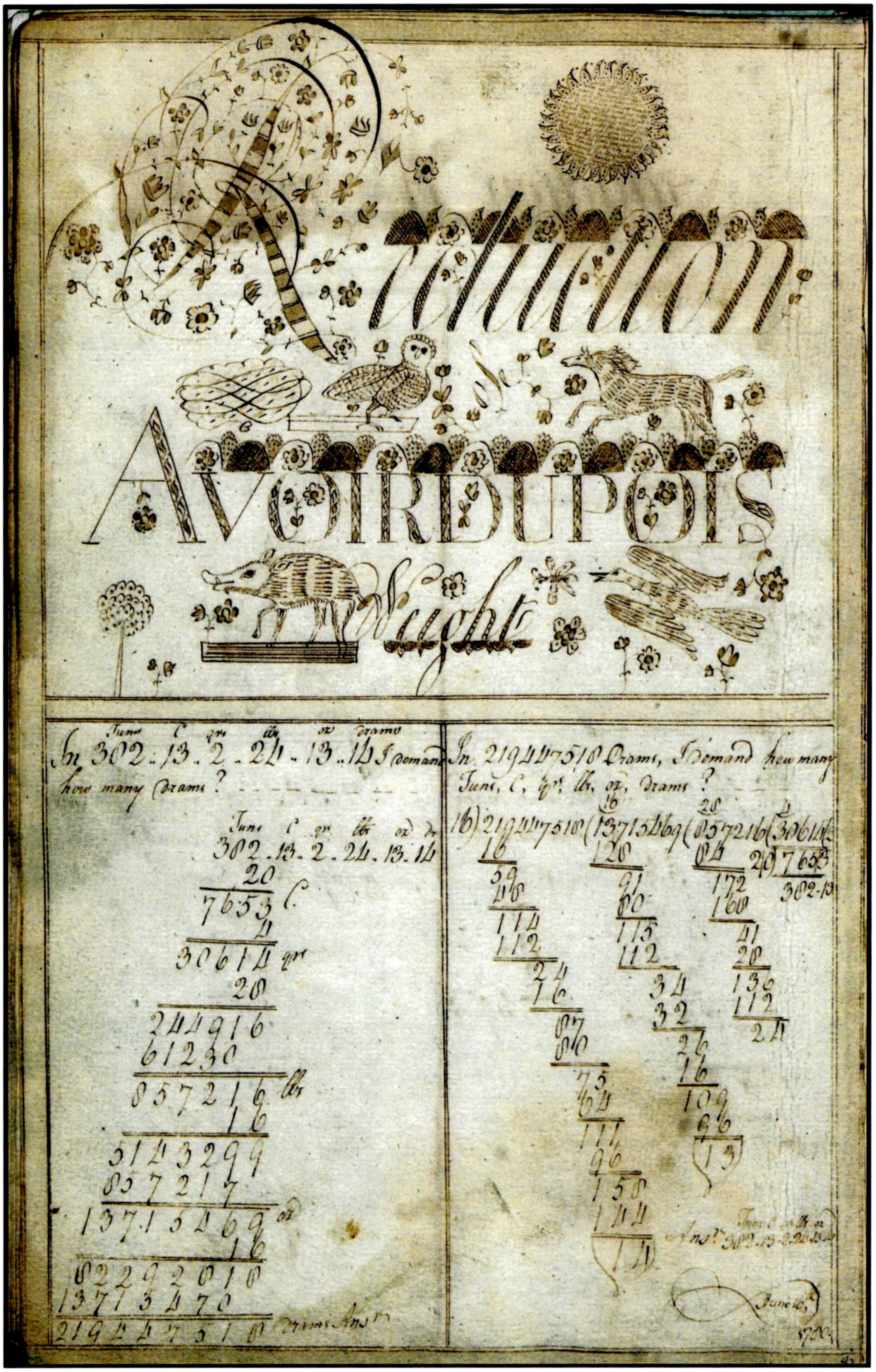

Figure 4.8. Elisabeth Ryan's page on "Reduction Avoirdupois Weight" (p. 81) (Southern Historical Collection, Wilson Library, University of North Carolina at Chapel Hill).

Reduction in the Composite Cyphering Book

Reduction was a vitally important component of *abbaco* arithmetic—particularly in the early European settlements in North America, before decimalization of currency occurred. Martha dedicated three pages (pp. 28–30) to reduction, and Elisabeth six pages (pp. 76–81). Figure 4.8 shows page 81 from Elisabeth's section. Both Martha (p. 28) and Elisabeth (p. 78) offered the following poem when introducing the topic:

Reduction shews how we of names in use,
May great to small and small to great reduce;
So that the answer which shall thence arise,
The given sum in value equalize.
Multiply or divide it back you must,
Which makes again your given number just.

The examples that Martha and Elisabeth used showed how reduction from both "great to small" and "small to great" could occur. Thus, for example, the first problem in Martha's section (p. 28) was: "In 36 pounds, I demand how many farthings."

The page in Elisabeth's section showing Figure 4.8 was dated June 16th, 1780. After spending a great deal of time on the elaborate heading, which incorporated pictures of at least two animals and two birds, Elisabeth showed how to apply reduction techniques to situations involving avoirdupois weight—a system used for weighing products not subject to apothecaries' weight (used for pharmaceutical items) or troy weight (for precious metals). At the bottom left of Figure 4.8, 382 tuns, 13 hundredweights (cwt) 2 quarters (qr) 24 pounds (lbs) 13 ounces (oz) and 14 drams (dr) were required to be reduced to drams. In order to do this, Elisabeth converted the 382 tuns 13 cwt to 7653 cwt, and then converted 7653 cwt and 2 qr to 30614 quarters. She then used long multiplication by 28 to convert the 30614 quarters and 24 lbs to 857216 lbs, and after multiplication by 16 (with 13 added), 13715469 ozs were obtained. Then, after multiplication by 16, with 14 added, an answer of 219447518 drams was obtained.

Finally, rather beautifully, the problem shown on the right side presented the "reverse" of the previous problem. This time, the challenge was to write 219447518 drams in tuns, hundredweight, quarters, pounds, ounces and drams. Whereas the first problem had required quite a few multiplications, this reverse problem required divisions. The inbuilt check was to return to the original amount, with the original units.

The problems in Figure 4.8 were concerned with avoirdupois weight. Earlier, a subtraction problem involving troy weight was considered. Of course, many other systems of units arose in the everyday operation of a plantation—apothecaries' weight, time, length, area, volume, capacity, angle, speed, and money, to name just a few. The conversion factors varied from nation to nation, state to state, and location to location. Was it likely that Martha or Elisabeth would ever need to be able to perform these reduction calculations? The answer is an unequivocal "Yes." Daily life on a plantation could involve the purchase and preparation of food, the cutting and sewing of cloth, the measuring of land, the weighing of quantities, etc. Someone had to have the responsibility for such matters, and also for budgeting and for financial planning. Often male owners of plantations insisted on keeping such matters to themselves, but some were pleased to hand on the responsibility to wives, sons, or daughters, who had the necessary knowledge and skills. That was particularly the case during the seven years of the Revolutionary War, when adult males, and youths, could be away at war (Sturtz, 2002).

Figure 4.9. Martha Ryan's celebration on reaching the rule of three direct (p. 31) (Southern Historical Collection, Wilson Library, University of North Carolina at Chapel Hill).

Rules of Three

Both Martha (p. 31—see Figure 4.9) and Elisabeth (p. 82) celebrated having reached the rule of three direct with full-page, highly decorative announcements. Martha's page included the drawings of five ships (The Privateer *General Washington*, the *Don Antonio*—a Portuguese prize, the Privateer *Freedom*, the Sloop *Nancy*, and the Schooner *Edward*), and an American flag. Martha's section included 17 pages and Elisabeth's section 18 pages on various rules of three. Thus, almost one-fourth of the pages comprising the composite cyphering book were concerned with the rules of three.

On the page following the celebration shown in Figure 4.9 there is a detailed account of what the rule of three direct is, and that description is clearly dated January 7, 1777, when Martha was just 11 or 12 years old. The announcement shown in Figure 4.9 includes the name of the ship *General Washington*, which was not built until 1779, indicating that the page shown in the Figure was prepared *after* Martha had completed at least one of her pages on the rules of three.

Page 33 in Martha's section shows eight problems which were solved using the rule of three direct. The handwriting in six of the eight problem statements and associated solutions is mature, but the handwriting for the last two problems, although neat, is quite immature. The page is initialed "H" at the bottom, and there appears to be little doubt that the first six problems were stated and solved by an adult tutor. The words for the last two problems were written by the same tutor, but the solutions were almost certainly entered by a young Martha, in 1777. Martha's page 38 was still concerned with the rule of three direct, and was clearly dated "Saturday morning, the first day of March, 1777." But Martha's page 40, which announced the rule of three reverse, was clearly dated January 4, 1781—in other words, almost four years had passed between the time when Martha grappled with rule of three direct problems and rule of three reverse problems. Then, between January 4, 1781 and January 31, 1781, Martha prepared 14 pages that would be included in the composite cyphering book. The entry for January 31 would be her last.

Not a few of the rule-of-three-direct problems that the girls solved might be described as practical, but nevertheless trivial. Take, for example, the following question which the tutor wrote down and solved on page 33 of Martha's cyphering book:

$$\text{If 11 herrings cost 11d, what will 1 \& } \frac{1}{2} \text{ cost at that rate?}$$

Obviously, 1 herring would have cost 1d. The answer was clearly $1^1/_2$d. But the tutor wrote:

Herrings

If 11 .... 11 .... $1\frac{1}{2}$

$\underline{2}$	$\underline{3}$ d	$\underline{2}$
22	22/33/1	3
	$\underline{22}$	
	11	
	$\underline{4}$	
	22/ $44/^1/_2$	
	$\underline{44}$	
	00	Answer $1^1/_2$d.

One wonders whether Martha comprehended what her tutor wrote.

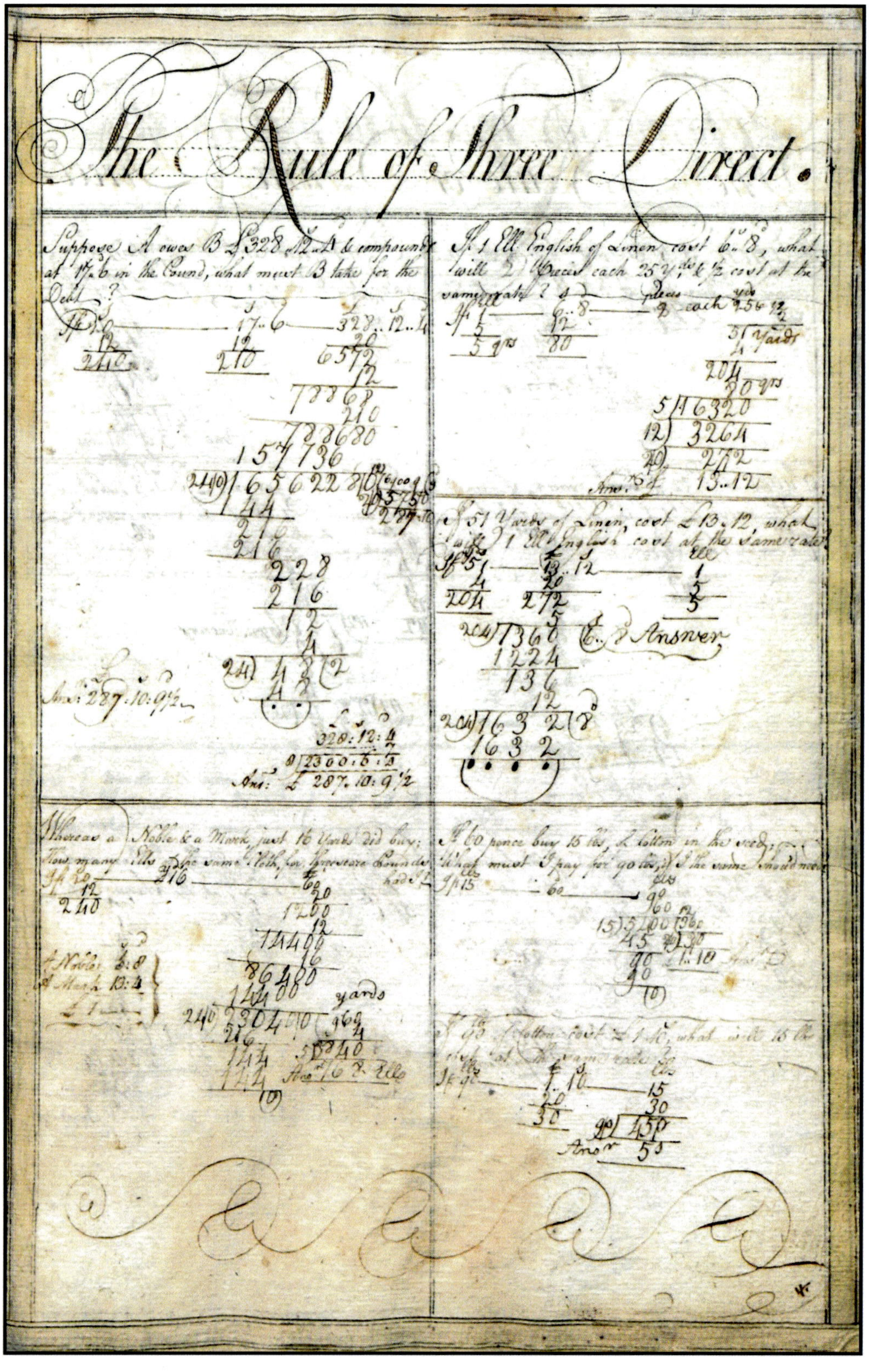
The Rule of Three Direct.

Suppose A owes B £328..12..4 & compounds at 17..6 in the Pound, what must B take for the Debt?

Ans. £287..10..9½

If 1 Ell English of Linen cost 6s..8d, what will 2 Pieces each 25 Yds & ½ cost at the same rate?

If 51 Yards of Linen, cost £13..12, what will 1 Ell English cost at the same rate?

Whereas a Noble & a Mark just 16 Yards did buy; How many Ells of the same Cloth, for Twenty Pounds had I?

If 60 pence buy 15 lbs of Cotton in the seed, What must I pay for 90 lbs, if I the same should need?

If 90 Cotton cost £1..10, what will 15 cost at the same rate?

Figure 4.10. Martha's solutions to five rule-of-three-direct tasks (p. 34) (Southern Historical Collection, Wilson Library, University of North Carolina at Chapel Hill).

More on the Direct Rule of Three

It is likely that Martha did not write any of the five direct-rule-of-three problem statements in Figure 4.10, which is reproduced from page 34 of the composite cyphering book. There is clearly a strong contrast between the maturity of the penmanship with the problem statements, and with the much more immature writing in the handwritten solutions. Presumably, a tutor (who included an "H" in the bottom right corner) wrote the questions on the page, and Martha wrote the solutions.

It will be instructive here to consider the six questions shown in Figure 4.10:

1. Suppose *A* owes *B* £ *328-12s-4d* and compounds at 17*s* *6d* in the pound, what must *B* take for the debt?
2. Whereas a Noble and a Mark just 16 yards did buy, how many Ells of the same cloth for three score Crowns had I?
3. If 1 Ell English of linen costs 6*s* *8d*, what will 2 pieces each 25 yards and a $\frac{1}{2}$ cost at the same rate?
4. If 51 yards of linen cost £ *13-12s,* what will 1 Ell English cost at the same rate?
5. If 60 pence buy 15 lbs of cotton in the seed, what must I pay for 90 lbs if I the same should need?
6. If 90 lbs of cotton cost *£1-10s,* what will 15 lbs cost at the same rate?

Each of Martha's solutions was set out using the traditional direct-rule-of-three format. The idea that with Question 5, for example, the cost would be 6 times 60 pence, or 360 pence (or *£1-10s*) did not seem to have occurred to Martha. Similarly, for Question 6, the answer could easily be obtained by dividing 30 shillings by 6, to get 5 shillings. But such thinking was not expected of 11- or 12-year-olds. Rather, the expectation was that they recognize that these were direct-rule-of-three tasks, for which a standard approach associated with the "golden rule" could be applied.

All of the questions could easily have come within the ambit of someone charged with the responsibility of keeping track of financial undertakings on a plantation. Martha's section of the cyphering book had many more problems similar to those shown in Figure 4.10. On the very next page (p. 35), Martha's tutor wrote the following six problems for her to tackle:

1. Linen at 4/- per yard, how many yards can I buy for £ *84 8s?*
2. Sugar at $6\frac{1}{2}$ d per pound. I demand the value of 1 cwt at that rate.
3. Bought a hogshead of molasses for £ 21-5. How many gallons did it contain, valuing the gallon at 2/6?
4. Oznabrigs at 2s 6d per yard, how many yards may I buy for £ 17-7-6?
5. If 1 cwt of sugar costs £ 3-0s8d, what is the value of 1 lb at the same rate?
6. Tobacco at $2\frac{1}{2}$ d per pound, I demand the value of 1 cwt at the same rate.

The solutions to all of these questions were obtained, clumsily, by the use of the direct rule of three. For example, with the first of these six questions, there did not seem to have been any thought given to dividing 4 into 1688, because £ 84 8s equaled 1688 shillings. That calculation appeared in Martha's application of the rule of three, but the simplicity of the direct proportion relationship inherent in the problem seemed to be lost.

Martha was being taught how to apply a rule without thinking much about meaning.

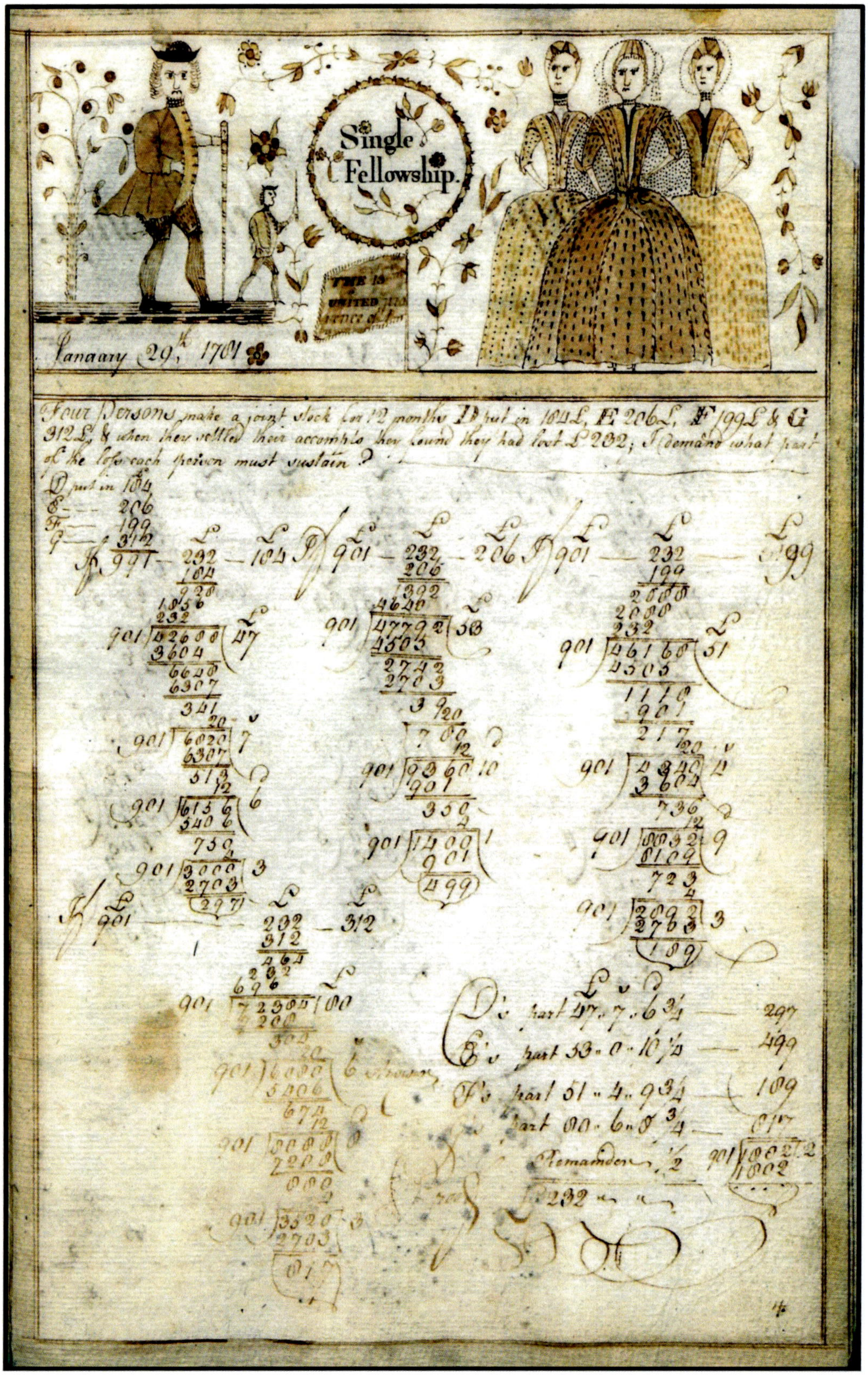

Figure 4.11. 15-year-old Martha Ryan, on Single Fellowship, January 29, 1781 (Southern Historical Collection, Wilson Library, University of North Carolina at Chapel Hill).

Single Fellowship in the Composite Cyphering Book

Both Martha and Elisabeth had entries in their sections on the topic "single fellowship." Martha's four pages on the topic (pp. 48–51) were completed between January 6th and January 29th, 1781. Figure 4.11 shows the entry in Martha's section for January 29th. The handwriting on the page is in the same mature hand found on most of the other pages of the composite cyphering book, and is probably that of Martha's tutor—although it is possible that the notes were entered by Martha herself. The same could be said of the beautiful illustrations and calligraphy in the decorative heading. Martha was probably 15 years old at the time. Elisabeth's section also had four pages on single fellowship (pp. 99–102). But dates on some of Elisabeth's earlier pages indicate that the entries on single fellowship in Elisabeth's section were made in the second half of 1780—that is to say, *before* Martha made her entries.

So far as the arithmetic covered on the pages was concerned, Martha's and Elisabeth's entries were almost identical. The same introductory notes were used, and same word problems stated and solved, in exactly the same manner. Furthermore, the writing would appear to be in the same, familiar, mature hand—which was probably that of the person who served as Elisabeth's and Martha's tutor (and who often included the symbol "H" at the bottom of a page).

The notes in Martha's and Elisabeth's sections introduced single fellowship in the same way. An unedited version is:

> The rule of fellowship is for merchants or other traders there are joint stocks in company to distribute unto every one his proportional part of the gain or loss according to his stock laid in his dividend into two parts commonly called the single and double rules of fellowship in the single rule having the particular gain or loss observe this general rule.
>
> As the total sum of the stocks
> Is the total gain or loss [*Note:* With Martha's version, the first word of this line was "To"]
> So is each man's particular stock
> To each man's particular gain or loss.

Then followed a model example, showing how the rule could be applied. The problem was:

> Two merchants *A* and *B* make a joint stock, *A* puts in 104*£*, *B*, 276*£*, they gained 127*£*;
> I demand, what part of it belongs to each? (From Elisabeth's section p. 99.)

The introductory notes made clear that, with single-fellowship scenarios, the gain or loss of each person contributing to the investment could be calculated by an application of the single rule of three (which is what was done in the application shown in Figure 4.11). In many textbooks the term "partnership" was preferred to "fellowship" (Jackson, 1906), but in cyphering books, "fellowship" was the term most frequently used.

The most striking feature of Figure 4.11 was the illustration in the banner heading. The patriotic spirit of the time is represented by the American flag. Why the three middle-class "ladies" on the right side were depicted as so indignant is not made clear, but there is the suggestion that they were being ill-treated by the British-background "rulers" on the left side. An analysis of Martha's and Elisabeth's composite cyphering book leaves one in no doubt that the sisters were, indeed, "daughters of the Revolution."

Figure 4.12. Martha's second-last entry (p. 52) in her cyphering book, January 30, 1781 (Southern Historical Collection, Wilson Library, University of North Carolina at Chapel Hill).

The Entries on "Double Fellowship" in Martha's Section

Martha Ryan's section included just two pages on double fellowship (pp. 52–53), whereas Elisabeth's section included four pages (pp. 103–106). The first two pages of Elisabeth's entries on double fellowship were almost identical with the two pages on that topic in Martha's section of the cyphering book.

The patriotic, defiant illustrations in Martha's banner headline in Figure 4.12 raise the question why so much of the interior part of the "D" in the word "Double" was shown as shrouded in black. Although we have no strong evidence to support our thesis, we wonder whether Martha knew that she was dying and that double fellowship was likely to be the last topic on which she would make entries in what had become her precious cyphering book. The D shaded in black was meant to represent imminent death. More will be said about why we believe that to be the case shortly.

Martha described double fellowship in the following terms (p. 52):

> It is called Double Fellowship when their gains are different not only with respect to their stocks, but in respect to the time of their continuance in company; & to resolve any questions in this rule, observe the following directions:
>
> Of the total sum of the product of each man's money multiplied by the time put in for, as to the total gain or loss; so is the particular product of each man's money multiplied by the time put in for, to each man's particular gain or loss.

The entry in Elisabeth's book (p. 103) introduced the topic with almost the same words. Then the entries for both girls stated the following word problem: "Two merchants *A* & *B* agree to trade together. *A* puts in 322£ for 4 months, and *B* 475£ for 7 months; they gained 520£. I demand each merchant's part?" Although the application of the single rule of three should have been expected to be straightforward, the fact that neither Martha nor Elisabeth had studied decimal fractions made the calculations quite lengthy.

In the example, the total sum of the product of each man's money multiplied by the time put in was 4613 (i.e., 1288 + 3325). Thus, according to the rule, *A*'s share of the profits, S_A pounds say, should be determined from:

$$4613 : 520 :: 1288 : S_A$$

and hence $$S_A = \frac{1288 \times 520}{4613}$$

But, S_A is in pounds, and hence a difficulty arises in determining what this last expression for S_A would give after conversion to pounds, shillings, pence and farthings. The result of the correct calculations on Martha's page (see Figure 4.12) was 145 pounds 3 shillings 9 pence and 2 farthings. A similar calculation should have arrived at the amount of profit that *B* should receive as 374 pounds 16 shillings 2 pence and 2 farthings, and these two amounts would have happily summed together to give the total gain of 520£. However, the calculation shown for *B* was a farthing out—which was probably explained by contemporary accounting practices taking a calculation for the amount of farthings to the lower whole number of farthings. Thus, for example, as in Figure 4.12, if 9660 farthings were divided by 4613, the answer would be taken to be 2 farthings; but if 8792 farthings were divided by 4613, the answer would be taken to be just 1 farthing.

So, this was Martha's last page of mathematics. She was just 15 years of age when it was prepared. Whether she copied it from Elisabeth's page on double fellowship is not clear. Neither is it clear whether most of the page was prepared for her by her tutor.

Interlude: Martha Ryan's Final Message

The two pages on double fellowship were not, in fact, the last contribution Martha made to the composite cyphering book. The last entry in Martha Ryan's section of the composite cyphering book, which was dated January 31st, 1781, appeared on page 41 of the book, 12 pages *before* Martha's first page on double fellowship. Her second page on double fellowship (p. 53) finished the quire she was using, and she looked for a blank page, earlier in the book, for her final message. She found such a page on the opposite side of an earlier leaf that had been highly decorated. The decorations had "bled through" the page, and probably earlier Martha had decided against making entries on such a page. But on January 31st, 1781, time was of the essence, and so that earlier, available page was chosen for her final message to any future reader. She drew a beautiful picture of a ship and signed off with the following poem:

The Brig *Standall*: Commanded by Captain Fearnot

1. Our Brig's prepared to go.
   And arm'd with men of might
   Our enemies to over throw
   In them we'll take delight.

2. We'll make our enemies to know
   Of them we're not afraid.
   Our sails shall spread to cruise the seas
   And to protect our trade.

3. So Fearnot is my name
   No danger will I dread.
   I'll make the tyrants bend to me
   Or soon I'll strike them dead ...

4. Farewell my friends to you
   Till I shall come again.
   My foes I will subdue
   No British tyrants reign.

Finis, January 31, 1781

Three little-known facts ought to be recorded here. First, the *Fearnot* was a largely Irish-crewed privateer vessel that attacked and captured British vessels during the period 1779–1782; second, the *Fearnot's* Captain was a certain Luke Ryan; and third, Ryan's exploits on the *Fearnot* were secretly supported by Benjamin Franklin (McMillin, 2011a, 2011b; O'Brien, 1920). Martha Ryan obviously knew of Luke Ryan's exploits. The name "Fearnot" seemed to be of significance to her—perhaps, she, herself, was "fearing not" as she faced death (for reasons unknown to the writers of this volume). And, the above poem, accompanied by Martha's beautiful image of a ship she named *Standall*, might have represented her final word, her final message of patriotism.

The next page in Martha's section of the composite cyphering book was dated January 4, 1781. We conjecture that Martha returned to her cyphering book in January 1781 knowing that she was about to die. Then, over a frenetic period of 27 days she prepared 14 full pages for her book. We would like to think that she created the above poem as her last message, her "finis," and placed it in the most beautiful and precious thing that she had ever created—her cyphering book. If that is true, then this chapter is dedicated to her.

It would be easy to romanticize Martha's story, but the fact is we do not have enough details of her life to know the full story. In spite of extensive searching, we could find no date of birth or death—or indeed any details about any aspect of Martha's life. She simply seemed to disappear. Fortunately, we discovered, from a subtraction exercise in her cyphering book, that she was born in 1765. There can be no doubt that the extraordinary cyphering book in which numerous pages carry her name was full of patriotism, and beautiful calligraphy, penmanship, drawings, and courage. And, the mathematics that was inscribed in it went further than the mathematics in almost all other cyphering books prepared by North American females in the eighteenth century.

More Advanced Forms of the Rule of Three Found in the Composite Cyphering Book

Both Martha and Elisabeth cyphered well beyond the direct rule of three. Martha devoted three pages to the "rule of three reverse," two pages to the "double rule of three direct," and two pages to the "double rule of three reverse." Elisabeth also paid attention to each of these same topics. She devoted nine pages to the direct rule of three, two pages to the rule of three reverse, two pages to the double rule of three direct," and two pages to the "double rule of three reverse."

In the following, an example of each of these variations on the rule of three is given. The examples are taken from pages in Elisabeth's section of the composite cyphering book.

Rule of three direct:	If 10 Ells English of cloth costs *£5-1s-3d*, what will 3 yards cost at the same rate? (p. 85)
Rule of three reverse:	If in 20 days, 6 labourers can do a piece of work, in how many days will 8 labourers do the same job? (p. 93)
Double rule of three direct:	If £100 principal in 12 months gain £6 interest, how much interest will £175 gain in 8 months? (p. 95)
Double rule of three reverse:	If £50 principal in 12 months gain £3 interest, what principal will gain £20 interest in 8 months? (p. 97)

For each rule, an explanatory statement and at least one model example were given. Thus for the last-named rule (the double rule of three reverse) the following explanation was given:

> Place your numbers in such order that your 2nd & 4th may be of one name and your 3rd and 5th; then the 3rd and 4th numbers must be multiplied together for a divisor, and the first number must be multiplied by the 2nd and that product by the 5th for the dividend, and the quotient will be the answer in the same name with your first number.

This was followed by the following "solution" to the problem "If £50 principal in 12 months gain £3 interest what principal will gain £20 interest in 8 months?"

	£	*m*	*£*	*m*	*£*
If	50	12	3	8	20
		50		3	
		600		24 Division	
		20 *£*			

24/ 12000 / 500 Answer

120

00

It goes without saying that the "explanation" would not have been extremely helpful for many young persons aged about 13—Elisabeth's age at the time. Although the model example might have helped learners to see what needed to be done, the main problem was whether the learner would ever be able to recognize which situations in real-life—in the case of Elisabeth around the plantation (and elsewhere)—would be associated with problems that could be solved using this double rule of three reverse. Of course, the same comment would apply for all the other rules. The approach adopted by the tutor may not have helped his (or her) young students to know what arithmetic to use, and how to do it, in appropriate real-life situations.

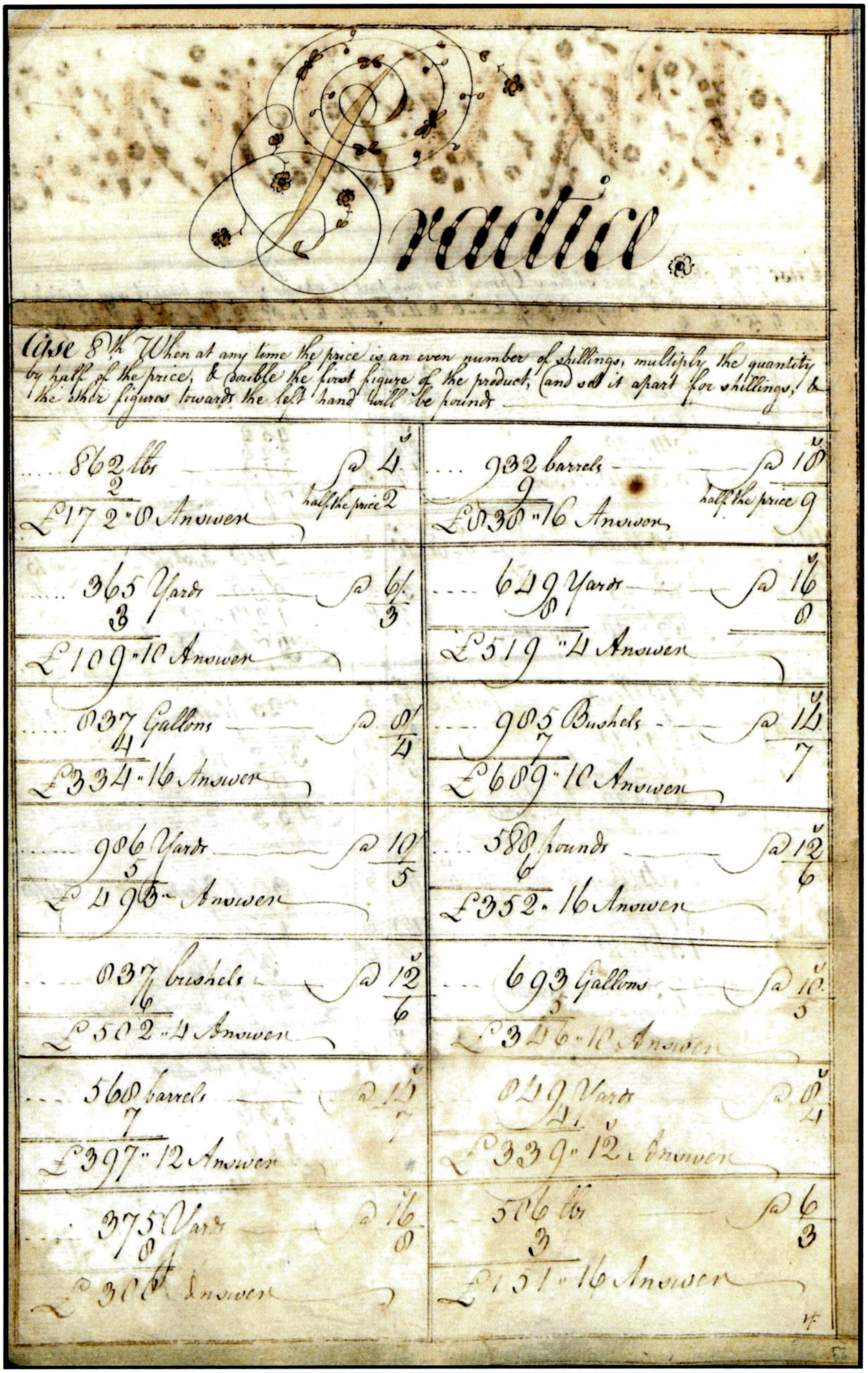

Figure 4.13. The fifth of 13 pages on "Practice" in Elisabeth's section (p. 112) (Southern Historical Collection, Wilson Library, University of North Carolina at Chapel Hill).

The Emphasis on "Practice" in Elisabeth's Section of the Composite Cyphering Book

Practice was an elementary part of the *abbaco* sequence and often appeared either immediately before or immediately after the rule of three and its various more advanced forms. Elisabeth devoted 13 pages to it. By contrast, there were no pages on it in Martha's section.

On her first page on "practice," Elisabeth introduced the topic in the following terms:

> These rules are most compendiously contrived for the speedy casting up of any sort of goods or merchandize, and therefore are of excellent use among merchants and tradesmen, etc., for their quick and elegant dispatch of business, and from their frequent use are called rules of practice; any question in the rule of three that hath a unit or 1 for its number may be much sooner done by these brief rules than by the method followed in that, as too much abounding in figures, in order for working the following tables are to be well understood and got perfectly by heart. (p. 108)

The method of practice relied upon knowing what would, in the 21st century, be regarded as unit fractions associated with quantities. Thus for example, the student was expected to learn that 6d, 4d, 3d, 2d, $1\frac{1}{2}$d, and 1d were $\frac{1}{2},\frac{1}{3},\frac{1}{4},\frac{1}{6},\frac{1}{8}$ *and* $\frac{1}{12}$ of a shilling, respectively. Similar "aliquot parts" of quantities (i.e., unit fractions for quantities) needed to be known for "parts of a tun" (e.g., 10 cwt was half a tun) and "parts of a cwt" (e.g., 14 lbs was half a quarter).

The note immediately below the decorative heading on the fifth of the Practice pages was denoted Case 5th. It stated: "When the given price is such a number of shillings, or shillings and pence as make an even part of a pound, divide the given quantity by that part, and the quotient will be pounds." The first exercise was concerned with "687 yards @ 10*s*" and the answer "£ 343-10*s*" was obtained, presumably by dividing £ 687 by 2. Numerous other similar examples were shown.

The 13 pages on practice had a very practical orientation, and the exercises were designed to enable the person doing them to cope with dignity in a society where calculations such as "687 yards @ 10*s*" might need to be done. Being able to do such calculations would have been helpful for anyone with accounting duties with respect to the financial calculations which might arise on a plantation in Bertie County, North Carolina, in the 1780s. But the rules and cases quickly got increasingly complicated. For example, Case 8th (see Figure 4.13) stated: "When at any time the price is an even number of shillings, multiply the quantity by half the price and double the first figure of the product and set it apart for shillings, and the other figures towards the left hand will be pounds." For the uninitiated this would have been difficult to comprehend. The first exercise showed what it meant—it was concerned with 862 lbs @ 4 *s*. So, according to the rule, all one needed to do was multiply 862 by half of 4, to get 1724, then double the 4 (which was the "first figure" mentioned in the statement of Case 8th), and the answer would be £ 172-8*s*. This was sheer magic, but it did give the correct answer.

And, so the magic continued. A few pages on, Elisabeth reached Case 12th: "If the quantity given has any odd weight annexed to it as $\frac{1}{4}$, $\frac{1}{2}$ or $\frac{3}{4}$ (after you have work'd as before taught for the whole numbers) then take $\frac{1}{4}$, $\frac{1}{2}$ or $\frac{3}{4}$ of the price, and add it to the other work" (p. 117). Each case had its own piece of magic. Unfortunately, in a real situation Elisabeth (or whoever else) would have had difficulty identifying which case, which piece of magic, was needed, And, even if the appropriate case was identified, then there was still the problem of remembering the magic.

Figure 4.14. Elisabeth's decorative, and patriotic, introduction to tare and tret (p. 121) (Southern Historical Collection, Wilson Library, University of North Carolina at Chapel Hill).

Elisabeth Ryan's Efforts with Tare and Tret

Ellerton and Clements (2012), in raising the question whether the Ryan sisters' composite cyphering book might be regarded as overly decorative, commented:

> It could be argued that although the extraordinarily beautiful cyphering book prepared in North Carolina around 1780 by Martha and Elisabeth Ryan represented a worthwhile artistic achievement, it said little about the quality of the sisters' mathematics learning. We would contend however, that that would be a 21st-century perspective, and that in the 1780s the idea that a cyphering book should be as attractive as possible prevailed. Furthermore, since the arithmetic in the Ryan sisters' cyphering book went well beyond the elementary, the book suggested that, yes, females could cope with complex ideas in arithmetic. In addition, the book served as a reference manual for the rest of the authors' lives. (p. 148)

The above argument, defending the Ryan sisters for giving as much attention to artistic form as to mathematical substance, is not without merit. But imagine the vast amount of time expended on producing the page shown in Figure 4.14 which, it could be argued, had very little to do with mathematics. Furthermore, our analysis of the sisters' composite cyphering book suggests that many of the entries—even the decorative, calligraphic entries like that shown in Figure 4.14—were prepared by a private tutor, and *not* by one of the sisters. That comment applies particularly to Elisabeth's section of the cyphering book—the 99 pages of which were prepared over a 14-month period from about April 1780 to June 1781. One must begin to question how much mathematics Elisabeth learned as a result of her cyphering experience.

Because of the huge mercantile influence on the *abbaco* curriculum, the topic "tare and tret" (often "tret" was written as "trett") was almost always to be found in elementary arithmetics published in Europe and in North America during the period 1500–1850. Merchants knew that importers and exporters valued the ideas and wanted their future employees to know about them.

Those dealing with packaging goods for export, or receiving them, quickly learned the distinctive language associated with tare-and-tret calculations. The meanings of terms like "neat," "gross," "cloff," "clough," and "suttle" needed to be known, and the associated arithmetic needed to become automatic. Consider, for example, "Rule 1," on the second of the three pages Elisabeth dedicated to tare and tret:

> *Rule 1:* The tare of any quantity of goods may be easily found, if at so much in the whole only by substracting the said allowance from the gross weight; if at so much per chest, &c, by multiplying the pounds tare by the number of chests &c, and substracting as before; & if at so much per cwt, by taking such part or parts of the gross weight, as the allowance is of a cwt. (p. 122)

Then, on the next page came Rule 2:

> *Rule 2:* Tret, always being 4 pounds per 104, the constant method of finding it is by taking the 26th part of the line it is to be deducted from (4 times 26 is 104). (p. 123)

Although the language was unusual, it was as clear as might be expected given the complexity of what was being described. Also, there was at least an effort to explain the rationale behind Rule 2. That was more than what could be said for the treatment of tare and tret by authors of some widely-used arithmetics (see, e.g., Pike, 1788). Nevertheless, there was little chance that Elisabeth would ever weigh or tax goods on the wharves—tare and tret was a part of the traditional *abbaco* sequence, and there seemed to be no other reason why Elisabeth was asked to study it.

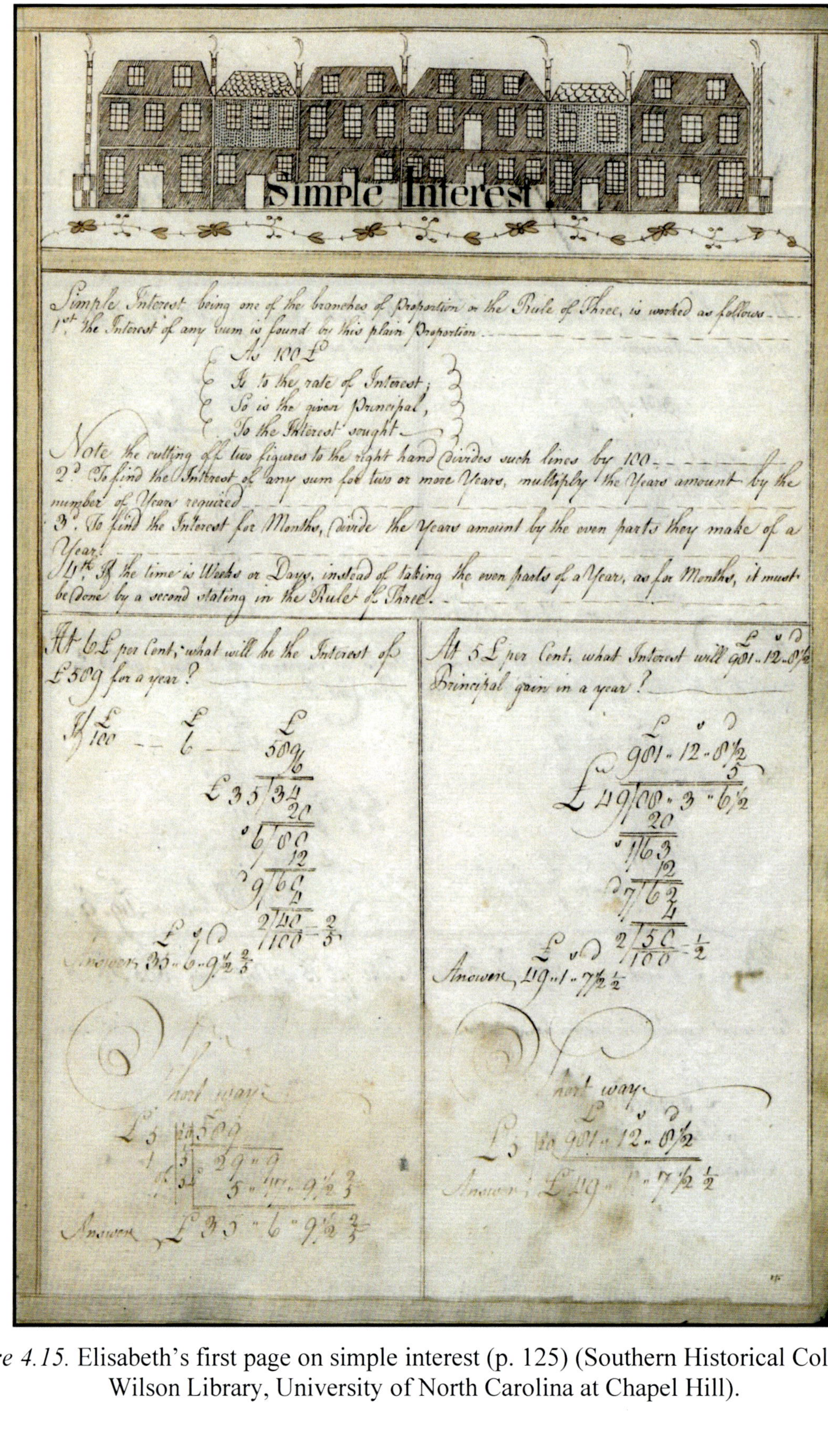

Figure 4.15. Elisabeth's first page on simple interest (p. 125) (Southern Historical Collection, Wilson Library, University of North Carolina at Chapel Hill).

Elisabeth's Entries on Simple Interest

Why Elisabeth should have shown a picture of what seems to be a group of row (or terrace) houses in her decorative heading is a matter for speculation—a possible explanation is that she was the daughter of an owner of a plantation, and investment in property was commonplace among those who belonged to that privileged class. Investment of capital drew interest payments, and hence it might have been appropriate for someone like Elisabeth to become competent in the arithmetic associated with simple and compound interest. That, and the fact that simple interest was an integral part of the *abbaco* sequence for elementary arithmetic, were probably why the tutor decided that Elisabeth should devote four pages to simple interest and two pages to compound interest.

Elisabeth's introduction to the topic made clear how it could be related to the direct rule of three. After pointing out that simple interest was one of the branches of proportion or the rule of three, she gave the rule for calculating simple interest in the following terms:

As 100 £ is to the rate of interest; So is the given principal, to the interest sought.

Immediately following this masterly summary of the fundamental nature of simple interest, Elisabeth's notes on the subject quickly became concerned with more mundane matters:

- Note the cutting off of two figures to the right hand divides such lines by 100.
- To find the interest of any sum for two or more years, multiply the year's amount by the number of years required.
- To find the interest for months, divide the year's amount by the even parts they make of a year.
- If the time is weeks or days, instead of taking the even parts of a year, as for months, it must be done by a second stating of the rule of three.

The first of these points arose because simple interest came before decimal fractions in the *abbaco* sequence. Effectively, the rule was giving a way of achieving an approximation to the correct amount of interest gained. The third of the points indicated that to find the interest for 7 months, say, add the interests for 6 months and 1 month (or for 4 months and 3 months).

The required approach was made clear through the first model example: "At 6 £ per cent, what will be the interest for £ 589 for a year?" The working was as follows:

If £	£	£
100	6	589
		6
		£ 35/34
		20
		6/80
		12
		9/60
		4
		2/40
		100 = 2/5

Answer: £35-6-9 $\frac{1}{2}$ $\frac{2}{5}$

In the calculation shown, the successive multiplications were applied to the remainders. Thus, for example,

£00/34 is multiplied by 20, to get 6/80 shillings, and then the 80 is multiplied by 12 to get 9/60 pence. Then, the 60 is multiplied by 4 to get 2/40 farthings, and /40 is 2/5 of a farthing. As stated, this had to be done without reference to decimal fractions, because decimals had not yet been formally studied.

At the bottom left of Figure 4.15, a "short way," involving a division by 20 and then two divisions by 5, was shown.

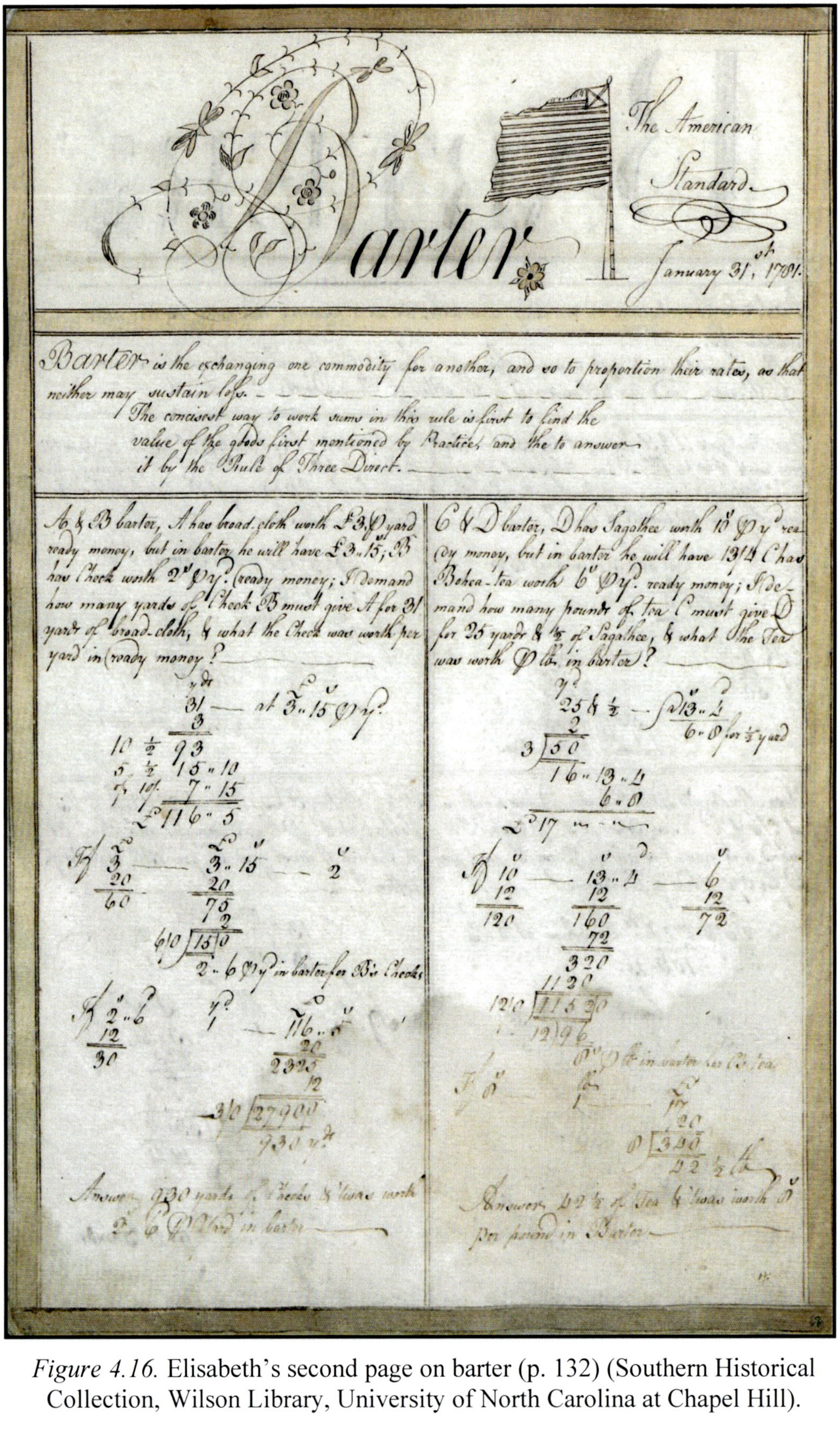

Figure 4.16. Elisabeth's second page on barter (p. 132) (Southern Historical Collection, Wilson Library, University of North Carolina at Chapel Hill).

Elisabeth's Entries on Barter

During the colonial and early Federal eras, bartering—that is to say, exchanging goods and services without any payment of actual monies—was a common phenomenon at all levels of society. On a merchant ship trading in the East Indies, for example, the three most powerful personnel on a ship were the captain, the navigator, and the supercargo. The principal task of the supercargo was to supervise exchanges of goods at ports so that by the end of the journey a suitable profit would be returned to investors (Henry & Kingston, 1878; Jackson, 1906). In early North Carolina much bartering occurred between plantation owners, and it would have been advantageous for a plantation to have within its ranks someone trained in bartering principles and calculations. There were four pages in Elisabeth's section of the composite cyphering book that were dedicated to the topic "barter," and Figure 4.16 shows the second of those pages.

The page began with a brief definition of barter—"Barter is the exchanging of one commodity for another, and so to proportion their rates, as that neither may sustain loss"). Such a definition might have sounded "good," but in practice it was often the case that both sides were trying to make a profit out of the exchange. Then came the statement:

> The concisest way to work sums in this rule is first to find the values of the goods first mentioned by Practice, and then to answer it by the rule of three direct. (p. 132)

Interestingly, exactly the same definition, and the same statement ("The concisest way …") appeared on the previous page.

On the second page, the definition and statement were followed by the following model example:

> Two merchants, *A* & *B* barter. *A* has 150 barrels of corn worth 16/8 per bbl; *B* has molasses at 4/2 per gallon. I demand how gallons of molasses *B* must give *A* for his 150 barrels of corn?

Any worthwhile supercargo would have known the answer to that question in an instant. Since 4/2 is one-fourth of 16/8, *B* would have to give *A* 600 gallons (i.e., 4 times 150 gallons) of molasses for the 150 barrels of corn. However, such a method required what might be called "number sense," something that was rarely emphasized in the *abbaco* tradition. In Elisabeth's book, the solution to the model example was set out as follows:

150 barrels of corn at 16/8
5
6/ 750
125 gall.

Here, the 150 was multiplied by 5 and the result divided by 6 because 3s 4d was 1/6 of a pound, and 16/8 was 5 times 3s 4d.

D	gall	£
If 4/2	1	125
		20
		2500
		12
		5/0 /3000/0
	Answer	600 gallons

The direct rule of three was applied, with the 125 pounds being converted to 30000 pence and the 4s 2d being converted to 50 pence.

Notice the use of the strokes when 30000 was divided by 50.

The model example presented a traditional method of carrying out barter—a method much honored in the *abbaco* tradition and one which would not be easily replaced.

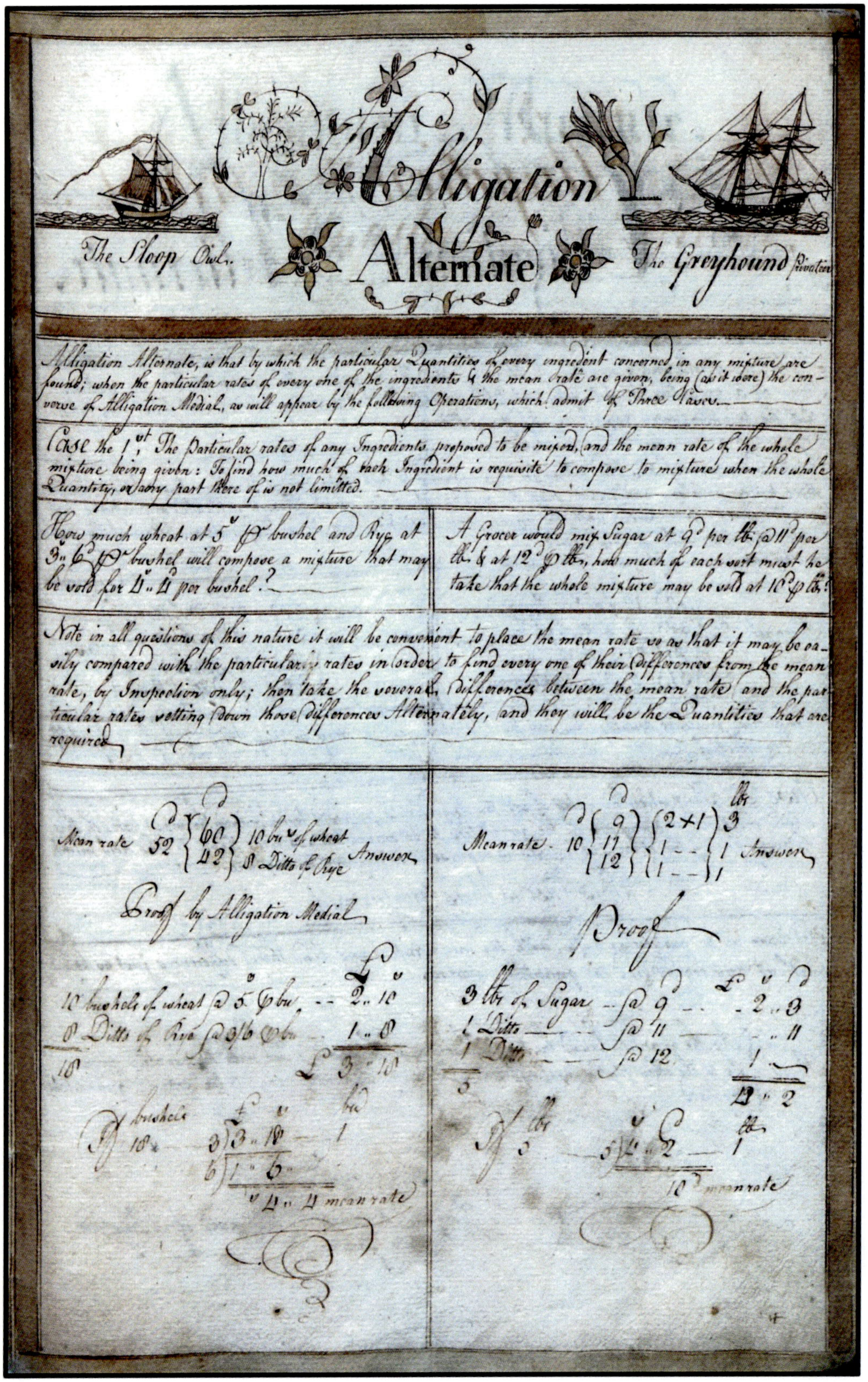

Figure 4.17. Elisabeth's first page on alligation alternate (p. 145) (Southern Historical Collection, Wilson Library, University of North Carolina at Chapel Hill).

Elisabeth's Entries on Alligation

"Alligation medial" and "alligation alternate" were standard middle-level topics in the traditional *abbaco* sequence (Jackson, 1906). They were usually placed after topics like practice, tare and tret, profit and loss, discount, and equation of payments. Elisabeth dedicated two pages to alligation medial (pp. 143–144) and three pages to alligation alternate (pp. 145–147).

As Figure 4.17 explains, alligation was concerned with the arithmetic associated with mixing quantities. In particular, alligation alternate was "that by which the particular quantities of every ingredient concerned in any mixture are found, when the particular rates of every one of the ingredients and the mean rate are given" (p. 145). Three cases were explained in Elisabeth's notes, the first being on page 145. One of the model examples on that page was:

> A grocer would mix sugar at 9d per lb, at 11d per lb, and at 12d per lb. How much of each sort must he take that the whole mixture may be set at 10d per lb?

This type of problem was summarized in "Case the 1st: The particular rates of any ingredients proposed to be mixed and the mean rate of the whole mixture being given: To find how much of each ingredient is requisite to compare the mixtures when the whole quantity or any part thereof is not limited" (p. 145). Then followed a statement setting out a procedure for solving such problems:

> Note in all questions of this nature it will be convenient to place the mean rate so as that it may be easily compared with the particular rates in order to find every one of their differences from the mean rate, by inspection only; then take the several differences between the mean rate and the particular rates setting down these differences alternately, and they will be the quantities that are required. (p. 145)

In solving the grocer's problem, stated above, Elisabeth's set out her working as follows:

		d		lbs	
		9	2 + 1	3	
Mean rate	10d	11		1	Answer
		12		1	

So, the answer to the original problem is 3 lbs of the 9d sugar can be mixed with 1 lb of the 11d sugar and 1 lb of the 12d sugar. Such a composite sugar mixture would be worth 10d per lb. The wording of the rule, as Elisabeth stated it, however, was not clear—although her model example did clarify what was intended. An outline of alligation algorithms for cases when more than three or four different mixtures are involved can be found in Pike (1788).

It is doubtful whether, in her later life, Elisabeth would have ever felt the need to make use of the principles of alligation, although there would have been contexts on a plantation in which mixing procedures might have been useful. But if, at some later time, Elisabeth did consult her notes on alligation there can be no guarantee that they would have been of much help to her. The rules for solving alligation problems, as they were outlined in her cyphering book, were difficult to comprehend. The application of those rules appeared to arrive at a solution somewhat magically. These rules had been worked out, over centuries, by so-called "algorists," and had long been accepted as unquestionably true.

Elisabeth referred to other types of alligation—medial, partial, and total—and gave details with respect to alligation medial (but not for alligation partial or alligation total).

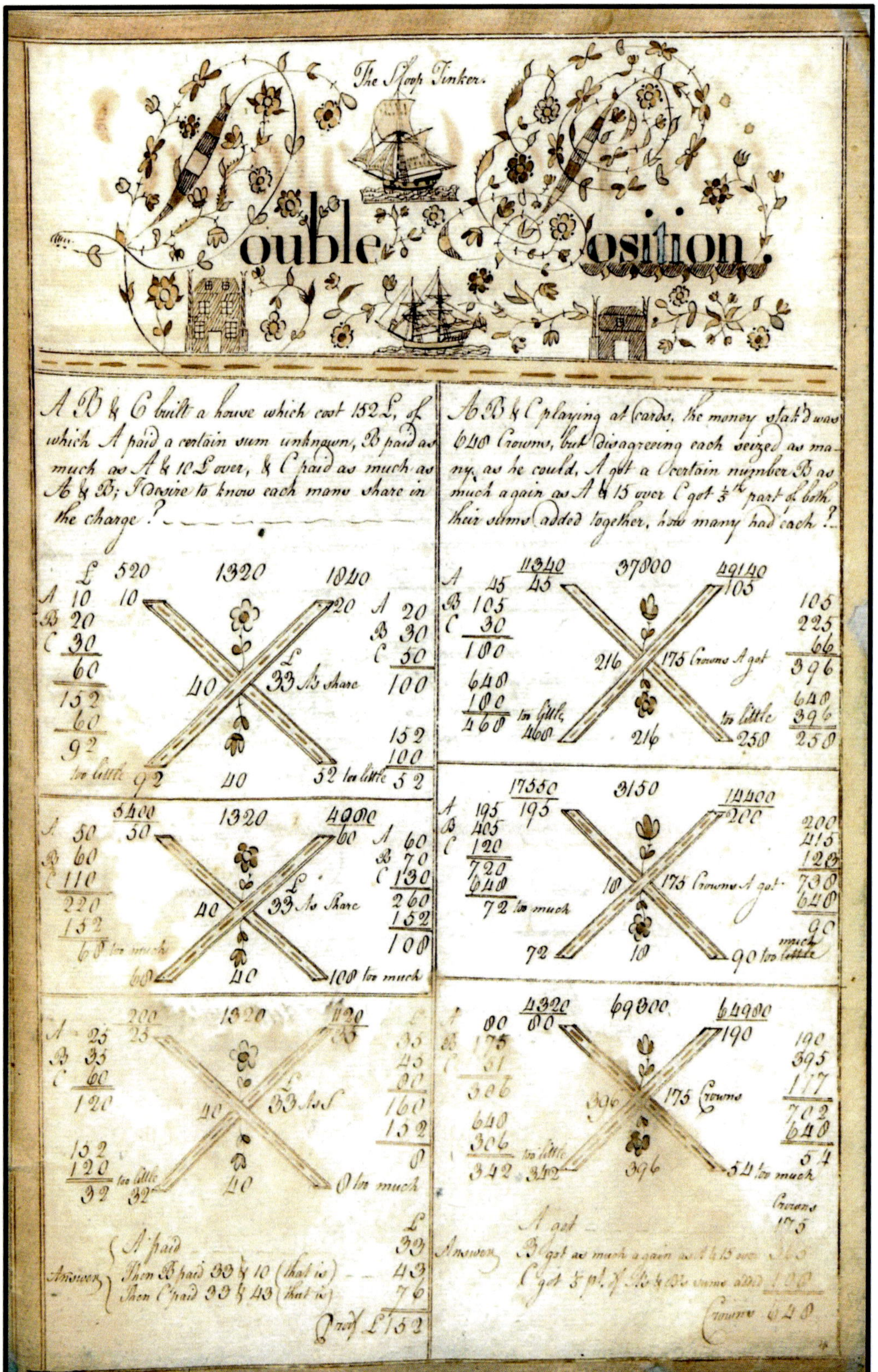

Figure 4.18. A page showing solutions to "double position" tasks (p. 151) (Southern Historical Collection, Wilson Library, University of North Carolina at Chapel Hill).

Elisabeth's Entries on Rules of False Position

The topic known as "false position" comprised rules for "single position" and for "double position." In Elisabeth's section of the cyphering book there were two pages (pp. 148–149) on single position, and two pages (pp. 150–151) on double position. Figure 4.18 shows the second of Elisabeth's two pages on double position.

With double position, a person working on a problem was expected to make two "intelligent guesses" at a solution, and from those guesses an algorithmic procedure could be used which would generate the right answer. The law of double position stated: "If the error be alike, i.e., both be too small or both be too long, divide the difference of the product by the difference of the error and the quotient will be the answer. If the error be unlike, that is, one too small and the other too large, divide the sum of the product by the sum of the error, and the quotient will be the answer."

On the first page for double false position (p. 150), the following notes were written in what was probably her tutor's handwriting—the page has the initial "H" at the bottom right-hand corner:

1st: The rule of double position is when two false positions are assumed to give a resolution to the question proposed;

2nd: When any question is stated in double position, make a cross, thus $\overset{a\ \ b}{\underset{e\ \ d}{\mathrm{X}}}$

3rd: Then make choice of any number you think may be convenient for your working. Call this your first position, and place it at the end of the cross at *a*, and then work with this position as if it were the true number sought, according to the nature of the question; then, having found out your error, either too much or too little, place it at the side of *d*. Then choose another number of the same denomination with the first position, which call your second position, and place it on the side of the cross at *b*; then work with this position as with the former and having found out your error, either too much or too little, place it at the side of the cross at *e*; and then the position will stand at the top of the cross and the errors at the bottom, each under his correspondent (*sic.*) position; and then multiply the errors into the position cross-wise, that is multiply the first position by the second error, and the second position by the first error, and put each product over its position.

4th: Then consider whether the errors are both alike. If they are, subtract the lesser product from the greater, and set the remainder for a dividend; then subtract the lesser error from the greater, and let the remainder be a dividend, and the quotient arising by this division is the answer.

5th: But if the errors are unlike, that is one too much and the other too little, then add the products of the positions and errors together, and their sums shall be a dividend; then add the errors together, and their sum shall be a divisor, and the quotient arising thence is the answer.

This was another piece of magic developed by algorists—no reason for the rule was put forward. In Figure 4.18, two problems were solved using the rule. The first was: "*A, B* and *C* built a house which cost 152*£*, of which *A* paid a certain sum unknown, *B* paid as much as *A* and 10*£* more, and *C* paid as much as *A* and *B*. I desire to know each man's share in the charge" (p. 151). The two positions taken (shown at the top of the first cross) were that *A* paid 10*£* and 20*£*, and these generated errors of 52*£* and 92*£* respectively (shown at the bottom of the cross). Then, after calculating 10×52 and 20×92, and subtracting 520 from 1840, to get 1320, the answer for *A*'s payment would be obtained by dividing 1320 by $(92 - 52)$. *A* would pay 33*£*, *B*, 43*£*, and *C*, 76*£*.

From a 21st-century perspective, that lengthy double-position solution is made to look extremely clumsy when compared with the algebraic solution based on the equation $x + (x + 10) + (2x + 10) = 152$, which gives $x = 33$ immediately. However, like most other eighteenth-century students, Elisabeth had never learned algebra, and so such a method was unavailable to her.

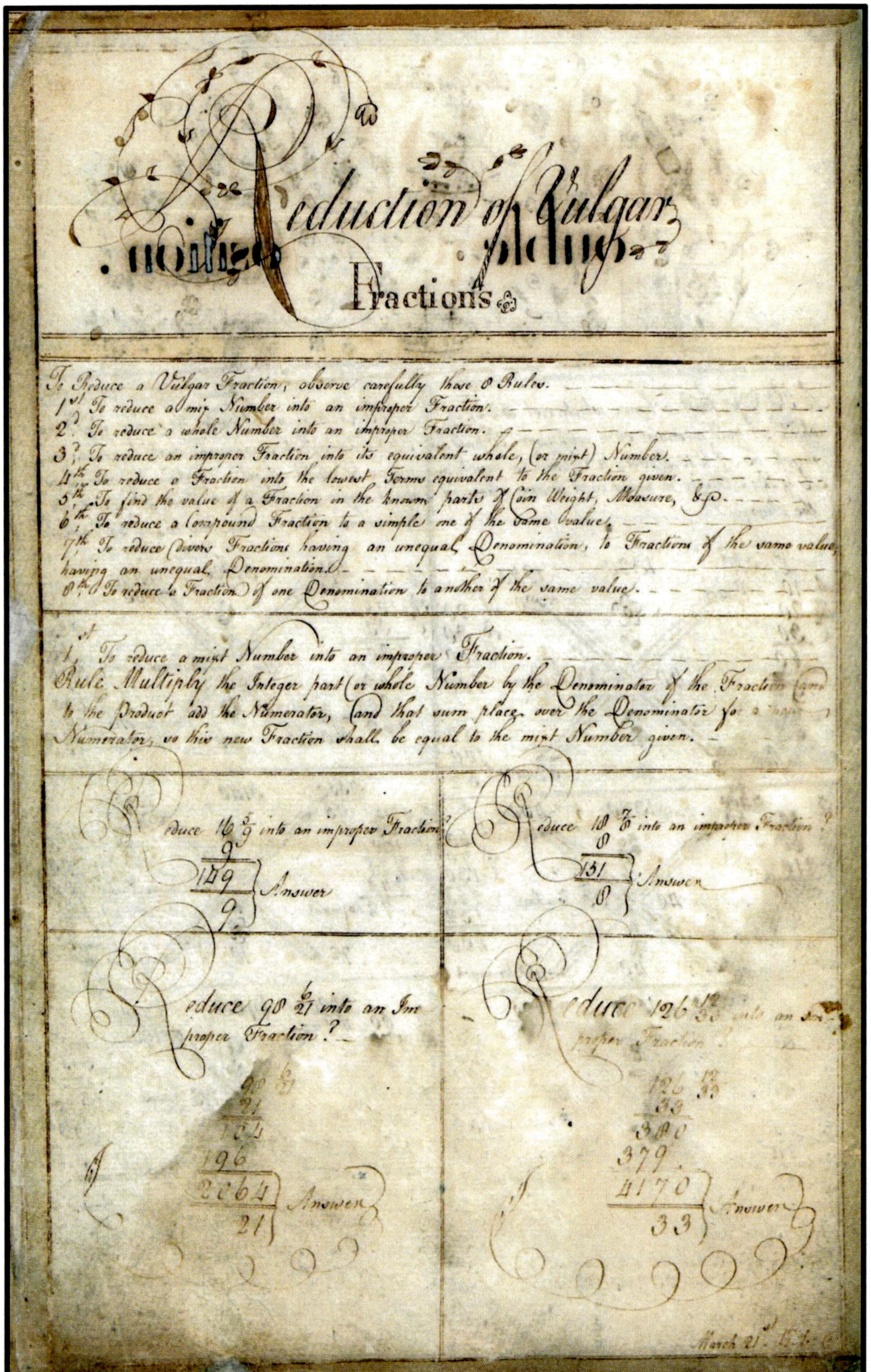

Figure 4.19. Elisabeth's single page on vulgar fractions (p. 152) (Southern Historical Collection, Wilson Library, University of North Carolina at Chapel Hill).

Elisabeth's Single Page on Vulgar Fractions

Figure 4.19 shows the only page in the composite cyphering book that formally deals with vulgar (or common) fractions. There are no pages at all, in the cyphering book, that deal formally with decimal fractions. Vulgar and decimal fractions concepts were well established among mathematicians in the eighteenth century, and appeared in many arithmetic textbooks aimed at schools. During the period 1607–1861 *implemented* school mathematics curricula in North America, as revealed by cyphering book data, corresponded more to perceived societal ethno-mathematical demands than to the *intended* curricula of textbook writers. Despite decimals having been formally described before the end of the sixteenth century, the first mention of decimals in our collection of 350 North American cyphering books was in 1766. Vulgar fractions were formally treated in less than half of those 350 manuscripts. In both North America and Great Britain, the lag-time between when mathematicians had developed sound ways of operating with vulgar and decimal fractions and when they came to be regarded as fundamental parts of the *implemented* school arithmetic curriculum was large. The topics might have appeared in textbooks that were, according to their titles or to claims in their prefaces, prepared for school use; but cyphering-book evidence makes it clear that both vulgar and decimal fractions were not widely studied by those attempting to learn arithmetic in British or North American schools (Ellerton & Clements, 2013).

As we have seen in this chapter, trying to solve problems arising from topics like practice, the various rules of three, simple and compound interest, discount, fellowship, and alligation without having formally studied vulgar or decimal fractions, or percentage, was more difficult than it should have been. Often students resorted to cases and rules which needed to be taken as "givens." Such an approach placed huge demands on memory, and on the ability to understand the meanings of sentences with obscure terminology and complex rules.

In Figure 4.19, eight rules for the reduction of vulgar fractions are specified (e.g., "To reduce a whole number into an improper fraction"). This was the final page in the composite cyphering book, and it seemed that time had run out—model examples were given for one rule only. The exception was the rule which stated that "to reduce a mixed number to an improper fraction, multiply the integer part (or whole number) by the denominator of the fraction and to the product add the numerator, and that sum place over the denominator for a new numerator, so this new fraction shall be equal to this mixt (sic.) number given" (p. 152). The last entries in this magnificent composite cyphering book were four model examples showing how this rule operated. Four mixed numbers were "reduced" to improper fractions. The given fractions were $16\frac{5}{9}$, $98\frac{16}{21}$, $18\frac{7}{8}$ and $126\frac{12}{33}$. The fractions seemed to have been arbitrarily chosen, and Elisabeth might have wondered which were the denominators and which the numerators.

Elisabeth's introduction to the theory of vulgar fractions appeared to be rushed, and there was a strong chance that she would have had difficulty following most, if not all, of the eight rules that she had noted on the final page of her cyphering book. There was no mention in her cyphering book of possible applications of fractions to real-life situations, and almost certainly she would not have been able to apply what she had learned about fractions to her daily circumstances. Furthermore, much of the *abbaco* arithmetic curriculum that she had studied could have been made simpler if Elisabeth had learned to use fractions at some earlier time. The *abbaco* curriculum was not particularly well sequenced, but it seemed that no-one was aware that that was the case.

The Historical Challenge

The Ryan sisters' composite cyphering book provides a glimpse of what mathematics education might have been like for well-to-do white children who were living on plantations within Southern states around the time of the Revolutionary War. In some cases, children of the privileged élite attended subscription schools, and some went to far-off boarding schools—but it was more likely that they would be educated at home by a resident tutor, or by an itinerant teacher. For both Martha and Elisabeth Ryan, the tutor seems to have been the person who initialed many of the pages in the cyphering book with the symbol "H."

"H" seemed to have a strong grasp of the main topics in the *abbaco* sequence which defined the intended mathematics curriculum for those North American children, aged 10 years or more, who prepared cyphering books (Ellerton & Clements, 2012). That curriculum started with numeration, and proceeded to operations on whole numbers, compound operations, reduction, the rules of three, practice, profit and loss, simple and compound interest, discount, equation of payments, fellowship, alligation, and false position. Elisabeth included pages on almost all of those topics in her cyphering book.

Martha did not have the time to do as much as Elisabeth. She started her cyphering book in 1776, but the next year both her father, James Ryan, and her maternal grandmother died. In 1778, 1779 and 1780 Martha probably did not prepare any cyphering-book pages—although there is a possibility she did so with another tutor and the pages were not included in the composite cyphering book. Elisabeth, two years younger than Martha, cyphered during 1780 and 1781. Both girls cyphered well beyond the rule of three—which was a very unusual achievement for females. Therefore, the sisters' cyphering book is important historically. The curriculum they followed probably included exactly the same topics as a boy would have followed if that boy had been taught by Tutor "H." Analysis of the questions that the girls were asked to solve suggests, however, that Tutor "H" may have had his eye on the sorts of arithmetical tasks that an intelligent wife of the owner of a plantation would have needed to solve. Many of the tasks on the direct rule of three, on practice, and on simple and compound interest, and discount, were of that kind.

One might like to think of the composite cyphering book as extraordinary because the penmanship, calligraphy and illustrations were of such high quality. Almost every page qualified as a wonderfully crafted piece of art. However, careful analysis casts doubt on whether the girls were responsible for this—the penmanship and calligraphy, and perhaps even some of the illustrations, could well have been done by Tutor "H." Furthermore, in the absence of hard data pointing to the contrary, one might easily conclude that the girls did not do much hard thinking, themselves, about the arithmetic in their cyphering book.

There can be no doubting, however, the incredible revolutionary spirit of the two girls—for that leapt out of almost every page. Martha and Elisabeth were, indeed, daughters of the Revolution. Even towards the end of her cyphering efforts (p. 150), Elisabeth wrote, at the foot of a page, the strong words: "Poor freedom is better than rich slavery. Feb^{ry} 9th, 1781." Just 10 days before that, Martha had penned her impressive little poem which showed she knew all about Captain Luke Ryan's amazing successes, in 1779 and 1780, as a privateer in helping to change the course of the Revolutionary War (McMillin, 2011a, 2011b). One has cause to wonder whether Luke Ryan was related to the Ryan sisters in any way.

Hopefully, this chapter will provide entrée for other researchers interested in the history of North American school mathematics to take up the challenge of investigating the teaching and learning of mathematics in "southern states" during the 17th, 18th and early 19th centuries.

Acknowledgments

This chapter could not have been completed without the help of the archivists and officers within the Southern Historical Collection in the Wilson Library at the University of North Carolina, at Chapel Hill. In particular, the help of Matthew Turi and the imaging technicians of the Research and Instructional Services Department of the Louis Round Wilson Special Collections Library is gratefully acknowledged.

The genealogical assistance of Barbara A. Day, especially in relation to finding some details about Martha and Elisabeth Ryan, and their parents, is also gratefully acknowledged.

References

Berkin, C. (2005). *Revolutionary mothers: Women in the struggle for America's independence.* New York, NY: Vintage Books.

Bethlehem Digital History Project. (2013). *Susan Shimer's cyphering book.* http://bdhp.moravian.edu/bethlehem/bethlehem.html, retrieved November 15, 2013.

Bickham, G. (c. 1740). *Youth's instructor in the art of numbers, a new cyphering book in which is shewn variety of penmanship, command of hand: Engraved for the use of schools.* London, United Kingdom: William & Cluer Dicey.

Butler, L. S. (1976). *North Carolina and the coming of the Revolution, 1763–1776.* Zebulon, NC: Theo. Davis Sons, Inc.

Cajori, F. (1890). *The teaching and history of mathematics in the United States* (Circular of Information No. 3, 1890). Washington, DC: Bureau of Education.

Clements, M. A., & Ellerton, N. F. (2012, October 27). *The curious case of duodecimals in North American cyphering books and early textbooks.* Paper presented to the History and Pedagogy of Mathematics (HPM) Conference, The University of California at Berkeley, California.

Cocker, E. (1664). *The tutor to writing and arithmetic.* London, United Kingdom. H. Tracy.

Cohen, P. C. (1982). *A calculating people: The spread of numeracy in early America.* Chicago, IL: University of Chicago Press.

Dilworth, T. (1773). *The schoolmasters assistant: Being a compendium of arithmetic, both practical and theoretical* (17th ed.). Philadelphia, PA: Joseph Crukshank.

Doar, A. K. (2006). *Cipher books in the Southern Historical Collection.* Master of Science thesis, Wilson Library, University of North Carolina at Chapel Hill.

Ellerton, N. F., & Clements, M. A. (2012). *Rewriting the history of school mathematics in North America 1607–1861: The central role of cyphering books.* New York, NY: Springer.

Ellerton, N. F., & Clements, M. A. (2013, September). *The mathematics of decimal fractions and their introduction into British and North America schools.* Paper presented to the Third International Conference on the History of Mathematics Education, held at Uppsala University Sweden, September 25–28, 2013.

Fisher, G. (1734). *Arithmetick in the plainest and most concise methods hitherto extant. With new improvements for dispatch of business in all the several rules. As also fractions, vulgar and decimals, wrought together after a new method, which renders both easy to be understood in their nature and use. ...* London, United Kingdom: Bettesworth and C. Hitch.

Grimes, J. B. (1910). *Abstract of North Carolina wills.* Raleigh, NC: E. M. Uzzell and Co.

Hathaway, J. R. B. (Ed.). (1900). *The North Carolina historical and genealogical register.* Edenton, NC: Clearfield.

Henry, W., & Kingston, W. (1878). *The two supercargoes.* London, United Kingdom: Sampson, Low, Marston, Searle & Rivington.

Jackson, L. L. (1906). *The educational significance of sixteenth century arithmetic from the point of view of the present time.* New York, NY: Teachers College Press

Jefferson, T. (1785). *Notes on the establishment of a money unit and of a coinage for the United States*. Paris, France: Author.

Linklater, A. (2002). *Measuring America: How the United States was shaped by the greatest land sale in history.* New York, NY: Plume.

McMillin, M. M. (2011a). *Gather the shadowmen: The lords of the ocean.* Richardson, TX: Hephaestus.

McMillin, M. M. (2011b). *Prince of the Atlantic: The war*. Richardson, TX: Hephaestus.

Middlekauff, R. (1963). *Ancients and axioms.* New Haven, CT: Yale University Press.

Monaghan, E. J. (2007). *Learning to read and write in colonial America.* Amherst, MA: University of Massachusetts Press.

Moravian Publication Office. (1876). *Historical sketch of the Moravian Seminary for Young Ladies.* Bethlehem, PA: Author.

Morley, V. (2002). *Irish opinion and the American Revolution 1760–1783.* Cambridge, United Kingdom; Cambridge University Press.

National Park Service. (1998). *National Registrar of Historic Places Registration Form: Jonathan Hill Jacocks House.* Washington, DC: United States Department of the Interior.

Newby, W. G. (1958). *Town of Hertford bi-centennial, 1758–1958: And historic data of Perquimans County, North Carolina.* New Bern, NC: Owen G. Dunn.

O'Brien, M. J. (1920). *A hidden phase of American history: Ireland's part in America's struggle for liberty.* New York, NY: Dodd, Mead and Company.

Pike, N. (1788). *The new complete system of arithmetic, composed for the use of the citizens of the United States.* Newburyport, MA: John Mycall.

Smallwood, M. B. (1964). *Some colonial and revolutionary families of North Carolina.* Washington, NC: Author.

Sturtz, L. (2002). *Within her power: Propertied women in colonial Virginia.* New York, NY: Routledge.

Thornton, T. P. (1996). *Handwriting in America: A cultural history.* New Haven, CT: Yale University Press.

Chapter 5

With a Tinge of Green: Mary Walters' Cyphering Book, 1820

Abstract: In 1820 Mary Walters was based in Baltimore, Maryland. She had emigrated from Ireland to North America and her cyphering book, prepared in 1820, provides evidence that in Maryland some Irish-heritage school teachers were responsible for engaging students in a form of mathematics education that had more than a tinge of green. Certainly, like all who followed the *abbaco* arithmetic curriculum, Mary was expected to learn how to add, subtract, multiply and divide counting numbers, and to apply elementary measurement principles in a range of everyday situations. When dealing with tasks involving the measurement of quantities, she needed to be able to change from one set of units to another. She also needed to learn to apply the direct rule of three in a wide range of real-life scenarios in which ratio and proportional thinking were needed. But the entries in Mary's cyphering book also made it clear that her teacher(s) wanted her to inject herself into problem-based scenarios by which she would learn to apply elementary arithmetic principles to tasks concerned with human-interest, life-and-death situations which often generated emotions like jealousy, love, and desire. In particular, she was expected to appreciate the power of stories, poetry and song to convey interesting problem scenarios.

Key Words: *Abbaco* curriculum; ciphering book; ciphering tradition; cyphering book; cyphering tradition; Irish ballads; Irish education; Irish in North America; Irish poetry; Irish mathematics education; problem-based learning.

When we first examined the 80-page, folio-sized cyphering book which is the focus of discussion in this chapter, we assumed that its main educational significance would be associated with it having been prepared, around 1820, by a girl who was proud that she had cyphered to the rule of three. Despite the fact that the manuscript had no covers, that its outside pages were grubby and tattered, that the corners of all pages were turned, and that all internal pages were age-toned, we were impressed by the consistently high level of penmanship and calligraphy, and by the beautiful 8" by 6" hand-painted rose (see Figure 5.1) on one of the pages. Around the rose were inscribed the words "The Rule of Three," and below it was an appropriate poem.

From a historical perspective, the fact that the book was prepared by a female who was proud of having cyphered to the rule of three was less significant than the fact that the book included many word problems expressed via the medium of poetry. We reached the conclusion that this genre was indelibly linked to the Irish heritage of the young woman who prepared the cyphering book in Baltimore, Maryland, around 1820. That young woman's name was Mary Walters.

Who Was Mary Walters, and When and Where Did She Prepare her Cyphering Book?

In order to find out more about who prepared the cyphering book we used the ancestry.com website to investigate persons with the name "Mary Walters"—because that was the name written prominently three times, on two adjacent pages of the cyphering book we were studying. We were not surprised to find that in the early 1800s there were many girls of that name who were of school age. We needed additional relevant data.

Nerida F. Ellerton and M. A. (Ken) Clements, *Abraham Lincoln's Cyphering Book and Ten other Extraordinary Cyphering Books*,
DOI 10.1007/978-3-319-02502-5_5,

Figure 5.1. Mary Walters' celebration, in 1820, on reaching the "rule of three."

The first piece of additional data was the fact that we purchased the manuscript from someone who had recently bought it at an estate sale in Oneida County, New York. Another piece of data was that the word "Baltimore" was written, in fluent calligraphy, on three pages.

Genealogical checking indicated that a "Mary Walters" who had been born in Kilkenny, Ireland, in about 1806 had emigrated to North America sometime between 1806 and 1818. This Mary Walters had married James Henry O'Connor (also from Kilkenny) in 1826, and had died in Taberg, Oneida County, New York, late in the nineteenth century.

Although no date at which the manuscript was prepared is specifically given in the cyphering book, the responses to two subtraction tasks in the book implied that Mary had begun to prepare her cyphering-book entries in 1820. Both of these tasks were in the form of word problems. The first stated that the "invention of gunpowder was discovered in the year 1302, and then asked: "How long is it since to the present year?" The calculation "1820 – 1302" was then shown, with the answer "518." The second task stated that the art of printing was discovered in the year 1449, and then asked, "How long is it since?" This time "1820 – 1449 = 371" was shown. So, there could be little doubt that whoever Mary Walters was, she prepared at least part of her cyphering book in 1820. A subtraction indicated that if, indeed, the Mary Walters who prepared this cyphering book had been born in 1806, then she would have been 13 or 14 years old in 1820. This calculation produced a result consistent with our judgment that the handwriting in the cyphering book was neat and consistent with what might be expected of a 13- or 14-year-old.

Where did Mary Walters Prepare her Cyphering Book?

We wondered whether it was possible that Mary had prepared her cyphering book somewhere in Great Britain, for although it included many problems requiring monetary calculations, all of these were in terms of sterling currency—pounds, shillings, pence and farthings—and not dollars and cents, the chief elements of the decimalized form of currency that had been introduced into the United States of America in the early 1790s. The issue was complicated by the fact that in 1820 different versions of sterling currency were still being used across the United States. Sterling currency was used in Ireland and in England, of course, but in Ireland the values of pounds, shillings, pence, etc., were not identical to those in England (Plowden, 1805; Yeats, 1928). Related, perhaps, to the absence of any emphasis on decimal currency in Mary's cyphering book was the fact that decimal fractions or common (i.e., vulgar) fractions were not formally dealt with in her book. Even with rule-of-three tasks which involved ratio and proportion concepts, decimal fractions were conspicuous be their absence.

When we checked leaves of Mary's cyphering book for watermarks, we were able to identify, unambiguously, an 1818 North American watermark—shown in Bidwell (2013). Furthermore, since "Baltimore" was written three times on one page of the manuscript, it seemed reasonable to assume that when, in 1820, Mary began her cyphering book, she was living in Maryland, in the United States of America. Maryland was a popular destination for Irish Catholics immigrating to the United States in the early decades of the nineteenth century (Doyle, 1980; Koos, 2001; Miller, 1985; Steinfeld, 1991).

From the above considerations, we concluded that the Mary Walters who prepared the cyphering book that we were examining had been born in Kilkenny, Ireland, in or around the year 1806. She had emigrated to the United States some time before 1818, had begun to prepare the cyphering book in Maryland in 1820, and had married James Henry O'Connor (also born in Kilkenny) in 1826. Genealogical data also revealed that Mary died in Taberg, Oneida County, New York, sometime after 1870—according to 1870 U.S. census data, she was 64 years old in 1870.

The Problem-Based, Poetic Genre in Mary Walters' Cyphering Book

Many word problems in Mary Walters' cyphering book were presented through the medium of poetry. The following poem, which is shown here in unedited form (but is also shown in Figure 5.2), was taken from the inside of the last leaf of Mary's book. It illustrates the genre.

Sound is found by exp (???) [*Words are missing*]
Feet in One second of Time (???) [*Words are missing*]
As I walkt out one cleare,
Twas on a summers Day;
Unto a mountain's highest Brow
There I did take my way.
And as I gaz'd the Country round,
Fine prospect there to see,
A Sportsman brave his game persued.
In a Wood Remtes [???] for me
The Distance I did wish to know,
Between that man and me;
A thought the thing it me be done,
If fire I could but see.
In a short time some game he found,
So much to my Desire;
As soon as up the game did rise,
He of his piece did fire.
Then with attention I observed,
As the Distace shou'd be found;
Twas just nine seconds after sight of fire,
Before I heard the sound.
Then take your Pen the distance show,
Between that [man] and me;
And you with Laurel shall be crouned,
And soon a schler be.

A copy of the handwritten version of the problem has been reproduced as Figure 5.2. The page was on the inside of the last leaf in Mary's cyphering book, and unfortunately the top right corner is no longer present—and several key words are missing.

Mary solved this problem by applying the so-called "direct rule of three." She obviously recognized the problem as equivalent to: "If, in 1 second, sound travels 1142 feet, how far would it travel in 9 seconds?" She applied a standard direct rule-of-three technique, multiplying 1142 feet by 9, to get 10278 feet. She then divided by 3 to get 3426 yards. On dividing this by 1760, she concluded that the sportsman was 1 mile and 1666 yards away.

According to Greg Koos, Director of the McLean County Museum of History, in Bloomington, Illinois, the above poem, and other poems in Mary's cyphering book, can be sung to the tune of traditional Irish songs (G. Koos, personal communication, August 13, 2013). In fact, in Mary Walters' cyphering book there were quite a few problems which were set as story-poems.

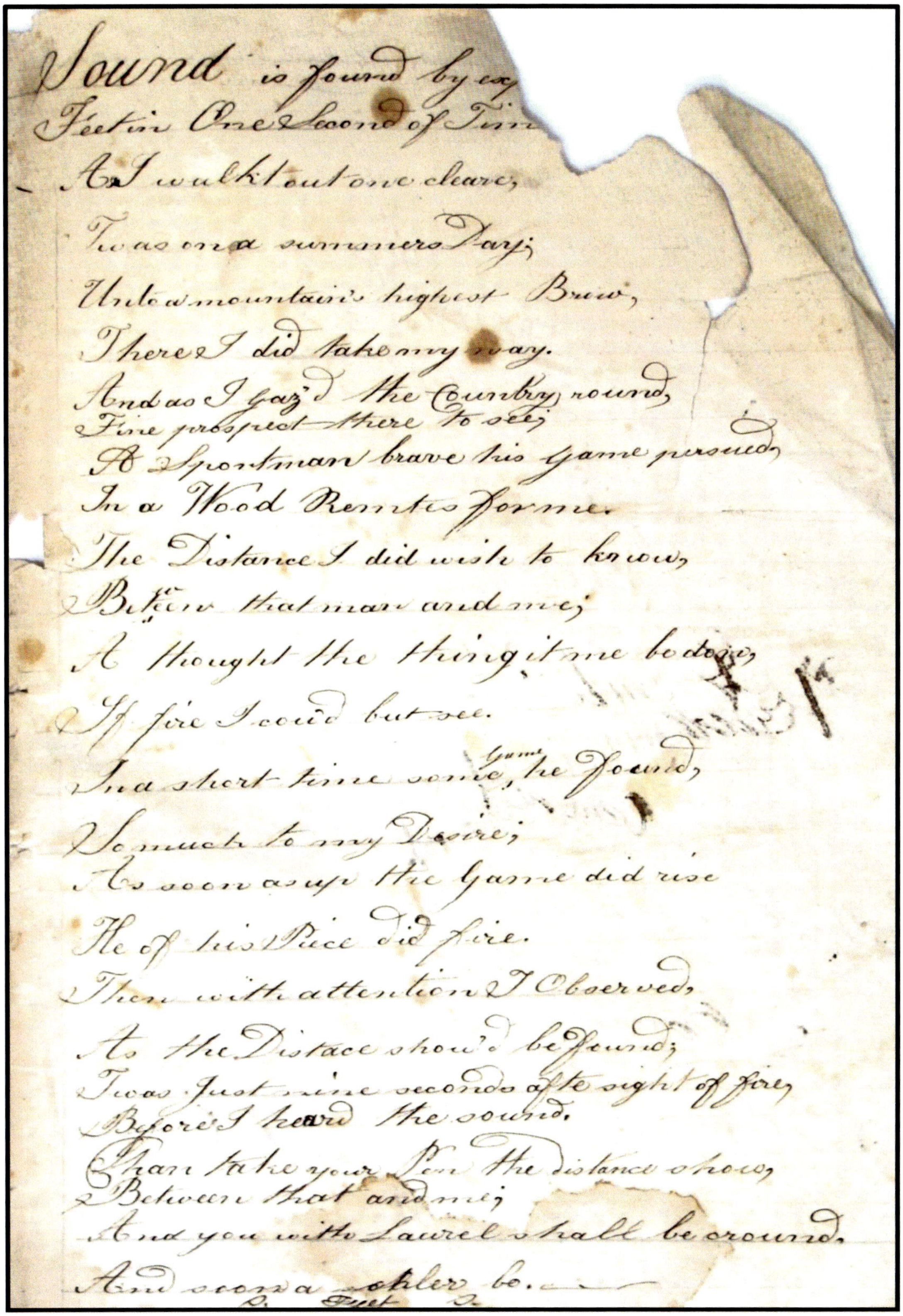
Sound is found by ex
Feet in One Second of Tim
As I walkt out one cleare,
Twas on a summers Day;
Unto a mountain's highest Brow,
There I did take my way.
And as I gaz'd the Country round,
Fine prospect there to see;
A Sportman brave his Game pursued,
In a Wood Remote from me.
The Distance I did wish to know,
Betwen that man and me;
I thought the thing it me be done,
If fire I could but see.
In a short time some Game he found,
So much to my Desire;
As soon as up the Game did rise
He of his Piece did fire.
Then with attention I Observed,
As the Distace shou'd be found;
Twas Just nine seconds afte sight of fire,
Before I heard the sound.
Than take your Pen the distance show,
Between that and me;
And you with Laurel shall be crownd.
And soon a schler be.

Figure 5.2. Using arithmetic to locate her man.

Yet, the *abbaco* arithmetic sequence was followed throughout the book, with poems carefully chosen to fit topics like "addition," "subtraction," "multiplication," "division," "reduction," and the "rule of three." In short, then, Mary (or Mary's teacher) adopted a problem-based approach to arithmetic, with word problems, often introducing contexts of considerable human interest, being presented in poetic form. Where might this problem-based, poetry-inspired approach to school arithmetic have originated? And from where, and from whom, did the poems originate?

Problem-based approaches to arithmetic through poetry were not totally unknown around 1820 (see, e.g., Hersee, 1997). Daniel C. Underhill (1825) published numerous editions of his *Tables of Arithmetic Made Easy* in which arithmetic was approached through poetry (Karpinski, 1980). Other North American authors of arithmetics who included problems in the form of poetry were Titus Bennett, John Bonnycastle, Nathan Daboll, Frederick Emerson, Zachariah Jess, Augustus M'Cord, Jason Mahan, Tobias Ostrander, and Benjamin Workman (Allen, 1970, 1990). But, when we compared the poems in texts by those authors with the poems in Mary Walters's cyphering book we did not find any poem that was more than remotely similar to poems in Mary's cyphering book. Furthermore, none of the textbook authors made the poems central to the intended learning process—that was true even with Ostrander's (1823) book, which included a total of 148 lines of verse, in 16 poems.

The most likely explanation of the approach found in Mary Walters's cyphering book was that either the poems were taken from some other text—like, for example, another cyphering book that used the poetic genre approach—or that they were dictated to her by a teacher who had committed many of the poems to memory (Cullen, 1988). It is also possible that the poems were actually composed by Mary's teacher for use with his (or her) students.

The following comments by Greg Koos (2001) on Irish-heritage schoolmasters in the United States appear to be pertinent to the matter under discussion:

> In 1804, a report by the Society for the Education of the Poor [in Ireland] noted that the lack of textbooks caused their replacement by "such romances and histories of profligate and daring adventures as have been handed down from generation to generation" ... Another important source for curricula were manuscript compilations. Schoolmasters were often the most literate persons of their communities. Among them were poets and translators. These men took up schools in the eighteenth century during the decline of the traditional patronage by leading families. (pp. 11–12)

Koos (2001) quoted an English-language version of a poem, "Letter to Father Fitzgerald," written in the 1780s by the great Gaelic poet, Eoghan Rua Ó Súilleabháin (1748–1782). In the poem, Ó Súilleabháin reflected on what his work as a teacher included:

> Reverend Sir
>
> Please to publish from the altar of your holy Mass
> That I will open school at Nocknagree Cross,
> Where the tender babes will be well off
> Fir it's there I will teach them their Criss Cross;
> Reverend Sir, you will by experience find
> All my endeavours to please mankind,
> For it's there I will teach them how to read and write;
> The catechism I will explain
> To teach young nymph and noble swain,

With all young ladies I'll engage
To forward them with speed and care,
With bookkeeping and mensuration,
Euclid's Elements and Navigation,
With Trigonometry and sound gauging
And English grammar with rhyme and reason.
With the grown-up youths I'll first agree
To instruct them well in the Rule of Three,
Such of them as are well able.
The cube root of me will learn
Such as are of tractable genius,
With compass and rule I will teach them,
Bills, bonds and informations,
Summons, warrants and supersedes,
Judgement tickets, good
Leases, receipts in full
And releases, short accounts.
With rhyme and reason,
And sweet love letters for the ladies. (Quoted in Carpenter, 1998, pp. 385–386)

Even today, Ó Súilleabháin's poetry is highly regarded (Dunleavy & Dunleavy, 1991).

Maryland, where Mary Walters prepared her cyphering book in 1820, was a likely destination for Irish schoolmasters being deported, or emigrating voluntarily, to North America (Koos, 2001). We know, for example, that in 1773–1774, at least six Irish schoolmasters sailed from English ports to North America and became indentured servants—three in Virginia and the other three in Maryland. Koos (2001) has claimed that "by the first quarter of the nineteenth century, Irish schoolmasters were participating in the settlement of the American frontier" (p. 18). At least one of Abraham Lincoln's teachers (Zachariah Riney, who taught the future President in Kentucky) was an Irish-heritage schoolmaster (Donald, 1995). These Irish schoolmasters tended to value highly the study of arithmetic (Brenan, 1935; Clarke, 1996; Porter, 1995). Indeed, according to Brenan (1935), the success of an Irish schoolmaster was often determined by his proficiency in mathematics—the reason being that "the ordinary people of Ireland would set no store by a school in which arithmetic did not figure prominently" (p. 81).

The poetry genre, so common in Mary Walters' book, could have been introduced to Mary by an Irish-heritage teacher who valued the use of poetry as a natural and vital part of his pedagogy. In such a case, he might have memorized many of the poems and dictated them to Mary. Alternatively, he might have written the poems in his own cyphering book and required Mary to copy them from his "parent" cyphering book into her own cyphering book. In Mary's book, even standard poems were given little twists:

Thirty days hath September
April, June and November.
February hath twenty-eight alone
And all the rest have thirty-one
Unless leap year, and then's the time
February hath twenty-nine.

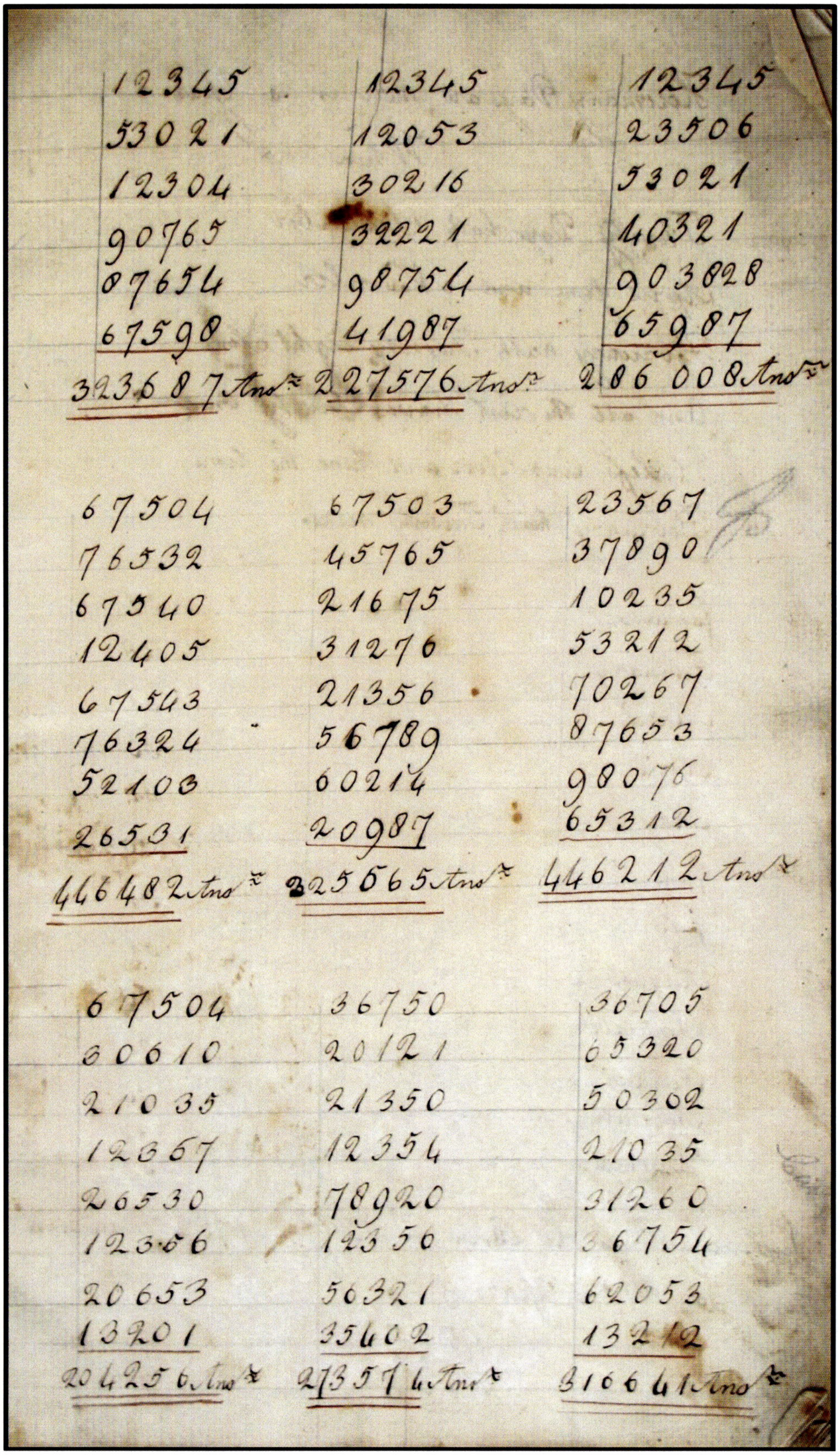

Figure 5.3. Mundane tasks requiring addition of whole numbers.

A Closer Look at the Problem-Based Poetry-Genre Approach to School Arithmetic in Mary Walters' Cyphering Book

Mary Walters seemed to follow a standard *abbaco* curriculum sequence for elementary arithmetic. She started with the four operations on whole numbers, taking addition, subtraction, multiplication and division in their turn.

Addition

Mary headed the first page of her cyphering book with a simple, but calligraphically impressive heading, "Addition." Then followed a standard, rule-bound written description of how counting numbers should be added:

1. Place the numbers under one another, so that units may stand under units, tens under tens, &c.
2. Add up the figures in the row of units, set down what remains above the even tens or if nothing remains a cypher, and for the tens carry as many ones to the next column.
3. Add up the other rows in the same manner, and in the last column put down the whole sum contained in it.

This was followed by 7½ pages of practice for addition tasks—including addition of pure numbers (see Figure 5.3), and addition of quantities, including money and time. From an educational perspective, none of this was particularly interesting—although Mary certainly maintained a high level of penmanship when committing her work to writing.

Then, all of a sudden, the poetry genre was introduced through several poems which invited addition of whole numbers. These poems were contextually interesting and would obviously have added spice to the application of addition.

An unedited version of the text for the first poem, concerning "Old Simon," is shown below. The poem, which occupied 34 lines in the cyphering book, was concerned with how a will was to be shared among claimants. However, this familiar context was described in sub-contexts which would have been of interest to a 14-year-old girl like Mary.

There are lines in this poem which featured a realistic and open approach to sexuality. The poem had its own structure, offered interesting and realistic scenarios, and demanded a solution. Mary's solution is shown in Figure 5.4. Mary then grappled with another addition task expressed in poetic form. Her solution to the task is shown next to the poem. Note that in Mary's cyphering book one finds classical Latin terms like "Tyro" being used in the poetry.

Old Simon

Old Simon aged ninety-two
At last did bid the world adieu
Some mouldy pelf he left behind
Which oft disturbed his craving mind.
Four sons he had and each now claims
His share, John, Simon, Ralph and James.
His wife b'ing dead, he'd daughters three
Called Susan, Ruth and Margery.
His will b'ing read young Simon found

He must receive three hundred pound.
Just eight more pounds too was Ralph's part
But John he bore an aching heart.
And of his fortune was beguil'd
Because he'd got their maid with child.
One shilling only was his share
But James, it seems, had better fare.
Of cash in pounds six score had he,
And Susan's part was ninety-three.
But Ruth unlucky girl had ran
Away with Joe the servant man.
And for the hasty crime she'd done
Was cut off just the same as John.
But Marg'ry's share amongst the rest
By calculation proved the best
For by the father's will we find
The chest which Simon left behind.
Had store of pelf—the treasure found
Amounted to four hundred pounds.
Of clothes and goods it did appear
Just thirty pound was made out clear.
All this was left to Madge because
She had obey'd her father's laws.
Now Tyro with ease may find
The fortune Simon left behind.

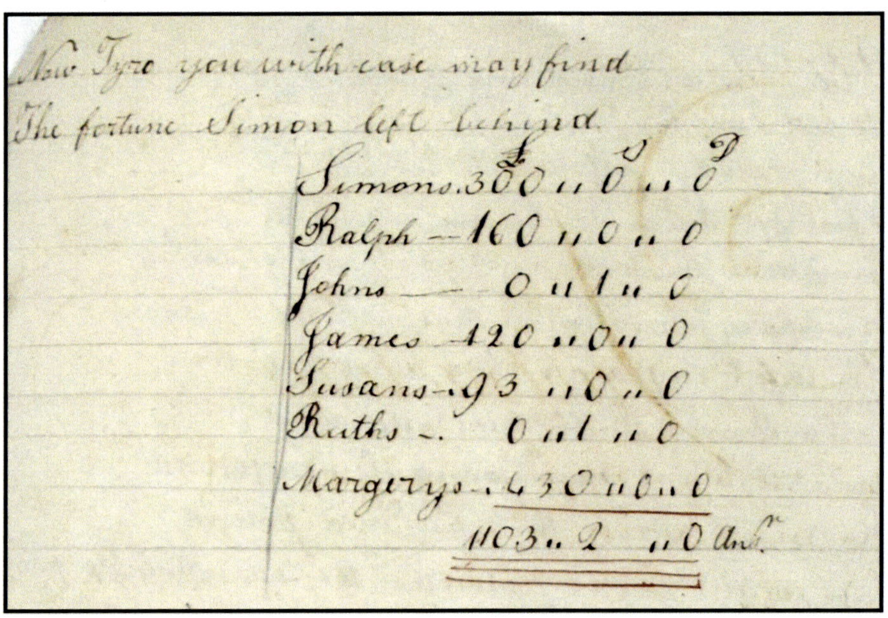
Now Tyro you with ease may find
The fortune Simon left behind

	£	s	d
Simons	300	0	0
Ralph	160	0	0
Johns	0	1	0
James	120	0	0
Susans	93	0	0
Ruths	0	1	0
Margerys	430	0	0
	1103	2	0 Ans.

Figure 5.4. Mary Walters' answer to the "Old Simon" task.

How Much Did He Borrow?

	£	s	d
John borrowed of his friend some cash			
In part he soon paid down	12	12	0
Twelve golden guineas still unpaid	0	6	8
A noble and a cround (*sic.*)	0	5	0
What did he borrow, tell me true	£13	3	8
In royal British coin			
This Tyro you may quickly do			
To make your learning shine?			

Subtraction

Immediately after the section on addition came a section on "Subtraction." It was structured in a similar way to the section on addition: a rule was given, practice "sums" were set and worked, and interesting problems based on possible real-life scenarios were presented through poetry.

The rule given for subtraction was the popular "equal additions" algorithm ("If the figure in the lower line be the greater, add ten to the upper one and then take the lower one from the sum, set down the remainder, and carry one to the next lower figure, and with which proceed as before."). Although this algorithm is rarely used in the United States of America today, it was widely used between 1500 and 1900 (Ellerton & Clements, 2012; Jackson, 1906).

Then, after devoting several pages to "sums" involving subtraction, Mary was introduced to real-life subtractive contexts, expressed in poetic form. A copy of the first task ("A Problem Posed by Master Jordan"), as it appeared in Mary's book, is shown in Figure 5.5, together with Mary's solution to it. The problem concerning Master Jordan is educationally interesting because in order to respond appropriately one needed to grasp how the semantic structure embedded in the real-life scenario related to the operation of subtraction. Experienced teachers, and mathematics education researchers, in the twenty-first century recognize that inability to cope with semantic structure in subtractive tasks can be the source of more errors than mistakes arising from incorrect application of algorithms (Clements, 1980).

An edited version of the second task ("How Much Remained for the Daughter?") is shown immediately after Figure 5.5, and Mary's solution to that problem is also indicated. With this task, the information that the lady decided to leave "Parson Wade" 700 pounds suggested that the person who wrote the poem was from a Protestant rather than a Roman Catholic background. Part of the value of embedding problems in real-life scenarios is that problem solvers need to work out what information is relevant and what is not. The fact that Parson Wade was a reverend did not affect the solution to the problem, but its inclusion in the problem pointed to values within Irish-based societies.

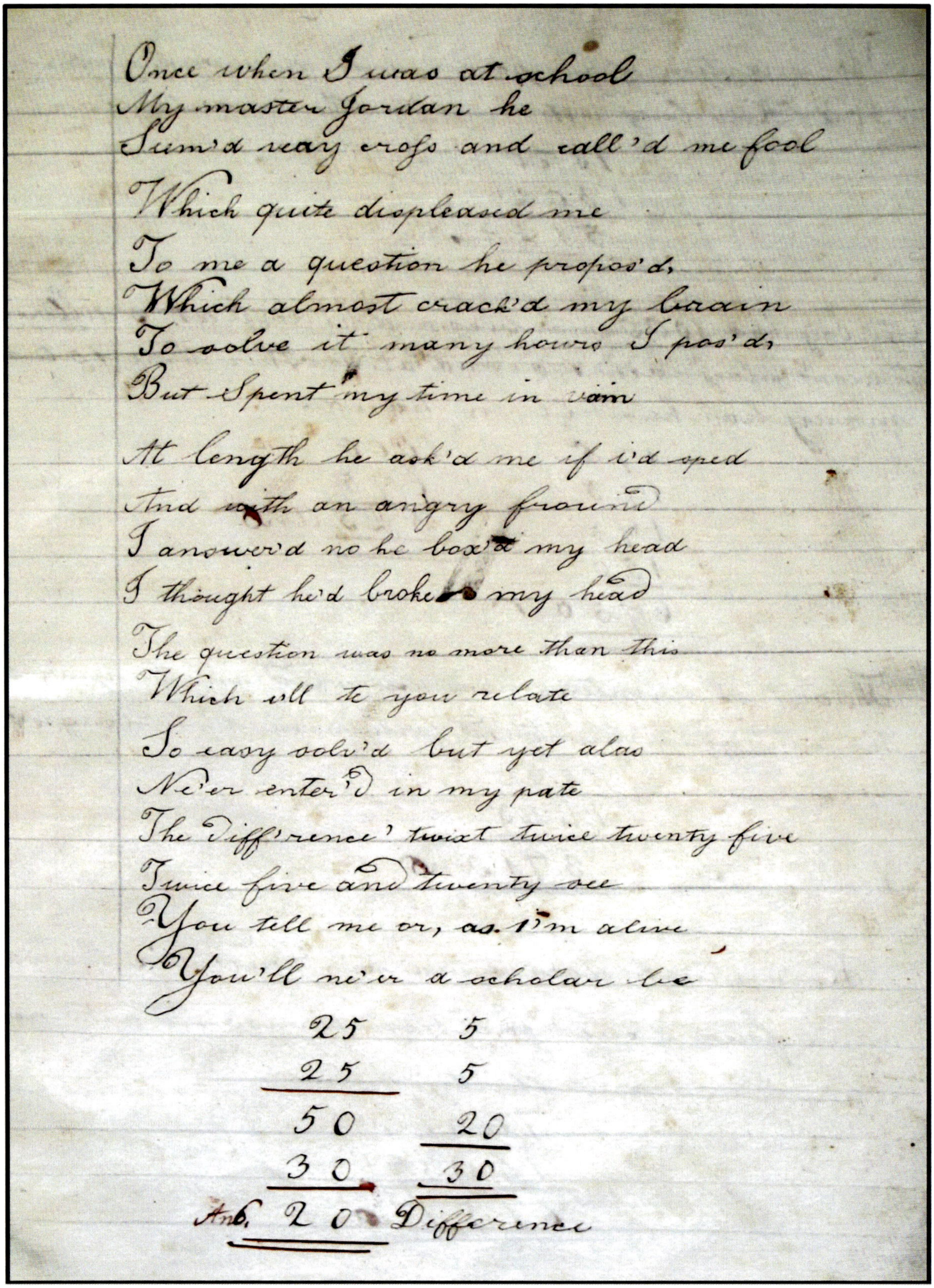

Once when I was at school
My master Jordan he
Seem'd very cross and call'd me fool
Which quite displeased me
To me a question he propos'd,
Which almost crack'd my brain
To solve it many hours I pas'd,
But spent my time in vain

At length he ask'd me if i'd sped
And with an angry frown
I answer'd no he box'd my head
I thought he'd broke my head
The question was no more than this
Which ill to you relate
So easy solv'd but yet alas
Ne'er enter'd in my pate
The Diff'rence 'twixt twice twenty five
Twice five and twenty see
You tell me or, as I'm alive
You'll ne'er a scholar be

25 5
25 5
50 20
30 30
Ans. 20 Difference

Figure 5.5. A subtraction problem posed by Master Jordan, and an answer.

How Much Remained for the Daughter? [*Edited*]

A lady left her daughter fair
Twelve thousand pound in gold,
To be distributed with care
To underneath is told.
First to a niece there must be paid
Just fourteen hundred pounds
And half that sum to Parson Wade
To make his glass go round.
And to her maid, Miss Nancy Hare
Three hundred pounds in cash
Who swells with pride and such an air
She apes any lady flash.
The steward and butler each must have
Just twice two hundred more,
And to a tenant farmer brave
In shining pounds, six scores.
The greasy cook, each other maid
B'ing three in number they
Had twenty guineas each one paid
To make them fine and gay.
The coachman Ralph and footman Dan,
Ten guineas and a crown apiece
Which made them toss about the can,
In ev'ry market town.
When all these legacies were paid,
What did remain behind
For Miss that blooming peerless maid
Whose virtues made her kind?

£	s	d	£	s	d	
Sum Left				12000	0	0
Niece	1400	0	0			
Parson	700	0	0			
Miss Nancy	300	0	0			
Steward and butler	400	0	0			
Farmer brave	120	0	0			
Cook	63	0	0			
Ralph and Footman Dan	21	10	0	3004	10	0
The daughter fair				8995	10	0

Multiplication and Division

The sections devoted to multiplication and division in Mary's cyphering book had the same structural features as the sections on addition and subtraction: first, a general description of the operation was given, followed by the statement of a rule for carrying out the operation; then sets of practice "sums" were stated and solved; and finally problems in context were presented through poetry. Here we shall present, in edited form, some of the poems in which real-life scenarios for multiplication, or division, were summarized.

The first of the poems is concerned with St James' Park, in London.

Find the Area (A Multiplication Scenario)

Near to St James's, in park you'll find
A fair canal delightful to mankind
Where sportive fishes gliding jump and play
When Phoebus warms them with refulgent ray.
The cheerful warblers in soft accents sing
And harmony encores the joys of spring.
The crowded mall in glitt'ring lustry hymn
And nature's scen'ry in itself combines.
The duke and earl, the star and gartered knight
And ladies glance along in silver white.
The area of this liquid space you'll find,
From what you see is underneath subjoin'd.

Length 2800 feet, breadth 100 feet; Ans. 280000 Area

Saving for a Cow (A Second Multiplication Scenario)

A careful maid had thirty hens,
Laid twenty eggs per day,
She took great care to search her pens,
Wherein her treasure lay.
The eggs she sold the cash put by
Still to increase her store.
Resolved was she to buy a cow
With five pounds and no more.
How many eggs must there be sold
To purchase Tydy say
At five for two pence pray unfold
The sum for Nancy Ray.

Mary set out her solution to this problem as follows:

100 shillings = 5 pounds
6 two pences in a shilling
600
5
Ans. 3000 eggs

Her way of thinking seemed to be as follows: 5 pounds was 100 shillings, which was 600 lots of 2 pence. Since 5 eggs sold for 2 pence, 600 by 5, or 3000, eggs would be needed.

Selling to Badger Fortongue (A Third Multiplication Scenario)

Old Fomer Careful fond of pelf,
To Badger Fortongue sold,
His whet and seems to hug himself,
And smile upon the gold,
A hundred bushels where *(sic.)* agreed,
The Badger straight to bring,
At nine and six pence true indeed,
It made old Careful sing.
What did the whole come to declare,
Which pleased Clod so well,
And made him hobble, sing and stare,
Ingenious Tyro tell.

Mary Walters' solution was:

£	s	d
0	9	6
		10
4	15	0
		10
47	10	0 Ans

To a modern reader, the definition of the problem might not be clear, but Mary's working revealed that all that was required was to multiply 9 shillings and 6 pence by 100, and then to show the answer in pounds, shillings and pence. Mary did this successfully, by multiplying by 10, and then by 10 again.

The Banquet (The First Division Scenario)

At a hunting suppose six score gentlemen be
All sons of fair freedom and dear liberty.
The banquet prepared with old Bonisace dine
To feast and carouse over punch ale and wine.
With punch bowl and ladle the field drink and sing
And soon to adore mighty Bacchus the king
For Bacchanals who favour the bottle and glass
Will smile in full bumpers wherever the pass.
If sixty-three pounds do the reckoning defray
Then what must each gentleman equally pay?

Mary Walters' solution was:

		£	s	d
120	12	63	0	0
	5	5		
		0	10	6 Answer

Mary's "solution" to the problem suggested that although she intended to divide 63-pounds by 120, she did not know how to do that. She seemed to want to divide the 63 pounds by 12, but could only get the 5 pounds part of that calculation. Then, instead of dividing by 10, she indicated she would divide by 5 (but did not carry out the division). Her answer of 10 shillings and 6 pence was correct, but it was not made clear how she got it.

Hodge to Grandmother Grannum (A Second Division Scenario)

Says Hodge to his Grandmother Grannum I see
That the money and purse which you've given to me
Is worth sixteen and eight pence, it's well it's no worse
For the cash is in value worth nine times the purse.
What sum then had Roger fair Ladies pre tell
What tickled his fancy and pleas'd him as well?

Mary Walters' solution was:

	s	d
10/	16	8
	1	8 (purse)
	15	0 (cash)

Mary's "solution" to the problem was correct but there was no indication given how the breaking down of the 16 shillings and 8 pence was carried out. Also, the introduction of "Roger" into the second-bottom line (whereas in the first line it was "Hodge") was curious—it may have had balladic origins, or may have been an error arising from dictation.

At the conclusion of the sections on the four operations a word problem, not presented in the form of a poem, was given. The problem, in unedited form, stated:

> One of the smarts in the accomptant's office making his addresses in an old lady's family who had five fine daughters she told him their father had made a whimsical will which might not soon be settled in chancery, and till then he must refrain his visits the young gentleman undertook to unravel the will which imported that the first four of her girls for times were together to make £25000; the four last £33000; the three last with the first £30000; and the three first with the last to make £28000; and the two last with the two first, £30000. Now sir if you can make appear what each is to have and as you like, seemingly my third daughter Charlotte, I am sure she will make you a great wife, and you are welcome what was Miss Charlotte in fortune.

Although the meaning of the question in this word problem is clear, the quality of the English expression leaves much to be desired. It seems likely that Mary Walters wrote down the question from a dictated statement, and did not copy it from a textbook or another cyphering book.

The solution to the problem was set out, in Mary's cyphering book, as follows:

£
25000
33000
30000
28000
32000
4/ 148000
37000 Sum of their fortunes

37000	37000	37000	37000	37000
25000	33000	30000	28000	32000
12000 Youngest	4000 Eldest	7000 Second	9000 Fourth	5000 Miss Charlotte's

Although no written explanation was given for why these calculations were made, it did seem that Mary knew what she doing—provided, of course, she was not given the working shown above in a dictation segment in which her teacher dictated the solution to her. Nevertheless, the context of the problem was likely to be of some interest to a 14-year-old girl like Mary.

Reduction, Exchange, and Units of Measurement

The introduction to this section in Mary's book stated: "In this and all the following rules all great names are brought into small by multiplication; on the contrary all small names into great by division." Following this statement, numerous examples were worked showing what the statement meant. Thus, for example, 130 shillings were converted, through multiplication, to 124800 farthings; 24192 farthings were converted, by division, to 24 guineas.

There were also questions in which an amount in one currency was to be converted to an equivalent amount in another currency. Thus, for example, one question asked how many pounds there were in 1178 dollars, with the rate being 1 dollar was equal to 4 shillings and 3 pence. This was the only time in Mary's cyphering book that there was any mention of U.S. currency. The approach to this problem in Mary's cyphering book was unusual. The 4 shillings and 3 pence was regarded as 17 units, with each unit equal to 3 pence. Then 1178 was multiplied by 17, and £250 6s and 6d was obtained. A rule for reducing "pounds sterling into foreign and English coins" was then stated: "Reduce the given pounds sterling and the given coin into one name that is if you can reduce them boath (*sic.*) into shillings, sixpences, four pences or 3 pences, &c. do so then divide one by the other and the quotient will be the answer."

Then followed 12 pages in which various kinds of units were listed (for troy weight, avoirdupois weight, apothecaries' weight, cloth measure, wine measure, ale and beer measure, dry measure, long measure, land measure, and time). For each category, unit equivalents were stated (e.g., for long measure, "3 barleycorns make 1 inch"), for these were needed in reduction exercises. Only one of the exercises was presented in the form of a poem, and that poem, together with the solution that Mary recorded, were:

	Mary Walters' Solution:		
To measure a neigbouring plain,	*a*	*r*	*r*
I took up my cross staff and	90	2	1
Having found the content of the whole,	4		
Ninety acre two roods and a pole.	362 roods		
What roods and what perches where there,	40		
Be pleased to make to appear.	14481 perches		

It is possible that the meter and the rhyme suffered from translation. Note that this aspect of the school mathematics curriculum, in which quantities with units were added, subtracted, multiplied or divided, was often referred to as "compound operations." It had an important place in the traditional *abbaco* curriculum sequence.

The Rule of Three

Having "cyphered to the rule of three" was regarded as an achievement, and the last 18 pages of Mary Walters' cyphering book were dedicated to situations and contexts in which ratio and proportion were involved. We have already seen (Figure 5.1) the beautiful, water-colored rose that Mary included in her book to celebrate her achievement of reaching the rule of three, and have examined the poem about measuring the distance from an observer to a hunter (see Figure 5.2). In fact, in Mary's book there were 11 poems within the 18 pages dedicated to the rule of three. However, only the *direct* rule of three was considered—no problems calling for either the *inverse* rule of three or the *double* rule of three were presented.

Early in the section, the direct rule of three the rule was stated—in the form of four numbered paragraphs (shown in Figure 5.6 and also in unedited form following Figure 5.6):

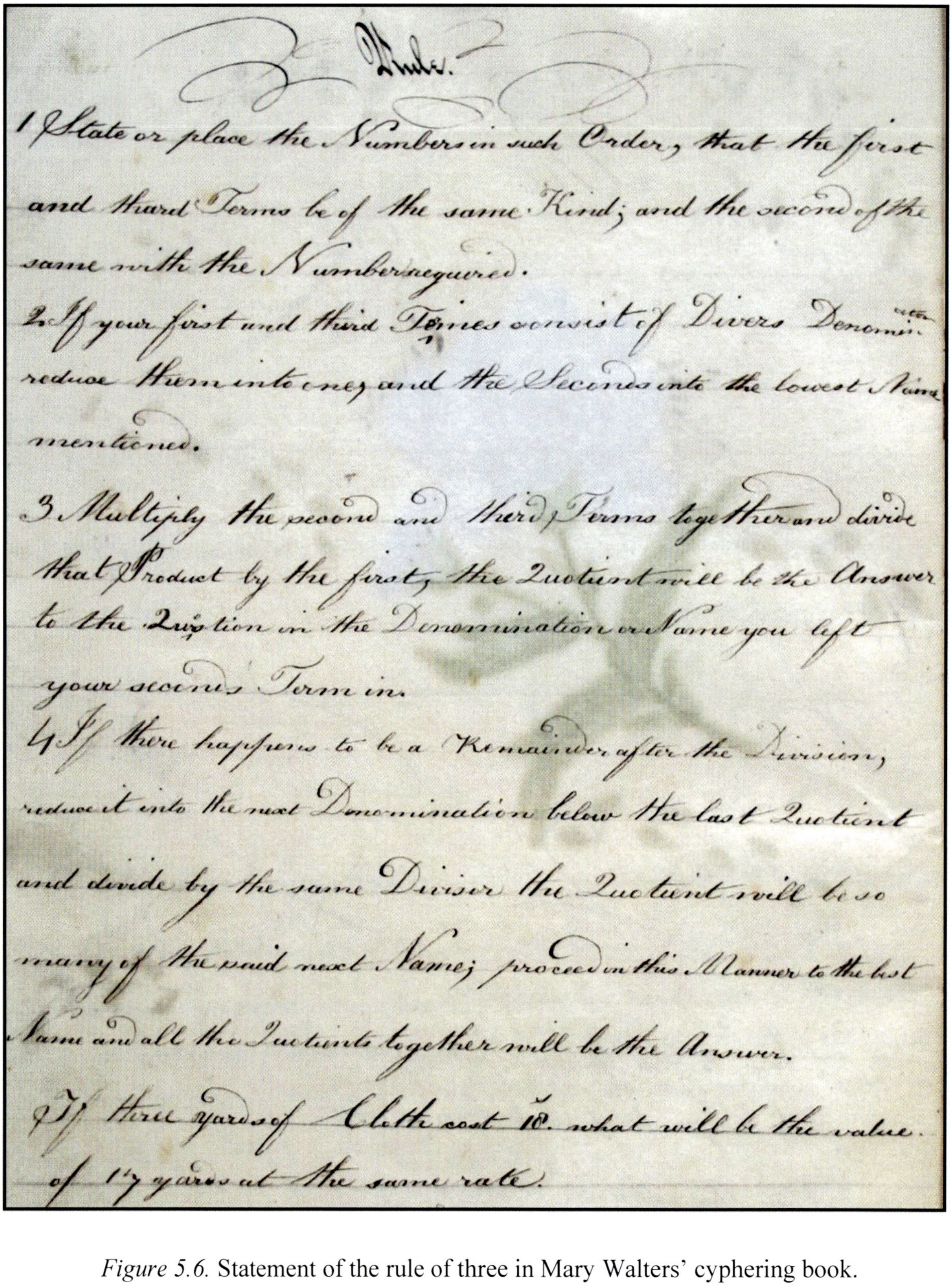
Rule.

1 State or place the Numbers in such Order, that the first and third Terms be of the same Kind; and the second of the same with the Number required.

2. If your first and third Termes consist of Divers Denominᵗⁿˢ reduce them into one; and the Seconds into the lowest Name mentioned.

3. Multiply the second and third Terms together and divide that Product by the first, the Quotient will be the Answer to the Question in the Denomination or Name you left your second Term in.

4 If there happens to be a Remainder after the Division, reduce it into the next Denomination below the last Quotient and divide by the same Divisor the Quotient will be so many of the said next Name; proceed in this Manner to the last Name and all the Quotients together will be the Answer.

Q If three yards of Cloth cost 10ˢ what will be the value of 17 yards at the same rate.

Figure 5.6. Statement of the rule of three in Mary Walters' cyphering book.

1. State or place the numbers in such order that the first and the third terms be of the same kind, and the second of the same with the number required.
2. If your first and third terms consist of divers denominations reduce them into one and the second into the lowest name mentioned.
3. Multiply the second and third terms together and divide that product by the first, the quotient will be the answer to the question in the denomination or name you left your second term in.
4. If there happens to be a remainder after the division, reduce it into the next denomination below the last quotient and divide by the same divisor the quotient will be so many of the unit next name; proceed in this manner to the last name and all the quotients together will be the answer.

Immediately after the statement of the rule of three (see Figure 5.6) came the example: "If three yards of cloth cost 18 shillings, what will be the value of 17 yards at the same rate. Mary's written response was set out as shown below. No explanation was given, for the calculations were intended to illustrate the points made in the four paragraphs by which the direct rule of three had been stated. Then followed other rule-of-three exercises (like, for example, "If twelve pounds of bacon cost me a Crown, For a flitch of sixscore what must I pay down?" and "Bought 36 ounces. of silver at the rate of 5 shillings and 4 pence per ounce. What does the wole (*sic.)* come to?").

yds : s :: yds
3 18 17
18
136
17
3/306
20/10|2
£5..2..0 Ans

This was a standard way of presenting solutions to rule-of-three tasks.

Only two of the poems found in the rule-of-three section of Mary Walters' book will be shown. The first is in Figure 5.7. An edited version (with punctuation and spelling corrections), together with Mary's solution to it, are now given:

A drover came riding a mare of good mettle,
Amongst the Welsh mountains to buy up some cattle,
When the Welshmen espy him they splutter and stare,
Saying, "Bless you, coot measter, come buy our fine ware.
They're sound wind and limb and as fat as a pig,
And plump as a codlin, and sleek as a snig (?);
By their horns you may see, they're young and all that,
Come bargain measter, you see they are fat."
Now the drover agrees for a score of their cows,
Which fed in a meadow close by the hay mons (?)
For seventy-five guineas, the sum was no more,
What each one was sold for be pleas'd to explore.

Mary Walters' Solution

C guineas :: C
20 : 75 :: 1
21
75
150
20/ 1575/ 78
14 £3. 18. 9
17
16
15
12
20/ 18/09

The Drover and the Welshmen (A Rule-of-Three Scenario)

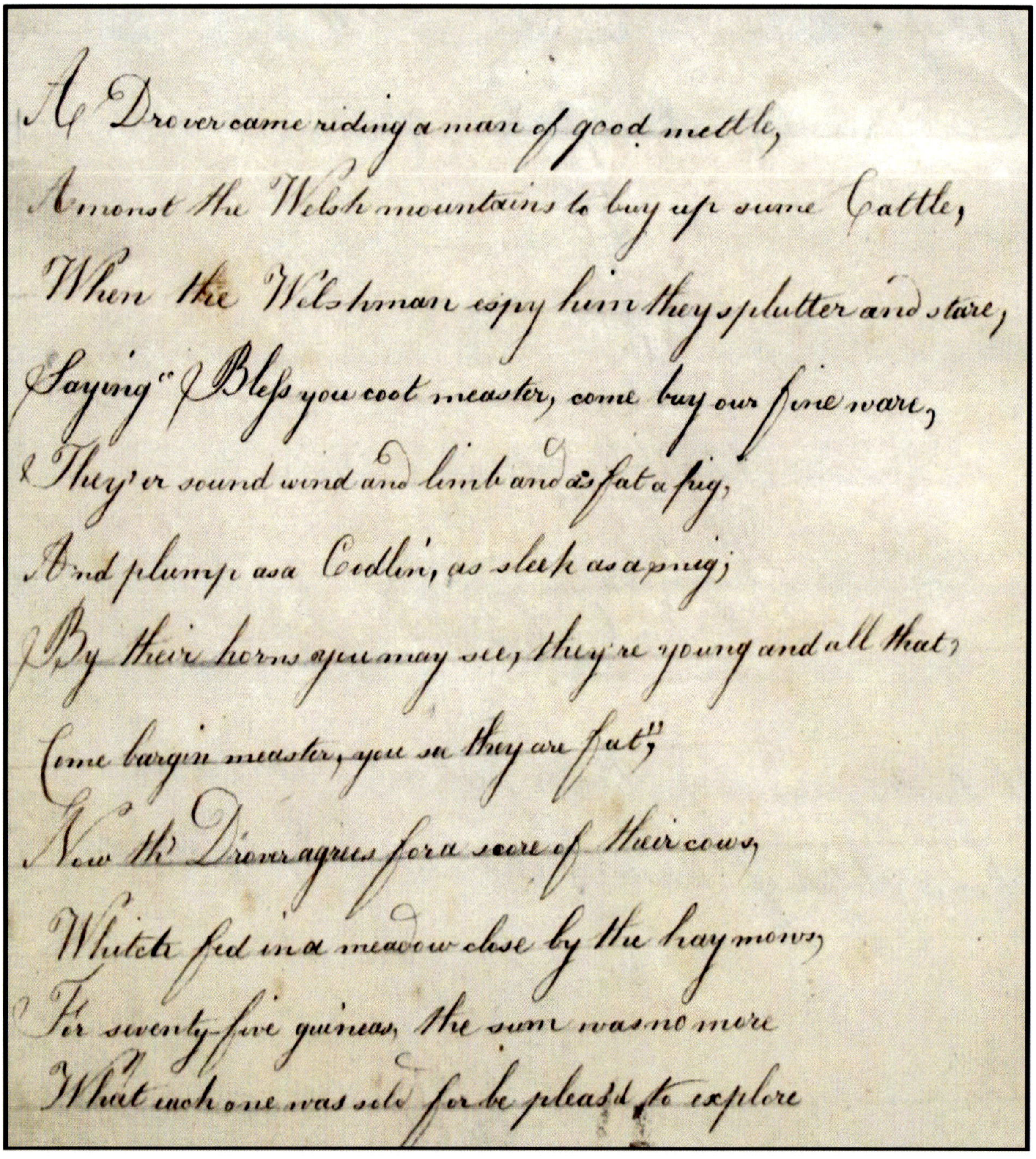
A Drover came riding a man of good mettle,
Amonst the Welsh mountains to buy up some Cattle,
When the Welshman espy him they splutter and stare,
Saying "Bless you coot measter, come buy our fine ware,
They're sound wind and limb and as fat a pig,
And plump as a Codlin, as sleek as a sprig;
By their horns you may see, they're young and all that,
Come bargin measter, you see they are fat"
Now the Drover agrees for a score of their cows,
Whitch fed in a meadow close by the haymows,
For seventy-five guineas, the sum was no more
What each one was sold for be pleased to explore

Figure 5.7. "The Drover and the Welshmen"—A poem in Mary Walters' cyphering book.

The handwritten version of the "Drover and the Welshmen" poem, shown in Figure 5.7, seemed to have a number of spelling errors: "amonst" should have been "amongst"; "Welshman" should have been "Welshmen"; "bargin" should have been "bargain"; and "whitch" should have been "which." Also, "as fat a pig" should have been "as fat as a pig." Similar spelling errors and word omissions can be found in most of the other poems—which suggests that the poems were dictated to Mary.

The second rule-of-three poem to be considered here concerned "Fair Sylvia." The poem is shown in unedited form.

Fair Sylvia (A Rule-of-Three Scenario)

Fair Sylvia blooming as the rose in June,
Oft met young Damon in the study bow'r,
Where warbling songsters chant a joyful tune
To chere the pair and bless their happy hours.
Oft very oft this lovely couple went
To pay a visit to the silent grove.
And mingle kisses sweet with true content
To seal the hours of constancy and love.
But oft embarrassed were the lovely pair
By a most rigid Father's stern decree
Who would not suffer any suitor there
Sworn foe, to love alike and unity
Should Damon near the ancient castle come.
He ne'er must have admittance to his dear
For the Father's answer was "she's not at home,
And really, Sir, you have no business here."
Such rude behaviour long this couple bore
With much anxiety and fretful pain,
Till they agreed at last to bear no more,
But hip to Scotland o'er the conquered plain
There to be linked in Hymen's sacred tye,
And crown their wishes by a speedy flight.
Favour'd by night, fair Sylvia forth did fly
With Damon her enamour'd heart's delight.
Soon in the morn ere Phoebus gilt the West,
The Father was informed of the plot,
And to pursue the couple thought it best.
So in an instant he on horseback got.
Just thirty miles the lovers were before
The Father when he began to ride.
Two miles he gained of them per hour or more
But happily the marriage knot was tyd
Before he did o'take the flying pair.
The rate they went was seven miles an hour.
How many hours would the Father be ye fair
Car he o'took them now to me explore.

Mary Walters' Solution:

m	*H*	::	*m*
2	1	::	30
			1
			2/30
			15

This poem made use of classical allusions, and was structurally similar to popular Irish poetry of the era (G. Koos, personal communication, August 13, 2013).

Educational Issues Arising from our Analysis of Mary Walters' Cyphering Book

It would be difficult to imagine a potentially more interesting scenario for 14-year-old Mary Walters than the one involving Fair Sylvia and Damon. Its use of an arithmetically-rich context to stimulate the learning of arithmetic might well have made Mary think that, yes, arithmetic could be relevant to interesting, real-life situations. Furthermore, the problem, though expressed in poetic form, was directly relevant to the curriculum that Mary's teacher appeared to be following, for it could be solved by applying the direct rule of three. However, it is unclear whether Mary's handwritten version of the scenario was copied from another text (another cyphering book, or perhaps a commercial text), or whether the poem was dictated to her by her teacher, or someone else. It is even possible, but admittedly unlikely, that Mary created the poem herself.

The solution that Mary recorded in her cyphering book occupied just five short lines, and no explanation of her approach was given. It seemed to be assumed that what she wrote was sufficient to communicate her thinking to others. We do not know if Mary was told—by her teacher, or by somebody else—how to tackle the problem. Perhaps, she worked out the method herself. Was her method approved by her teacher?

Similar comments could be made with respect to word problems in other parts of Mary's cyphering book. Consider, for instance, Mary's approach to the routine rule-of-three task, "If 2 lb of sugar costs 1 shilling and 6½ pence, what will 24 of the same cost?" (see Figure 5.8). It appears to have been the case that Mary, like many other students of her own generation, and students in many previous and future generations, did not think much about the mathematics demanded by word problems. The fact that she did not include the unit (lb.) after the "24" in her handwritten statement of the problem could have been due to carelessness—but it could also have been the result of her mindlessly writing down something that was being dictated to her.

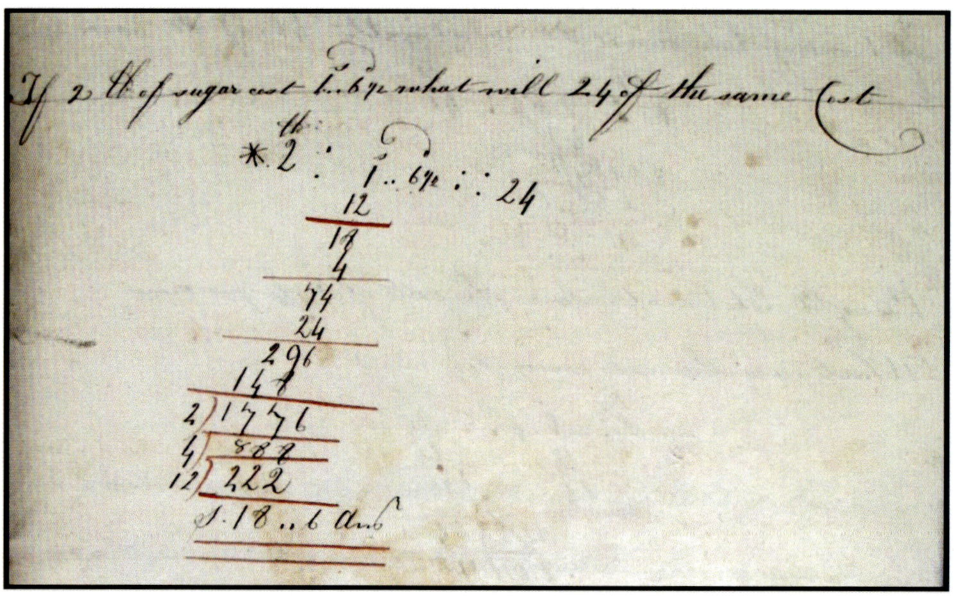

Figure 5.8. "If 2 lbs of sugar cost 1 s. 6 ½ d, what will 24 of the same cost?"

Mary converted the 1 shilling and 6½ pence to 74 farthings and then, using the rule of three, multiplied this 74 by 24 before dividing by 2. This gave her 888, which, it seems, she presumed to be in farthings. She divided this by 4 (to convert to pence) and then divided 222 by 12 to get her answer in shillings—obtaining 18 shillings and 6 pence as her answer. What she never seemed to realize was that if she had thought about the meaning of the problem, she could simply have multiplied 1 shilling and 6½ pence by 12 to get 18 shillings and 6 pence. In her working shown in Figure 5.8, Mary unthinkingly applied the rule-of-three algorithm ("multiply second by third and then divide by first").

We do not know much about the circumstances surrounding Mary when she studied arithmetic in 1820, but it is obvious that it was intended that she learn to add, subtract, multiply and divide counting numbers effectively, that she learn to apply elementary measurement principles in a range of everyday situations, that she learn to change from one set of units to another, and that she learn how to apply the direct rule of three in a wide range of everyday situations. In short, it was expected that she follow the same kind of elementary *abbaco* sequence that was widely taught and learned in many parts of the world at that time (Ellerton & Clements, 2012).

Intended, Implemented and Attained Curricula

The entries in Mary's cyphering book also made it clear that the *intended* curriculum was more than a mere acquaintance with elementary *abbaco* topics—she was expected to learn the elementary *abbaco* content in a very special way. Her teacher wanted her to inject herself into problem-based scenarios by which she would learn to apply elementary arithmetic principles to tasks that were of intense interest to her. Such tasks involved human interest situations. There was no attempt to divorce her arithmetic studies from contexts involving all kinds of human, emotionally-charged, dramatic relationships. In particular, she was expected to appreciate the power of poetry to convey interesting problem scenarios. In that sense, Mary's intended curriculum was something more than a set of arid arithmetical skills.

Although entries in the cyphering book carry unmistakable messages about the *implemented* curriculum, they are nevertheless silent on the teaching methods adopted by Mary's teacher in Baltimore in 1820. Did Mary's teacher use "blab-school" methods whereby Mary spoke aloud, and sang, to herself and to others, throughout the school day? Such methods were common in schools in Baltimore in the nineteenth century (Johnson, 1994). And, was her learning checked during recitation sessions—when she was able to speak individually with her teacher(s)? Such an approach was often to be found in schools conducted by Irish-heritage teachers (Koos, 2001).

We do not know exactly how Mary was expected to learn arithmetic, but we do know that she prepared a remarkable 80-page cyphering book which included numerous topic-relevant poems. Was she expected to learn the poems by heart, to further a form of pedagogy and an oral tradition by which students would learn curriculum content while simultaneously committing ballads, songs and poems to memory? We simply do not know the answer to that question.

We do know, though, that in the poems in Mary's cyphering book it was frequently the case that words were poorly spelt, sentences poorly constructed, and appropriate punctuations either not used or used incorrectly. There is a strong suggestion that Mary often had notes, including poems, dictated to her. If that was the case, then how well did she learn the mathematics? That question is concerned with the *attained* curriculum.

The fact is, we do not know how well Mary learned the arithmetic she recorded in her cyphering book. We would like to think that the problem-based, poetry-inspired pedagogy evident in what she recorded in her cyphering book prepared her well to survive, arithmetically, with dignity in the United States of America during the period of her life from 1820 onwards. But would she have learned better, and understood more, if she had been involved with a form of pedagogy which did not make use of ballads, poetry and song?

We regard Mary Walters' cyphering book as extraordinary, and believe that it provides evidence that some Irish-heritage school teachers in the United States of America in the first half of the nineteenth century were responsible for giving school arithmetic in the United States more than a tinge of green. But, as we have seen, from a historical perspective, Mary's cyphering book raises more questions than we have been able to answer in this chapter. We have seen that various authors of arithmetic in the United States (e.g., Ostrander, 1823) often posed problems through verse, and the same was true of authors of arithmetics in Great Britain (Hersee, 1997).

The young Abraham Lincoln attended several blab schools—in Kentucky, and in Indiana—which were probably conducted by Irish-heritage schoolmasters. But his cyphering book (see Chapter 6 of this book) was qualitatively different from Mary Walters' book—in Abraham Lincoln's book we did not find anything like the problem-based, poetry-inspired approach to arithmetic that we found in Mary Walters' cyphering book.

Did the Irish schoolmasters use different approaches to their teaching of arithmetic when dealing with boys as opposed to when they were dealing with girls? We know that slightly less than 20 percent of extant North American cyphering books from the period 1607–1861 were prepared by females, it being more common for girls to prepare sewing samplers than cyphering books (Ellerton & Clements, 2012). The ratio of extant female to male cyphering books suggests that there was an attitude among parents that, for girls, time spent at school on sewing samplers could more easily be justified than time spent on cyphering. Yet, as Clarke (1996) has pointed out, Irish schoolmasters inherited an education tradition in which, at all levels, and for both male and female students, arithmetic was an integral component of the school curriculum.

Only a small proportion of those girls in North America who stayed on at school after gaining a very basic education studied arithmetic. Was the proportion of girls who continued with arithmetic higher among Irish-heritage and Continental-European-background families than among those from England, Scotland or Wales (Koos, 2001; Northend, 1917; Ring, 1993; Smith, 2005)? That is an important and researchable question. We know that there were many emigrants from Ireland who worked as school teachers in the British colonies in America, and in the United States of America, throughout the period 1607-1850 (Linehan & Murray, 1898)—a worthy theme for research would be how these teachers viewed arithmetic within the overall curriculum, and whether they taught arithmetic differently from teachers with non-Irish backgrounds.

There is therefore a sense in which our examination of Mary Walters' extraordinary cyphering book has set an agenda for future researchers.

Acknowledgment

We would like to thank Greg Koos, Director of the McLean County Museum of History, in Bloomington, Illinois, for reading and commenting on a draft of this chapter. He emphasized that in the early nineteenth century the use of ballads, poetry and songs was part of an Irish pedagogical tradition, and we have attempted to incorporate that idea throughout the chapter. Thanks are also due to Dr. William H.A. Williams for his help in analyzing and interpreting the poems.

References

Allen, H. D. (1970). The verse problems of early American arithmetics. *The Journal of Rutgers University Libraries, 33*(2), 49–62.

Allen, H. D. (1990). Classic "measurement" problems feature practical fanciful. *The Mathematical Log, 34*(1), 1, 4, 6.

Bidwell, J. (2013). *American paper mills: A directory of the paper trade with notes on products, watermarks, distribution methods and manufacturing techniques.* Hanover, NH: University Press of New England.

Brenan, M. (1935). *Schools of Kildare and Leighlin, A.D.1775–1835.* Dublin, Ireland: Gill and Son.

Carpenter, A. (1998). *Verse in English from eighteenth-century Ireland.* Cork, Ireland: Cork University Press.

Clarke, P. (1996). A glimpse at Irish accounting history. *Irish Accounting Review, 3*(2), 23–40.

Clements, M. A. (1980). Analyzing children's errors on written mathematical tasks. *Educational Studies in Mathematics, 11*(1), 1–21.

Cullen, L. (1988). *The hidden Ireland: Reassessment of a concept.* Dublin, Ireland: Lilliput.

Donald, D. H. (1995). *Lincoln.* New York, NY: Simon and Schuster.

Doyle, D. N. (1980). *Ireland, Irishmen and revolutionary America.* Cork, Ireland: Mercier.

Dunleavy, J., & Dunleavy, G. (1991). *Douglas Hyde—A maker of modern Ireland.* Berkeley, CA: University of California Press.

Ellerton, N. F., & Clements, M. A. (2012). *Rewriting the history of school mathematics in North America 1607–1861.* New York, NY: Springer.

Hersee, J. (1997). Multiplication is vexation. *Paradigm, 24,* 2–10.

Jackson, L. L. (1906). *The educational significance of sixteenth century arithmetic.* New York, NY: Teachers College Press.

Johnson, W. R. (1994). "Chanting choristers": Simultaneous recitation in Baltimore's nineteenth century primary schools. *History of Education Quarterly, 34*(1), 1–23.

Karpinski, L. C. (1980). *Bibliography of mathematical works printed in America through 1850* (2nd ed.). New York, NY: Arno Press.

Koos, G. (2001). The Irish hedge schoolmaster in the American backcountry. *New Hibernia Review, 5*(2), 9–26.

Linehan, J. C., & Murray, T. H. (1898). *Irish schoolmasters in the American colonies, 1640–1775, with a continuation of the subject during and after the War of Revolution.* Washington, DC: American-Irish Historical Society.

Miller, K. A. (1985). *Immigrants and exiles: Ireland and the Irish exodus to North America.* New York, NY: Oxford University Press.

Northend, M. H. (1917). *Memories of old Salem: Drawn from letters of a great-grandmother.* New York, NY: Moffatt, Yard and Company.

Ostrander, T. (1823). *The elements of numbers, or, easy instructor; being a new and comprehensive system of practical arithmetic...* Canandaigua, NY: J. D. Bemis and Co.

Plowden, F. (1805). *An historical review of the State of Ireland from the invasion of that country under Henry II to Its union with Great Britain on the First of January 1801.* Philadelphia, PA: W. F. McLaughlin and Bartholomew Graves.

Porter, T. (1995). *Trust in numbers: The pursuit of objectivity in science and public life,* Princeton, NJ: Princeton University Press

Ring, B. (1993). *Girlhood embroidery: American samplers and pictorial needlework, 1650–1850.* New York, NY: Knopf Publishers.

Smith, E. B. (2005). "Let them compleatly learn": Manuscript clues about early women's educational practices. In *A manuscript miscellany*, Summer 2005 Institute. Manchester, United Kingdom: Manchester University Press.

Steinfeld, R. J. (1991). *The invention of free labor: The employment relation in English and American law and culture 1350–1870.* Chapel Hill, NC: University of North Carolina.

Underhill, D. C. (1825). *Tables of arithmetic made easy.* New York, NY: Mahlon Day.

Yeats, W. B. (1928). *Coinage of Saorstát Éireann.* Dublin, Ireland. The Stationery Office.

Chapter 6

He would be Good: Abraham Lincoln's Early Mathematics, 1819–1826

Nerida F. Ellerton, Valeria Aguirre Holguín, and M. A. (Ken) Clements

Abstract: This chapter, jointly written by Nerida Ellerton, Valeria Aguirre Holguín, and Ken Clements, offers a comprehensive analysis of the earliest extant handwritten manuscript of Abraham Lincoln—22 pages (11 leaves), from a cyphering book that Lincoln prepared between 1819 and 1826. The pages are examined from historical, mathematical and educational perspectives, and an order in which they were written is conjectured, with justifications provided. This is the first time all 22 pages have been examined, it having become received tradition that there were only 20 surviving pages. The authors argue that if the corpus of all extant U.S. cyphering books were to be studied, then the future President's cyphering book would not be particularly outstanding from calligraphic, penmanship, or abstract mathematical points of view. Analysis reveals, nevertheless, that Lincoln succeeded in his quest to prepare an attractive cyphering book in which almost all entries were arithmetically correct—despite his having to cope with rough frontier circumstances.

Keywords: Abraham Lincoln; Abraham Lincoln in Indiana; Abraham Lincoln's schooling; arithmetic; ciphering book; copybook; cyphering book; decimal fractions; history of education; history of mathematics; history of school mathematics; rule of three; sum book.

This chapter analyzes the 22 extant pages of the cyphering book written by Abraham Lincoln. All previous commentators on Lincoln's cyphering book have examined, at most, 20 pages, but two of the authors (Ellerton and Clements) located an additional leaf (two pages) held within the Houghton Library at Harvard University, and so in this work 22 pages, not 20, have been studied.

These 22 pages comprise the oldest surviving manuscript in the future President's handwriting. They were were prepared in Indiana, at various times between 1819 and 1826, and Lloyd A. Dunlap, an assistant editor of *The Collected Works of Abraham Lincoln* (Basler, Pratt & Dunlap, 1953), commented that there is no reason to accept or reject Thomas L. D. Johnston's statement that Lincoln's cyphering book "represented five or six years' work" (Dunlap, 1959, p. 6).

The analysis in this chapter throws new light on Abraham Lincoln himself, and on the history of school mathematics in the United States of America. That said, it should not be imagined that Abraham Lincoln's cyphering book was very different from cyphering books prepared by other North American students in the 1820s. We shall argue that what made Lincoln's book "extraordinary" were the conditions under which it was prepared.

A Note on Terminology

Ellerton and Clements (2012) have made it clear that the most appropriate term for the handwritten manuscripts on arithmetic prepared by North American schoolchildren between 1607 and 1861 is either "cyphering book" or "ciphering book," and definitely not "cypher book" or "cipher book," or "sum book."

Nerida F. Ellerton and M. A. (Ken) Clements, *Abraham Lincoln's Cyphering Book and Ten other Extraordinary Cyphering Books*,
DOI 10.1007/978-3-319-02502-5_6,

Throughout their book, Ellerton and Clements (2012) used the word "cyphering," and in what follows, when referring to Abraham Lincoln's manuscript, we shall use that term. Notice, however, that in 1859 Lincoln himself used an "*i*" when he wrote that when he was at school no qualifications were required of a teacher in Indiana "beyond readin', writin' and cipherin' " (quoted in Basler et al., 1953, Vol. 3, p. 511). But Dennis Hanks, who lived with the Lincolns during the future President's school days, used a "*y*" when, in 1891, he wrote, of Lincoln: "He was always reading, writing, cyphering, writing poetry" (quoted in Arnold, 1891, p. 24).

The Structure of this Chapter

In this chapter we analyze all 22 extant pages of Abraham Lincoln's cyphering book. Although we have attempted to correct faulty, or romanticized, interpretations of Lincoln's cyphering efforts, we do not expect to have the final word on some of the issues that we raise. Before outlining the structure, it will be useful to point to three special contributions that we see this chapter making.

The first, and most important contribution, is that we have been the first to point out that there are 11, and not 10, surviving leaves from Lincoln's manuscript. A leaf on the "double rule of three," held by Harvard University since 1954, has never before been taken into account by Lincoln scholars (see, e.g., Basler et al., 1953). The two pages of that leaf show that claims that Lincoln never studied the double rule of three were wrong.

Two of the authors (Ellerton and Clements) have actually handled and photographed this eleventh leaf, and there can be no doubt that it was prepared by Lincoln. We should add, parenthetically, that when Ellerton and Clements were examining cyphering books held by Harvard University they came across another cyphering book prepared by another Abraham Lincoln—this was prepared in the mid-1750s by a great half-uncle of the future President. We will not comment further on this manuscript by the other Abraham Lincoln, except to say that its existence emphasizes that the lineage of the young Abraham Lincoln who lived in Spencer County, Indiana, in the early 1820s did include at least one person who had prepared a cyphering book.

A second contribution in this chapter is the re-ordering of the pages. The editors of *The Collected Works of Abraham Lincoln* (Basler et al., 1953) reproduced the 20 pages of which they were aware in a particular order, but did not claim that their order was the one in which Lincoln had prepared the pages. Both Louis Warren (1959) and Maurice Dorfman (1966a) did, in fact, attempt to order them, with Dorfman (1966a) stating that "the order in which Lincoln wrote them, based on the arithmetic books used and also on imprints from other pages, is quite conclusive" (p. 62).

In the first column of Table 6.1 our order for the pages is shown, and in the second, third and fourth columns the orders put forward by Dorfman (1966a), Warren (1959) and Basler et al. (1953), respectively, are shown. The final column shows the names of authors of textbooks to which material on particular pages of Lincoln's manuscript can be unambiguously linked. Our order for the pages, based on the published arithmetic texts that can be linked to pages, on imprints from other pages, on comparing patterns in which edges of pages are torn, on visual evidence of early stitching holes, and on our knowledge of the sequence in the standard *abbaco* arithmetic curriculum, differs in important ways from the orders provided by Basler et al. (1953), Warren (1959) and Dorfman (1966a). Although we are confident that our order is the best so far, we are not prepared to say that there is conclusive evidence that our order is totally correct.

Table 6.1
Suggested Orders in which the 22 Surviving Pages from Lincoln's Cyphering Book were Prepared

NFE/VAH /MAC	Dorfman (1966a)	Warren (1959)	CWAL (1953)	Description of the Page	Current Owner (2013)	Linked with a Textbook?
1	5	2	6	Simple Subtraction	Shapell Manuscript Foundation	
2	6	1	5	Compound Subtraction 1	Shapell Manuscript Foundation	
3	1	5	9	Multiplication 1	Columbia University	
4	2	6	10	Multiplication 2	Columbia University	
5	3	7	2	Division 1	ALPLM (Springfield, IL)	
6	4	8	1	Division 2	ALPLM (Springfield, IL)	
7	7	3	18	Compound Addition	Library of Congress	Dilworth (1806), p. 13
8	8	4	17	Compound Subtraction 2	Library of Congress	Dilworth (1806), p. 35
9	9	9	3	Compound Multiplication	Indiana Historical Society	Dilworth (1806) p. 43
10	10	10	4	Compound Division	Indiana Historical Society	Dilworth (1806), p. 44 and p. 51
11	11	19	8	Single Rule of Three 1	Brown University University of Chicago	Dilworth (1806), p. 65
12	12	20	7	Single Rule of Three 2	Brown University University of Chicago	Dilworth (1806), p. 66
13	Not Listed	Not Listed	Not Listed	Double Rule of Three 1	Harvard University	S. Pike (1822), p. 84
14	Not Listed	Not Listed	Not Listed	Double Rule of Three 2	Harvard University	S. Pike (1822), p. 85
15	13	11	12	Simple Interest 1	Indiana University	S. Pike (1822), p. 102
16	14	12	11	Simple Interest 2	Indiana University	S. Pike (1822), p. 103
17	15	13	14	Simple Interest 3	Yale University	S. Pike (1822), p. 108
18	16	14	13	Simple Interest 4	Yale University	S. Pike (1822), p. 109
19	17	15	19	Compound Interest 1	Private	S. Pike (1822), p. 111
20	18	16	20	Compound Interest 2	Private	S. Pike (1822), p. 111
21	19	17	16	Discount 1	Chicago History Museum	S. Pike (1822), p. 112; Jess (1821) p. 106; Walkingame (1801), p. 63
22	20	18	15	Discount 2	Chicago History Museum	S. Pike (1822), p. 113; Jess (1821) p. 107; Walkingame (1801), p. 63

Our third important contribution arises from the fact that all three authors are mathematics educators, trained not only in mathematics but also in education. In addition, Ellerton and Clements have published extensively on the history of school mathematics around the world (see, e.g., Clements, Keitel, Bishop, Kilpatrick & Leung, 2013; Singh & Ellerton, 2013). The kind of overview made possible by the authors' backgrounds was not something available to previous writers on Lincoln's cyphering book.

Following this introductory section, the 22 extant pages will be analyzed in the following way. On the left side, a page from Lincoln's cyphering book will be reproduced; and, on the right side, comments on aspects of that page deemed to be educationally, mathematically, or historically significant will be made.

Although it might be thought that such an approach would lead to a dry mathematical analysis, the authors hope that readers will gain much from being able to place each page into a wider context, especially a context that takes into account the history of early school mathematics, not only in the United States of America but also in other nations. As far as we know, this method of analysis has never before been used with respect to Lincoln's cyphering book.

Stories associated with the early Kentucky and Indiana years of Abraham Lincoln abound (see, e.g., Arnold, 1891; Bartelt, 2008; Beveridge, 1958; Herndon & Weik, 1936; Tarbell & Davis, 1896), and will not be repeated in this book unless they are obviously relevant to the discussion. We believe that our analysis of the cyphering efforts of Lincoln provides convincing evidence that he was extremely intelligent, hard-working and persevering, someone who was rarely satisfied with second best.

One would find it hard to read through the 22 pages of Abraham Lincoln's cyphering book carefully and not conclude that he took pride in the fact that he *owned* the book, and that what went into it was *his*. Despite some rather derogatory comments he made about the schools he attended, and about his teachers, in an autobiographical statement to Jessie Fell in 1859—"There were some schools, so called; but no qualification was ever required of a teacher" (quoted in Basler et al., 1953, Vol. 3, p. 511)—we gained the impression, when studying the 22 pages, that he sensed that this work on arithmetic would prepare him well for the future. We will, however, focus our discussion on Abraham's cyphering prowess rather than on any of his future efforts in mathematics. We will not elaborate on his efforts to learn surveying, or on his later enthusiasm for mastering *Euclid's Elements.* Nor will we analyze examples where he applied ideas and logic from Euclid to real-life situations—for details on these matters, see Dorfman (1966b) and Hirsch and Van Haften (2010).

Under What Circumstances did Abraham Lincoln Prepare his Cyphering Book?

We argue that when one considers the circumstances in which Abraham Lincoln prepared his cyphering book, the 22 extant pages become worthy of the adjective "extraordinary." What were some of those circumstances? Throughout the period when Abraham was preparing his book, his home was a log cabin (with dimensions about 20 ft by 18 ft), originally built by his father, Thomas. In 1819, after Thomas had returned from Kentucky with his second wife, Sarah Bush (Johnston) Lincoln, the cabin was updated, so that it was given a floor, windows and a door (Bartelt, 2008; Beveridge, 1958; Herndon & Weik, 1936). Water had to be carried from a stream over a mile away.

Tom Lincoln, Sarah Bush Johnston Lincoln, Sarah ("Sally," Abraham's sister), Abraham, Sarah Bush Lincoln's three children from her previous marriage (John, Sarah and Alexander Johnston), and Dennis Hanks lived in this house, together (Beveridge, 1958). Sometimes, there were others. Figure 6.1 depicts a young Abraham Lincoln in the act of splitting rails—something for which he acquired a reputation.

Figure 6.1. From a plate titled *Lincoln, the Rail-Splitter*, in Courtenay (1917); Illustrations by A. M. Turner and Harriet Kaucher (public domain).

Conditions in the schools that Lincoln attended were not much different from those in his home. Freezing midwestern winters had to be endured, with heating provided by an open fire. Lincoln's cyphering book pages were prepared by writing with a quill taken from a bird, using home-made ink. The very few printed books that were available were expensive. Paper was scarce, and also expensive. Preparing a cyphering book in New York, Philadelphia, or Boston was an altogether different thing from preparing one in rural Indiana. Furthermore, out in the midwest it was difficult to find teachers capable of teaching arithmetic beyond the four operations on whole numbers (Ellerton & Clements, 2012; Tarbell & Davis, 1896).

Textbook Considerations

Herndon and Weik (1936) stated that all students at Lincoln's schools learned to cypher "and afterwards used Pike's *Arithmetic*" (p. 34). Many early commentators (e.g., Houser, 1943) assumed that this referred to Nicolas Pike's (1788) famous early arithmetic, but we were unable to find any problems taken from that arithmetic in Lincoln's cyphering book. The dilemma seemed to be resolved when later scholars (e.g., Dorfman, 1966a) recognized that there were problems in Lincoln's cyphering book from a popular arithmetic by *Stephen* Pike (1811). That finding suggested that Stephen Pike's book was used, and not Nicolas Pike's. That was especially likely to be the case since during the period between 1815 and 1825, Stephen Pike's *Arithmetic* was more widely used in schools than Nicolas Pike's (Karpinski, 1980). After scholars had also noticed that Lincoln's cyphering book included problems that were identical to, or almost identical to, problems in arithmetics authored by Thomas Dilworth (1806) and Zachariah Jess (1821), it was assumed that Lincoln must have had access to arithmetics by Dilworth, Jess and Stephen Pike. But the question remains—in what ways were these books used by Lincoln's teachers and by Lincoln himself?

Despite the comments in the last paragraph, we would emphasize that we do not dismiss the possibility that the "Pike" to whom Herndon and Weik (1936) referred was, indeed, Nicolas Pike—who was the first U.S citizen to write a major arithmetic textbook. Although many editions of Nicolas Pike's arithmetics were published between 1788 and 1825 (Karpinski, 1980), the text was too complex for many school students, and also for many teachers (Ellerton & Clements, 2012). It is possible, though, that an arithmetic by Nicolas Pike was available to Abraham Lincoln, but his teachers used examples from one or more of Thomas Dilworth's and Stephen Pike's arithmetics.

Abraham Lincoln did not throw much light on the question of whether he used an arithmetic by Nicolas Pike or one by Stephen Pike when, in 1858, he informed Leonard Swett:

> My father … had suffered greatly for the want of an education, and he determined at an early age that I should be well educated. And what do you think his ideas of a good education were? We had an old dog-eared arithmetic in our house, and father determined that somehow, or somehow else, I should cipher clear through that book. (Quoted in Swett, 1889, p. 456)

The possibility that the "old dog-eared arithmetic" had been written by Nicolas Pike is consistent with a claim made by Francis Van Natter (1963) that Lincoln's initial interest in surveying was sparked by his studying the small section on the "mensuration of superficies" placed toward the end of Nicolas Pike's (1788) book. It was unusual for a school arithmetic to have a section on such matters—Nicolas Pike's book did, and Stephen Pike's did not. Furthermore, it was more likely that an arithmetic written by Nicolas Pike would have been "old" and "dog-eared" in the early 1820s than one by Stephen Pike, because the first edition of Stephen Pike's *Arithmetic* had appeared in 1811, whereas the first edition of Nicolas Pike's had been published in 1788 (Karpinski, 1980).

Thus, the possibility that Nicolas Pike's book was used cannot be ruled out. Nevertheless, there is evidence that Stephen Pike's *Arithmetic* was also commonly referred to as "Pike's arithmetic." David Banta (1906), picking up an old arithmetic textbook which, he said, he had used at school in Indiana in the 1840s, referred to it as "Pike's *Arithmetic*" (p. 135). He then read aloud the definition of long division that he had learned many years earlier. The book from which he read was Stephen Pike's *Arithmetic*. According to Banta, this *Arithmetic*, by *Stephen* Pike, was used far more often than any other arithmetic in early Indiana schools.

The second point that needs to be made about textbooks is that our page-by-page analysis of Abraham Lincoln's cyphering book will suggest that often Lincoln was taking down details dictated to him by a teacher (and was not copying directly from a textbook). Thus, for example, the words from Dilworth's (1806) statement of the rule of three included the word "supposition" (p. 65), but this was misspelled in Lincoln's cyphering book as "suposeitsn." There were other similar errors which pointed strongly to the teacher dictating at least some of the notes.

There have even been suggestions that single pages from arithmetics like those authored by Stephen Pike, Thomas Dilworth and Zachariah Jess could have been passed around to students (Dunlap, 1959). The fact is, little is known about the actual classroom environments, the material resources available, or the teaching practices adopted, in Lincoln's schools.

Dunlap (1959), wondering why there was not a single source for all pages of Lincoln's cyphering book, asked himself a series of questions which, almost as if talking to himself, he proceeded to answer:

> If Dilworth provided the text for three pages, why not for all? It is possible that one of Lincoln's three teachers used Dilworth as his text, another used Jess. But, if that be the case, why is it only part of a page from Dilworth and not the remainder of it? If Lincoln were copying from a book in front of him, would he not have taken all relevant material? And, recalling Lincoln's statement concerning the qualifications of his teachers, it is conceding much to believe that one of them employed two textbooks. (p. 10)

Indeed, the fact that Abraham's book includes questions from Dilworth (1806), Jess (1821) and Pike (1822) does not negate the possibility that his teachers had their own handwritten cyphering books in which were written questions from Dilworth, Jess and Pike. Such questions, and text surrounding the questions, could have been dictated to Lincoln. Ellerton and Clements (2012) cited many early nineteenth-century North American commentators on education who indicated that commercial arithmetics were not readily available, and were not used, in most schools.

With respect to Thomas Dilworth's arithmetic, it should be noted that 57 different American editions were published, by various publishers, between 1773 and 1825 (Karpinski, 1980). However, between 1819 and 1826, when Lincoln prepared his cyphering book, the influence of Dilworth's text on U.S. school mathematics had waned, with the works of U.S.-born authors such as Daniel Adams (1822), Warren Colburn (1822), Nathan Daboll (1820), Stephen Pike (1822) and Michael Walsh (1820) being preferred. When citing Dilworth we shall mainly refer to an 1806 version (published in New York by G. Jansen). Note, however, that we do not know which actual edition of Dilworth's *Arithmetic* was the source of the numerous "Dilworth examples" in Lincoln's cyphering book.

Similarly, Karpinski (1980) listed 10 different editions of Stephen Pike's *Arithmetic*, and 4 keys, published between 1811 and 1826. Karpinski (1980) also listed 17 different editions of Jess's *Arithmetic* and 2 keys, published between 1799 and 1821. When referring to Stephen Pike's *Arithmetic* we shall mainly refer to the 1822 Philadelphia edition, published by Benjamin Warner; and when we refer to Zachariah Jess's *Arithmetic*, we shall refer to the 1821 edition published in Philadelphia by McCarty and Davis. But the truth is, we do not know which editions provided the inspiration for the "Pike" and "Jess" examples in Lincoln's cyphering book.

Penmanship and Calligraphy

The quality of Lincoln's calligraphy, in headings, and general penmanship throughout the 22 extant pages was noteworthy. Indeed, given the rough conditions, the absence of smooth, flat surfaces, and the scarcity of paper, one might say that the future President did a remarkable job keeping up such high standards of presentation. This was *his* book and he wanted to be proud of it. He also wanted his family to be proud of it, and in that he succeeded, for his step-mother would guard Lincoln's cyphering book for many years. Next to the Bible, a cyphering book was often regarded as the most important book a family could have (Wickersham, 1886).

That said, in comparison with the penmanship and calligraphy found in other cyphering books across the United States before 1850, Lincoln's book was not particularly special. Ellerton and Clements (2012), who have studied about 1000 North American cyphering books, have established that, when cyphering, most students were prepared to spend much time on calligraphic headings and careful penmanship. But relatively few of the other students had to cope with conditions like those in Spencer County, Indiana, where Lincoln prepared his book.

In North America, it became an accepted tradition that a beautiful and mostly correct cyphering book would serve as a passport to desirable employment, especially if the arithmetic went as far as, or beyond, the rule of three. In a similar vein, a student who wanted to proceed to college would be expected to show his cyphering book to college authorities who interviewed him. From a teacher's point of view, at the end of a term there would be an evening when students would be examined through oral questioning by a local committee and, after the formalities, over supper, cyphering books would be on display. Teachers knew that their future employment could hinge on the appearance of their students' cyphering books.

For How Long Did Lincoln Attend Schools in Indiana, and Who Were his Teachers?

When preparing this chapter we became increasingly perplexed by the conflicting statements made in relation to Abraham Lincoln's school education. One of the difficulties is that Lincoln made a number of autobiographical statements—aimed at political audiences—in which he said he attended school for less than a year in total, and named three teachers. Subsequent interpretations of these statements could have generated inaccurate impressions. We now elaborate on that point.

In June 1860, Lincoln wrote that he "went to A B C schools by littles, kept by Andrew Crawford, [James] Sweeney [Swaney], and Azel W. Dorsey" (quoted in Basler et al., Vol. 4, p. 62). He also stated that "the aggregate of all his schooling did not amount to one year" (p. 62). Burlingame (2008), Donald (1995), Funk (1969), and Warren (1931) agreed with the order in which Lincoln said he had his teachers, but others have stated the order to be Crawford, Dorsey and Swaney (e.g., Grigsby, 1865; Wilson, 1940), or Dorsey, Crawford and Swaney (e.g., Turnham, 1865). Funk (1969) provided additional details, stating that:

> The first school was opened in 1819–1820. Andrew Crawford, who had moved into the community the year before, taught school for a session of two or three months. … Lincoln's second school was taught by James Swaney … This second school was probably held in 1821–22 … James Swaney was the son of a Michael Swaney who had moved to Indiana from Hardin County, Kentucky, just before the Lincolns. … The parents died in 1817 and James and his sister, Charott, were left orphans. The two children were bound over to Azel W. Dorsey of Spencer County who was appointed their legal guardian. … In 1820, James reached twenty-one and having learned to "read, write and cipher through the rule of three," he was ready to teach in the pioneer school. He was hired for the 1821–22 school term. The school stood on the Hoskins farm in Jackson Township of Spencer County. The school was four-and-one-half miles southwest of the Lincoln farm. (pp. 150–151)

Funk (1969) noted that, although the school building in which Swaney taught was larger than the Crawford school, and had two fireplaces, the distance from home probably explains why Abraham attended Swaney's school for only a brief period. Warren (1931) agreed with that explanation.

It seems likely that Abraham's third school was in the 1824–1825 or the 1825–1826 winter term with Azel Dorsey (Funk, 1969; Warren, 1931). This school "stood on Section 22 near the Lincoln home" (Goodspeed Bros and Co, 1885, p. 410). Azel Dorsey was much respected in the community and was reputed to be the best of Abraham's teachers (Funk, 1969; Warren, 1931).

In our research, we found reference to a fourth Indiana teacher, James Davis Bryant (or Briant). Bryant "taught all through this portion of the county, and Abe Lincoln was one of his pupils" (Goodspeed Bros. & Co., 1885, p. 412). Bryant was born in 1800, probably in Kentucky where his parents Zachariah and Tabitha had moved from Virginia. It appears that he was an itinerant teacher, and taught in several schools in Spencer County in the early 1820s, including a log-cabin school erected near Buffalo about 1822 where it is said that "Azel W. Dorsey was one of the first teachers" (Goodspeed Bros. & Co., 1885, p. 412).

The fact that there is no other reference to James Bryant as one of Abraham's teachers begs comment. Azel Dorsey was highly regarded both as a teacher and as a prominent citizen in the community (Azel W. Dorsey, Lincoln Pedagogue. 1930). As his popularity within the local school context, as well as in the wider community, grew, so the demands on his time became greater. We conjecture that Dorsey may have used the opportunity to have Bryant share some of the teaching burden, while he (Dorsey) still ran the school and attended to other matters in town. The possible presence of a fourth teacher would go some of the way toward explaining the appearance, in 1826, of problems from Walkingame (1802) and Jess (1821) on the last pages of Abraham's cyphering book.

We believe that Lincoln cyphered with all four of his Indiana teachers. Dennis Hanks (1865—see page 105) claimed that he purchased the sheets of paper that formed Abrahams's cyphering book, and we conjecture that Dennis bought that paper when the future President attended Andrew Crawford's school during the 1819–1820 winter. It is possible that the subtraction tasks on the page that we regard as the first of the extant pages of Lincoln's cyphering book were prepared at Crawford's school. Traditionally, in the United States, those students who would prepare cyphering books began to cypher at age 10 (Ellerton & Clements, 2012). Wilson (1940) claimed that Abraham learned to cypher to the rule of three while attending Crawford's school.

So far as Lincoln's statement that the aggregate of his formal schooling did not extend beyond 12 months was concerned, we raise the question "what does 12 months mean in that context?" Nathaniel Grigsby (1865), who went to school with the future President, stated that Lincoln attended schools in Indiana for a total of about 18 months. Grigsby also alluded to Lincoln attending schools in Kentucky before his family moved to Indiana and, in an interview with William Herndon, Dennis Hanks (1865) confirmed that that was indeed the case. Another who went to school with Lincoln, David Turnham, stated in an 1865 interview with William Herndon that Lincoln attended schools kept by Andrew Crawford and Azel Dorsey for a total of about four winters (Turnham, 1865). Wilson (1940) maintained that Lincoln attended school for at least two years. A conservative estimate would have Lincoln attending school for a total of at least three winter terms, and possibly more than that. We believe that the weight of evidence is against Dennis Hanks's claim that Lincoln went to school for a total of six months only, even though Hanks lived with the Lincoln family during the time that Lincoln went to school.

The statements relating to different periods of time can be fairly easily reconciled. In the early nineteenth century, most schools in the northern and mid-western states of the United States operated over two terms, a winter term and a non-winter, or fall, term. Each term lasted about three months. Typically, older boys were expected to work to support the family during the non-winter months, and they did not attend school during those months. If a person attended both winter and non-winter terms, then this was often considered to be attending school for "a year" (whereas, in fact, the actual time at school was only about six months). Indeed, according to Burlingame (2008), "the school year was short, extending from the close of the fall harvest to the planting season" (Vol. 1, p. 31).

If someone went to school for the winter term only, then that person was sometimes said to have gone to school for half a year (whereas in reality it was for only three to four months), and some might even have said for the "whole school year." It seems that Lincoln attended schools in Indiana for at least three winter terms (with Crawford, Dorsey and Bryant). He also attended Swaney's school (at Gentryville, four miles from Lincoln's home) and schools in Kentucky. So, depending on the way one counts, the estimates of how long Lincoln attended school (by Lincoln himself and by Dennis Hanks, on the one hand, and by Nathaniel Grigsby and David Turnham on the other hand), can be partly reconciled. That is hardly surprising, for each of the four informants had first-hand experience of events and times surrounding Lincoln's schooling.

We believe, therefore, that the evidence indicates that Lincoln attended school for about two years, where "two years" is taken to mean four "terms" or about 12 calendar months in total. We should add that if that is true, then the amount of time that Lincoln spent at school was above the average for children in the United States, especially for children in the midwest (Clements et al., 2013; Ellerton & Clements, 2012).

There is evidence that Lincoln attended Dorsey's school for six months. A plaque—erected by the Illinois Historical Society and still standing in Huntsville, Illinois—states that Azel Waters Dorsey, who lived between 1784 and 1858, moved to Huntsville after conducting a school in Spencer County, Indiana. The plaque indicates that Abraham Lincoln attended Dorsey's school for six months in 1824. Goodspeed Bros. and Co (1885) noted that Azel Dorsey opened his school in 1821 or 1822, and that it operated for about 3 years. Two of Lincoln's cyphering-book pages were dated 1826, and it could have been the case that these were prepared when James Bryant was assisting at Dorsey's school. A stone from Lincoln's tomb in Springfield, Illinois, was erected over Dorsey's grave in 1932, and bears the following inscription: "Erected by Logan Settles in memory of Azel Dorsey, school teacher of Abraham Lincoln 1826."

The following matter-of-fact statement was made by Nathaniel Grigsby (1865) to William Herndon in September 1865, not long after Lincoln's death. Grigsby's brother, Aaron, married Abraham's sister, Sarah (Sally) Lincoln.

> My father came from Kentucky in the fall of 1815 and settled in what is now Spencer County. ... [Abraham's father, Thomas, and his mother, Nancy] moved to the State in the year 1816 or 1817. Abraham, Sally and myself all went to school. We first went to school to Andy Crawford in the year 1818 in the winter, the same year that Mrs Lincoln died, she having died in October. Abe went to school nearly a year, say nine months. I was going to school this time and saw Lincoln there most, if not all, the time. The second schoolmaster we went to was Mr Azel Dorsey. Abraham Lincoln went to school to Azel about six months. Lincoln did not go to school to him all the time. We had to go about two miles to school. The third time we went to school was to a Mr Swaney, who taught six months. Lincoln did not go to school to him all the time. Lincoln had to walk about four miles. Lincoln was about the first school nine or ten years of age. The second school, he was about fourteen or fifteen, and the third school he was about sixteen or eighteen. ... I think Lincoln ciphered at Crawford's school, Dorsey's and Swaney's. He used Pike's

> *Arithmetic*. Ray's was sometimes used. We only wrote, spelled and ciphered. ... While other boys were idling away their time, Lincoln was at home studying hard, would cipher on boards, wooden fire shovels, etc., by the light of the fire that burnt in the heath; had a slate sometimes, but if not handy would use boards. He would shave boards bright and cipher on them, dirty them, reshave them. Abe would sit up late reading and rise early doing the same.
>
> (Nathaniel Grigsby to William Herndon, September 12, 1865, p. 112)

The sequence Grigsby gave for Abraham's teachers was Crawford, Dorsey, Swaney, and the possibility remains that Abraham returned briefly to Dorsey's school late in 1825 and early in 1826. Much of Nathaniel Grigsby's (1865) statement to Herndon has been corroborated by the evidence of others. For example, there can be little doubt that Lincoln checked his calculations on shovels (see Burlingame, 2008, p. 34), before entering what he regarded as a totally correct solution into his cyphering book. Clearly, though, Grigsby was wrong about Lincoln having used, at school, an arithmetic authored by Joseph Ray, because no book written by Ray was published before 1834 (Karpinski, 1980). Perhaps, Grigsby confused Ray with Dilworth—Ray and Dilworth were the two largest-selling authors of arithmetics used in U.S. schools during the period 1774–1865.

A Different Form of Instruction, and the *Abbaco* Arithmetic Curriculum

Mathematics instruction, and the mathematics curriculum, in the United States of America are very different now from what they were early in the nineteenth century when Abraham Lincoln attended schools in Kentucky and Indiana. Much has been made of the small amount of time, overall, that Lincoln attended school, but rarely does one ever find a statement indicating the fact that in the early 1800s many U.S. school-age children rarely attended school (Clements et al., 2013; Ellerton & Clements, 2012). Of those who did attend, many (especially boys in New England, New York, Pennsylvania, and the emerging midwestern states) attended during winter months only, and did not study any mathematics beyond elementary arithmetic. As late as the 1870s, the hero of Mark Twain's *Adventures of Huckleberry Finn* went to school only "three or four months" during which time he learnt the rudiments of reading, writing, spelling, and "the multiplication table up to six times seven is thirty-five" (Clemens, 1885, p. 18).

Two centuries ago, American teachers did not stand at the front of the room and teach, and most students, even those studying mathematics, did not own a mathematics textbook. Written examinations of any kind were not used. Most teachers of any branch of mathematics did not have formal qualifications in mathematics.

This is not an appropriate platform from which to give details of the individual, recitation style of instruction and learning that prevailed in the schools Lincoln attended, but we would urge those attempting to make sense of Lincoln's early school education to acquaint themselves with historical scholarship on what U.S. rural schools were like in the early 1800s. It should be added that at least two of the schools which Lincoln attended—those conducted by Andrew Crawford and Azel Dorsey—were "blab" schools in which students were expected to talk aloud, mainly to themselves, throughout the school day (Beveridge, 1958; Donald, 1995; Koos, 2001, Warren, 1931). At a blab school the only time during a school day when a student was not expected to talk continuously was when he, or she, was called to the teacher for individual recitation. Although some scholars have suggested that the recitation process did not include arithmetic (e.g., Banta, 1906), it is undeniably the case that recitation and cyphering were linked in early North American schools (Ellerton & Clements, 2012). Furthermore, it is likely that most of the entries in Abraham Lincoln's cyphering book were made during school hours, because the school furniture included a long writing desk made from "a broad board upon pins or supporters driven into the wall" (Smith, 1867, p. 35). This writing desk was located immediately below a hole in the wall created to provide light (Smith, 1867). Such writing desks were often found in common schools (Chapman, 1895), and offered a reasonably flat surface for students attempting to produce their best efforts in calligraphy and penmanship.

Another matter that greatly affected Lincoln's arithmetical learning was the sort of mathematics curriculum that he was expected to follow during his time at school. Lincoln studied what historians are now calling the *abbaco* curriculum. This had its origins in Continental Europe, and had a strong commercial flavor to it (Ellerton & Clements, 2012). Of most relevance here is that our analyses of Lincoln's cyphering book will reveal that he was expected to learn topics like simple and compound interest, and discount, before he had formally studied common fractions (often called "vulgar fractions") or decimal fractions, or percentage.

One of the most forthright statements about Lincoln's early schooling was penned by the Reverend C. J. Schwarz (1920):

> The statement that Abraham Lincoln had hardly any schooling, and only of the rudest sort, and that he made well-nigh superhuman efforts to reach the nearest school, is fiction. ... Here in Indiana, from 1817 to the close of 1819, no school was available for young Abe, the nearest one being at Troy, thirteen miles distant. However, a large settlement was made in the vicinity about 1820, and soon there were nine schools within a radius of from one to four miles from the Lincoln farm. (p. 50)

Schwarz went on to state that one of the early schools "stood near the Lincoln home, on the same tract of land, where probably Abe was a pupil for a number of terms. Azel W. Dorsey was one of his teachers" (p. 50). The other teachers mentioned in Schwartz's article were Bryant, Crawford, and Adam Shoemaker. Reference to Shoemaker as a teacher of Abraham Lincoln has also been made by the Indiana Department of Public Instruction (1917), and by De La Hunt (1916). Schwarz (1920) noted that Adam Shoemaker, the teacher, was a visiting minister at the Pigeon Creek Baptist Church, and that Abraham Lincoln may have been a pupil of Shoemaker's in 1828 at Troy, when Lincoln operated a ferry there for a short time.

In addition, Schwarz (1920) described the nature of education in these early schools, noting that the teachers were "educated men and capable teachers for those days" (p. 50). His description of the schools' curricula at that time is pertinent here:

> These schools taught more than is generally supposed. The thorough and persistent drilling in the fundamentals, reading, writing, and arithmetic, and the progressive application of same, with bearings on the actual concerns of life, gave the pupil facility and pleasure for more advanced studies. (p. 50)

Teaching focused on individual instruction, whereby "more thorough and additional instruction could be given in accordance with each scholar's advancement" (Schwarz, 1920, p. 50). Lincoln's cyphering book provides strong evidence of the power of such individual instruction.

Jefferson Davis, the former President of the Confederate States, was less positive than Schwarz about learning which took place when students prepared cyphering books. In a biography of her husband's life, Varina Davis (1890) reported his description of how students worked out their "sums," and then, "after a sufficient amount of attention" merely copied what they had written into their cyphering books. Davis commented that "after some time a bright boy could repeat all the rules; but if you asked him to explain why … he could give the rule but no reason for it" (p. 19). That was the way Jefferson Davis interpreted his memories of his cyphering days at school, but his was just one case. Not surprisingly, this interpretation has been endorsed by some historians (see, e.g., Miller, 2006, p. 56). Ellerton and Clements (2012), however, rejected such a point of view as unfair to a cyphering tradition that endured and prospered for over 500 years.

A Page-by-Page Analysis of Abraham Lincoln's Cyphering Book

Please note that, in the following commentaries on the 22 extant pages of Abraham Lincoln's cyphering book, the future President will usually be referred to as "Abraham."

The pages have been reproduced, with the permission of the owners of the 11 leaves—which are located in different places across the United States of America. The photographic images which are shown were provided by the owners. We gratefully acknowledge the support of those who assisted us during the process of obtaining permission to reproduce the pages in this book. In the figure caption for each image, appropriate acknowledgement is given in the manner requested by each owner.

Every effort has been made to present authentic images of these 22 extant pages. The pages are old and worn, and each leaf has a different history of where and how it has been preserved. It has been a privilege to be able to include all of these pages together, for the first time since they were penned by Abraham Lincoln. Through these images, we hope that readers will not only feel a sense of the age and the history associated with each leaf, but will also be able to glimpse something of the mathematical growth of the lad who would one day become President of the United States of America.

The First of the Extant Pages in Lincoln's Cyphering Book

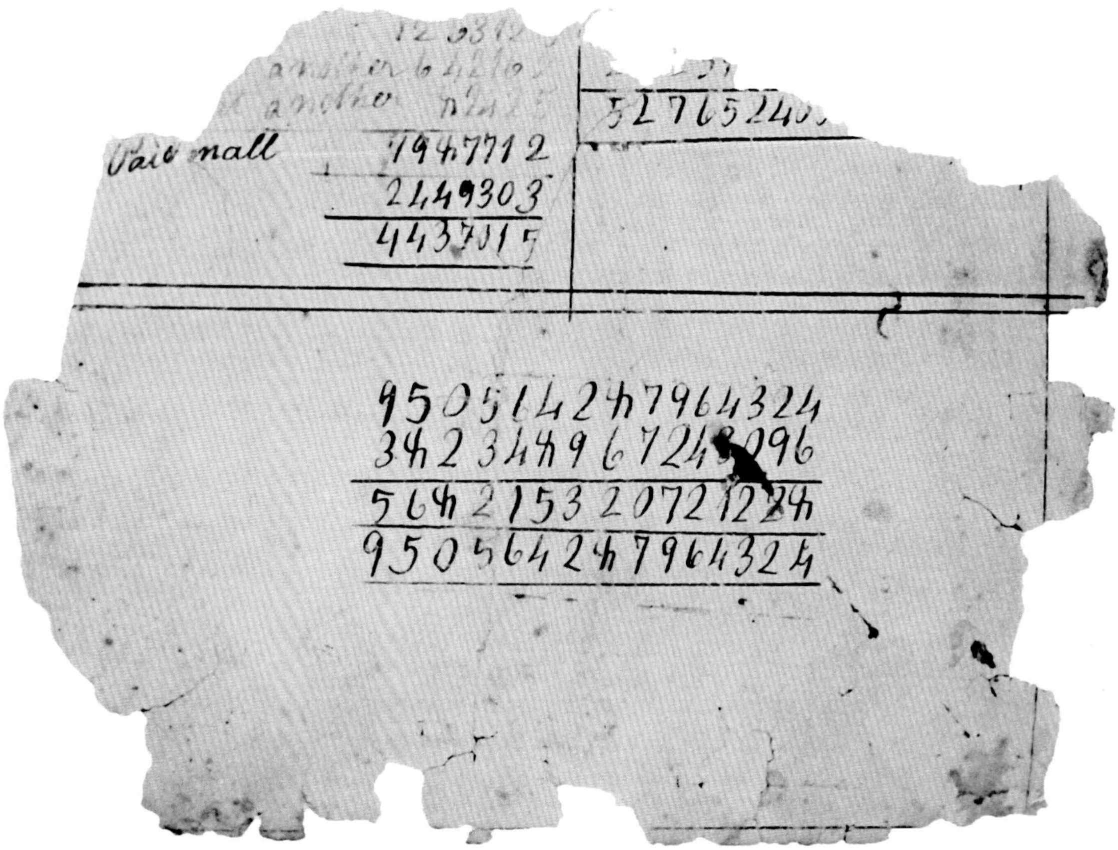

Figure 6.2. "Simple Subtraction" (fragment from the first of the 22 pages). By permission of the Shapell legacy partnership.

It is with some trepidation that we claim that of the 22 extant pages from Abraham's cyphering book, the page fragment shown in Figure 6.2 was the first to be prepared by Abraham. Although the editors of the 1953 *Collected Works of Abraham Lincoln* (Basler, Pratt, & Dunlap, 1953) did not claim that the order in which they reproduced the 20 pages—not 22 pages, because they were not aware of the existence of the leaf currently held at Harvard University—was the original order, they obviously placed the pages in an order that they thought could be correct. They thought that the page shown in Figure 6.3 was the sixth page. Warren (1959) listed it as the second page, and Dorfman (1966a) listed it as the fifth.

Why do we believe that the page shown in Figure 6.2 was prepared before all of the other extant pages? Before we answer that question we wish to comment that there may well have been pages before this one on numeration and addition—but these are long since gone. Furthermore, since Lincoln attended several different schools with different teachers, it is possible that he repeated some topics, and that therefore the pages in his original cyphering book (which were cumulatively prepared at the different schools) may not have been in strict sequence from a mathematical point of view.

Two of the authors of this chapter (Ellerton and Clements) own a large collection of North American cyphering books (comprising about 350 manuscripts). They also own about 3000 copies of North American school mathematics textbooks (mostly arithmetics) published between 1788 and

1900—these include about 200 arithmetics published in the United States before 1826. Ellerton and Clements have immersed themselves in these books over a period of about nine years, and are in no doubt that the cyphering (*abbaco*) tradition controlled how school mathematics took place in North American schools up to about 1840. In the *abbaco* tradition, simple multiplication and division would *never* have been studied before simple addition and subtraction—and yet that is what was implied by the sequences put forward by the *CWAL* editors and by Dorfman (1966a).

On the fragment shown in Figure 6.2 a striking visual effect is provided by the long subtraction task, 950564287964324 – 382348967243096. After Abraham had done this correctly, he then added the answer (568215320721228) to the second number, to obtain the first number. This process of adding was often found in cyphering books and in commercial arithmetics of the time. Although it was simply a "check," during the eighteenth and nineteenth centuries authors of arithmetics and of cyphering books usually referred to this as a "proof" (see, e.g., Jess, 1821, p. 8; Pike, 1811, p. 10). Today's mathematicians would certainly not approve of such a loose use of the word "proof."

The subtraction task on the page has not been found in any commercial arithmetic of the period, and it is likely that it was created by one of Abraham's teachers, or by Abraham himself. Certainly, lengthy subtractions like the one shown were commonly found in many books, but the subtraction concept was sufficiently simple that even if a teacher did not have a strong grasp of mathematics he or she would have felt capable of posing such tasks.

At the top left of the fragment of the page shown in Figure 6.2, part of another subtraction task is shown. This task was probably concerned with adding 1263125, 642165, and 82425, and then subtracting the result from 4437015. At one stage, Abraham wrote "Paid in all" suggesting that the original task had been set in some kind of monetary context.

As stated above, it is possible that this page was not the first of the pages that Abraham prepared for his cyphering book. There could be missing pages, and sometimes compound addition tasks appeared before simple subtraction tasks in commercial arithmetics. That was, in fact, the case with Thomas Dilworth's (1792, 1806) popular arithmetics and, as we shall see, several questions in Abraham's cyphering book were from one of Dilworth's texts. However, none of the tasks on the pages we have ordered as first and second appeared in any of Dilworth's arithmetics.

We conjecture that the page reproduced in Figure 6.2 was prepared by Abraham late in 1819 or early in 1820, when he was attending a blab school conducted by Andrew Crawford. It is possible that Abraham's original cyphering book contained at least two earlier pages, one dealing with numeration and one with addition—and, perhaps, there was a third additional page in which rules for subtraction were outlined. However, in Spencer County, Indiana, around 1820, paper was scarce, and expensive, and it could have been the case that Abraham decided that since he had already learned about numeration and addition (perhaps when he attended school in Kentucky, before his family moved to Indiana), he should not "waste" precious pages in his cyphering book with such elementary material.

The conjecture that this was the first of the extant pages that Abraham prepared is also supported by the space that Abraham allowed himself when writing on the page. There is no suggestion of material being crammed together, or that Abraham felt that the writing itself needed to be made smaller than what he would normally write. On many of the later pages, Abraham's writing became smaller, and more of the pages were fully covered with the writing. It seems that as he prepared the page of which part is reproduced in Figure 6.2, Abraham was not very concerned about conserving space within the blank pages that he had recently acquired.

The Second of the Extant Pages in Lincoln's Cyphering Book

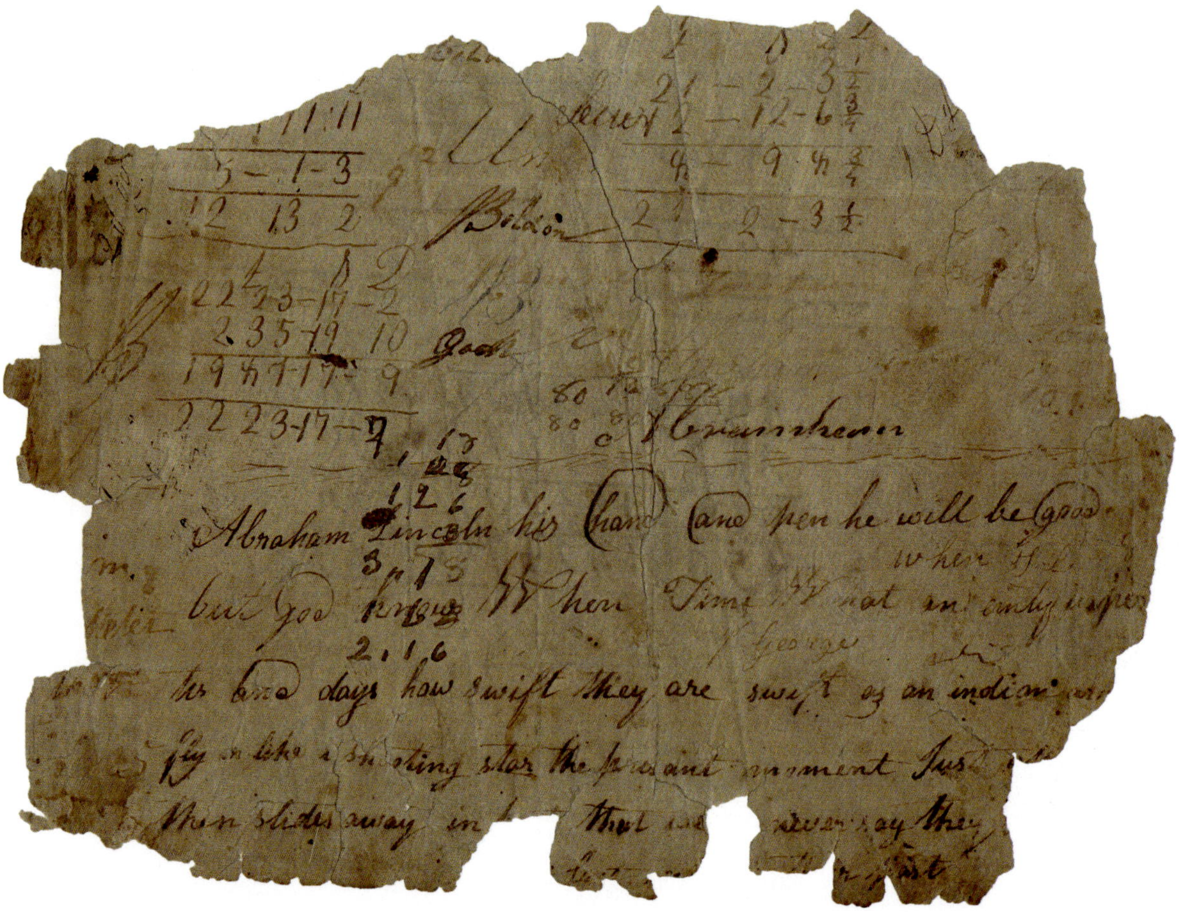

Figure 6.3. "Compound Subtraction 1" (fragment from the second of the 22 pages). By permission of the Shapell legacy partnership.

The fragment shown in Figure 6.3 is on the back of the fragment shown in Figure 6.2. The editors of *CWAL* (Basler et al., 1953) had this as the fifth page on their list, Warren (1959) had it as the first page, and Dorfman (1966a) as the sixth. We believe that this page was prepared by Abraham late in 1819 or early in 1820, when he was attending the school conducted by Andrew Crawford.

It was fairly common, but not universally the case—an exception was Nicolas Pike's (1788) text—for textbook authors to ask students to do "subtraction of money" tasks (i.e., tasks requiring subtraction of amounts of English money, or Federal money) before simple multiplication and division tasks (see. e.g., Daboll, 1820, p. 27; Dilworth, 1792, p. 22; Jess, 1821, p. 9; Pike, 1811, p.13). It is unlikely that Abraham would have prepared this page after he had completed multiplication or division tasks. That adds further support to our claim that this fragment was prepared before pages which dealt with multiplication and division.

It has not been possible to trace any of the three subtraction tasks on the fragment in Figure 6.3 to any commercial arithmetic of the period. As with the first subtraction task, it is likely that the tasks were created by Lincoln's teacher. Certainly, subtractions of quantities of English money like those shown were commonly found in many arithmetics, and the concepts involved were sufficiently simple that most teachers would have felt capable of posing such tasks.

On this fragment, Abraham was concerned with subtraction of English money. One such task (see the top right corner of the fragment), written under the heading *£ S D*, was "*£21-2-3½*" less "*£12-12-6¾*." The farthings were expressed as vulgar fractions, ½ and ¾, despite the likelihood that Abraham had never formally studied common (i.e., vulgar) fractions.

Such a task could be dealt with in the following way: "3 farthings from 2 farthings you can't do; therefore, since 3 from 4 is 1, add 1 and 2 to get 3 farthings; pay back the 1 pence that was added (in the form of 4 farthings), to get 7 pence from 3 pence, which you can't do—so, 7 pence from 12 pence is 5 pence, and add that to 3 pence to get 8 pence; pay back the 1 shilling (i.e., 12 pence) that was added, to get 13 shillings from 2 shillings, which you can't do—so, 13 shillings from 20 shillings is 7 shillings, and add that to 2 shillings to get 9 shillings; and finally, pay back the 1 pound (i.e., 20 shillings) that was added, to get 13 pounds, which from 21 pounds is 8 pounds. Hence the answer is 8 pounds 9 shillings 8 pence and 3 farthings." That was Abraham's answer. He then did a check by adding that answer to *£12-12-6¾*, which returned him to *£21-2-3½*.

Almost 50 years after the Declaration of Independence (in 1776), children in U.S. schools were still learning to calculate in English money (with vulgar fractions being used as notation for farthings). They also learned to calculate with Federal money (with decimal fractions). This dual system continued for many years—to the 1840s, at least.

After the subtraction tasks Abraham included the following (unedited) little poem (quoted in *CWAL*, Basler et al, 1953, pp. 1–2). The bracketed words were supplied in the text by Herndon and Weik (1936). The unedited text was "taken down before the original manuscript became badly frayed" (Basler et al., 1953, p. 2). The part of the text from "Time what an emty vaper tis ..." came from the first two stanzas of a hymn by Isaac Watts (Wilson & Davis, 1998—p. 135, see footnote 2).

Abraham Lincoln
his hand and pen, he will be good,
but God knows When. Time what an emty vaper
tis and days how swift they are swift as an indian arr[ow]
fly on like a shooting star the presant moment Just [is here]
then slides away in h[as]te that we [can] never say they['re ours],
but [only say] th[ey]'re past.

Unlike the poetry in Mary Walter's manuscript (see Chapter 5), Abraham's poem was not concerned with mathematics. Many students who prepared cyphering books personalized their efforts by adding doggerel or small poems, and often they were pleased to point out that the books were theirs. The comment "This is my book" was often to be found. It is likely, too, that Abraham added the poem at some later date. It is well known (Herndon & Weik, 1936) that Abraham was often trying to find paper on which he could record poems, doggerel, etc., which appealed to him, and on this page his writing for the poem appears to be more mature than for the currency calculations.

The partial leaf whose sides have been reproduced in Figures 6.2 and 6.3 is very worn, and it is probably the case that the remainder of the leaf no longer exists. The fact that part of the original leaf is missing, and that what remains has uneven edges around its boundary, support the conjecture that this remnant was from an early leaf in the original manuscript. As such, it is likely that it would have been subjected to more wear and tear than other leaves, and that could have resulted in the original being torn.

The Third of the Extant Pages in Lincoln's Cyphering Book

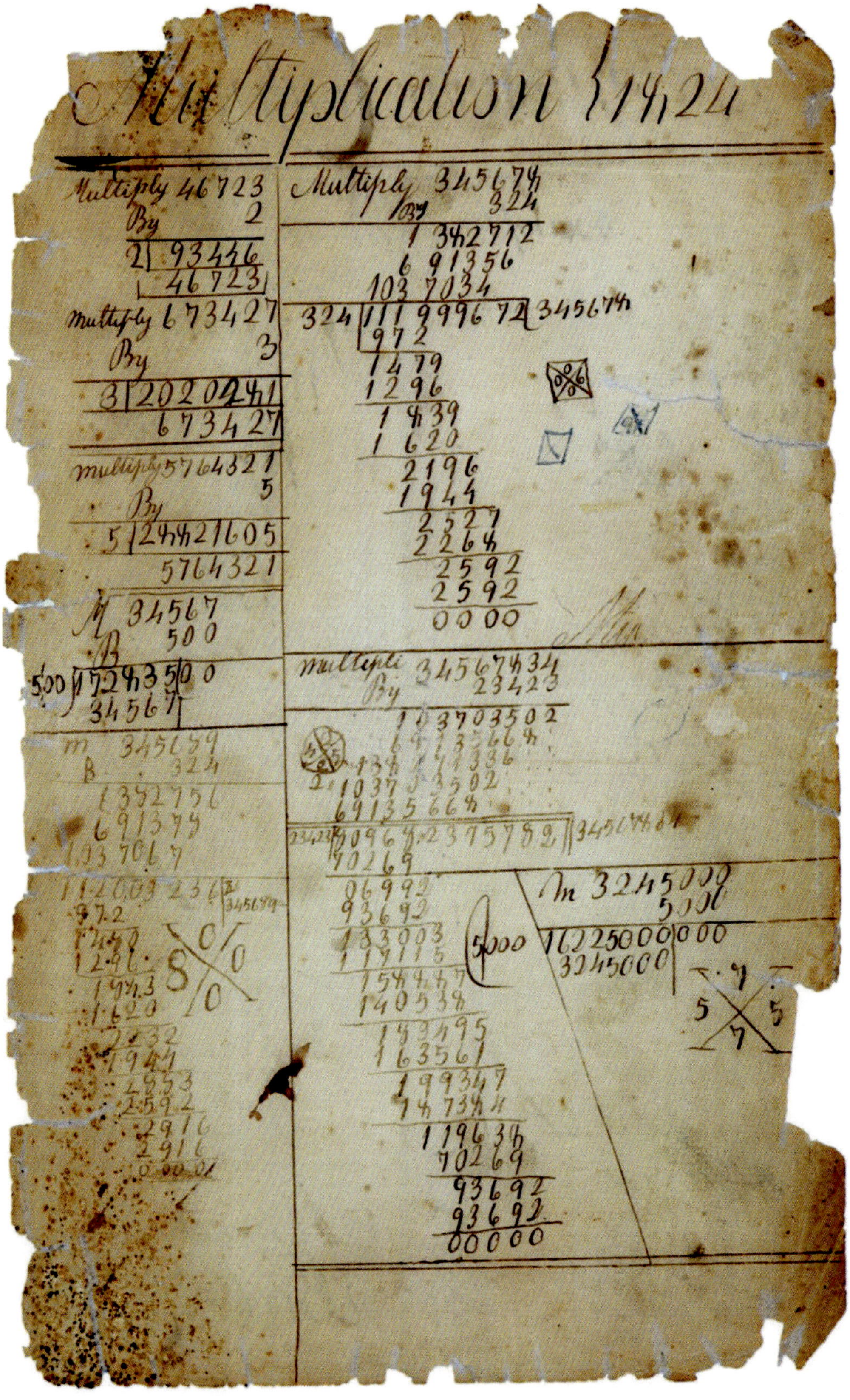

Figure 6.4. "Multiplication 1" (third of the 22 pages). Rare Book & Manuscript Library, Columbia University in the City of New York.

The page shown in Figure 6.4 was identified as the first of the extant pages by Dorfman (1966a), but fifth by Warren (1959), and ninth by Basler et al. (1953). There can be little doubt that this "multiplication" page would have preceded the two "division" pages. The page that is shown was dated 1824 by Abraham, and is currently held in the Butler Library at Columbia University. Most likely, the page was prepared by Abraham at Azel Dorsey's school.

The page starts with simple multiplications—by 2 ("46723 by 2"), by 3 ("673427 by 3"), and by 5 ("5764321 by 5")—and quickly proceeds to long multiplication ("345678 by 324").

Abraham checked all of his calculations by division—which, given that division had not yet been formally covered, might be regarded as strange. However, Abraham had previously attended Andrew Crawford's school (Beveridge, 1958; Herndon & Weik, 1936), and it is likely he learned to carry out the four operations on whole numbers at that school.

An interesting feature is that all multiplication calculations were not only checked by division, but also by a method known as "casting out nines." We shall now describe this method of checking.

Take for example the calculation for multiplying 345678 by 324. If one adds the digits for 345678 one gets 33, which has 6 as remainder if it is divided by 9; then, if one adds the digits for 324 one gets 9, which has 0 as remainder if divided by 9; so, one places these remainders within a cross, with 6 on the right and 0 on the left; then, if these two remainders are multiplied together, and the result divided by 9, the remainder is 0, and that remainder is inserted at the top of the cross; finally, if the sum of the digits in the product (i.e., 111999672) is obtained, one gets 45, which has 0 as remainder if divided by 9. This 0 is inserted at the bottom of the cross. The check is that the entry at the top of the cross (i.e., 0) should equal the entry at the bottom (in this case, also 0). Since these are equal, it can be assumed that the original calculations were correct. Although this method is not fool-proof it does provide a quick and useful check, one which has been honored by its use over centuries (see, e.g., Daboll, 1820; Pike, 1788; Swetz, 1987; Wilson, 1935).

Dorfman (1966a) maintained that Abraham got his casting-out-of-nines calculations "all wrong" (p. 65), but in fact, he got nearly all of them all correct. Clearly, Dorfman was not able to identify how Abraham had cast out nines. Perhaps the reason why this occurred was because Dorfman relied on Martin's (1800) text (see Dorfman, 1966a, pp. 68–69), and Martin's method of casting out nines was different from Abraham's. Martin (1800) also stated that $94 \div 4789 = 50\frac{89}{94}$, and $564 \div 7649 = 135\frac{352}{564}$ (p. 14), which could only be correct if notation for division in Scotland (where Martin lived and had his book published) was different from notation for division in the United States of America and England. It is little wonder that Martin's book was never published in the United States (Karpinski, 1980). As far as the authors could ascertain, the only copy of Martin's book in the United States is the one held at the University of California at Berkeley's Library—that was the copy consulted by Dorfman (1966a).

From what we have said, it is clear that Dorfman's claim (p. 63) that four of the pages in Abraham's cyphering book were based on Martin's book is not supported by the available evidence.

The page shown in Figure 6.4 is full of calculations, and we are persuaded that Abraham did them all himself. These were not copied out of a textbook. Herndon and Weik (1936) told how Abraham's step-mother recalled that Abraham used to write over any flat wooden surfaces in the family home and would then plane the surfaces and start again. Part of the cyphering tradition was that only entries which were definitely correct should be entered in a cyphering book.

The Fourth of the Extant Pages in Lincoln's Cyphering Book

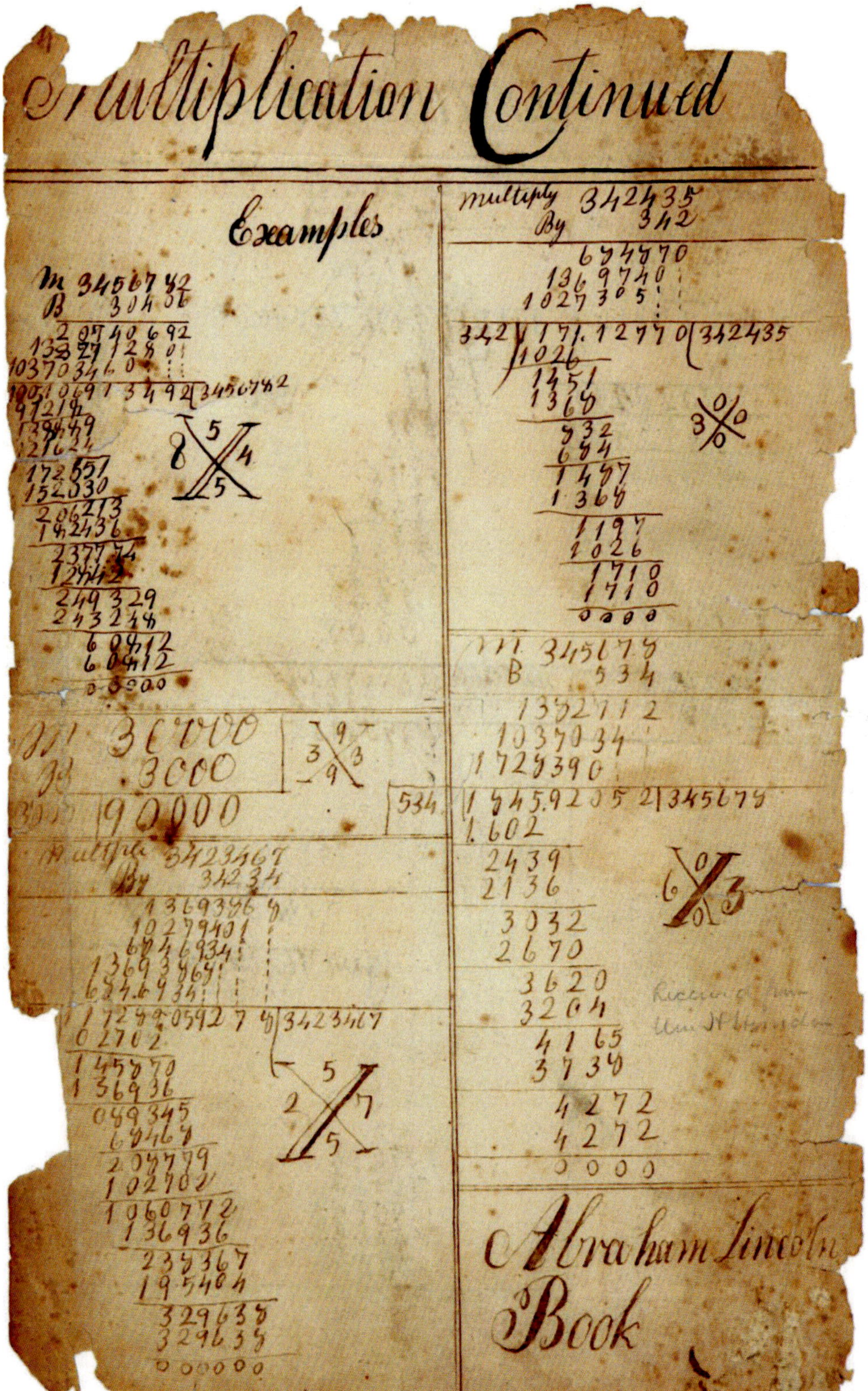

Figure 6.5. "Multiplication 2" (fourth of the 22 pages). Rare Book & Manuscript Library, Columbia University in the City of New York.

Dorfman (1966a) claimed that the page shown in Figure 6.5 was the second of the extant pages. Warren (1959) listed it as the sixth, and Basler et al. (1953) as the tenth. It was headed "Multiplication Continued." Most likely, the page was prepared by Abraham when he was attending Azel Dorsey's school. The calculations it shows were not found in any of the arithmetics published before 1824 that we examined. The leaf is currently held by Columbia University.

The headings on the page displayed mid-level calligraphy, of a standard very typical of U.S. cyphering books of the time. In the bottom right corner Abraham wrote "Abraham Lincoln's Book," also in reasonable quality calligraphy. Given the rough circumstances in which Abraham lived, both at home and at school, the attention he gave to making this cyphering book the best he could make it is noteworthy. This book was important to him, and as William Herndon would find out, this was the only book prepared by Abraham that his step-mother would keep for many years (Herndon & Weik, 1936). This pride in a cyphering book was not unusual—as Ellerton and Clements (2012) have pointed out, often the two most valued books that a family owned were the Bible and a cyphering book.

Around the middle of the left column in Figure 6.5 the handwritten numerals seem to show the task 30000 by 3000, and the answer obtained appears to have been an incorrect 90000. Immediately before the 30000 there is an upper-case *M,* and before the 3000 is an upper-case *B.* It was not made clear what the *M* and the *B* represented. The numerals for this task were probably written by someone other than Abraham—perhaps by Abraham's teacher at the time. It is interesting that the casting-out-nines check that was shown included two 9s. Normally, zeros (0s) would appear instead of the 9s. This is the only example in the cyphering book when the casting-out-nines check was incorrect, and it is likely that the error was not Abraham's.

In an age when all calculations, except for the most simple, had to be written down, these calculations pointed to a young man who had thoroughly learned his multiplication tables, and who had a solid grasp of numeration. Young children did not usually begin to prepare a cyphering book until they were 10 years of age, and the first thing they would enter in their books was a numeration table, which depicted numbers from units to trillions.

The two pages on multiplication (see Figure 6.4 and Figure 6.5) included only tasks which involved pure calculation. There are no word problems on either page. Textbooks such as those by Thomas Dilworth (1806) and Stephen Pike (1822) included, in their sections on multiplication, many word problems incorporating real-world scenarios. It is possible that Abraham solved multiplication word problems on pages of his cyphering book that have been lost. It is not clear where the tasks shown in Figures 6.4 and 6.5 came from, and whether Abraham's calculations were approved by a teacher.

Although cyphering books rarely showed any indication that the pages had been checked by teachers, in fact most of them had been checked. The recitation system employed in the schools meant that the teacher checked a student's accuracy for a calculation task before the working was entered into the cyphering book. Solutions would be written on scrap pieces of paper, on bark, on slates, or indeed on any surface that might be suitable. These solutions would be shown to teachers, who were supposed to check that they were correct, and that the students who had prepared them understood what they had written. That was the ideal, but the practice often fell short of that. Nevertheless, many of the arithmetics used in the United States in the nineteenth century have pictures on the front cover showing teachers in recitation with individual students, or with small groups of students. That was how teaching and learning took place up to about 1850. Before 1850, the modern practice of a teacher teaching all students in a class, simultaneously, was not to be found in most U.S. schools.

The Fifth of the Extant Pages in Lincoln's Cyphering Book

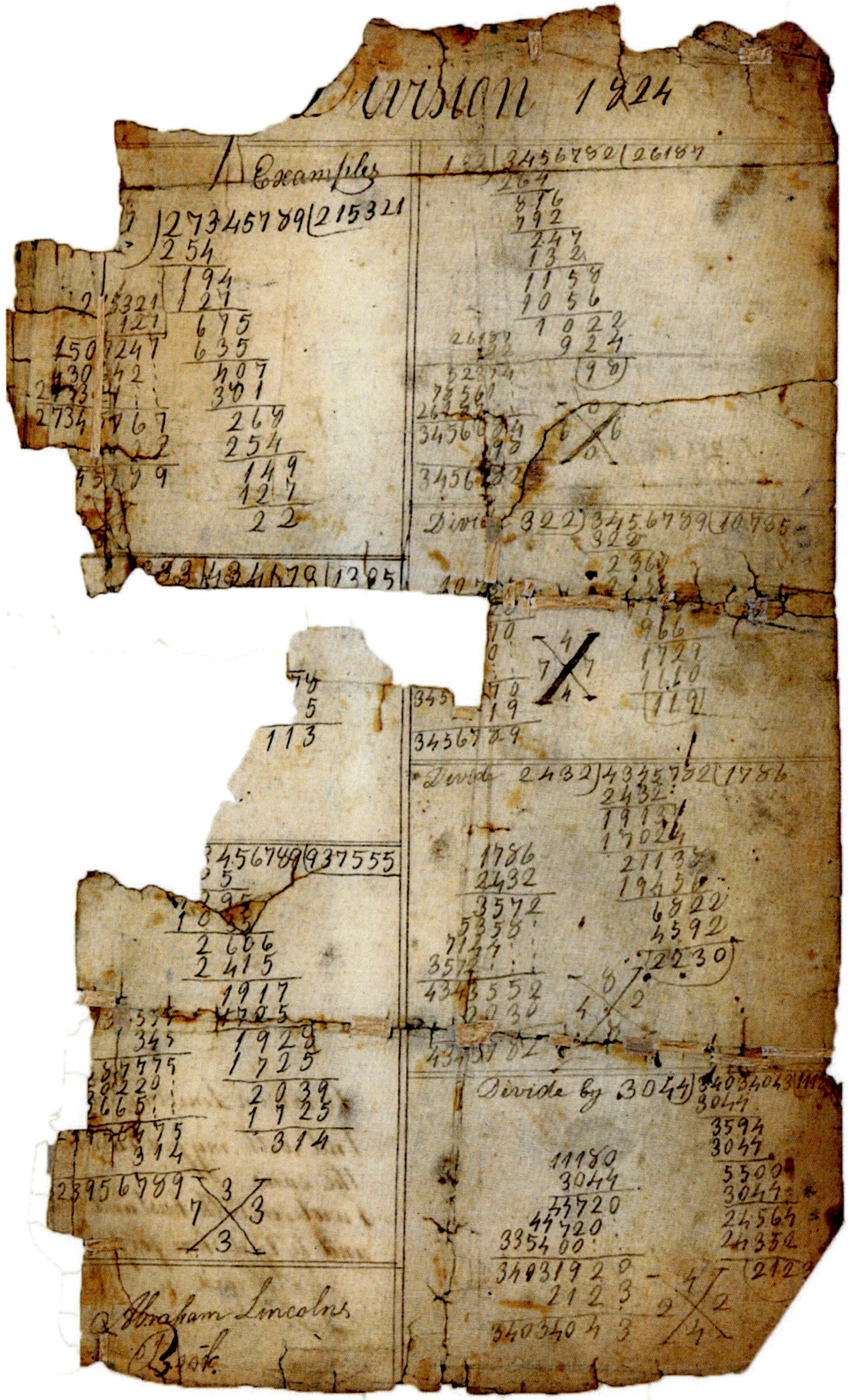

Figure 6.6. "Division 1" (fifth of the 22 pages).
Abraham Lincoln Presidential Library & Museum (ALPLM).

The page shown in Figure 6.7 was listed as the first of the extant pages by Basler et al. (1953), as eighth by Warren (1959), and as fourth by Dorfman (1966a). Abraham headed it "Long Division C ..." (with the "C" presumably standing for "Continued"). It was not dated, but since Abraham clearly dated the page that preceded it as 1824, this page was probably completed in 1824. The authors could not find any of the long-division tasks and the associated calculations in an arithmetic published before 1824. Despite Dorfman's (1966a) suggestion to the contrary, it is almost certain that these calculations were not based on material in Martin (1800). We agree with Dorfman and Warren, though, that this page would have followed the "Long Division 1824" page in its sequence, and cannot understand how Basler et al. (1953) might have thought otherwise. This leaf is currently held by the Abraham Lincoln Presidential Library and Museum, in Springfield, Illinois.

At the bottom right corner of this page is the piece of doggerel, "Abraham Lincoln is my name; And with my pen I write the same; I wrote in both haste and speed; and left it here for fools to read." The year 1826 was indicated immediately after this "poem," suggesting that two years after he completed the long-division calculations, Abraham returned to his cyphering book, looking for any blank space in which he might amuse himself by writing a poem. Whether this poem was actually created by Abraham is not known.

The two pages on division were concerned solely with practicing the long division algorithm—word problems which required division were not to be found. It is possible that such problems were on a page that has been lost. There is no hint that Abraham was introduced to the galley method of division that was used by Thomas Prust for division (see Chapter 3), and there is no statement of rules or cases. In fact, the division operation does have some mathematically interesting properties—thus, for example, whereas $(12 \times 4) \times 2$ is equal to $12 \times (4 \times 2)$, a similar property is not true for division. $(12 \div 4) \div 2$ does not equal $12 \div (4 \div 2)$. There is no evidence that such matters were ever considered by Abraham.

Furthermore, many elementary arithmetic topics normally covered in the standard *abbaco* cyphering curriculum were not dealt with in Abraham's cyphering book—among the topics that one might have expected Abraham to study were: numeration, addition, reduction, decimal fractions, vulgar fractions, and practice. Only 22 pages of Abraham's cyphering book are known to exist, but it is possible that the original cyphering book that Abraham left in the safe-keeping of his step-mother had substantially more pages—Lloyd A. Dunlap (1959) quoted Thomas L. D. Johnston as saying that "the latter years of the book is gone" (p. 6).

Although it could be argued that the original book had between 12 and 16 leaves (that is to say, between 24 and 32 pages), we believe that there may have been more. In the Ellerton-Clements collection of about 350 North American cyphering books, cyphering books that went beyond the rule of three almost always comprised more than 32 pages. Such manuscripts usually occupied at least 50 pages. We believe that some of the pages in Abraham's original cyphering book have been lost—probably for ever.

Lincoln did say, in a brief autobiographical note written to Jesse Fell in 1859, that he had cyphered only to the "rule of three" (Herndon & Weik, 1936), but in the traditional *abbaco* curriculum sequence the double rule of three, simple interest, compound interest, and discount were all normally studied after the rule of the three. Lincoln dealt with all of those topics—undoubtedly, he cyphered *beyond* the rule of three.

The Seventh of the Extant Pages in Lincoln's Cyphering Book

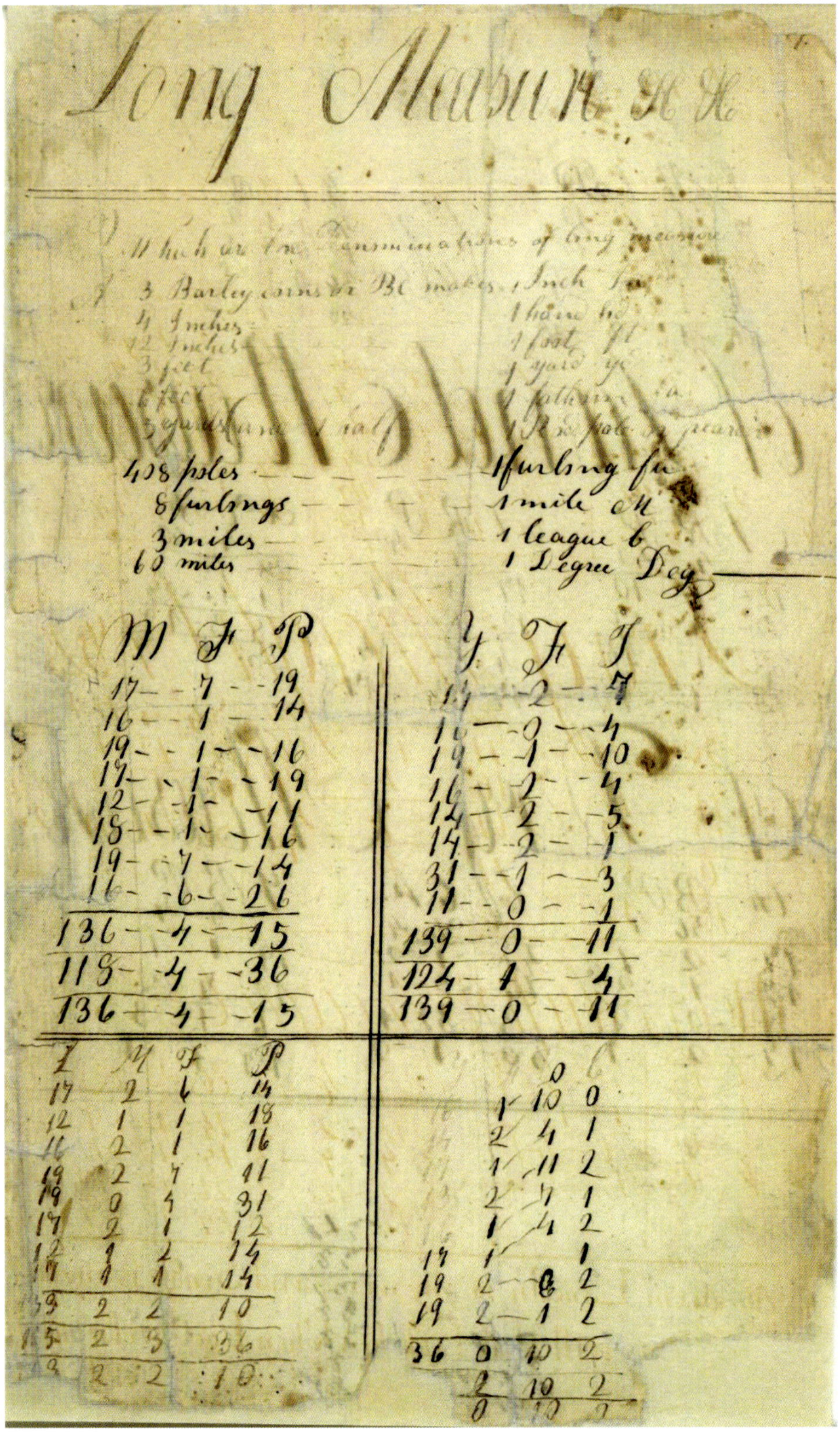

Figure 6.8. "Compound Addition" (seventh of the 22 pages).
Library of Congress, Herndon-Weik Collection of Lincolniana.

The page shown in Figure 6.8 is one side of the leaf currently held in the Library of Congress, in Washington, DC. It is concerned with the addition of length measures.

The term "compound operations" was used by authors of early arithmetics to refer to tasks which invited the use of addition, subtraction, multiplication or division of quantities. The page shown can be linked to a section on "compound addition" in Thomas Dilworth's (1792, 1806) New York editions of *The Schoolmasters Assistant; Being a Compendium of Arithmetic both Practical and Theoretical*. This section appeared in those editions before anything dealing directly with subtraction, multiplication, or division was introduced. One would reasonably have expected material like that shown in Figure 6.8 to appear very early in a cyphering book. The editors of *CWAL* (Basler et al., 1953) showed the "compound addition" page as the eighteenth page in their list, Warren listed it as third, and Dorfman (1966a) as seventh. Both Basler et al. and Dorfman included pages involving multiplication and division before the compound addition page.

There are two errors on Abraham's version of the page—first, he indicated that there were 408 poles in a furlong, when the correct correspondence was 40 poles in a furlong; and second, the sum for the "miles" column of the first problem should be 137 and not 136. Abraham's method of checking was, apparently, to sum the columns and, having obtained the answer to subtract the first entry from the answer. Then, the first entry would be added to this, and the answer should once again be obtained. This was not a safe method of checking, and it failed to reveal Abraham's error in his first set of calculations. This method of checking was not shown in Dilworth's (1806) *Schoolmasters Assistant*, and was probably the product of the thinking of Abraham or of his teacher.

It is unlikely that Abraham copied what he wrote on this page directly from Dilworth's text. We say that because there are small differences between what Abraham wrote and what appeared in the Dilworth text. Although the "408 poles makes 1 furlong" error could have been a copying error, it could also have been an error made as a result of dictation by a teacher to a student. In the "Yards-Feet-Inches" addition task, Abraham entered 31 yards-1 foot-3 inches, but in both the 1792 and 1806 editions of Dilworth, the entry read 11 yards-1 foot-3 inches. The answer in a commercially-published key to Dilworth's arithmetics (A Teacher of Mathematics, 1812) assumed that the entry was 11 yards-1 foot-3 inches, and so it is obvious that Abraham did not check his answers against the key. Although the difference between what Abraham wrote and what appeared in Dilworth's arithmetic could have been the result of an error arising from dictation, it could also have been the will of the teacher to make the problems slightly different from those set by Dilworth.

Dilworth's 1792 and 1806 New York editions included 18 pages and 19 pages, respectively, on compound addition. There was compound addition of money, of troy weight, of avoirdupois weight, of apothecaries' weight, of long measure, of cloth measure, of land measure, of liquid measure, of wine measure, of Winchester measure, of motion. Most elementary cyphering books included entries on most of these measures—for addition, subtraction, multiplication and division of the measures. Compound operations were regarded as an elementary part of arithmetic and therefore it was expected that they would be dealt with very early in a cyphering curriculum. It is likely that there were quite a few additional pages—possibly as many as 10—in Abraham's original cyphering book which dealt with aspects of compound operations not dealt with in the pages which remain. Seen from that perspective, it is likely that Abraham's original cyphering book was considerably longer than the 22 pages which remain.

We seriously contemplated placing the compound addition page as first in our list, but decided to place it seventh, behind the page which featured a lengthy, pure subtraction task, and behind pages focusing on simple multiplication and division. Some arithmetics included simple subtraction, multiplication and division tasks before compound arithmetic tasks, but others did not.

The Eighth of the Extant Pages in Lincoln's Cyphering Book

Figure 6.9. "Compound Subtraction 2" (eighth of the 22 pages). Library of Congress, Herndon-Weik Collection of Lincolniana.

The page shown in Figure 6.9—on "land measure" and "dry measure"—is held in the Library of Congress, Washington, DC. The same tasks can be found in the "compound subtraction" sections of Thomas Dilworth's (1792, 1806) New York editions of *The Schoolmasters Assistant; Being a Compendium of Arithmetic both Practical and Theoretical* (see pages 24 and 25 in the 1792 edition, and pages 35 and 36 in the 1806 edition). The editors of *CWAL* (Basler et al., 1953) showed this "compound subtraction" page as the seventeenth page in their list, and Warren (1959) and Dorfman (1966a) as the fourth page and eighth page, respectively, in their lists. Basler et al. included a page involving compound multiplication and a page involving compound division before the compound addition page, but it is unlikely that that would have been the case in any cyphering book.

The two questions at the top of Figure 6.9 were concerned with subtraction of long measure. For the first one, a student needed to know that there were 40 poles in a furlong, 8 furlongs in a mile and 3 miles in a league; and for the second, 3 barleycorns in an inch, 12 inches in a foot, and 3 feet in a yard. All of this was taxing on a student's memory. The student had to be careful when carrying out the subtractions, because at each step, different relationships between units had to be considered. Abraham made very few errors in his cyphering book, but he got the second of these long-measure subtractions wrong—his answer should have been 35 yards 2 feet 10 inches and 1 barleycorn. His check was obviously inadequate, for it failed to reveal that he had made a mistake.

With the land-measure subtractions tasks, the student needed to use the relationships 40 poles in length by 1 pole in width make 1 rood, and 4 roods make 1 acre. With the dry-measure subtraction tasks, there were 4 pecks in 1 bushel, and 36 bushels in 1 cauldron (of coal). All of Abraham's calculations were correct for these six dry-measure tasks.

At the bottom of the page, Abraham wrote, in a mature hand, "Abraham Lincoln, his hand and pen. he (sic.) will be good. But God knows when." He also performed some calculations with fractions. The poem and the fractions calculations were probably done at a later time, when Abraham was looking for writing space.

The tasks on this leaf were not pitched at a high level. It should not be assumed that because some of the tasks were in Thomas Dilworth's popular arithmetic it follows that Abraham either owned or had ready access to Dilworth's arithmetic. Lloyd A. Dunlap (1959), after a careful analysis, reached a different conclusion. He argued that although Abraham had grappled with tasks from Dilworth's *Arithmetic*, he did not have direct access to Dilworth's text:

> It seems probable that Lincoln never used any arithmetic text. His sum book was a reproduction of earlier manuscript sum books compiled by his teachers, years before, under the direction of their now nameless tutors. The source was not Pike, nor Dilworth, nor Jess, but Andrew Crawford, James Swaney, Azel Dorsey and the anonymous men who taught them. (p. 10)

Any of Lincoln's early teachers—Crawford, Swaney, Dorsey, and Bryant from Indiana, and his two early teachers from Kentucky, Zachariah Riney and Caleb Hazel—may have compiled their own cyphering books when they were students (Barrett, 1860; Miller, 2006; Nicolay, 1904).

The Ninth of the Extant Pages in Lincoln's Cyphering Book

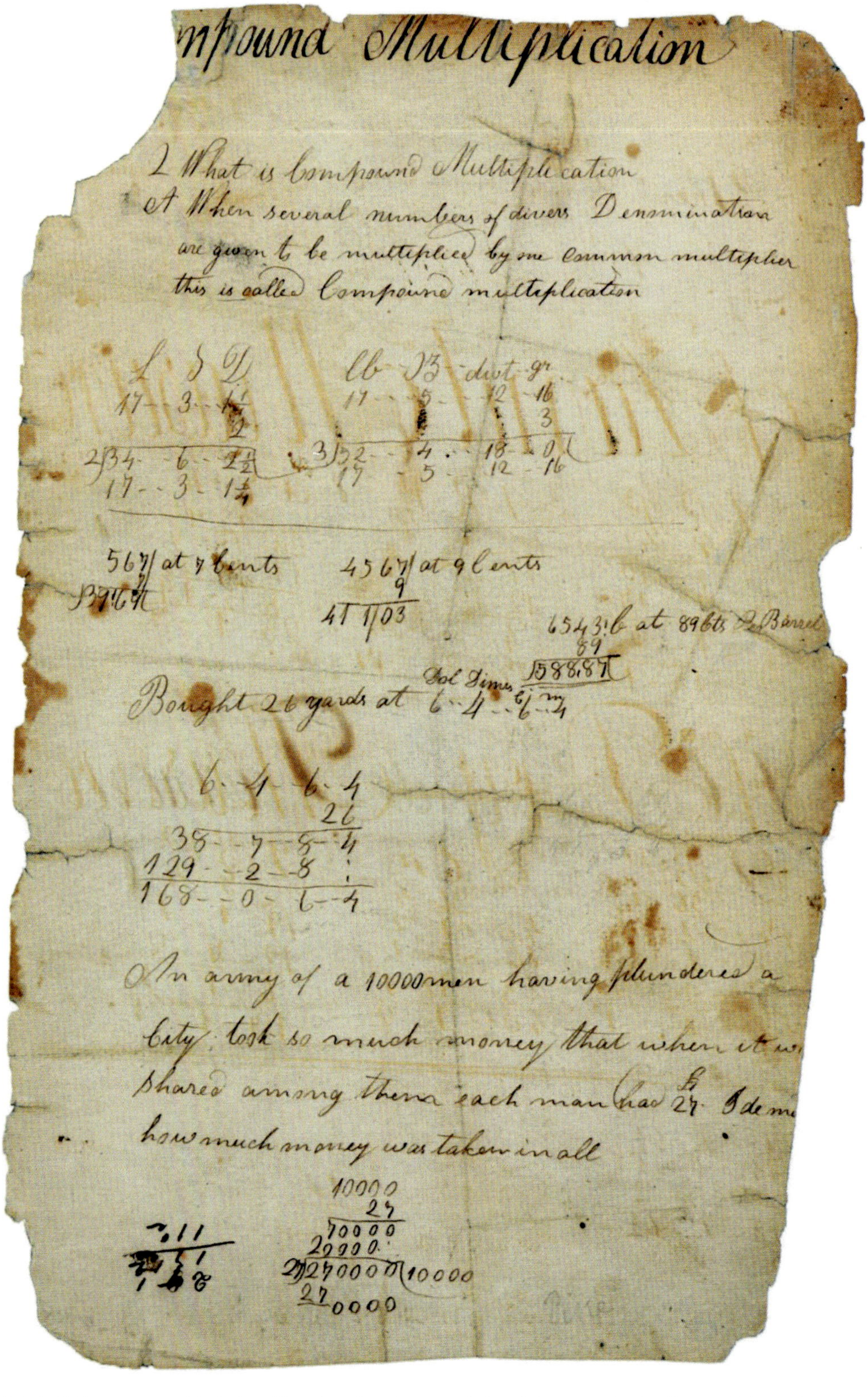

Figure 6.10. "Compound Multiplication" (ninth of the 22 pages). Indiana Historical Society.

The page shown in Figure 6.10 was placed third by Basler et al. (1953), but ninth by both Warren (1959) and Dorfman (1966a). The reverse-imaged words "Of Land Measure" and "Of Dry Measure," which can be seen across the page, confirm that this page was originally opposite the page previously shown in Figure 6.9. The page shown in Figure 6.10 was not dated but was probably completed by Abraham in 1824. It is held by the Indiana Historical Society.

This page, and the one which follows it, can be linked with the North American editions of Thomas Dilworth's *The Schoolmasters Assistant; Being a Compendium of Arithmetic Both Practical and Theoretical*. The exact placements of the relevant passages in that book depend on the year of publication. For this particular "compound multiplication" page, the description of compound multiplication given at the top of the page corresponds word for word to the description in the New York edition of Dilworth published by Hugh Gaine in 1792 (p. 30), and in the 1806 New York edition published by G. Jansen (pp. 42–43). Most of the examples—including the one concerning an army of 10000 "plundering men"—also appeared, word for word, in Dilworth's *Arithmetic*.

In the question asking for the cost of 6543 lbs at 89 cents per lb, Abraham did not complete the calculations—he apparently forgot to multiply by the 9. Otherwise, his calculations were correct. In his solution to the "plundering men" problem, he showed his calculations but did not give his answer. The genre in which he showed his working was, once again, PCA ("Problem-Calculation-Answer"). This PCA style of responding to a written task can also be seen with the working set out in the middle of the page, where Abraham multiplied 6 dollars 4 dimes 6 cents and 4 mills by 26. This could have been answered by multiplying the decimal 6.464 by 26.

English currency notation, involving pounds, shillings, pence and farthings, was still widely used in the United States (different states had different exchange rates for their versions of pounds, shillings, and pence). Notice that for the task from Dilworth (1806, p. 43), showing 17 pounds 3 shillings and 1¼ pence multiplied by 2, on this page, Abraham "checked" his answer by dividing by 2. Although he was probably 14 or 15 years old, and lived in a nation which had used decimal currency since 1792, he had not yet formally studied decimal fractions, and consequently set out his working in the "compound multiplication" manner shown. However, on this page he also shows calculations for "567 at 9 cents," "4567 at 9 cents," and "6543 lb at 89 cts." These seem to have done by a more direct approach that was similar to, but not exactly the same as, decimal multiplication.

The compound operations (addition, subtraction, multiplication and division) tasks were of high relevance in a nation in which many different systems of currency and weights and measures were constantly being used. Patricia Cline Cohen (2003) maintained that the cyphering approach "involved transcribing formulaic rules out of printed textbooks word for word into the pages of individual copybooks, together with the exact examples offered by the text to illustrate each rule" (p. 44). Ellerton and Clements (2012), however, disagreed with Cohen's assessment of the way the cyphering tradition worked itself out in North America during the colonial and early Federal periods (up to about 1850). On this page, for example, although a young man (Abraham Lincoln) responded to problems that had appeared in a commercial text (authored by Dilworth), it is obvious that he did not merely copy solutions out of the text. Certainly, the definition of compound multiplication given at the top of this page was precisely that shown in Dilworth (1806, pp. 44–45), but none of the examples were worked in Dilworth's text, and no answers for these examples were provided by Dilworth. It is true that, a key to Dilworth's *Arithmetic* had been published (A Teacher of Arithmetic, 1812), but Abraham would not have had access to this. On the other hand, although it is clear that Abraham did not copy from Dilworth's text (Dunlap, 1959), it is likely that he often wrote words that were dictated to him by a teacher.

The Tenth of the Extant Pages in Lincoln's Cyphering Book

To Exercise Multiplication

There were 40 men concerned in payment a sum of money and each man paid 1271£ how much was paid in all—

1271
40
40)50840(
1271

If 1 foot contain 12 inches I demand how many there are in 126 feet

126
12
252
126
12)1512(
126

Of Compound Division

Q What is compound Division

A When several numbers of Divers Denomination are given to be divided by 1 common divisor this called Compound Division—

Abraham Lincoln His Book

Figure 6.11. "Compound Division" (tenth of the 22 pages). Indiana Historical Society.

The page shown in Figure 6.11 was listed fourth by Basler et al. (1953), but tenth by both Warren (1959) and Dorfman (1966a). It was not dated but was probably prepared by Abraham in 1824. It is currently held by the Indiana Historical Society.

Again, the PCA ("Problem-Calculation-Answer") genre is evident. As was the case on the preceding compound multiplication page, there is no direct application of decimal fractions, despite the fact that in 1824 the United States had been using decimal currency for more than 30 years.

The page, and the one which precedes it, can be linked with Dilworth's *Arithmetic* although the exact placements of the relevant passages depend on the year of publication. For this particular "compound division" page, the description of compound division given toward the bottom of the page corresponds, word for word, to the description in the 1806 New York edition of Dilworth published by G. Jansen (see p. 51). The page is reproduced in Figure 6.12 below. The two exercises at the top of Abraham's page—concerning 40 men each paying 1271 pounds, and the number of inches in 126 feet—also appeared, word for word, in this same edition of Dilworth (1806, p. 44, Questions 5 and 6).

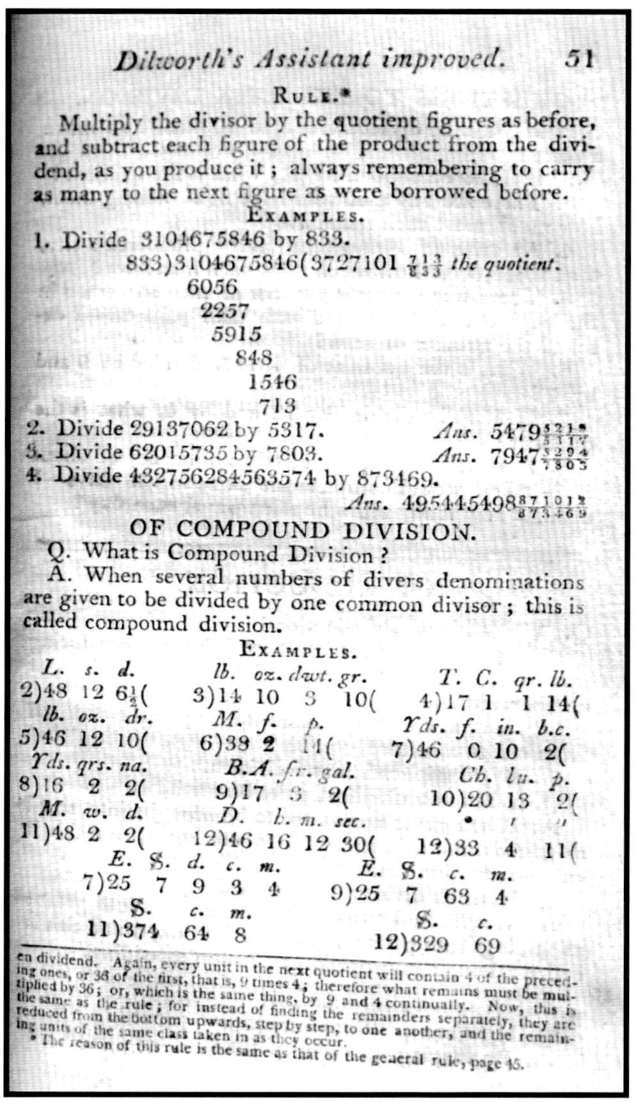
Dilworth's Assistant improved. 51

RULE.*

Multiply the divisor by the quotient figures as before, and subtract each figure of the product from the dividend, as you produce it ; always remembering to carry as many to the next figure as were borrowed before.

EXAMPLES.

1. Divide 3104675846 by 833.

833)3104675846(3727101 $\frac{713}{833}$ *the quotient.*
6056
2257
5915
848
1546
713

2. Divide 29137062 by 5317. *Ans.* 5479$\frac{5219}{5317}$
3. Divide 62015735 by 7803. *Ans.* 7947$\frac{5294}{7803}$
4. Divide 432756284563574 by 873469. *Ans.* 495445498$\frac{871012}{873469}$

OF COMPOUND DIVISION.

Q. What is Compound Division ?

A. When several numbers of divers denominations are given to be divided by one common divisor ; this is called compound division.

EXAMPLES.

L. s. d. 2)48 12 6½(*lb. oz. dwt. gr.* 3)14 10 3 10(*T. C. qr. lb.* 4)17 1 1 14(
lb. oz. dr. 5)46 12 10(*M. f. p.* 6)38 2 14(*Yds. f. in. b.c.* 7)46 0 10 2(
Yds. qrs. na. 8)16 2 2(*B.A. fr. gal.* 9)17 3 2(*Ch. lu. p.* 10)20 13 2(
M. w. d. 11)48 2 2(*D. h. m. sec.* 12)46 16 12 30(° ' " 12)33 4 11(
E. $. d. c. m. 7)25 7 9 3 4 *E. $. c. m.* 9)25 7 63 4
$. c. m. 11)374 64 8 *$. c.* 12)329 69

en dividend. Again, every unit in the next quotient will contain 4 of the preceding ones, or 36 of the first, that is, 9 times 4; therefore what remains must be multiplied by 36; or, which is the same thing, by 9 and 4 continually. Now, this is the same as the rule; for instead of finding the remainders separately, they are reduced from the bottom upwards, step by step, to one another, and the remaining units of the same class taken in as they occur.

* The reason of this rule is the same as that of the general rule, page 45.

Figure 6.12. Page 51 from Dilworth (1806).

The Eleventh of the Extant Pages in Lincoln's Cyphering Book

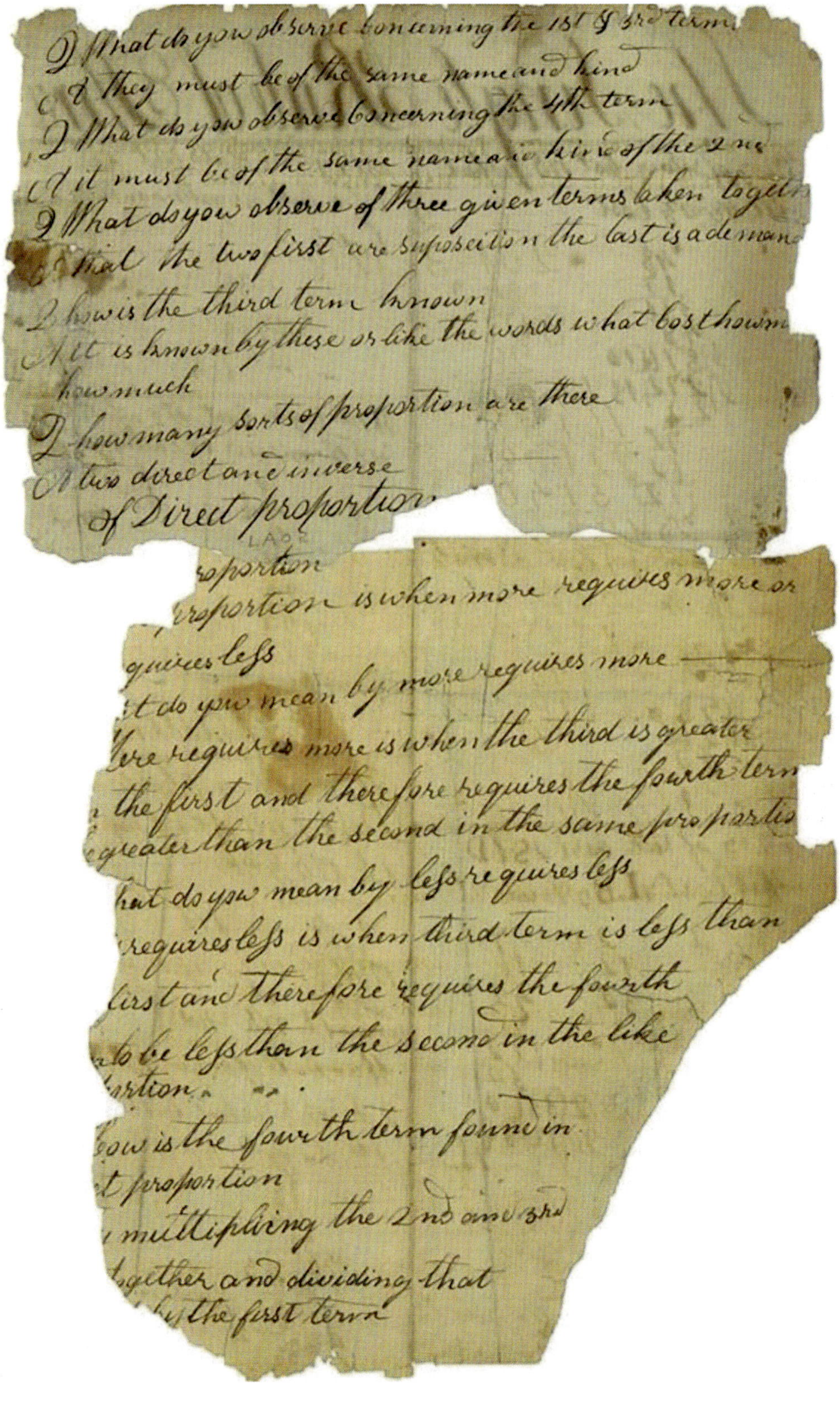
Q What do you observe concerning the 1st & 3rd term
A They must be of the same name and kind
Q What do you observe concerning the 4th term
A it must be of the same name and kind of the 2nd
Q What do you observe of three given terms taken togeth
A that the two first are supposition the last is a deman
Q how is the third term known
A it is known by these or like the words what cost how m
how much
Q how many sorts of proportion are there
A two direct and inverse
of Direct proportion

roportion
roportion is when more requires more or
quires less
t do you mean by more requires more
re requires more is when the third is greater
the first and therefore requires the fourth term
greater than the second in the same proportio
hat do you mean by less requires less
requires less is when third term is less than
first and therefore requires the fourth
to be less than the second in the like
rtion.
ow is the fourth term found in
t proportion
multiplying the 2nd and 3rd
gether and dividing that
y the first term

Figure 6.13. "Rule of Three 1" (eleventh of the 22 pages).
Upper section of the leaf: McLellan Lincoln Collection, John Hay Library, Brown University.
Lower section of the leaf: Special Collections Research Center, University of Chicago Library.

The upper fragment shown in Figure 6.13 is currently held by Brown University. It was listed eighth by Basler et al (1953), nineteenth by Warren (1959), and eleventh by Dorfman (1966a). None of those authors knew of the lower fragment—now held at the University of Chicago. It was only in 2010 that Brown University, the University of Chicago, and the Abraham Lincoln Presidential Library jointly announced that the upper and lower fragments were originally part of the same leaf (Reynolds, 2010). Abraham probably prepared the writing on both fragments in 1824 or 1825.

On the fragments shown, the celebrated "rule of three" is described and notes on direct proportion are given. Although the wording was taken from Dilworth (1806, pp. 65–66), there are differences, between Dilworth's and Abraham's texts. For example, in the top line Abraham used the abbreviations "1st & 3rd" but Dilworth wrote "first and third." Similarly, in the next line, Abraham wrote "4th" and Dilworth, "fourth"; and, in the next line Abraham used "2nd" and Dilworth "second." Then, Abraham had difficulty in spelling "supposition." Abraham replaced Dilworth's explanation for the third term ("It is known by these, or the like words, what cost? how many? how much?") with "It is known by these or like the words what cost how many how much"). Almost certainly, the passage was dictated to Abraham, and he was unable to cope with the linguistic demands of at least one sentence. It is unlikely that Abraham copied this directly from Dilworth's *Arithmetic* (or from a single page removed from Dilworth's text).

Later in life, Abraham recalled that while at school he cyphered to the rule of three (Herndon & Weik, 1936). In fact, he cyphered beyond the rule of three, but in the late 1850s the politically astute Presidential contender knew that having "cyphered to the rule of three" was a claim that carried weight at all levels of society. The rule of three was often referred to as the "golden rule" and, throughout the period 1200–1850, in Europe and in America, it provided the standard approach to the crucially important study of proportionality.

When, in the late 1850s, Abraham Lincoln became a Presidential contender, his support team set out to project a "rail-splitter" image for Lincoln—the public should come to believe that he was an intelligent person who knew what it was to struggle against the odds, and win (Plummer, 2001). The image had to be that he had grown up in extremely poor circumstances, only barely supported in his desire to be educated by a father who himself was uneducated. But, the image also had to be that he was extremely intelligent and hard-working, and because of his sheer intelligence and his tenacity had been able to become a successful lawyer and a very able politician. Given the desired image, a claim that he cyphered "to the rule of three" told the public that he was intelligent, but any suggestion that he cyphered "beyond the rule of three" had the danger of suggesting that he had had more educational opportunities as a youth than was consistent with the desired rail-splitting image.

The previously unreported leaf located in the Houghton Library at Harvard University—which will be discussed shortly—was concerned with the double rule of three. The contents logically and arithmetically followed upon the single rule of three, and can be dated as 1825. These present fragments, giving Dilworth's statement of the single rule of three, would have been completed before the Harvard leaf. The Harvard leaf was based on pages from Stephen Pike's (1822) *Arithmetic*, and not on text in Dilworth's *Arithmetic*.

With rules on one side and their applications on the other side of the fragments, Abraham had completed the first, fundamental phase of his mathematics education. He had cyphered to the rule of three, and that constituted *rite de passage* to higher forms of learning. Young Abraham was ambitious and he would return to study more mathematics. Indeed, in later years, he would seriously take up the study and application of surveying, and would immerse himself in the logical beauty of *Euclid's Elements*. We believe that during his school days, young Abraham gained an inkling of the power and beauty of mathematics (Dorfman, 1966b; Hirsch & Van Haften, 2010).

The Twelfth of the Extant Pages in Lincoln's Cyphering Book

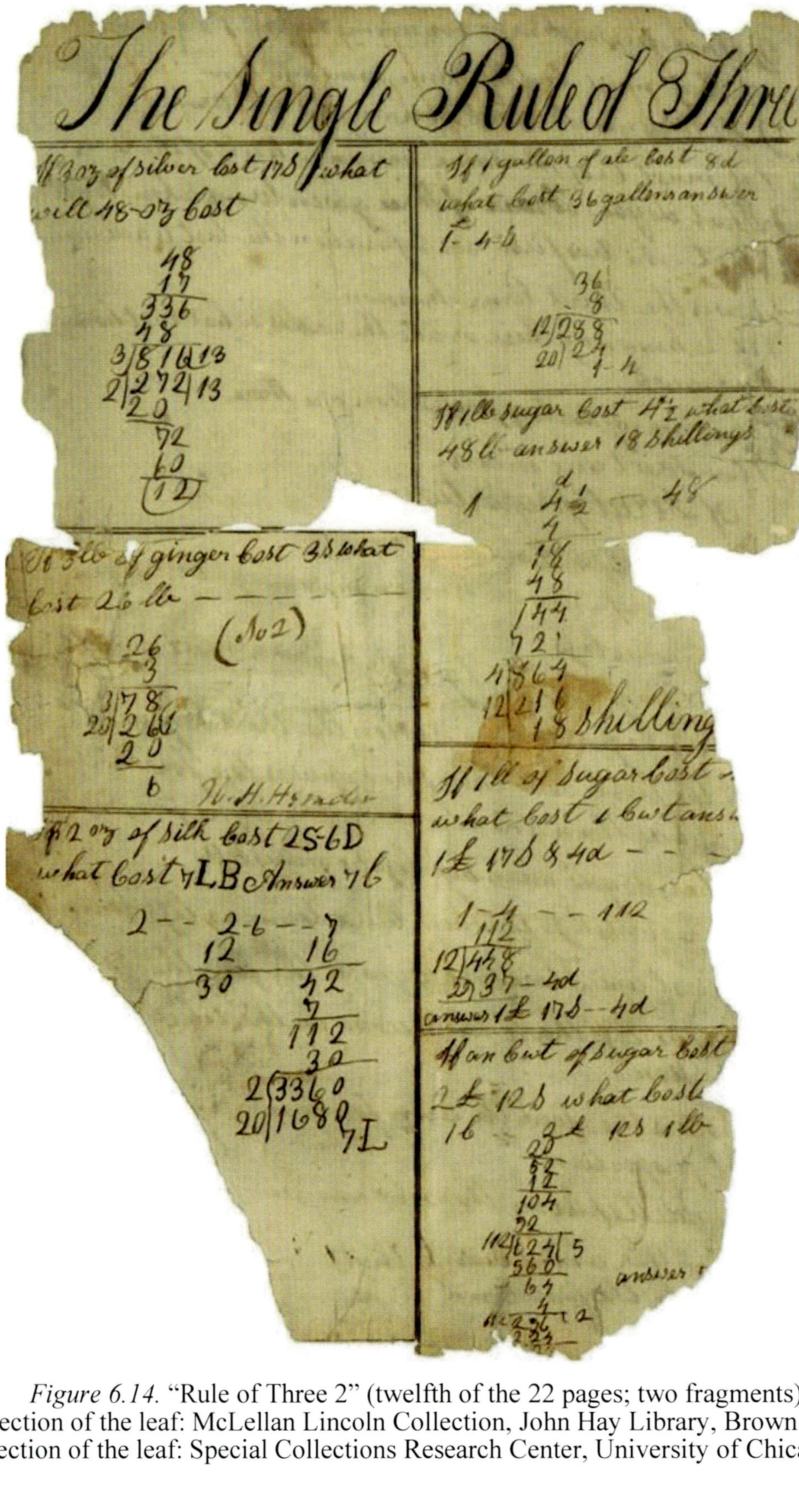
The Single Rule of Three

If 3 oz of silver cost 17s what will 48 oz cost

If 1 gallon of ale cost 8d what cost 36 gallons answer £1-4-0

If 1 lb sugar cost $4\frac{1}{2}$ what cost 48 lb answer 18 shillings

If 3 lb of ginger cost 3s what cost 26 lb

(No 2)

W. H. Herndon

If 2 oz of silk cost 2s 6d what cost 7 LB Answer 7 16

If 1 lb of sugar cost what cost 1 cwt answer £1 17s & 4d

answer £1 17s 4d

If an cwt of sugar cost 2£ 12s what cost 16

Figure 6.14. "Rule of Three 2" (twelfth of the 22 pages; two fragments).
Upper section of the leaf: McLellan Lincoln Collection, John Hay Library, Brown University.
Lower section of the leaf: Special Collections Research Center, University of Chicago Library.

The upper fragment, held by Brown University, was listed seventh by Basler et al (1953), but twentieth by Warren (1959), and twelfth by Dorfman (1966a). Although not dated, it was probably prepared in 1824 or in 1825. The lower fragment is now held by the University of Chicago.

These fragments show seven examples of how the single rule of three was used. All seven are on page 66 of Dilworth's (1806) *Arithmetic*. A solution to the first problem ("*If 3 Oz. of silver cost 17s what will 48 Oz. cost*") was set out as a model example by Dilworth, who wrote (on p. 66):

	Oz.		*s.*		*oz.*		
As	3	:	17	::	48		
					17		
						2/0	*£.* *s.*
					3] 816 [	27/2 [	13 12

With no further explanation, Dilworth stated the answer was *13£ 12s.*

This solution assumed an understanding of how Dilworth used symbols. But Abraham preferred to do things his own way. He followed the golden rule and "multiplied the third by the second" to get (48)(17) = 816 (he used a long multiplication procedure). Then he "divided by the first ingeniously" to get $816 \div 3 = 272$. He realized that this answer was in shillings, so he divided by 20 to get 13 pounds and 12 shillings—although he did not actually state his answer.

The procedure with the rule of three was regarded as very important, and something not to be questioned. It probably did not occur to Abraham that all he needed to do was to multiply 17 shillings by 16, and then convert to pounds and shillings.

With the "gallon of ale" question, Abraham attempted to follow the strictures of the golden rule. He "multiplied the third by the second" to get 288 pence, and then divided by 1 (the "first"). He converted the 288 pence to 1 pound and 4 shillings by dividing by 12, then 20.

With the first "sugar" question, Abraham did not indicate that the 4½ referred to pence, and he did not proceed immediately to multiply 4½ by 48—probably because he did not know how to multiply 4½ by 48. He converted the 4½ pence to 18 farthings, and multiplied by 48, to get 864 farthings. Then, after dividing by 4 and by 12, he obtained 18 shillings as his answer. PCA genre was evident in his response to this problem, as it was in his responses to all of the problems on this page. The second "sugar" question asked: "*If 1 lb of sugar cost 4d what cost 1 cwt.*" Abraham's working showed that he needed to think of 1 cwt as 112 lb. The 4 by 112 gave 448 which would be in pence. Dividing by 12 gave 37 shillings and 4 pence, which he wrote as "*1£ 17S - - 4d.*"

Abraham solved the "ginger" problem ("*If 3 lb of ginger cost 3S what cost 26 pounds*") with a straightforward application of the rule of three—but, of course, the answer could easily have been obtained without recourse to the rule of three. Interestingly, for some unknown reason, William Herndon, felt the need to attach his signature to Abraham's solution to this problem. The "silk" problem ("*If 2 oz of silk cost 2S 6D, what cost 7lb*") was slightly more challenging. Abraham converted the *2S 6D* to 30 pence, and *7lb* to 112 oz. Then 30 and 112 were multiplied to get 3360; this was divided 2 ("the first") and then Abraham asserted that the resulting 1680 pence was "7*£*."

For the third sugar problem ("*If an cwt of sugar cost 2£ 12S, what cost 1 lb?*") Abraham converted *2£ 12S* to *52S* and then *624D*. He also converted *1 cwt* to *112 lb* and, applying the rule of three, divided 624 by 112, to get 5 pence and 64 remaining. He converted the 64 pence into 256 farthings by multiplying by 4, and then he divided the 256 by 112, and got 2 and 32 remainder. The fragment does not allow the reader to see what Abraham did after that.

It is clear that all of this was Abraham's work. He was not copying from a textbook, or from someone else's work. Some of his thinking seemed to have been reasonably complex.

The Thirteenth of the Extant Pages in Lincoln's Cyphering Book

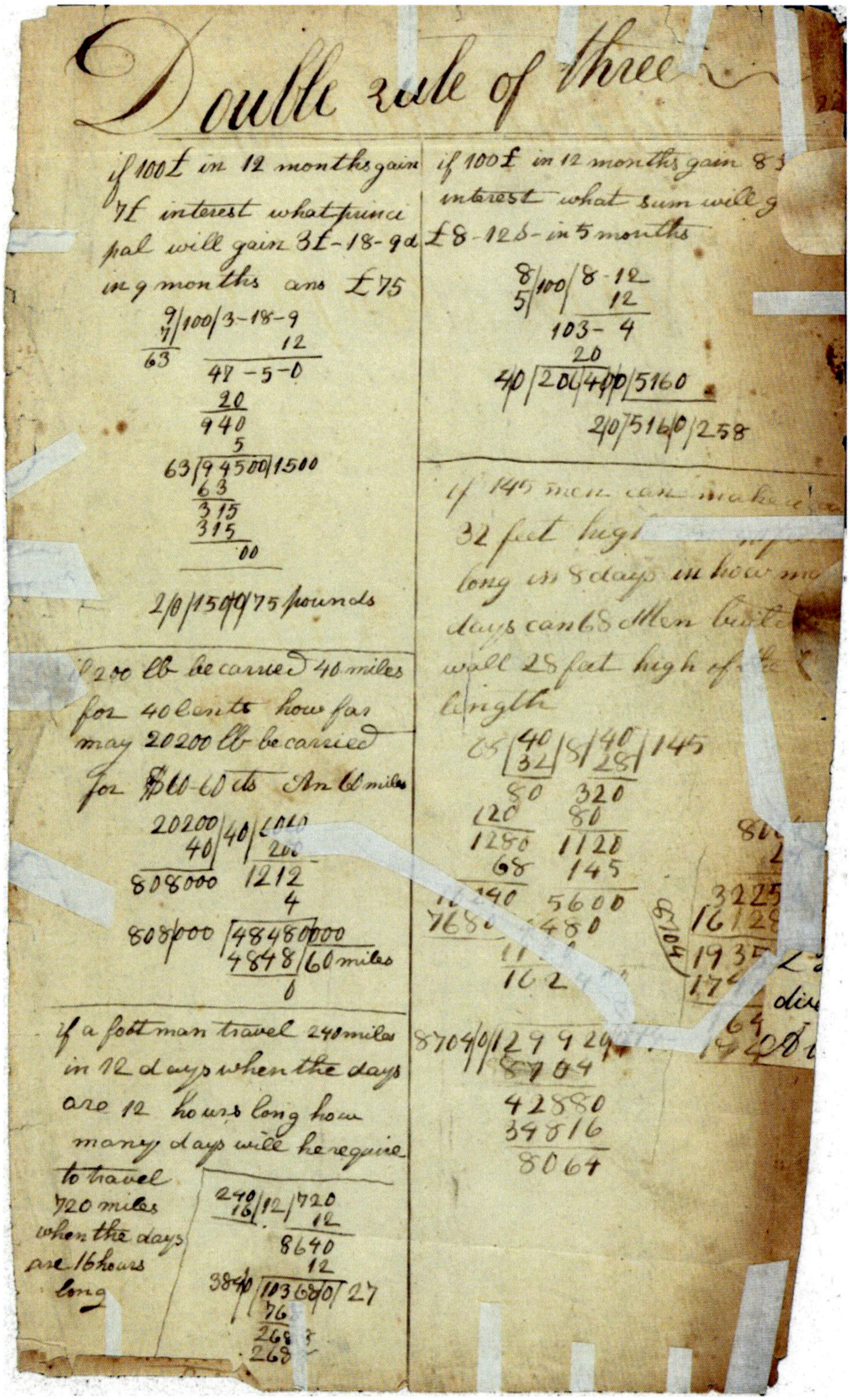
Double rule of three

if 100£ in 12 months gain 7£ interest what princi pal will gain 3£-18-9d in 9 months ans £75

if 100£ in 12 months gain 8 interest what sum will g £8-12s- in 5 months

if 200 lb be carried 40 miles for 40 cents how far may 20200 lb be carried for $60-60 cts An 60 miles

if 145 men can make 32 feet high long in 8 days in how m days can 68 men build wall 28 feet high of the length

if a foot man travel 240 miles in 12 days when the days are 12 hours long how many days will he require to travel 720 miles when the days are 16 hours long

Figure 6.15. "Double Rule of Three 1" (thirteenth of the 22 pages).
Page from Abraham Lincoln's mathematical exercise book, c. 1825. MS Am 1326.
Gift of Christian A. Zabriskie, 1954. Houghton Library, Harvard University.

The page shown in Figure 6.15 is held in the Houghton Library at Harvard University. Basler et al (1953), Warren (1959), and Dorfman (1966a) were unaware of its existence. Houghton Library holds a letter, dated December 7, 1875, from William H. Herndon to someone whose name is hard to decipher, indicating that he was giving the page to the recipient of the letter, and stating that the page was definitely prepared by Abraham Lincoln in 1825. Although the noted Herndon expert, Douglas Wilson, agreed that our initial conjecture that Herndon's letter was addressed to Willard Cutting Flagg (1828–1878) was tenable, subsequent investigations by Wilson have not found evidence to support that conjecture (D. L. Wilson, personal communication, September 24, 2013). Further research is needed to identify the recipient of Herndon's letter.

In pages immediately before that shown in Figure 6.15, Abraham answered problems that could be found in a text by Thomas Dilworth, but for this page, and for all the remaining pages, none of the questions were from Dilworth's *Arithmetic*. The main source of his exercises now became an arithmetic authored by Stephen Pike—we do not know if it was the 1822 edition of Pike's *Arithmetic*, but we will assume it was. It was probably the 1824–1825 winter term, and Abraham was now attending Azel W. Dorsey's school, about two miles from his home. Like Andrew Crawford, Dorsey conducted school as a "blab" school, in the Irish tradition.

Four of the five questions on the page shown are identical to, or very similar to, questions on page 84 or page 85 of Stephen Pike's (1822) *Arithmetic*. Thus, for instance, the first problem that Abraham addressed stated, "*If 100£ in 12 months gain 7£ interest what principal will gain 3£-18S-9d in 9 months*," and Question 5 in Pike (1822) stated, "If 100*L* in 12 months gain 6*L* interest, what sum will gain 3*L*. 7s. 6d. in 9 months?" (p. 84). Some of the numbers differed slightly, but the structure was the same.

Some of the double-rule-of-three problems involved inverse rather than direct proportional relationships. Pike (1822) stated the rule for inverse proportion problems as: "Transpose the inverse extremes; that is, set that which is in the first place under the third; and that which is in the third place under the first; then work as in Direct Proportion" (p. 83). Abraham worked out what that meant and was able to apply the rule in his own way. His answers were correct. And although his solutions were presented using PCA genre, anyone who understood Pike's method of solving such problems would easily have recognized what Abraham was doing.

The fact that so many of the questions came from Stephen Pike's *Arithmetic* could explain why Augustus H. Chapman told William Herndon that Abraham obtained a copy of "Pike's Arithmetic" when working on the double rule of three (Dorfman, 1966a; Dunlap, 1959). However, as we argued earlier, the possibility that Abraham had had, and continued to have, access to one of Nicolas Pike's arithmetics should not be ruled out.

Abraham was beginning to think more deeply about mathematical relationships, and about how useful mathematics could be. He was 16 years old, and was living within a family of seven in a 20 ft by 18 ft cabin. Tradition has it that Thomas Lincoln, Abraham's father, was not altogether happy that his strong son, now coming toward manhood, did not give himself full-time to the task of helping the family to survive—but Abraham's step-mother encouraged him to go on with his education (Herndon & Weik, 1936). Further research is needed to determine whether that tradition is unfair to Thomas Lincoln.

In 1825 it made little sense for Abraham to attend school other than during winter months. During non-winter months his labor was needed to generate food and income that would help support a family in need. But, during the winter he could not work outdoors, and therefore going to school could be justified. Even so, now that he had succeeded in cyphering "to the rule of three," the time had come for him to think seriously about what the future might hold for him.

The Fourteenth of the Extant Pages in Lincoln's Cyphering Book

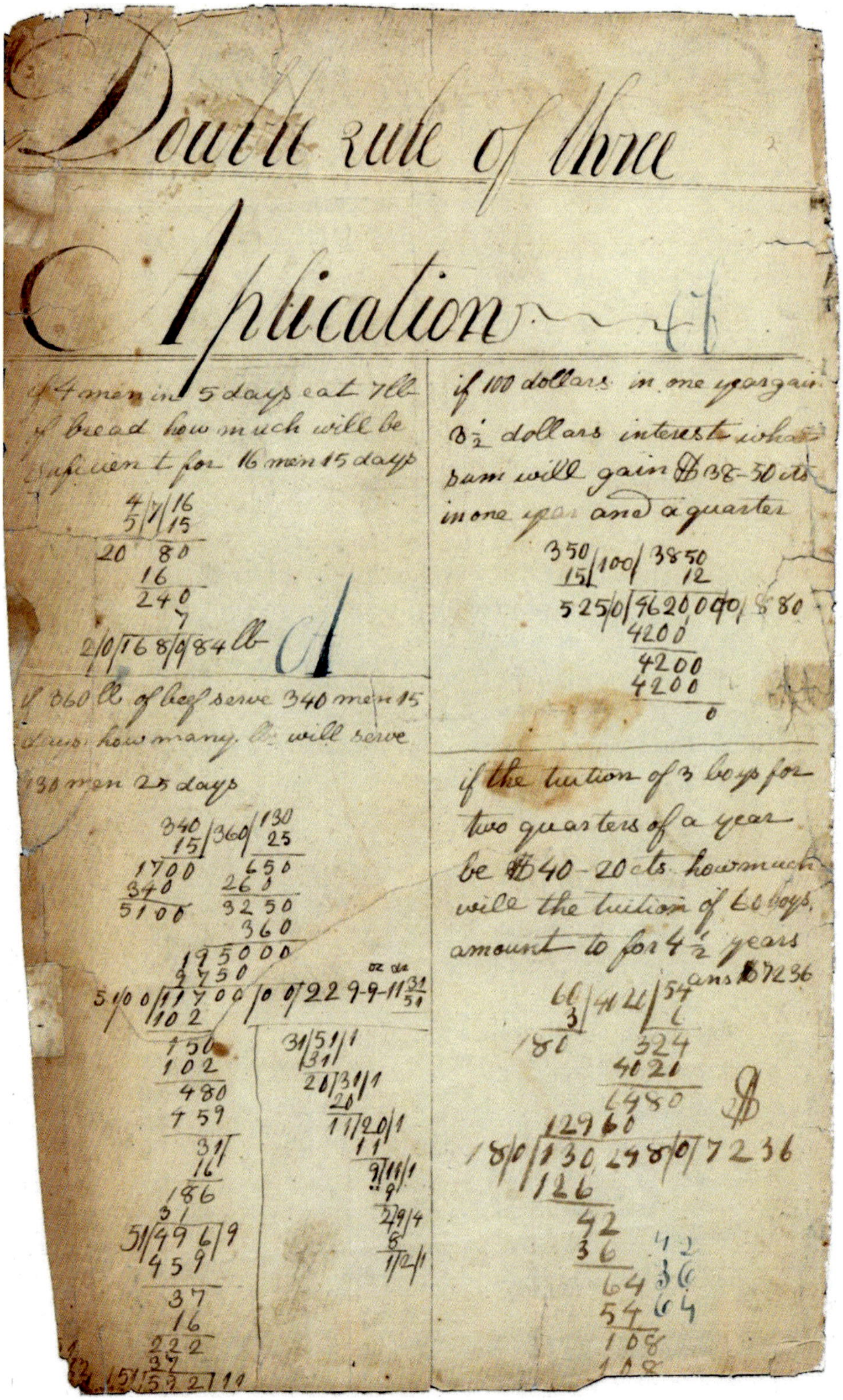

Figure 6.16. "Double Rule of Three 2" (fourteenth of the 22 pages).
Page from Abraham Lincoln's mathematical exercise book, c. 1825. MS Am 1326.
Gift of Christian A. Zabriskie, 1954. Houghton Library, Harvard University.

The page shown in Figure 6.16 is on one side of a leaf held in the Houghton Library at Harvard University. Like all previous writers who have commented on Abraham's studies in mathematics at school, Basler et al (1953), Warren (1959), and Dorfman (1966a) were unaware of the existence of this leaf.

Abraham was usually a good speller, but he misspelled "Application" in the calligraphic heading at the top of this page. The example at the bottom right of the page ("*If the tuition of 3 boys for two quarters of a year be $40-20 cts how much will the tuition of 60 boys amount to for 4½ years*") can be found in an arithmetic by Jacob Willetts (1822, p. 92). Although it is the only question from that source that we have discovered in the 22 extant pages, it is true, nonetheless, that during the final 10 extant pages, problems from a number of different textbooks can be found. The most likely explanation is that at least one of the two teachers whom Abraham had during that period owned a cyphering book which included the problems, and that either Abraham copied problems from that cyphering book, or they were dictated to him. The "tuition of 3 boys" problem from Willetts (1822) is also interesting because it is suggestive of the vast range of real-life situations that could be modeled by rule-of-three and double-rule-of-three arithmetic.

The two examples at the top of this page ("*If 4 men in 5 days eat 7 lb of bread, how much will be sufficient for 16 men in 15 days*" and "*If 100 dollars in one year gain 3½ dollars interest, what sum will gain $38.50 cents in one year and a quarter*") were from Stephen Pike's (1822) *Arithmetic* (p. 85). These examples were in a section which Pike headed "Promiscuous Examples." The problems in such sets belonged to major classes of problems which had been dealt with earlier in the book. A student's task was to decide, for each problem, which major class, which sub-class, etc., was relevant, and then to solve the problem by the "best" method. In other words, the student was invited to reflect on which method was most appropriate to solve a problem. Some writers used the term "promiscuous" problems (e.g., in Bennett, 1815; Butler, 1819), but others called them "miscellaneous problems" (e.g., Sterry & Sterry, 1795, pp. 115–117). Such sets of problems were to be found in many arithmetics used in the eighteenth and nineteenth centuries. They were also in many students' cyphering books (Ellerton & Clements, 2012). It was assumed that developing and writing down correct solutions to problems compelled students to "reflect" on problem structures, and meant that they were more likely to recognize and be able to solve structurally similar problems in the future (Van Egmond, 1980).

Most of the double-rule-of-three problems on the leaf held at Harvard University did not involve *direct* proportions. Rather, they involved *inverse* proportions that required students to think carefully and relationally about the mathematics. Statements of rules for tackling rule-of-three problems were to be found in many cyphering books and in almost every printed arithmetic published up to the 1850s (see, e.g., Dilworth, 1806; Pike, 1788). Helping students to learn to apply the rule of three, and associated rules, was difficult because the same relationships embodying ratio and proportion concepts arose in such a wide range of different real-life contexts (Ellerton & Clements, 2012).

Although many students solved rule-of-three problems without appreciating why the rule was appropriate, we did not get an impression that Abraham applied rules mindlessly. Nevertheless, the PCA genre he employed resulted in solutions being presented as a maze of numbers and calculations. Be that as it may, a thoughtful examination of Abraham's solutions always tended to suggest that this was, indeed, Abraham Lincoln's work. Every solution was worked out carefully before being written in the cyphering book. There was hardly anything crossed out. This was to be a book for life, and Abraham did not intend that it be riddled with errors. It had to be right.

The Fifteenth of the Extant Pages in Lincoln's Cyphering Book

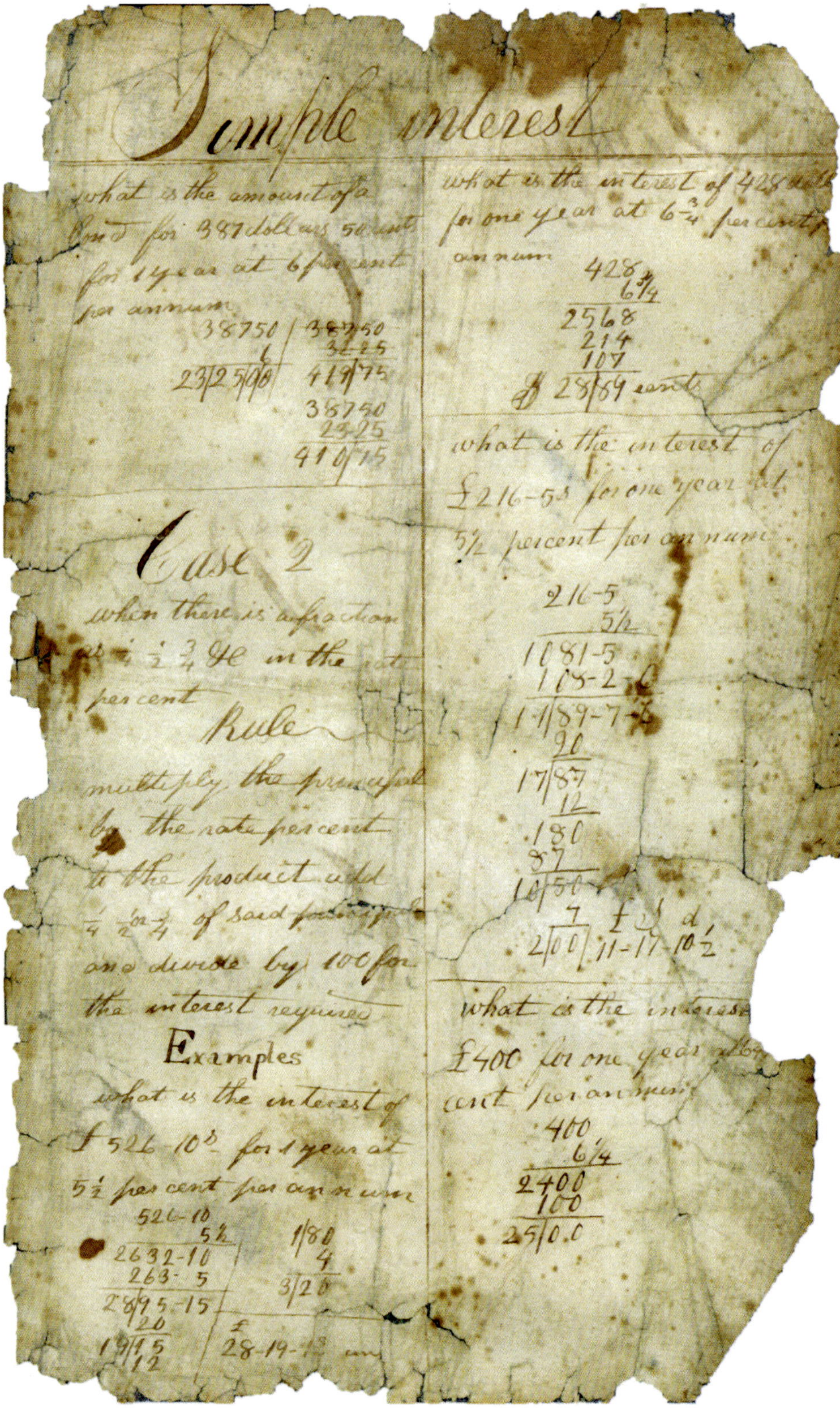

Simple interest

what is the amount of a bond for 387 dollars 50 cents for 1 year at 6 per cent per annum

38750 | 38750
6 | 2325
23|25|00 | 41|0|75

38750
2325
41|0|75

Case 2

when there is a fraction as 1/4 1/2 3/4 &c in the rate per cent

Rule

multiply the principal by the rate per cent to the product add 1/4 1/2 or 3/4 of said principal and divide by 100 for the interest required

Examples

what is the interest of £526 10s for 1 year at 5½ per cent per annum

526-10
5½
2632-10
263-5
28|95-15
20
19|15
12

1|80
4
3|20

£
28-19-1s

what is the interest of 428 dollars for one year at 6¾ per cent per annum

428
6¾
2568
214
107
$ 28|89 cents

what is the interest of £216-5s for one year at 5½ percent per annum

216-5
5½
1081-5
108-2-6
11|89-7-6
20
17|87
12
10|50
4
2|00

£ s d
11-17-10½

what is the interest of £400 for one year at 6¼ per cent per annum

400
6¼
2400
100
25|0.0

Figure 6.17. "Simple Interest 1" (fifteenth of the 22 pages).
Courtesy Lilly Library, Indiana University, Bloomington, Indiana.

The page shown in Figure 6.17 is held at Indiana University. It was the first of four pages to be dedicated to the topic "simple interest." Dorfman (1966a) listed it as the thirteenth of the extant pages, Warren (1959) as the eleventh, and Basler et al. (1953) as the twelfth. There was probably at least one earlier page on simple interest in Abraham's cyphering book. We say that because this page does not include an introduction to the topic, and the first case mentioned is "Case 2."

Under Case 2 a rule is stated. Ellerton and Clements (2012) identified two distinct genres characteristically present in most cyphering books. The first was the already-discussed PCA ("Problem-Calculation-Answer") genre, which applied particularly to written solutions to problems. The second was the IRCEE ("Introduction-Rule-Case-Example-Exercise") genre, which was used when a topic was being introduced. Most textbook authors adopted these two genres, and so did students preparing cyphering books. Throughout Stephen Pike's (1822) textbook, in particular, both the PCA and IRCEE genres were often to be found.

Five of the six problems on this page are identical to, or very similar to, problems on page 102 in Stephen Pike's (1822) *Arithmetic.* From a curriculum perspective, there were educational issues associated with introducing simple interest to Abraham, for the topic introduced concepts like "percent" and "decimal fractions," neither of which Abraham had formally studied. The first problem, in the top left corner, asked: "*What is the amount of a bond for 387 dollars and 50 cents for 1 year at 6 per cent per annum.*" This is similar to Problem 16 on page 102 of Pike (1822). Abraham's calculations show that he multiplied 387|50 by 6, and then added 23|25 to 387|50. A stroke was used to indicate where the whole number component of the calculation finished.

The example at the bottom left corner was not found in Pike (1822). It involved pounds, shillings and pence, and 5½ per cent. This required careful thought, but Abraham appeared to get the right answer after converting what he wrote as 28|95-15 pounds to 28 pounds 19 shillings and "something." He multiplied the |15 by 12 to get pence, and obtained 1|80, and then converted the |80 to 3|20 farthings by multiplying by 4.

Four of the five problems on the page involved a mixed number as the percentage to denote the rate of interest (the rates being 5½ per cent, 6¾ per cent, 5½ per cent, and 6¼ per cent). For a student who had never been taught vulgar fractions formally, such problems could have caused difficulty. A rule, for Case 2, was given to help students: "Multiply the principal by the rate percent [and] to the product add ¼, ½, ¾ etc. of said product and divide by 100 for the interest required." Abraham managed to follow and apply that rule successfully.

In Spencer County where Abraham and his family lived in Indiana it is likely that knowledge of how to calculate simple interest would have been a distinct advantage. Such problems will be discussed in our analyses of the next three extant pages.

Abraham was expected to deal with decimal fractions, vulgar fractions, and percent, even though he had never studied these topics. His teacher was probably Azel Waters Dorsey, who, we think, liked to use Stephen Pike's *Arithmetic*—or owned a cyphering book in which examples from Pike's book had been written. A careful examination of the intended curriculum, as it was evidenced by the names of chapters in Pike's (1822) *Arithmetic,* reveals that Simple Interest was the name for Chapter 17, Vulgar Fractions for Chapter 26, and Decimal Fractions for Chapter 27. Percent was not the name for any chapter. The sequencing of topics within the curriculum was flawed. At that time, however, North America did not have curriculum experts who could deal with such issues and formal teacher-training courses were non-existent (Ellerton & Clements, 2012). Furthermore, rural Indiana was a long way from Boston and New York, and it was hard to find anyone who might help a keen student to fill in the gaps left by an inadequately structured curriculum. Considering the difficulties, Abraham did remarkably well arithmetically.

The Sixteenth of the Extant Pages in Lincoln's Cyphering Book

Simple Interest

what is the interest of £855-17-6 for one year at 5¾ per cent per annum

855-17-6
5¾

Case 3

when the interest of a sum given for several years is required

Rule

Multiply the interest of the given for one year by the number of years

what is the interest of ... for one year at 6¼ per cent per annum

Examples

what is the interest of £246 18s for 5 years at 4¼ per cent per annum

what is the interest of $426-18 for one year at 4½ per cent per annum

Figure 6.18. "Simple Interest 2" (sixteenth of the 22 pages).
Courtesy Lilly Library, Indiana University, Bloomington, Indiana.

The page shown in Figure 6.18 is held at Indiana University. It was the second of four pages to be dedicated to the topic "simple interest." Dorfman (1966a) listed it as the fourteenth of the extant pages, Warren (1959) as the twelfth, and Basler et al. (1953) as the eleventh. It is surprising that Basler et al. did not recognize that this page, which elaborated upon Case 3, would definitely have followed the previous page, which dealt with Case 2.

As with the first page on simple interest, the PCA and IRCEE genres are strikingly identifiable on this page. Two of the four problems on the page could be found in Stephen Pike's (1822) *Arithmetic*. The question at the top left ("*What is the interest of £855-17-6 pence for one year at 5¾ percent per annum*") is almost identical to Question 6 on page 102 of Pike (1822). It will be instructive to try to reconstruct Abraham's method when he attempted to solve this problem.

Abraham first multiplied the principal (£855-17-6) by 5 (to get £4279-7-6), then he halved the principal (he erroneously obtained £427-13-6, instead of £427-18-9) and then halved what he obtained (he erroneously got £213-16-3, instead of £213-16-9). He then added his subtotals and got 49|20-17-3, inserting a vertical stroke after the "49" to indicate that 49 pounds were in the answer. From a modern perspective, the vertical stroke was used to indicate what would occur after division by 100 (for, the problem was about percent). But, what was Abraham to do with the "20" after the 49? He multiplied by 20 (to convert from pounds to shillings), and added in the 17 from the previous line, to get 417. He then used another vertical stroke after the 4 (4|17) to indicate that the 4 represented the number of shillings in the answer. Then followed 17 multiplied by 12 (to convert from shillings to pence), and with the 3 from two lines above now added in, he got 207. Another vertical stroke was used (this time after the 2, so 2|07), and after multiplying the 07 by 4 and getting 28, the 2 was taken to be the pence in the answer. Then Abraham stated his answer as *£49-4-2*. Although this was almost correct (the correct answer was *£49-4-3*), Abraham had made two calculation slips along the way.

The method Abraham used was very similar to that shown by Pike (1822) in a model example on page 102. Interestingly, it was also very similar to the method shown for exactly the same problem on page 120 of McKenney's (1825) *Key* to *Pike's Arithmetic*. But McKenney made no calculation errors, whereas Abraham made two. So it is almost certain that Abraham was not copying from the *Key*.

We would reiterate that it is unlikely that Abraham had ever formally studied decimal fractions or vulgar fractions or percentage. This particular problem involved those three topics, and Abraham's method was very similar to that shown in a model example in Pike's (1822) *Arithmetic* and in McKenney's (1825) *Key* to that *Arithmetic*. We still have the impression that Abraham did all of this by himself, perhaps with some guidance from his teacher (who probably would have accepted the answer Abraham obtained as "near enough"). It is interesting that Abraham used exactly the same method for the example at the bottom right of the page—that example could not be found in Pike (1822), or in any other textbook we consulted.

The words used for Case 3, were identical to those found in Pike (1822, p. 103), but the rule that Abraham wrote following upon Case 3 provided an abbreviated form of the corresponding rule given by Pike.

The evidence is strong that Azel W. Dorsey, with the aid of problems from Stephen Pike's (1822) *Arithmetic*, was able to take Abraham to another level in mathematics. Dorsey did not stay in Spencer County, Indiana, but moved on to Schuyler County, Illinois, where he would open another school.

The Seventeenth of the Extant Pages in Lincoln's Cyphering Book

Figure 6.19. "Simple Interest 3" (seventeenth of the 22 pages).
Abraham Lincoln Collection, General Collection,
Beinecke Rare Book and Manuscript Library, Yale University.

The page shown in Figure 6.19 is held at Yale University. It was the third of four pages dedicated to the topic "simple interest." It was stated to be the fifteenth of the extant pages by Dorfman (1966a), the thirteenth by Warren (1959), and the fourteenth by Basler et al. (1953). It includes a statement of "Case 6" and "Rule 5," but although Abraham's wording of Case 6 is identical to Pike's (1822, p. 108), his wording for Rule 5 is quite different from Pike's (p. 108). Almost certainly, there was a leaf before this one which included Cases 4 and 5, but that leaf has long since disappeared.

The calligraphy at the top of the page is unusual. In particular, the upper-case "E" at the front of "Example" is not like any of the calligraphy on other pages of Abraham's cyphering book. That is probably the reason why William Herndon felt the need to write "Genuine W. H. Herndon." Despite Herndon's assertion, we still have reservations about the authenticity of this page. The size of the writing is unusually small, and some of the letters are not formed in the same way as corresponding letters on other pages (compare, for example, the heading "Simple Interest" on this page with the same words on the previous page). A handwriting expert is needed to see if our concerns are justified. Note that we have no doubt that this leaf was within the cyphering book passed on to Sarah Bush Lincoln, and then passed on to William Herndon. We *are* questioning, though, whether this page was originally penned by Abraham.

There is some evidence that this page was taken down from dictation, because "C~a~s~e~~S~i~x~~" was written, whereas in Pike's (1822) *Arithmetic* this heading was written simply as "CASE 6" (p. 108).

Abraham (assuming he was the writer) worked through five simple interest exercises on this page related to Case 5 (stated on page 107 of Pike, 1822). Four of these were identical to exercises in Pike (1822, pp. 108–109). These questions, although apparently simple (e.g., Question 1, p. 108, "*What principal at interest for five years, at six per cent per annum, will amount to £650*"), were linguistically difficult and conceptually demanding. With Question 1, Abraham multiplied by 6 to get 30 (which represents a 30 percent growth over five years) and added this 30 to 100 to get 130. After stating a rule-of-three relationship, he then divided 130 into 65000 to get 500, and gave the answer as 500£ (which was the answer Stephen Pike gave to the same problem). Although this method was correct it was not transparent.

Once again, on this page the influence of Stephen Pike's (1822) *Arithmetic* was pervasive. It seems that Azel Waters Dorsey—or whoever Abraham's teacher was at the time—relied heavily on Pike's approach to such problems, and expected his students to do the same.

Note that on this page there are five examples, all requiring monetary calculations, and of the five, four were concerned with English money (pounds, shillings, pence and farthings) and only one involved official Federal money (dollars, dimes, cents and mills). That, despite the fact that the official U.S. currency was of decimal denomination and had been minted in the United States since 1792. Well into the nineteenth century the various states continued to use the names for English money. The fact that the values for the same unit would vary from state to state. arithmetically challenged not only schoolchildren but also those who happened to travel from one state to another, or carry out business transactions with those who lived in states other than their own. There is no record of Abraham ever formally studying decimal fractions, or decimal currency, while attending school, but he seemed to cope well with money calculations that involved the rule of three, simple and compound interest, and discount—irrespective of whether these calculations involved Federal currency or English money.

The Eighteenth of the Extant Pages in Lincoln's Cyphering Book

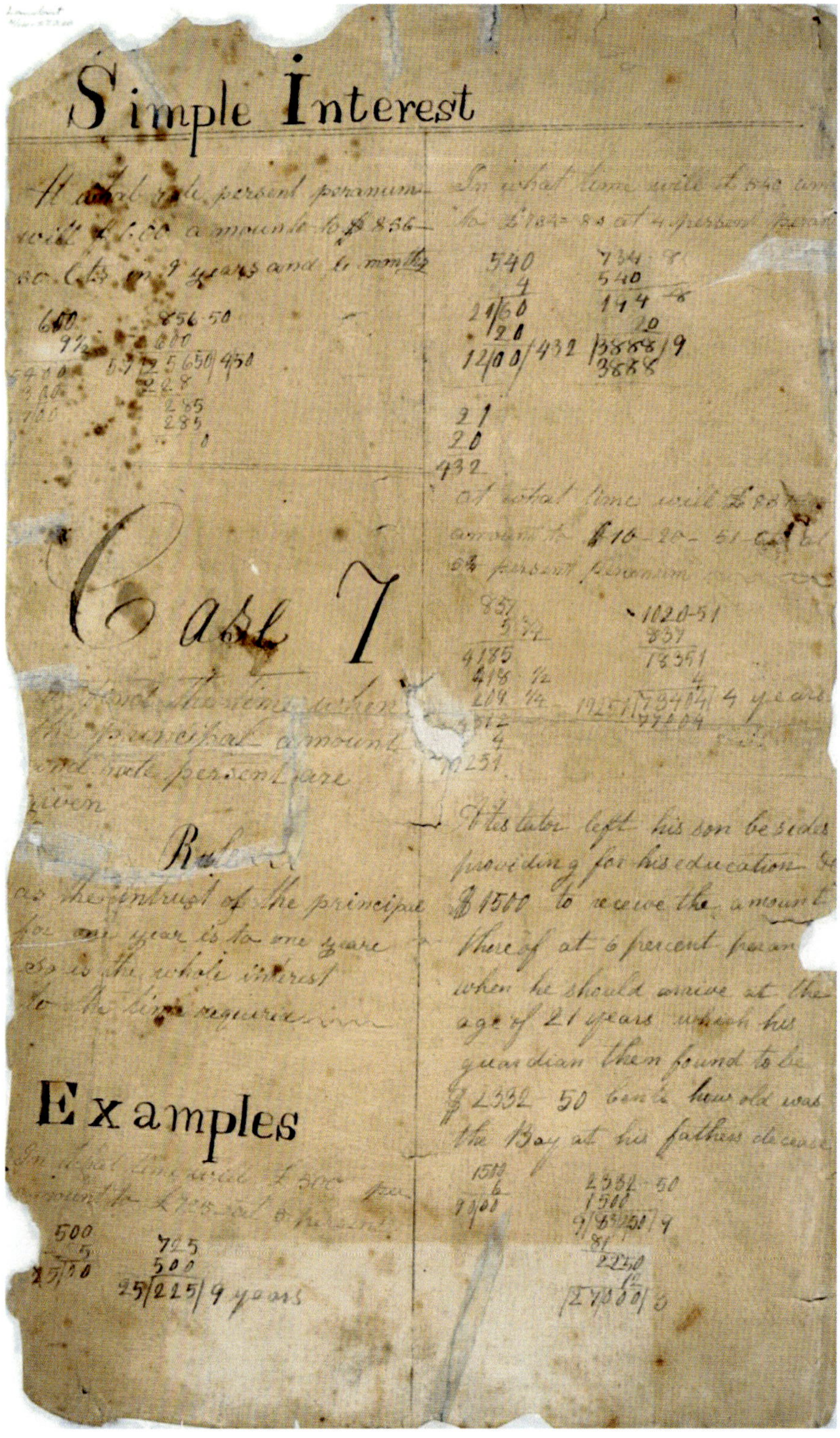

Figure 6.20. "Simple Interest 4" (eighteenth of the 22 pages). Abraham Lincoln Collection, General Collection, Beinecke Rare Book and Manuscript Library, Yale University.

The page shown in Figure 6.20 is held at Yale University. It is the fourth and last page dealing with the topic "simple interest." Of the extant pages, Dorfman (1966a) placed it sixteenth, Warren (1959) fourteenth, and Basler et al. (1953) thirteenth. It includes a statement of "Case 7."

It has not been possible to find a question like the one at the bottom right of the page (about a tutor leaving his son some money) in Pike's (1822) *Arithmetic.* However, a very similar problem was found in the popular arithmetic by Zachariah Jess (1821). In fact, Jess's text seems to have been influential with respect to the material on this page. The wording of Case 7, and of the rule following that, is exactly the same as wording on page 101 of Jess (1821). A similar statement is true for four of the examples or exercises. It seems that all of a sudden, Jess's book has become the main background influence. Dorfman (1966a) also drew attention to this, and claimed that the information and questions on this page were "definitely from Zachariah Jess's book" (p. 71). We would not go quite so far as that, for the first example in Pike (1822) after Case 7 and the rule is exactly the same as the example after Case 7 and the rule on this page. Jess's and Pike's treatments of simple interest were very similar.

The extreme form of PCA genre on the page warrants comment. Within the solutions to five examples (or exercises) that are shown, there is neither a logical connective (such as "therefore") nor an equals sign. The setting out for each exercise is based on algorithms that were applied—and probably not understood (because there is no evidence that Abraham had ever studied vulgar or decimal fractions).

Take, for example, the question at the top right of the page, "*In what time will L540 amount to L734 8s at 4 percent per year?*—which is identical to Question 2 on page 109 of Pike (1822). A twenty-first century teacher of mathematics would be excused for thinking that the vertical strokes at various places in the solution statement stood for decimal points, but Abraham had not studied decimal fractions. So what do the vertical strokes represent other than responses to the imperative "this is what you need to do to get the right answer"?

The "21" and the "12" on the left side of the solution are meant to represent "21 pounds 12 shillings," The 540 pounds has been subtracted from the 734 pounds and 8 shillings to get 194 pounds and 8 shillings, and after multiplying the 194 by 20 and adding 8, 3888 was obtained. Finally, 3888 was divided by 432 (because 21 pounds 12 shillings equals 432 shillings), and the correct answer obtained was 9 (years). In fact, Abraham seemed to know precisely what he was doing all the way, but he was not learning to set out his problem-solving efforts in a way that would communicate readily to others (and, possibly, after some time, even to himself). Abraham was being enculturated into a system of mathematics education that, at the subconscious level at least, emphasized the need to be able to get the right answer using as few as words as possible. He knew what he was doing, but he did not write down details of his logical processes.

The Nineteenth of the Extant Pages in Lincoln's Cyphering Book

Figure 6.21. "Compound Interest 1" (nineteenth of the 22 pages). Private Collection, courtesy of Seth Kaller, Inc.

This page is currently held by a private owner. Dorfman (1966a) listed it as the seventeenth of the extant pages, Warren (1959) as the fifteenth, and Basler et al. (1953) as the nineteenth. Both of the pages on the leaf are concerned with compound interest. The leaf was originally passed on, by William Herndon, to Peter H. Brandt, and it would seem that Brandt, or some later owner, folded it—which probably accounts for its obviously poor condition.

Stephen Pike (1822) felt the need to devote only one page (p. 111) to compound interest, and the treatment of the topic in Abraham's cyphering book is almost identical to that in Pike's text. This page repeats the rule stated by Pike, and all exercises are from Pike. The page begins in the IRCEE genre with a description of the concept of compound interest; that is immediately followed by a rule for calculating compound interest. The introduction and the rule were similar to, but not identical with, statements by Pike (1822). Then followed an example which, in fact, was the same example as that presented as a model example by Pike (1822, p. 111): "*What is the compound interest of £500 for 3 years at 5 per cent?*"

Pike (1822), using year-by-year calculations, arrived at the answer 78 pounds 16 shillings and 3 pence. His setting out was unusually full, with PCA genre not being adopted. By contrast, Abraham used an algorithmic approach, and his setting out was in the PCA genre—he did not explain *why* he did his calculations. Nevertheless, he quickly arrived at the correct answer.

In the twenty-first century this task could be completed quickly by using an electronic calculator to find the number 500 less than the value of 500 times 1.05^3. But in 1824, the task would have been time-consuming, even though conceptually it required little more than three successive applications of simple interest.

Probably, both Abraham and his teachers had never studied algebra formally, and therefore did not have access to appropriate compound interest formulae. He conceived of the problem as requiring three separate simple interest problems, and doggedly proceeded with the rather messy calculations. Only 13 out of 350 manuscripts in the Ellerton-Clements cyphering-book collection focus on algebra. In the 1820s, high schools were in their infancy in the United States and hardly anyone outside of universities studied algebra (Kilpatrick & Izsák, 2008; Simons, 1924). Furthermore, Abraham would have been faced with the same previously-mentioned difficulties arising from his never having formally studied vulgar or decimal fractions, or percent. Nevertheless, he coped, and his setting out suggested that the problems did not cause him much trouble.

The word problems on compound interest that Abraham copied into his cyphering book, and solved, were to be found, word-for-word, in Stephen Pike's (1822) *Arithmetic* (see p. 111). Each involved calculations in English and not Federal money. How that could have been happening more than 30 years after the introduction of decimalized currency into the United States of America did not seem to worry anyone. All over the United States of America students worked on simple and compound interest tasks, as well as on so-called "reduction" tasks and "currency exchange" tasks.

Certainly, Abraham, in his future life, would be likely to have to make decisions with respect to compound interest scenarios that would affect his financial well-being. But one wonders why a curriculum had not been developed in which almost all of the calculations required were in Federal dollars and cents. Then, again, teachers in Spencer County, Indiana, around 1825 dealt with classes of children of widely different ages and arithmetical backgrounds. It would have been unreasonable to expect any of Abraham's teachers to create a curriculum relevant to the needs of Abraham and the other children—the teachers had not been trained to do this, and in any case their knowledge of arithmetic was not sufficiently strong to meet such a challenge.

The Twentieth of the Extant Pages in Lincoln's Cyphering Book

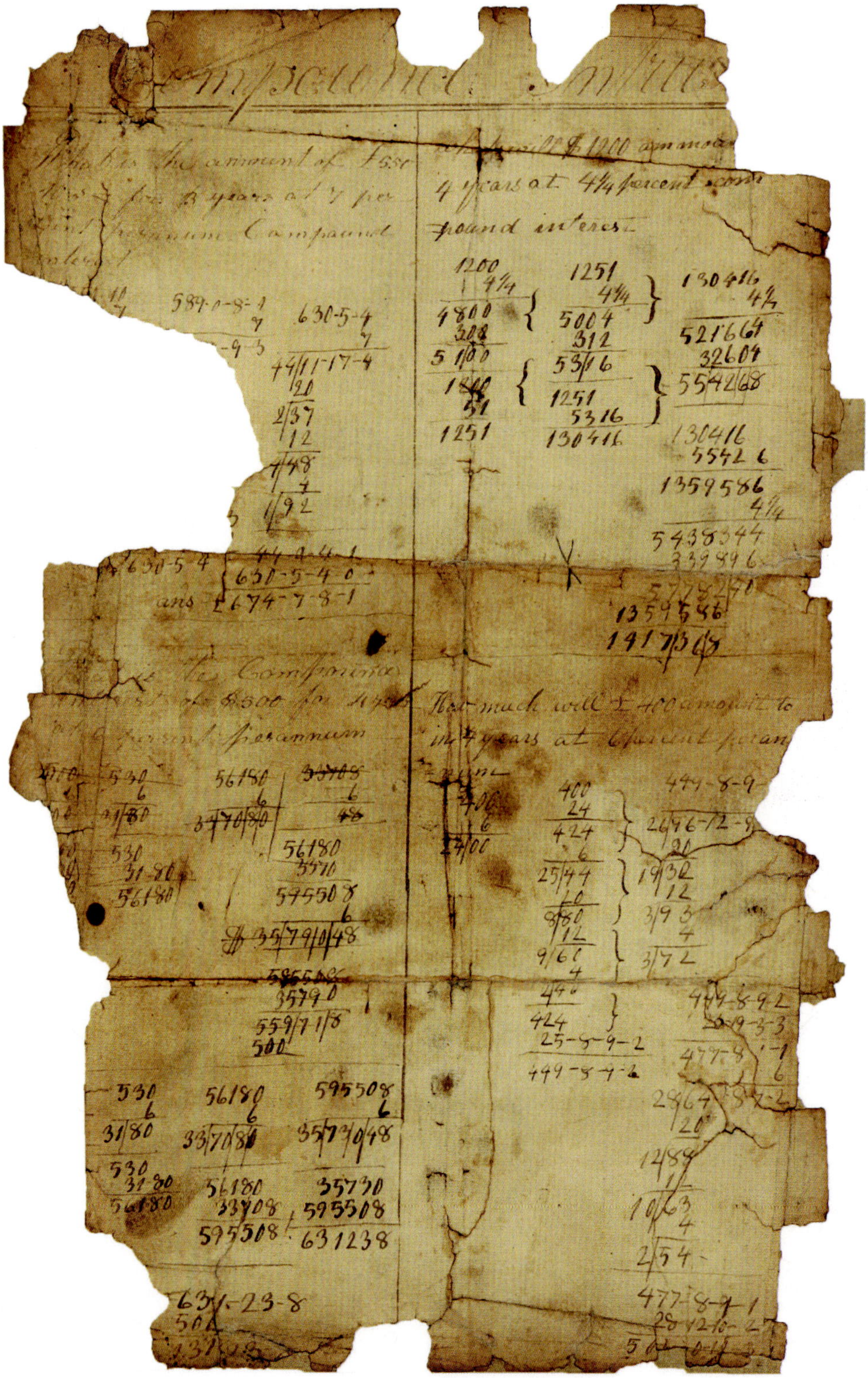

Figure 6.22. "Compound Interest 2" (twentieth of the 22 pages). Private Collection, courtesy of Seth Kaller, Inc.

This page is currently held by a private owner. Dorfman (1966a) listed it as the eighteenth of the extant pages, Warren (1959) as the sixteenth, and Basler et al. (1953) as the twentieth. The arithmetic on both of the pages of the leaf was concerned with compound interest.

There are four exercises on compound interest on this page, and three of those are either identical to, or very similar to, questions in the single page on the topic in Stephen Pike's (1822) *Arithmetic* (see p. 111). The three questions on this page similar to those in Pike (1822) were:

[*Top left*] *What is the amount of £550 ?? shillings ?? pence for 3 years at 7 per cent per annum compound interest.*

Pike (1822) *Question 2:* What is the compound interest of 450L for 3 years at 5 per cent per annum?

[*Bottom left*] *What is the Compound Interest of \$500 for 4 years at 6 per cent per annum*

Pike (1822) *Question 4:* What is the compound interest of 500 dollars for 4 years, at 6 per cent per annum?

[*Bottom right*] *How much will £400 amount to in 4 years at 6 per cent per annum?*

Pike (1822) *Question 5:* How much will 400L amount to in 4 years, at 6 per cent per annum?

On the top right of the page, Lincoln also worked on the following task:

> *What will \$1200 ammount (sic.) to [for] 4 years at 4¼ percent compound interest.*

That question could not be found in Pike (1822) or in any other arithmetic that we consulted.

It can be seen that two of the four questions involved Federal money. Consideration of the question not linked to Pike (1822) will be instructive because the interest rate involved a vulgar fraction (4¼ percent). In his calculations to determine 4¼ percent of \$1200 for the first year, Abraham used both decimal and vulgar fractions concepts—despite the likelihood that he did not consciously know the mathematical language for what he was doing. It seems that he knew that to get ¼ of \$1200 he needed to divide the 1200 by 4. Then, having added the 300 to 4800 to get 5100, he knew, somehow, that to get percent you had to put a vertical stroke to the right of the second number from the left, and that the number on the left of the vertical stroke represented the amount of interest. That led him to add \$51 to the original principal (\$1200), to get the "new" principal of \$1251.

In his calculations for the second year, Abraham still had to deal with decimal and vulgar fractions. He did not seem to know enough about fractions to realize that when he divided the 1251 by 4, he got 312¾, and so, truncated to the nearest dollar, his interest for the second year would be \$53.16, with ¾ of a cent "left over." Apparently, it was standard accounting practice to "round down," and so the interest for the second year would remain \$53.16 (and not be "rounded up" to \$53.17). He would perform similar calculations and make similar truncations, for his third- and fourth-year calculations. The final answer presented by Abraham was 1417|36|8. He did not interpret the meaning of 1417|36|8 in terms of dollars, cents and mills.

With respect to the other three solutions that Abraham showed on the page, his method of presentation differed considerably from the method shown in Pike's (1822) model example for compound interest. Our analysis of Abraham's work on this page indicated that, from an arithmetical perspective, at least, Abraham was his own person. He wanted to understand things his way.

The Twenty-First of the Extant Pages in Lincoln's Cyphering Book

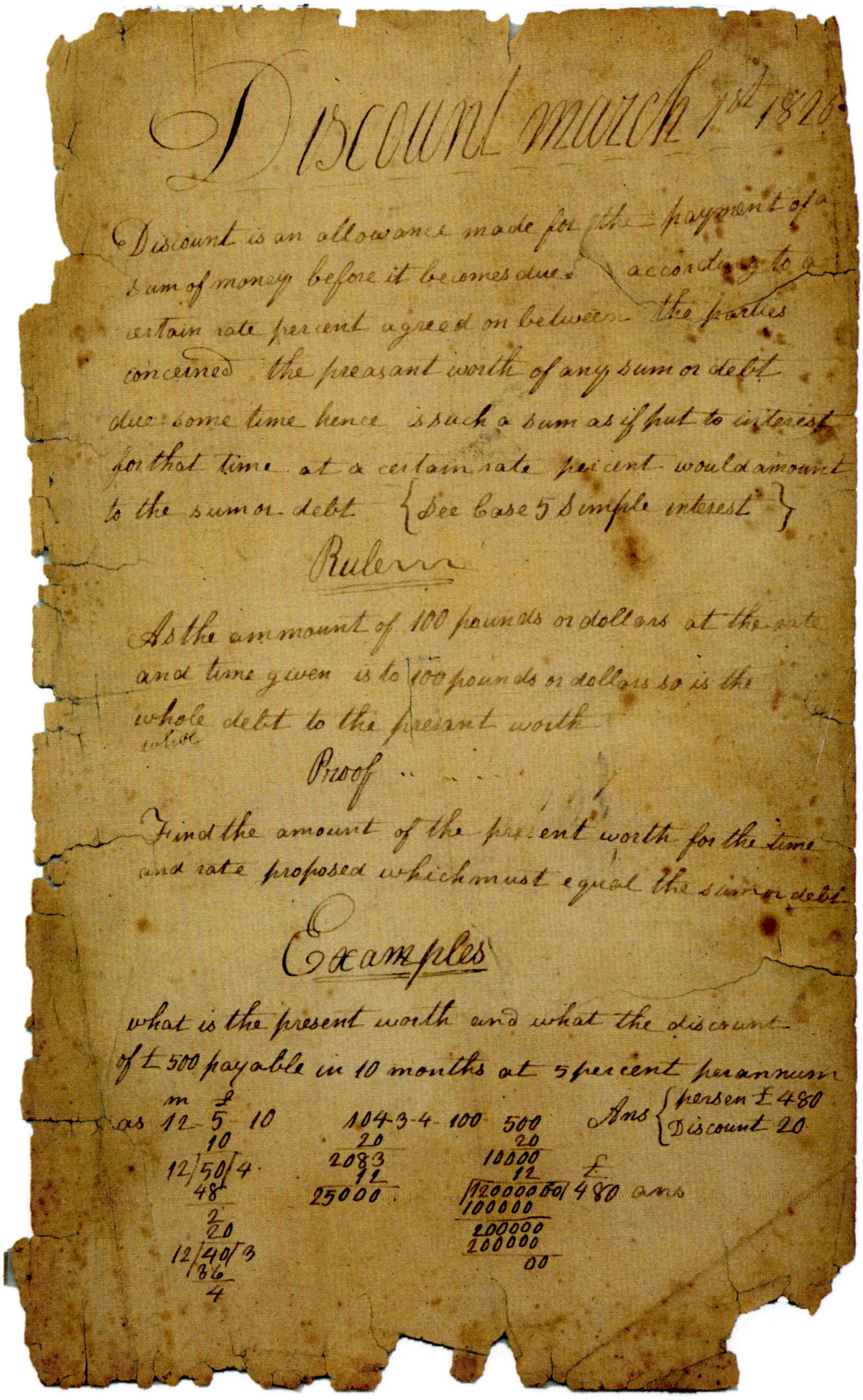

Discount march 1st 1826

Discount is an allowance made for the payment of a sum of money before it becomes due according to a certain rate percent agreed on between the parties concerned the preasant worth of any sum or debt due some time hence is such a sum as if put to interest for that time at a certain rate percent would amount to the sum or debt {See Case 5 Simple interest}

Rule

As the ammount of 100 pounds or dollars at the rate and time given is to 100 pounds or dollars so is the whole debt to the present worth

Proof

Find the amount of the present worth for the time and rate proposed which must equal the sum or debt

Examples

what is the present worth and what the discount of £500 payable in 10 months at 5 percent per annum

Ans {present £480, Discount 20}

£ 480 ans

Figure 6.23. "Discount 1" (twenty-first of the 22 pages).
Chicago History Museum, ICHi-52426.

This page is currently held by the Chicago History Museum. Dorfman (1966a) listed this as the nineteenth of the extant pages, Warren (1959) as the seventeenth, and Basler et al. (1953) as the sixteenth. The arithmetic on this page is concerned with discount, and we believe that Abraham's teacher at the time this page was prepared—March 1st 1826—was Azel Waters Dorsey, possibly assisted by James Bryant.

Our checking of possible background texts from which this page could have been developed led us to make the following three conclusions:

1. The seven-line introductory statement written immediately after the heading ("Discount is an allowance made for the payment of a sum of money before it becomes due ... would amount to a sum equal to the sum or debt") is to be found, word for word, in Pike (1822, p. 112).
2. The wording of the rule ("As for the ammount (sic.) of 100 pounds or dollars ... to the present worth"), and of the "proof" ("Find the amount of the present worth for the time and rate proposed which must equal the sum or debt") is to be found, almost word for word, in Jess (1821, p, 106).
3. The example at the bottom of the page ("What is the present worth ...") is identical to Question 8 on page 64 of Francis Walkingame's (1802) *The Tutor's Assistant, Being a Compendium of Arithmetic, and a Complete Question-Book*, which had been published in London. However, Abraham's solution did not resemble the method used in a model example shown by Walkingame (1802) on page 63.

It appears that Abraham, or Abraham's teacher, had Walkingame's *Arithmetic* as a new resource—. which might have been introduced by James Bryant. Although Walkingame's *Arithmetic* was never published in the United States—it *was* published in Canada—it was easily the best-selling arithmetic in England during the period 1750–1850 (Denniss, 2012; Michael, 1993; Wallis, 1963).

The forms of words used for the definition of discount, for the rule, and for the "proof" for discount calculations were linguistically complex. As an aside, it should be noted the meaning of the word "discount" was not the same as its meaning in the 21st century.

The problem at the bottom of the page ("*What is the present worth and what the discount of L500 payable in 10 months at 5 percent per annum?*") was to be associated with the rule "*As the amount of 100 pounds or dollars at the rate and time given is to 100 pounds or dollars, so is the whole debt to the present worth.*" The calculations on the left side of the page were concerned with determining the amount of interest on 100 pounds that should be paid for 10 months if the rate of interest was 5 per cent per annum. To do this, Abraham found ten-twelfths of 5 pounds, by multiplying by 10 and dividing by 12. He obtained 4 pounds 3 shillings and 4 pence, and seemed to reason that therefore 100 pounds would accumulate to 104 pounds 3 shillings and 4 pence (i.e., 25000 pence) over 10 months. Then, since 500 pounds equaled 120000 pence (obtained by multiplying by 20, then 12), he used the direct rule of three. He multiplied 120000 by 100, to get 12000000, and then divided 12000000 by 25000 to get 480. Hence the "present worth" would be 480 pounds, and so, as Abraham stated, the discount should be 20 pounds.

Although discount might have been an important concept in the lives of midwestern frontier families (who were often buying and selling land, etc.), this method of conceiving and calculating discount would not have been transparent to most people. Be that as it may, Abraham employed the standard *abbaco* method for tackling rebate/discount problems of this kind.

The Twenty-Second of the Extant Pages in Lincoln's Cyphering Book

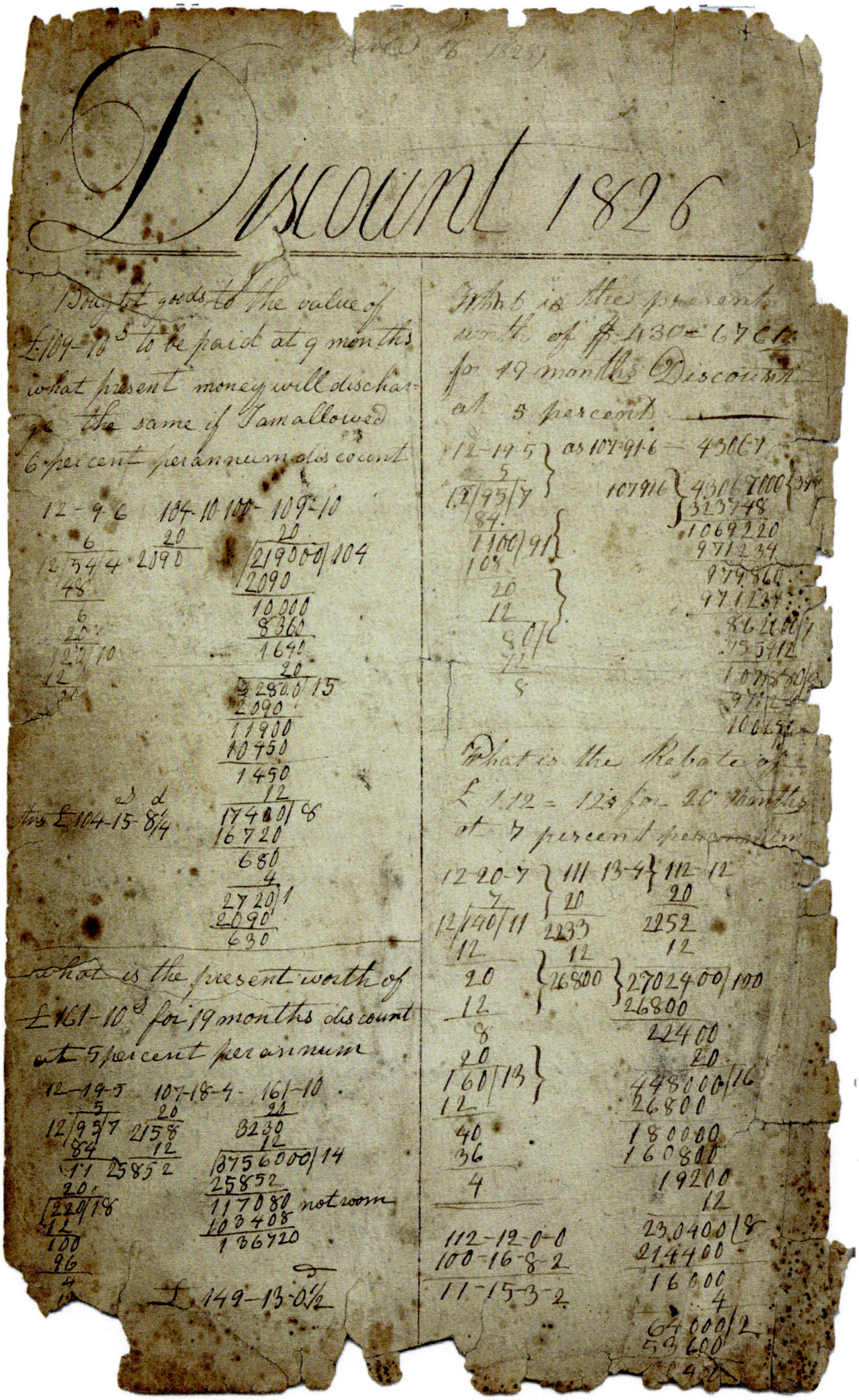

Figure 6.24. "Discount 2" (twenty-second of the 22 pages).
Chicago History Museum, ISHIi-52427.

This page, currently held by the Chicago History Museum, was from our perspective, the final (i.e., the twenty-second) page of Abraham's cyphering book. Dorfman (1966a) listed it as the twentieth (and last) page, Warren (1959) placed it eighteenth, and Basler et al. (1953), fifteenth.

The arithmetic on this page was, once again, concerned with discount, and we believe that Azel Dorsey, possibly assisted by James Bryant, was Abraham's teacher at the time—almost certainly in March 1826. The fact that the writing on this page goes close to both margins is consistent with this having been the very back page of the cyphering book.

On this page there are four questions on discount and Abraham's handwritten responses to them are shown. Once again, it is instructive to consider the sources of the questions:

1. The first question (top left corner) is identical to Question 4 on page 63 of Walkingame (1802). Abraham's correct answer is the same as that given by Walkingame—but Walkingame did not show his working. Abraham's working is structurally similar to what he showed in the final example on the previous page.
2. The second question (bottom left corner) is identical to Question 2 on page 107 of Jess (1821). In Abraham's written response to this question we find two unusual aspects: the first was, he made two errors; the second was, he did not complete the problem, but wrote "no room." Then he gave the answer 149 pounds 13 shillings and a halfpenny. That was the correct answer, but it was shown, immediately after the statement of the question, on page 107 in Jess (1821), and perhaps Abraham copied from that source. Because of his two errors, Abraham would not have arrived at the correct answer if he had completed his working.
3. The third question (top right corner) is identical to Question 5 on page 113 of Pike (1822). Once again, Abraham's solution included minor errors, and it is not clear that he actually completed a solution. An interesting feature of this problem is that it involved Federal, not English, money. That change necessitated a different pattern of thought (if one were to answer the problem correctly).
4. The fourth problem (bottom right) is identical to Question 6 on page 113 of Stephen Pike's (1822) *Arithmetic*. Abraham did a lot of work trying to achieve a correct solution to this problem, but did not express his answer clearly.

For the first time we sensed a level of exasperation in Abraham's attitude. For three of the four problems on this page he did not give a correct answer.

This was probably Abraham's final page of cyphering at school. He was now 17 years old, and textbooks like Jess (1821), Pike (1822) and Walkingame (1802) had many more problems, and many more topics that went well beyond "discount." Discount was only Chapter 20 in Stephen Pike's (1822) textbook and, after that, there were chapters dealing with equation of payments, barter, loss and gain, fellowship, exchange, vulgar fractions, decimal fractions, involution, square root, cube root, alligation, position, arithmetical and geometrical progressions, etc., through to topics like annuities, permutations, combinations and duodecimals.

Abraham was now ready to move on to a wider world. Perhaps, neither Dorsey nor Bryant was able to provide the assistance Abraham felt he needed in order to go further in his mathematical studies. He was now struggling, sometimes, to get answers shown in the textbooks (or in the cyphering book) that the teacher had. But mathematics had cast its spell over Abraham, and he would return to it in a few years' time to study logarithms, trigonometry, surveying and geometry (Dorfman, 1966b). Later, he would invest much time in memorizing and applying *Euclid's Elements* (Hirsch & Van Haften, 2010).

Reflections on an Extraordinary Cyphering Book

In a cartoon which appeared in the *Evansville Courier* of January 29, 1922 (reproduced in Erekson, 2012, p. 135) a figure representing "the public" is shown entering an exhibition called "The Land of Romance-Sentiment-Knowledge." This cartoon associated the Southwestern Indiana Historical Society's launch of the "Lincoln Inquiry" with that Society's quest to reclaim Southern Indiana's role in the development of the character and heritage of President Abraham Lincoln by examining Lincoln's life in and around Pigeon Creek during the period 1816 through 1830.

Our investigation into Abraham Lincoln's cyphering book has brought into sharp focus the fact that so little genuine historical scholarship has been applied to the 22 extant pages. Twenty of the pages were reproduced in *The Collected Works of Abraham Lincoln* (Basler et al., 1953), and in 1959 one of the associate editors of the *Collected Works*, Lloyd A. Dunlap, wrote a brief, but balanced, commentary on the pages. Louis Warren (1959) provided a good initial statement on the order in which Lincoln originally prepared the pages, and in 1966 Maurice Dorfman (1966a) had an article published in the *Lincoln Herald* which, from our perspective, offered a chronological ordering of the pages that was more accurate than Warren's. Dorfman followed this up with a speculative article that looked at the role of mathematics in Lincoln's later life (Dorfman, 1966b).

Dorfman (1966a) claimed that Lincoln, or one of his Spencer County teachers, had access to an arithmetic written by Nathaniel Martin. That claim seemed to be based solely on his observation that Martin's text used and explained the casting-out-nines checks that Lincoln displayed on some of the early pages of his cyphering book. However, none of the multiplication and division tasks that Lincoln dealt with could be found in Martin's (1800) text. Although Dorfman's (1966a) claims about the influence of Martin's text on Lincoln's arithmetic education were, almost certainly, wrong, very few historians or others interested in Lincoln knew enough about the history of mathematics education in general, or about school mathematics in particular, to be in a position to challenge what Dorfman (1966a) wrote. Like Nathaniel Martin, and Maurice Dorfman, David H. Donald (1995) claimed that Lincoln did not "get the idea of 'casting out nines'" (p. 31). As our analysis shows, Lincoln *did* know how to cast out nines correctly.

Slowly, the expression "sum book" began to be used in relation to the future President's cyphering book (see, e.g., Donald, 1995), even though that expression was never used, anywhere, at the time Lincoln was creating his book. Thus, for example, on the cover of the Fall 1966 *Lincoln Herald* an impressive drawing by Lloyd Ostendorf depicted William H. Herndon and Sarah Bush Johnston Lincoln looking at Lincoln's "sum book" on September 8, 1865. It is well known that a meeting between the two actually took place on that day, and so the impression was given that Lincoln's cyphering book had "Abraham Lincoln's sum book" written on its cover. Almost certainly it did not, yet pictures such as Ostendorf's, and verbal or written statements by others, have resulted in the term "sum book" being used often with respect to Lincoln's cyphering book.

Does it matter? "Surely," one might say, "Lincoln's cyphering book was full of 'sums,' in the sense that that word has been used for centuries, so why not call his book a 'sum book'"? We believe it does matter, simply because the use of "sum book" has meant that the correct term, "cyphering book" (or "ciphering book"), has not been used. Thus, attention has not been focused on the importance of the cyphering tradition in the history of education.

A long, and educationally important, cyphering tradition had been translated from Europe to North America. The influence of that tradition was peaking in the 1820s (Ellerton & Clements, 2012), with Lincoln's manuscript being one of thousands of similar cyphering books prepared in the United States at that time. In schools, students aged 10 years or more who wished to study mathematics were expected to learn that subject through the cyphering approach. Teachers who had studied mathematics at school—and, certainly, not every teacher was in that category—had created their personal cyphering books and, as Dunlap (1959) and Ellerton and Clements (2012) have pointed out, most U.S. school students in the 1820s did not own an arithmetic textbook. If, when studying arithmetic, a student worked from a book, then that book was more likely to have been his or her teacher's cyphering book, or father's old cyphering book, etc., than an arithmetic textbook.

From our analysis, and from our knowledge of what North American cyphering books were like, we can say confidently that, from a school mathematics perspective, there was not much that was very special about Abraham Lincoln's cyphering book. The calligraphy and penmanship were fair, but certainly not better than average for cyphering books. Like Abraham Lincoln, students who prepared cyphering books often wrote little poems, and doggerel, in their books. Like Abraham, they often signed their names on pages throughout their book. Most of the solutions to problems in cyphering books were correct because that was part of the cyphering tradition—students might be punished if they entered "solutions" into their books that were incorrect, or untidy.

The order of the topics in Abraham Lincoln's cyphering book was typical of the order in other cyphering books—the four operations on whole numbers would be followed by "compound operations" (i.e., applying the four operations to quantities like money, time, weight, etc.). Then would follow "reduction" and, finally, the rule of three would be reached—which was an achievement to be remembered. Having cyphered to the rule of three was a definite qualification, one that was taken seriously by students, teachers, parents, school evaluation committees, prospective employers, and by college admissions officers looking for new students.

But Abraham Lincoln cyphered beyond the rule of three. That has not been well known, for three reasons. First, Lincoln himself said that he cyphered to the rule of three but did not mention that he cyphered beyond that rule. Second, most historians have not been aware of the leaf from Lincoln's cyphering book held by the Houghton Library, at Harvard University, which deals with the "double rule of three." And third, we are not aware of any Lincoln historian who has taken account of what is known as the *abbaco* arithmetic curriculum, which was an integral part of the cyphering tradition that controlled school mathematics. With that curriculum, simple interest, compound interest, and discount—all topics that Lincoln addressed in his cyphering book—were dealt with *after* the rule of three.

The *abbaco* curriculum, which had developed over many centuries (Ellerton & Clements, 2012; Van Egmond, 1980), found its expression in the order of treatment of topics in school mathematics (mainly arithmetic) textbooks and cyphering books. Because the extension of the Hindu-Arabic numeral system to decimal fractions had come rather late (after about 1600), decimal fractions came late in the *abbaco* sequence. Traditionally, common (or vulgar) fractions also came late in the *abbaco* sequence, probably because students found common fractions difficult. Simple interest, compound interest, and discount came earlier than decimal fractions and common fractions. Our analysis of young Abraham's efforts with simple and compound interest, and with discount, revealed that his lack of knowledge of decimal and common fractions forced him to deal with problems by using algorithmic procedures that any young learner would have found hard to understand.

We suspect that difficulties Abraham experienced when attempting to solve discount problems persuaded him that enough was enough. His last leaf was dated "*March 1st 1826*." "There were fences to build, grubbing to do, lands to clear, and the thousand-and-one other things to do, to prepare the land for crop" (quoted in Burlingame, 2008, p. 108). But such duties were not attractive to young Abraham Lincoln—the time was coming when he would move on from this segment of his education to a world that reached far beyond Pigeon Creek.

It is because of our knowledge of the cyphering tradition that we feel that we have been able to establish a more likely "order" for the pages which remain of Lincoln's cyphering book. We do not claim the last word on this sequencing task, and hope that analysis of the remaining pages of the manuscript, perhaps using modern technology, will throw further light on the order in which they were originally prepared.

There is a saying that it only takes a spark to set a fire going, and there is evidence that Abraham's early mathematical studies provided that spark. He would return to mathematics in later life, and recent analyses suggest that as a lawyer, and as President of the United States of America, he was profoundly influenced by the structure of reason that he found in mathematics (Hirsch & Van Haften, 2010).

Why is Abraham Lincoln's Cyphering Book Extraordinary?

If Abraham Lincoln's cyphering book was so ordinary, then why do we say that it was "extraordinary"? Our reason for doing so is that Abraham prepared his cyphering book in difficult circumstances, and his achievement in preparing a book that went a fair way beyond the rule of three was extraordinary. Over one-half of the 1000 or so cyphering books that Ellerton and Clements have examined were prepared in Massachusetts, or New York or Pennsylvania. Preparing a cyphering book in Boston or New York City, or Philadelphia—where there was a ready supply of paper, and flat, smooth desks and tables, and where qualified tutors were available to assist with difficult topics—was an altogether different thing from preparing a cyphering book on the uneven, crudely crafted surfaces in the school houses near Pigeon Creek. By the 1820s, relatively few students who prepared cyphering books in the Eastern states lived in small log cabins.

Of the 350, or so, North American cyphering books in the Ellerton-Clements collection only two were prepared in Indiana. Three were prepared in Tennessee, one in Kentucky, and none in Illinois. It is from that perspective that we recognize that Abraham Lincoln's cyphering book was remarkable. Our analysis has revealed that he made a huge effort to understand what he was doing for every problem he included in "his book." Sometimes, his teachers may have dictated problems to him, but it seems to us that he always solved them himself. We have linked many of the problems he solved to questions in published arithmetics (before our analysis, others had discovered that some questions could be found in arithmetics by Stephen Pike, Zachariah Jess, and Thomas Dilworth, and we have added the arithmetic by Francis Walkingame).

Sarah Bush Lincoln guarded and cherished her step-son's cyphering book for years. What a shame that this extraordinary book, prepared by a remarkable person, would be divided, and scattered to the winds. We trust this chapter goes some of the way to establishing why Abraham Lincoln's cyphering book represented an extraordinary achievement. That book did indeed represent his "hand and pen," and he would "be good."

References

A Teacher of Arithmetic. (1812). *A key to Dilworth's Arithmetic; wherein every question is worked out at full length, in a short and comprehensive manner.* New York, NY: Smith & Forman.

Adams, D. (1822). *The scholar's arithmetic—Or Federal accountant.* Keene, NH: John Prentiss.

Arnold, I. N. (1891). *The life of Abraham Lincoln.* Chicago, IL: A. C. McClurg & Company.

Azel W. Dorsey, Lincoln Pedagogue (1930, July 7). *Lincoln Lore, 65*, 1–2.

Banta, D. D. (1906). The early schools of Indiana: Third installment from the papers of D. D. Banta first published in the *Indianapolis News* in 1892. *The Indiana Quarterly Magazine of History, 2*(1), 131–138.

Barrett, J. H. (1860). *Life of Abraham Lincoln, of Illinois.* Cincinnati, OH: Moore, Wilstach, Keys & Co.

Bartelt, W. E. (2008). *"There I grew up": Remembering Abraham Lincoln's Indiana youth.* Indianapolis, IN: Indiana Historical Society Press.

Basler, R. P., Pratt, M. D., & Dunlap, L. A. (Eds.). (1953). *The collected works of Abraham Lincoln.* New Brunswick, NJ: Rutgers University Press.

Bennett, T. (1815). *New system of practical arithmetic peculiarly calculated for the use of schools in the United States* (5th ed.). Philadelphia, PA: Bennett & Walton.

Beveridge, A. J. (1958). *Abe Lincoln in Indiana.* Fort Wayne, IN: Public Board for Allen County.

Burlingame, M. (2008). *Abraham Lincoln: A life* (2 vols). Baltimore, MD: The Johns Hopkins University Press.

Butler, W. (1819). *An introduction to arithmetic designed for the use of young ladies* (3rd ed.). London, United Kingdom: Simpkin, Marshall & Co.

Chapman, T. J. (1895). Some early schools of Pennsylvania. *Pennsylvania School Journal, 44*, 205–209.

Clemens, S. (Mark Twain). (1885). *Adventures of Huckleberry Finn.* New York, NY: Charles L. Webster.

Clements, M. A., Keitel, C., Bishop, A. J., Kilpatrick, J., & Leung, F. (2013). From the few to the many: Historical perspectives on who should learn mathematics. In M. A. Clements, A. J. Bishop, C. Keitel, J. Kilpatrick, & F. Leung (Eds.), *Third international handbook of mathematics education* (pp. 7–40). New York, NY: Springer.

Cohen, P. C. (2003). Numeracy in 19th-century America. In G. M. A. Stanic & J. Kilpatrick (Eds.), *A history of school mathematics* (Vol. 1, pp. 43–76). Reston, VA: National Council of Teachers of Mathematics.

Colburn, W. (1822). *Arithmetic upon the inductive method of instruction being a sequel to intellectual arithmetic*. Boston, MA: Cummings and Hilliard

Courtenay, C. M., with illustrations by A. M. Turner & H. Kaucher (1917). *Abraham Lincoln*. New York, NY: Sam'l Gabriel Sons & Company.

Daboll, N. (1820). *Daboll's schoolmaster's assistant; improved and enlarged; being a plain practical system of arithmetic adapted to the United States*. Auburn, NY: Thomas M. Skinner.

Davis, V. J. (1890). *Jefferson Davis: Ex-President of the Confederate States of America* (Vol. 1). New York, NY: Belford Company.

De La Hunt, T. J. (1916). *Perry County: A history*. Indianapolis, IN: The W. K. Stewart Co.

Denniss, J. (2012). *Figuring it out: Children's arithmetical manuscripts 1680–1880*. Oxford, United Kingdom: Huxley Scientific Press

Dilworth, T. (1792). *The schoolmasters assistant; being a compendium of arithmetic both practical and theoretical*. New York, NY: Hugh Gaine.

Dilworth, T. (1806). *The schoolmasters assistant; being a compendium of arithmetic both practical and theoretical*. New York, NY: G. Jansen.

Donald, D. H. (1995). *Lincoln.* New York, NY: Simon & Schuster.

Dorfman, M. (1966a). Lincoln's arithmetic education: Influence on his life. *Lincoln Herald, 68*(2), 61–80.

Dorfman, M. (1966b). Lincoln's arithmetic education: Influence on his life: Part II. *Lincoln Herald, 68*(3), 108–120.

Dunlap, L. A. (1959). Lincoln's sum book. *Lincoln Herald, 61*(1), 6–10.

Ellerton, N. F., & Clements, M. A. (2012). *Rewriting the history of school mathematics in North America 1607–1861.* New York, NY: Springer.

Erekson, K. A. (2012). *Everybody's history: Indiana's Lincoln Inquiry and the quest to reclaim a President's past.* Amherst, MA: University of Massachusetts Press.

Funk, A. L. (1969). *A sketchbook of Indiana history*. Rochester, IN: Christian Book Press.

Goodspeed Bros. & Co. (1885). *History of Warwick, Spencer and Perry Counties, Indiana*. Chicago, IL: Author.

Grigsby, N. (1865, June 13). Interview with Nathaniel Grigsby. Reproduced in D. L. Wilson & R. O. Davis (Eds.). (1998). *Herndon's informants: Letters, interviews and statements about Abraham Lincoln* (pp. 111–115). Urbana, IL: University of Illinois Press.

Hanks, D. (1865, June 13). Interviews with Dennis Hanks. Reproduced in D. L. Wilson & R. O. Davis (Eds.). (1998). *Herndon's informants: Letters, interviews and statements about Abraham Lincoln* (pp. 38–42, and pp. 103–106). Urbana, IL: University of Illinois Press.

Herndon, W. H., & Weik, J. W. (1936). *Herndon's life of Lincoln: The history and personal recollections of Abraham Lincoln.* New York, NY: Albert & Charles Boni.

Hirsch, D., & Van Haften, D. (2010). *Abraham Lincoln and the structure of reason.* New York, NY: Savas Beatie LLC.

Houser, M. L. (1943). *Young Abraham Lincoln mathematician*. Peoria, IL: L. O. Schriver.

Høyrup, J. (2005). Leonardo Fibonacci and *abbaco* culture: A proposal to invert the roles. *Revue d'Histoire des Mathématiques, 11*, 23–56.

Indiana Department of Public Instruction. (1917). *Twenty-eighth biennial report of the State Superintendent of Public Instruction for the school years ending July 31, 1915 and July 31, 1916*. Fort Wayne, IN: Fort Wayne Printing Co.

Jess, Z. (1821). *The American tutor's assistant, improved; or, a compendious system of decimal, practical arithmetic.* Philadelphia, PA: M'Carty & Davis.

Karpinski, L. C. (1980). *Bibliography of mathematical works printed in America through 1850* (2nd ed.). New York, NY: Arno Press.

Kilpatrick, J., & Izsák, A. (2008). A history of algebra in the school curriculum. In C. E. Greenes & R. Rubenstein (Eds.), *Algebra and algebraic thinking in school mathematics: Seventieth yearbook* (pp. 3–18). Reston, VA: National Council of Teachers of Mathematics.

Koos, G. (2001). The Irish hedge schoolmaster in the American backcountry. *New Hibernia Review, 5*, 9–16.

Martin, N. (1800). *A new system of arithmetic, to which is added, a form of book-keeping by single entry.* Air, Scotland: J. & P. Wilson.

McKenney, F. (1825). *A complete key to the Teacher's Assistant, or System of Practical Arithmetic, compiled by Stephen Pike.* Philadelphia, PA: McCarty & Davis.

Miller, R. L. (2006). *Lincoln and his world: The early years.* Mechanicsburg, PA: Stackpole Books.

Nicolay, J. G. (1904). *A short life of Abraham Lincoln.* New York, NY: The Century Co.

Pike, N. (1788). *A new and complete system of arithmetic, composed for the use of citizens of the United States.* Newbury-Port, MA: John Mycall.

Pike, S. (1811). *The teachers assistant; or a system of practical arithmetic; wherein the several rules of that useful science, are illustrated by a variety of examples, a large proportion of which are in Federal money. The whole is designed to abridge the labour of teachers, and to facilitate the instruction of youth.* Philadelphia, PA: Johnstonn (sic.) and Warner.

Pike, S. (1822). *The teachers assistant or a system of practical arithmetic; wherein the several rules of that useful science, are illustrated by a variety of examples, a large proportion of which are in Federal money. The whole is designed to abridge the labour of teachers, and to facilitate the instruction of youth.* Philadelphia, PA: Benjamin Warner.

Plummer, M. A. (2001). *Lincoln's rail-splitter: Governor Richard J. Oglesby.* Urbana, IL: University of Illinois Press.

Reynolds, J. (2010, April 14). Reunited set of math problems gives clues to how Lincoln learned. *The State Journal-Register* (Springfield, IL), 1–2.

Schwarz, C. J. (1920). Lincoln's schooling and pioneer schools. *The Fortnightly Review, 27*(8), 50–51.

Simons, L. G. (1924). *Introduction of algebra into American schools in the 18th century.* Washington, DC: Department of the Interior Bureau of Education.

Singh, P., & Ellerton, N. F. (2013). International collaborative studies in mathematics education. In M. A. Clements, A. J. Bishop, C. Keitel, J. Kilpatrick, & F. Leung (Eds.), *Third international handbook of mathematics education* (pp. 827–860). New York, NY: Springer.

Smith, W. C. (1867). *Indiana miscellany: Consisting of sketches of Indian life, the early settlement, customs and hardships of the people.* Cincinnati, OH: R. P. Thompson.

Sterry, C., & Sterry, J. (1795). *A complete exercise book, in arithmetic: Designed for the use of schools in the United States.* Norwich, CT: John Sterry & Co.

Swett, L. (1889). Mr Lincoln's story of his own life. In A. T. Rice (Ed.), *Reminiscences of Abraham Lincoln by distinguished men of his* time (8th ed.) (pp. 455–468). New York, NY: North American Review.

Swetz, F. J. (1987). *Capitalism & arithmetic: The new math of the 15th century.* La Salle, IL: Open Court.

Tarbell, I. M., & Davis, J. M. (1896). *The early life of Abraham Lincoln containing many unpublished documents and unpublished reminiscences of Lincoln's early friends.* New York, NY: S. S. McClure.

Turnham, D. (1865). Interview with David Turnham. Reproduced in D. L. Wilson & R. O. Davis (Eds.), *Herndon's informants: Letters, interviews and statements about Abraham Lincoln* (1998: pp. 120–122). Urbana, IL: University of Illinois Press.

Van Egmond, W. (1980). *Practical mathematics in the Italian Renaissance: A catalog of Italian abbacus manuscripts and printed books to 1600.* Firenze, Italy: Istituto E Museo di Storia Della Scienza.

Van Natter, F. M. (1963). *Lincoln's boyhood: A chronicle of his Indiana years.* Washington, DC: Public Affairs Press.

Walkingame, F. (1802). *The tutor's assistant, being a compendium of arithmetic, and a complete question-book.* London, United Kingdom: C. Whittingham.

Wallis, P. J. (1963). An early best seller: Francis Walkingame's "The Tutor's Assistant." *The Mathematical Gazette, 47*(361), 199–208.

Walsh, M. (1820). *A new system of mercantile arithmetic: Adapted to the commerce of the United States, in its domestic and foreign relations; with forms of accounts, and other writings usually occurring in trade* (4th ed.). Salem, MA: John D. Cushing.

Warren, L. A. (1931). Lincoln's Hoosier schoolmasters. *Indiana Magazine of History, 27*(2), 104–118.

Warren, L. A. (1959). *Lincoln's youth: Indiana years seven to twenty-one 1816–1830.* New York, NY: Appletom.

Wickersham, J. P. (1886). *A history of education in Pennsylvania.* Lancaster, PA: Inquirer Publishing Company.

Willetts, J. (1822). *The scholar's arithmetic designed for the use of schools in the United States.* Poughkeepsie, NY: Paraclete Potter.

Wilson, D. K. (1935). *The history of mathematical teaching in Scotland to the end of the eighteenth century*. London, United Kingdom: University of London Press.

Wilson, D. L., & Davis, R. O. (Eds.). (1998). *Herndon's informants: Letters, interviews and statements about Abraham Lincoln.* Urbana, IL: University of Illinois Press.

Wilson, W. E. (1940). *The Wabash.* New York, NY: Farrar & Rinehart.

Chapter 7
Sacrobosco's Heritage: Thomas Dixson's Cyphering Book, 1630–1632

Abstract: The 302-page handwritten manuscript by Thomas Dixson which is the focus of attention in this chapter is 46 years older than the oldest of the British cyphering books mentioned by John Denniss (2012). The analysis presented here suggests that the curriculum sequence put forward by Johannes de Sacrobosco early in the thirteenth century, for the study of arithmetic, may have been more influential in seventeenth-century England than Leonardo of Pisa's *abbaco* curriculum sequence. Some attention in this chapter is given to the so-called "galley" algorithm for division, which Dixson used. In keeping with the standard Sacrobosco curriculum sequence, Thomas Dixson included, early in his arithmetic studies, sections on arithmetical and geometrical progressions, and on common fractions—those topics would have come much later if the *abbaco* curriculum had been followed.

Key Words: Abbaco curriculum; arithmetical progressions; ciphering book; common fractions; cyphering book; galley method for division; geometrical progressions; Leonardo of Pisa; problem solving; Robert Recorde; Sacrobosco; vulgar fractions.

The oldest of the extraordinary cyphering books that we will study was prepared in England between 1630 and 1632. The author was a certain Thomas Dixson (who sometimes spelled his family name as Dixon). This cyphering book has 302 folio-sized pages and is currently held within the Plimpton Collection in the Butler Library at Columbia University, New York (where it is located within the post-1600 collection, and has call number MS 511.1630). It is in excellent condition.

The topics dealt with in this cyphering book were, in order, addition, subtraction, multiplication, and division (including applications to money, time, weight), reduction (including applications to money, time, weight), arithmetical and geometrical progressions, single and double rule of three (including with common fractions), numeration of fractions (including unit fractions, and fractions of the form m/n when $m \neq 1$), abbreviating (i.e., canceling) fractions, reduction of fractions, addition, subtraction, multiplication and division of common fractions, fellowship, and loss and gain. The last few pages were prepared by a John Dixon in 1701—presumably, John was a descendant of Thomas, and by 1701 the spelling of Dixson had been relaxed to Dixon.

The approach adopted by Thomas Dixson when preparing his manuscript is well illustrated in Figure 7.1, which shows a page concerned with the topic "Reduction." This topic, which would be found in cyphering books throughout the period 1200–1850, called for a quantity expressed in a relatively "large" unit (e.g., a day) to be reduced to the same quantity only expressed in a "smaller" unit (e.g., an hour). In some cyphering books (but not in Thomas Dixson's) some reduction tasks were also concerned with the reverse direction. The example at the top of the page shown in Figure 7.1 asked: "In one year, how mani days," and the example at the bottom of the page asked: "In one year how mani days, hours and minits." With respect to Figure 7.1, the attention of the reader is directed to the following:

- The high standard of calligraphy for the first main heading;
- The "old-English" spelling of "tyme" and of other words such as "mani" and "minits;"
- The absence of question marks;

Nerida F. Ellerton and M. A. (Ken) Clements, *Abraham Lincoln's Cyphering Book and Ten other Extraordinary Cyphering Books*, DOI 10.1007/978-3-319-02502-5_7,

- How the page was carefully divided into sections by double vertical, horizontal and oblique line segments. It appears that this setting out was designed before any writing on the page occurred.
- Thomas Dixson indicated that this was his work—"Thomas Dixson booke 1630." This indication of ownership can be found on many pages of the manuscript.
- With both Julian and Gregorian calendars there were 365 days in a year. Multiplication by 365 was achieved by multiplying by 13, then by 4, then by 7, and finally adding 1. In both examples, the multiplications were checked by galley division (discussed later in this chapter).

The next question in the Dixson manuscript stated: "In 12 years ½ how many months, weeks, days and hours." The answers were checked using a sequence of galley divisions. Then followed a carefully graded set of exercises on time. After that, Thomas Dixson moved on to reduction of "waight," the units being pounds and "ounzes," and reduction tasks dealing with lengths, etc. Altogether, 23 pages were devoted to reduction.

The standard of calligraphy and penmanship in the Dixson manuscript was high, and each page was beautifully ruled so that what was written on a page was well-arranged. There is an impressive amount of arithmetic recorded, and when reading Thomas's entries in the cyphering book we did not get the impression that he was merely copying from a textbook or from another cyphering book. Although we can never know for certain, we believe that he personally completed each of the many calculations that are shown. Almost all of the calculations are correct.

There are no explanations given directly, and in that sense the genre adopted by Thomas Dickson was different from the "Introduction-Rule-Case-Example-Exercise" (or IRCEE) genre found in many early North American cyphering books (Ellerton & Clements, 2012).

But why do we regard this cyphering book as "extraordinary," and worthy of special attention? After all, we do not know much about the circumstances surrounding the preparation of the book, or where Thomas Dixson was living when he prepared his manuscript. None of Thomas's teachers was named, and the place where he prepared "his booke" (his preferred spelling) was never stated. In the sequence of topics studied, arithmetical and geometrical progressions appeared earlier than usual in the *abbaco* tradition. There are no multi-colored, or even black-and-white ink diagrams.

This Dixson cyphering book is "extraordinary" not only because it was prepared in England in the early 1630s, but also because its 302 pages covered a wide range of topics in considerable detail. It was prepared more than 40 years before a 14-page 1676 cyphering book, which is the earliest of the British cyphering books noted by Denniss (2012).

The Dixson cyphering book appears to have been prepared with the kind of confidence that points to an established tradition. The pages in the manuscript were dated, on numerous occasions, between 1630 and 1632, and the very neat entries indicate that Thomas was confident that he knew what he was doing. It seems unlikely that his cyphering book was one of a kind. In fact, it could offer a model for the state of mathematics education which existed in England when Jamestown was settled in Virginia in 1607, when the Pilgrims established the Plymouth colony in 1620, and when the Puritans began to develop the Massachusetts Bay Colony around 1630. From a historical perspective, the book is, undoubtedly, worthy of careful study.

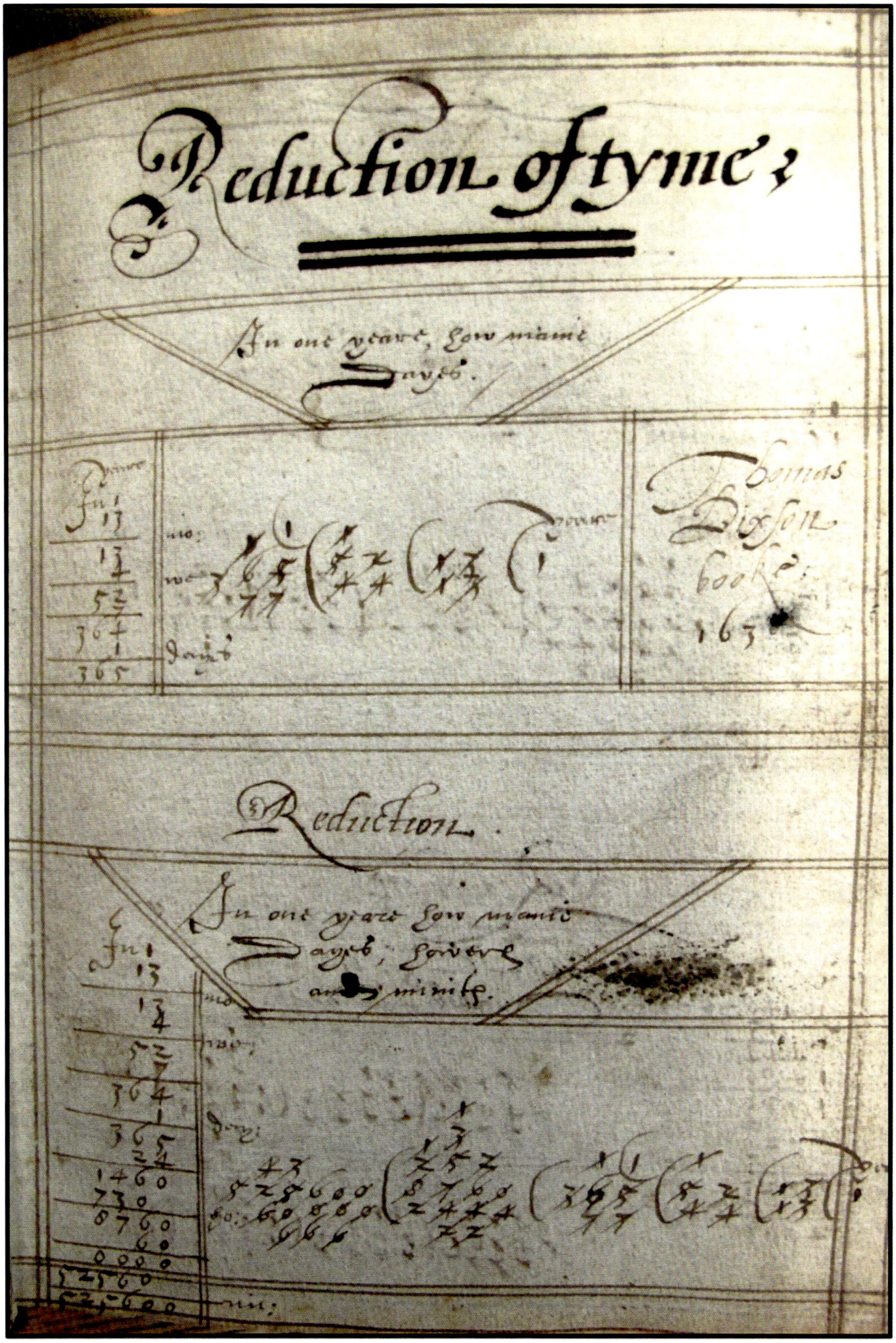

Figure 7.1. "Reduction of tyme" from Thomas Dixson's cyphering book (George A. Plimpton Collection, Rare Book & Manuscript Library, Columbia University in the City of New York).

As far as we know, there has been no scholarly analysis of cyphering books prepared in England during the first half of the seventeenth century. According to John Denniss (2012), there are two large collections of cyphering books in England. The second largest is held by Denniss himself and the largest, which was originally put together by John Hersee, is held by the Mathematical Association. The Denniss collection comprises about 125 manuscripts and the Hersee collection about 200 manuscripts (Stedall, 2012). The oldest of the cyphering books in those collections is dated 1684, but Denniss has a facsimile copy of a 14-page manuscript dated 1676. We (Ellerton and Clements) own 100 British cyphering books but all of ours were prepared between 1749 and 1888. Yet, Thomas Dixson's cyphering book, held at Columbia University, New York, is clearly dated (on numerous pages) between 1630 and 1632. Its very existence provides the opportunity to investigate questions about mathematics education in England during the first half of the seventeenth century.

It has become received tradition among British historians (e.g., Howson, 1982; Stedall, 2012) that, during the first half of the seventeenth century, very little mathematics worthy of that name was studied in English schools. However, we would argue that in the first half of the seventeenth century in England, much arithmetic was studied and recorded within the context of business. We would not deny that relatively few upper-class English schools at that time provided a form of mathematics education which went beyond what was offered in the so-called "dame schools"—those small, private institutions that cared for children up to about the age of 10 (Cubberley, 1920; Monroe, 1917). Most upper-class schools concentrated almost entirely on Latin grammar, with hardly any mathematics being taught (Howson, 1982). According to tradition, there had developed an attitude among British genteel classes that the study of arithmetic should be confined to future clerks (Jones, 1954). Indeed, Eva Taylor (1954) claimed that around 1600 the upper classes in England regarded mathematics as worthless, something to be spoken of in the same breath as astrology and demonology. Thus, it is not surprising that some historians (e.g., Howson, 1982) have claimed that around 1600 Great Britain was well behind other leading European nations—such as the Netherlands and France—in its provision of mathematics education.

Robert Recorde (1549), the famous author of *The Grounde of the Artes,* maintained that the quality of mathematics education in England in the first half of the seventeenth century was well below that in other leading European nations. After lamenting "the unfortunate condition of England, seeing so many great clerks to arise in sundry other parts of the world, and so few to appear in this our nation," Recorde (1549) went on to say that in England there was a "contempt," or disregard of learning, and that anyone who tried to teach mathematics to others was likely to be "derided and scorned, and so utterly discouraged to take in hand any like enterprise again" (p. 212). Yet, despite what Recorde stated, it appears to have been the case that by the end of the 16th century there had emerged a group of merchant adventurers in London who were vitally interested in arithmetic procedures of the *abbaco* variety (Baker, 1568, 1687; Davis, 1960).

Not all scholars have agreed with Recorde's summary of the situation. Florence Yeldham (1926), for example, drew attention to the influence of what she called "Arabic arithmetic in England" during the period 1200–1650. She offered a long list of English scholars who were "known to have taught or written upon numbers" (p. 92). She paid particular attention to the influence of John Holywood—or, as he was better known, Johannes de Sacrobosco. Sacrobosco may or may not have been born in England, but it is certain that he taught at the University of Paris during the first half of the thirteenth century—a time when Leonardo of Pisa's (Fibonacci's) text *Liber Abbaci* was making Hindu-Arabic numerals and operations known to European merchants (Franci, 2009; Long, McGee & Stahl, 2009; Van Egmond, 1976, 1980).

Lambert Jackson (1906) and Frank Swetz (1987) are among many scholars to have claimed that during the fifteenth century four different types of mathematics vied for recognition within

Western Europe: "theoretical tracts, *algorisms*, *abacus* arithmetic, and *computi*" (Swetz, 1987, p. 27). For Swetz, "theoretical tracts" were arithmetic derived from the writings of Anicius Manlius Severinus Boethius (c. 475–524), which focused on logical connections. That kind of mathematics held sway in European universities as part of the so-called "quadrivium." In this arithmetic, numbers were often still represented by Roman numerals (Ifrah, 2000; Howson, 1982). By contrast, European scholars like Leonardo of Pisa, Johannes de Sacrobosco, and Roger Bacon, called for less attention to be given to the theoretical tracts, and more to calculations based on *algorisms* (or algorithms) developed with respect to the ten Hindu-Arabic numerals, 0, 1, 2, …, and 9 (Pedersen, 1985; Smith, 1908; Yeldham, 1926). They de-emphasized the importance of arithmetic associated with the use of abacuses or other physical counting aids that had been commonly used at "counting tables," and of *computi,* which were almanac-like calculations associated with calculations for determining dates of movable religious occasions (like Easter) (Reynolds, 1993; Smith, 1908; Swetz, 1987). The *algorism* approach was consistent with the *abbaco* tradition that increasingly held sway among the business communities of Europe (Ellerton & Clements, 2012; Van Egmond, 1976, 1980).

We have argued elsewhere (Ellerton & Clements, 2012) that the cyphering tradition in Europe probably originated with Leonardo of Pisa (or Fibonacci), and that over time a well-established sequence of topics could be found in most cyphering books and in most arithmetic textbooks. Table 7.1 is based on the analysis of the *abbaco* sequence of topics that we presented in Ellerton and Clements (2012).

Table 7.1
The Abbaco Sequence of Arithmetic Topics 1200–1850

Elementary *Abbaco* Topics	Intermediate *Abbaco* Topics	More Advanced *Abbaco* Topics
Numeration	Double rule of three	Alligation (mixing quantities)
Addition	Inverse rule of three	Fellowship (the arithmetic of partnerships)
Subtraction	Tare and tret	Single and double position
Multiplication	Loss and gain	Involution (square and cube roots, etc.)
Division	Simple interest	Evolution (squaring, cubing, etc.)
Compound operations (four operations on quantities)	Compound interest	Arithmetical progressions
Reduction	Discount	Geometrical progressions
Practice	Vulgar (i.e., common) fractions	Permutations and combinations
Direct rule of three	Decimal fractions (after about 1650)	Mensuration of superficies
	Annuities	Logarithms (after about 1650)
	Equation of payments	Fortification (occasionally)
	Gauging	

We have examined about 1000 North American cyphering books, and are prepared to say that in the 18th century and the first half of the 19th century, the order implied by Table 2.1 was fairly rigidly adhered to by most North American students who studied arithmetic. Sometimes, the order was not totally consistent with that shown in Table 2.1, but variations were only minor. Yet, from a mathematical sequencing point of view, the order was not ideal—note, for example, that simple and compound interest, and discount, were usually studied before common and decimal fractions.

But the order of topics in the Dixson cyphering book departs significantly from the order shown in Table 2.1 because "arithmetical progressions" and "geometrical progressions" were dealt with immediately after "reduction." Then, immediately after the rule of three came a detailed treatment of common (or vulgar) fractions, occupying over 100 pages. This was followed by a reconsideration of the rule of three, but now fractions were involved. Finally "fellowship" problems, involving fractions, were solved.

These differences prompted us to look carefully at the order of treatment of topics in five popular early English arithmetics—by Recorde (1549), Wingate (1630), Cocker (1678), Dilworth (1784), and Walkingame (1785). We found that chapters on numeration, addition, subtraction, multiplication, division, and reduction were always the first chapters in a book—occasionally, there was an additional chapter, concerned with what was called "compound operations," when the operations were applied to quantities (money, lengths, areas, volumes, capacities, weights and measures), but usually such applications were dealt with in the chapters on numeration, the four operations, and reduction. Of the five authors, only Recorde treated progressions before the various rules of three. Common (or "vulgar") fractions were dealt with quite early by Recorde, but much later by Walkingame and Dilworth. In all of the texts, the applied topics of loss and gain, equations of payment, simple and compound interest, discount, barter, exchange, fellowship, alligation, and false position, came late in the textbook. No textbook had a chapter titled "Percentage," even though four of them—the exception was Cocker—had chapters on simple and compound interest. In Recorde's and Wingate's texts, chapters on common fractions appeared before chapters on simple and compound interest but that was not the case with Walkingame's and Dilworth's texts.

Elementary counting and operations on small numbers characterized arithmetic in "dame schools," or with privately-employed governesses or tutors, and moving on from there marked an important educational stage for a young child. Typically a 10-year-old child—usually a boy—would triumphantly begin his cyphering book by elaborately decorating a title page (Burton, 1833). He would then devote the next page to the topic of numeration, and then would follow addition, subtraction, multiplication, division, compound operations, reduction, practice and the direct rule of three. Reaching the rule of three represented a watershed, and if the child progressed beyond this unofficial marker then he (and in most cases it was a "he") would proceed to study topics like the double rule of three, the inverse rule of three, tare and tret, loss and gain, simple and compound interest, discount, vulgar (or common) fractions, annuities, and equation of payments. This represented the second marker. After the mathematical basis for decimal fractions had been more carefully elaborated in the seventeenth century, some students would also study decimal fractions at this intermediate stage. Those who moved on to the more advanced forms of arithmetic would study the topics shown in the third column of Table 2.1.

We have not been able to find many North American or British cyphering books prepared in the 17th century, and so our analysis of the Dixson manuscript, as well as of the small manuscript, probably prepared in Maine (New England) in the 1660s—which was examined in Chapter 2 of this book—can throw some light on the extent to which the *abbaco* curriculum sequence was followed at that time.

By the end of the 16th century the *algorism*-type of arithmetic, with its emphasis on Hindu-Arabic numeration and the use of standard procedures for addition, subtraction, multiplication and division for whole numbers, fractions, and compound numbers (i.e., the arithmetic associated with quantities such as time, weight, length, area, volume, etc.), had won the day among the business communities of Europe (Davis, 1960; Howson, 1982; Ifrah, 2000; Jackson, 1906; Knox, 1788; Swetz, 1987; Yeldham, 1926, 1936). Business houses employed skilled reckoners to develop special rules and cases that would enable them to solve numerical problems associated with partnerships, with mixing quantities, with calculating taxes, and with loss and gain, barter, simple and compound interest, discount, and proportion. Those responsible for training future reckoners developed treatises in which problem types, and associated rules and cases, were set out clearly. These would be recorded in cyphering books—usually prepared by chosen boys, but sometimes by girls, who had reached the age of 10—and it was emphasized to the students that those books would become their calculational guidebooks for life. Merchants recognized that it was profitable to employ competent arithmeticians of the *algorism* variety because they could use numbers to quantify what had happened, what was happening, and what might happen in virtually every area of business.

According to Yeldham (1926, 1936), in England in 1600, "sheer usefulness had won the day, and the older theoretical arithmetic of Boethius, heavily handicapped from the time *algorism* was introduced, was left almost unread" (Yeldham, 1936, p. 12). Employers sought out potential employees who had been trained by reputable reckoning masters in market-place arithmetic and who could represent them in their dealings with merchants and with reckoning masters. Therefore, at the beginning of the 17th century the demand for *algorism* practitioners increased rapidly right across Western Europe—even in nations as isolated as Iceland (Bjarnadóttir, 2009).

People with inherited property and influence in high society were not always convinced of the worth of this "new" mathematics (Davis, 1960; Ifrah, 2000; Recorde, 1549; Swetz, 1987). For many of them, anyone who worked on or with mathematics was to be regarded as a peddler of the purely mechanical kind (Money, 1993; Taylor, 1954). Such attitudes were lurking within the world-views of religiously-oriented settlers, and even of some businessmen of English origins, who became part of the European communities in the far-flung North American settlements.

The Galley Method of Division

We begin our analysis of the Dixson cyphering book by looking at how Thomas Dixson used what was known as the "galley" method for division. This was so called because some thought that the visual pattern after applying the algorithm often looked like a ship (or galley—see Figure 7.2, which shows an embellished version of the kind of written record an application of the method would produce).

Sometimes the galley method was called the "scratch method" because digits were scratched out and replaced with new ones during the application of the algorithm. According to Karpinski (1925), this method is thought to have originated with Chinese or Indian mathematicians, and often the temporary recording of a calculation occurred by making scratch marks in the sand (Lam, 1966; Lambert, 1906). This galley or scratch method was used for centuries for the purpose of carrying out division calculations, especially in India and in Arab nations, Europe, and in colonial settlements in North and South America. In the 17th century, and even into the 18th century, it was still being used in many European nations—including England, Germany, and Spain (see Long, McGee & Stahl, 2009; Recorde, 1549; Smith, 1908; Swetz, 1987)—and in North and South America.

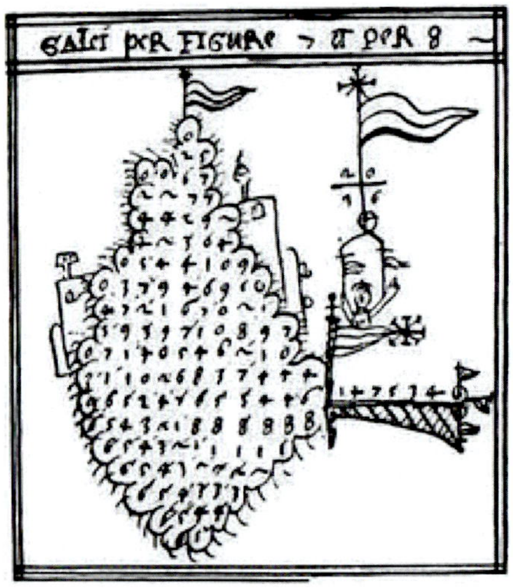

Figure 7.2. This illustration showing a decorated form of galley division (this diagram, reputedly drawn by Honoratus, a Venetian monk, has been reproduced from Smith (1908, p. 487)).

The following example of galley division (Figure 7.3), with 494865 divided by 58, is taken from Thomas Dixson's manuscript.

		~~1~~								
		~~5~~	~~1~~	~~0~~						
	~~3~~	~~0~~	~~3~~	~~2~~	9					
4	~~9~~	4	~~8~~	~~6~~	~~5~~	/	8	5	3	2
	5	8	8	8	8					
		~~5~~	~~5~~	~~5~~						

Figure 7.3. Thomas Dixson's use of the galley method for dividing 494865 by 58.

Those familiar with this algorithm would quickly recognize that

$$\frac{494865}{58} = 8532 \text{ and 9 remainder, or } \frac{494865}{58} = 8532\,\frac{9}{58}$$

The actual setting out by Thomas Dixson is shown in Figure 7.4, which also shows how Thomas showed two checks for his calculations—one check used the famous casting-out-nines procedure and the other used long multiplication.

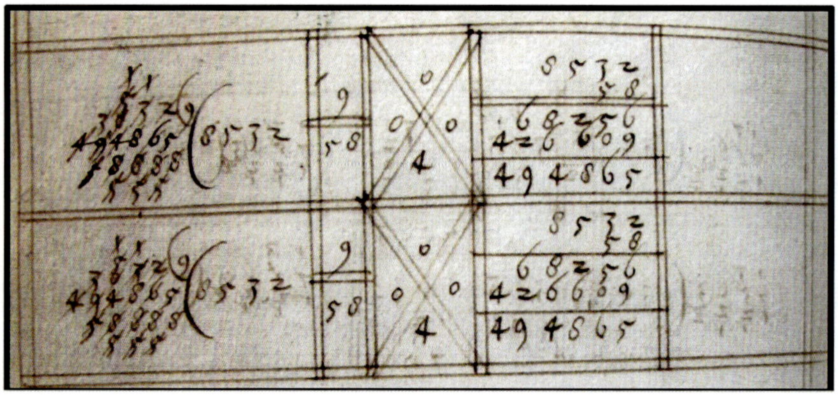

Figure 7.4. Thomas Dixson's entry for dividing 494865 by 58 (George A. Plimpton Collection, Rare Book & Manuscript Library, Columbia University in the City of New York).

A 21st-century reader might wonder what Thomas was thinking when he did these calculations. In Figure 7.3, we first wrote the dividend (i.e., 494865), followed by a bar. Then below the dividend, and one space in from the left side of the dividend, we wrote the divisor (i.e., 58). After the bar (next to the dividend), we will write the whole-number part of the quotient (i.e., 8532) as that emerges during the calculations. Thus:

4	9	4	8	6	5	/
	5	8				

Now, as in the modern algorithm, we divide 494 by 58. The whole-number part of the quotient is 8, so we write that in the space for the quotient at the right of the bar. Thus:

~~4x~~	9	4	8	6	5	/	8
	~~5x~~	8	8				
		5					

Next, we multiply each digit of the divisor by the quotient digit, 8, one number at a time. First, we say, 8 times 5 is 40, and 40 from 49 is 9. We cross out the 4 and the 5, and leave the 9—the crossing out is indicated by an x and a horizontal bar. Then we say 8 times 8 is 64, and subtract that from 94, leaving 30, which is placed above the 94 (with the 8 and the 94 being crossed out). We then rewrite the divisor, 58, with the 5 one line below where the 58 was before, only moved one space to the right, and the 8 moved up one. Thus:

	3	0						
~~4x~~	~~9x~~	~~4x~~	8	6	5	/	8	5
	~~5x~~	~~8x~~	8					
		5						

Then, as in the modern algorithm, we divide 308 by 58. The whole-number part of the quotient is 5, so we write that to the right of the 8 in the space for the whole number part of the quotient at the right of the bar. We calculate 5 times 58, and subtract the product (290) from 308, leaving 18. The divisor (58) is re-inserted, and overall the display is changed as shown:

		1						
	~~3x~~	~~0x~~						
~~4x~~	~~9x~~	~~4x~~	8	6	5	/	8	5
	~~5x~~	~~8x~~	~~8x~~	8				
		~~5x~~	5					

Once again, as in the modern algorithm, we divide 186 by 58. The whole-number part of the quotient is 3, which is written to the right of the 8 and the 5 in the space for the whole number part of the quotient at the right of the bar. We calculate 3 times 58, and subtract the product (174) from 186, leaving 12. The divisor (58) is re-inserted, and overall the display is changed as shown:

		~~1x~~	1							
	~~3x~~	~~0x~~	~~3~~	2						
~~4x~~	~~9x~~	~~4x~~	~~8x~~	~~6x~~	5	/	8	5	3	2
	~~5x~~	~~8x~~	~~8x~~	~~8x~~	8					
		~~5x~~	~~5x~~	5						

Finally, as in the modern algorithm, we divide 125 by 58. The whole-number part of the quotient is 2, which is written to the right of the 8, the 5, and the 3 in the space for the whole number part of the quotient, and overall the display is changed as shown:

		~~1~~	~~1~~	0						
	~~3~~	~~0~~	~~3~~	~~2~~	9					
~~4~~	~~9~~	~~4~~	~~8~~	~~6~~	~~5~~	/	8	5	3	2
	~~5~~	~~8~~	~~8~~	~~8~~	~~8~~					
		~~5~~	~~5~~	~~5~~						

We can conclude that 4984865 divided by 58 equals 8532 and 9 remainder.

We do not know whether Thomas Dixson wondered *why* the galley method worked. He probably did not, and was happy to learn merely to use and apply the algorithm, in example after example. A 21st century, mathematically-inclined person, however, might wonder, "What's going on?" What is the mathematical basis for this galley algorithm?

Essentially, what we did was:

- Divide 464000 by 58, and then subtract the 464000 from 494865 (to get 30865);
- Divide 29000 by 58, and then subtract the 29000 from 30865 (to get 1865);
- Divide 1740 by 58, and then subtract the 1740 from 1865 (to get 125);
- Divide 116 by 58, and then subtract the 116 from 125 (giving 9 remainder).

Notice that $\dfrac{464000}{58}=8000$, $\dfrac{29000}{58}=500$, $\dfrac{1740}{58}=30$, and $\dfrac{116}{58}=2$.

Although Thomas Dixson never used algebra, it is interesting to recognize that algebraically the galley procedure relied on the identity:

$$\frac{a+b+c+d+\ldots}{K}=\frac{a}{K}+\frac{b}{K}+\frac{c}{K}+\frac{d}{K}+\ldots$$

The idea is to find *a, b, c, d, ...* which are successive multiples of 10^nK, the *n*'s being integers.

Figure 7.5 shows how 9685236 could be divided, galley-style, by 5239. The dividend 9685236 has its digits crossed out, the divisor 5239 is the first four digits of the entries immediately below the dividend, the whole-number part of the quotient (1848) is to the right of the dividend, and the remainder is made up of digits that were not crossed out (3564, to the right of the number above the dividend—start at 35 on the top, go down to 6, and then to 4):

		~~4~~	3	5							
		~~5~~	~~5~~	~~6~~							
		~~6~~	~~7~~	~~8~~							
		~~8~~	~~5~~	~~4~~	6						
	~~2~~	~~4~~	~~2~~	~~8~~	~~3~~						
~~4~~	~~4~~	~~5~~	~~6~~	~~0~~	~~7~~	4					
~~9~~	~~6~~	~~8~~	~~5~~	~~2~~	~~3~~	~~6~~	/	1	8	4	8
~~5~~	~~2~~	~~3~~	~~9~~	~~9~~	~~9~~	~~9~~					
	~~5~~	~~2~~	~~3~~	~~3~~	~~3~~						
		~~5~~	~~2~~	~~2~~							
			~~5~~								

Figure 7.5. Applying the galley algorithm, to find the value of 9685236 divided by 5239.

It followed that

$$\frac{9685236}{5239} = 1848 \text{ and } 3564 \text{ remainder, or } \frac{9685236}{5239} = 1848\,\frac{3564}{5239}.$$

Thomas Dixson's setting out for the same task is shown in Figure 7.6 (in which Thomas has also shown a casting-out-nines check and a long-multiplication check).

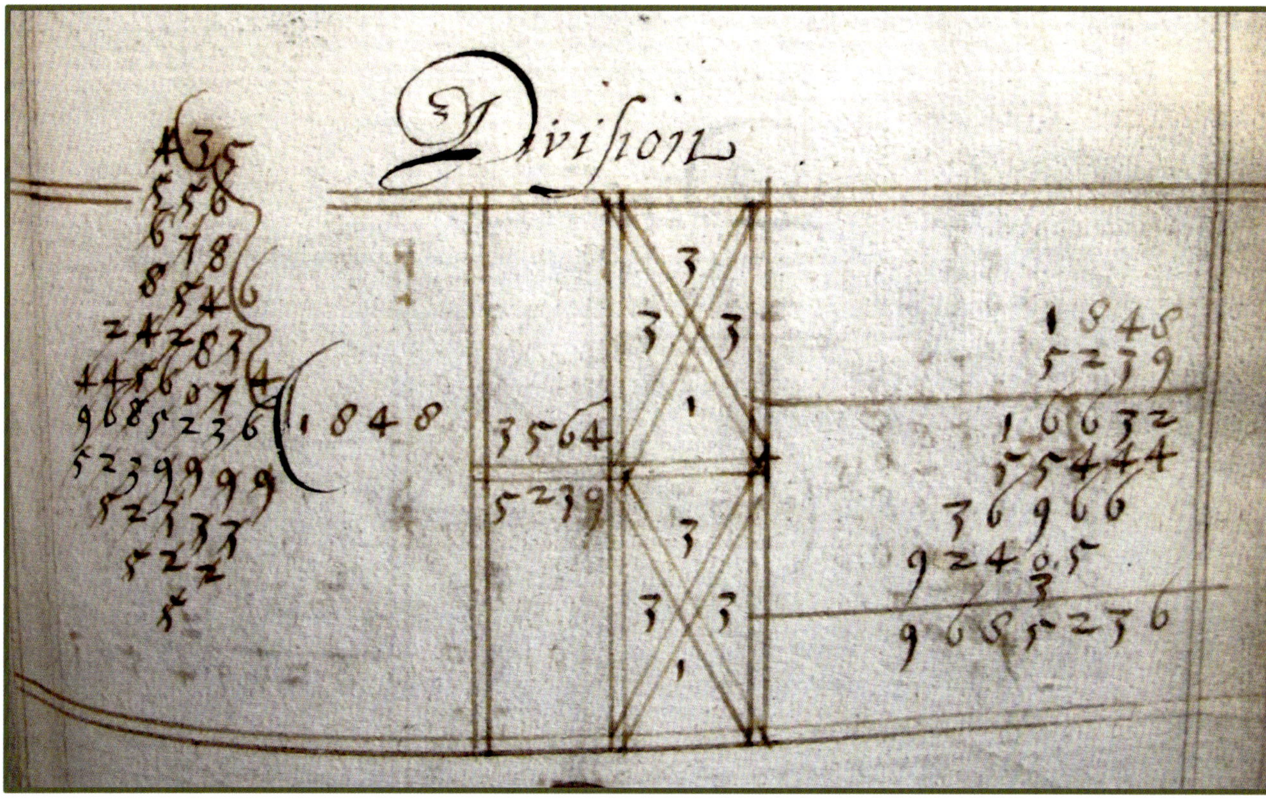

Figure 7.6. Thomas Dixson's galley-division for 9685236 divided by 5239 (George A. Plimpton Collection, Rare Book & Manuscript Library, Columbia University in the City of New York).

Thomas Dixson and Common Fractions

Common or Vulgar Fractions

On at least 90 of the 302 pages in Thomas Dixson's cyphering book there are calculations in which common fractions are involved. Many of the pages are concerned with the four operations for fractions—addition, subtraction, multiplication and division—and some are concerned with "abbreviation" of fractions (that is to say, reducing them to their "lowest form"). Some are concerned with writing fractions which have different denominators so that they will have a common denominator.

The treatment of fractions is through what might be called a problem-solving approach, although the contexts of the problems were not "real-life." Thus, for example, Figure 7.7 shows Thomas's response to the problem "If $\frac{5}{8}$ required $\frac{45}{64}$, what requireth $\frac{2}{3}$ after the same proportion?"

Thomas approached this problem via the rule of three. He multiplied $\frac{45}{64}$ by $\frac{2}{3}$, and then divided that product by $\frac{5}{8}$. The answer was obtained by multiplying 45 by 2 by 8 to get 720, then 64 by 3 by 5 to get 960, and then simplifying the fraction $\frac{720}{960}$, to get $\frac{3}{4}$. That was the correct answer, but how Thomas learned the algorithm which he applied is, of course, something that will never be known. Nevertheless, Thomas felt sufficiently proud of what he had just done to be moved to write: "Thomas Dixson his Ciphering booke Anno Domini 1630."

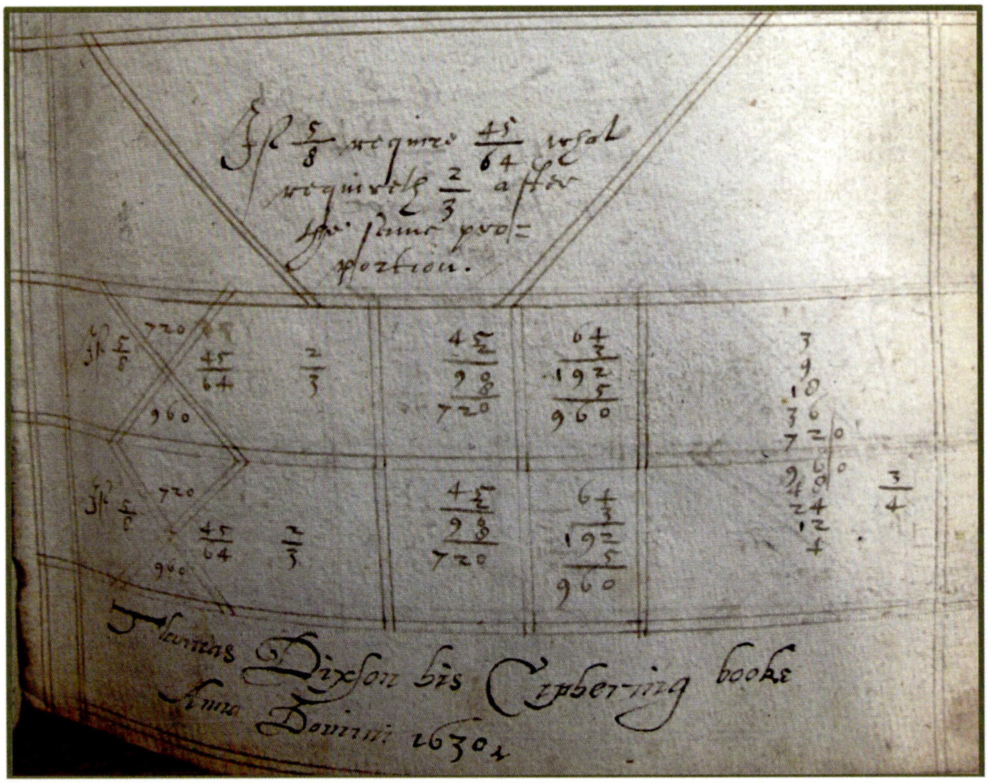

Figure 7.7. "If $\frac{5}{8}$ required $\frac{45}{64}$, what requireth $\frac{2}{3}$ after the same proportion?" (George A. Plimpton Collection, Rare Book & Manuscript Library, Columbia University in the City of New York).

Thomas used the same algorithm to find answers to other context-free tasks like "If $\frac{6}{7}$ admitt $\frac{8}{15}$, what admitteth $\frac{2}{3}$ after the same proportion?" He also applied the algorithm to solve slightly more contextual problems like: "If $\frac{3}{4}$ of the same goods cost 15 pounds, how much shall $7\frac{5}{6}$ cost at the same rate?" Thomas recorded his solutions to problem after problem, and almost always his working was correct. We shall never know whether he could apply the algorithms in real, out-of-school situations, but he learned how to respond correctly to such tasks on the pages of his cyphering book. On reading his manuscript we did not get the impression he merely copied, or applied "rules without reason," but we cannot be sure of that.

On no page in Thomas Dixson's 302-page book is there any mention of decimal fractions. It would be surprising if it had been otherwise for it was only in 1585 that Simon Steven had formalized the concept of a decimal fraction and had shown how decimals could be useful in everyday practices (De Morgan, 1847). The lag time from 1585 to 1630 was too short for decimal fractions to have been incorporated into elementary mathematics education.

Whose System of Arithmetic Topics Held Sway in Mathematics Education in Early 17th-Century England—Sacrobosco's or Leonardo of Pisa's?

The life of Johannes de Sacrobosco (or Ioannis de Sacro Bosco, c.1195–c.1256) is shrouded in mystery. Although one cannot be sure exactly when or where he was born, or where he died (Pedersen, 1985), many British scholars, including Augustus de Morgan (1847) and Florence Yeldham (1926), have been adamant that he was born somewhere in Great Britain. Some scholars believe he was born in England, some in Scotland, and some in Ireland. He probably studied at Oxford University, and he definitely taught at the University of Paris in the 1220s.

In the short introduction to *Algorismus Vulgaris*, or *Communis*, Sacrobosco provided a popular introduction to the Hindu-Arabic numeration system. Many manuscript copies of *Algorismus Vulgaris* survive today (Yeldham, 1926).

Of special relevance to this chapter is that Sacrobosco introduced the Hindu-Arabic numeration system in a slightly different way from Leonardo of Pisa, especially in relation to the order of topics after numeration, the four operations on integers, and reduction. It was Sacrobosco's order that is to be found in Thomas Dixson's cyphering book—and also in Robert Recorde's (1549) *The Ground of the Arts*. In particular, Thomas Dixson, and Recorde followed immediately upon their sections on reduction with a section on progressions. An examination of the entries in Table 2.1 will show that this was much earlier than would be expected under the standard ordering of *abbaco* arithmetic topics.

Thomas Dixson's first entry on arithmetical progressions is shown in Figure 7.8. Although we have not been able to decipher all of the words in Figure 7.7, it is clear that the "totall" of 1 + 2 + 3 + 4 + 5 + 6 + 7 + 8 + 9 was required. The indication that a multiplication by 5 occurred suggests that Thomas found the sum of the nine digits by calculating (1 + 8) + (2 + 7) + (3 + 6) + (4 + 5) + 9. This interpretation of an arithmetical progression being linked to the sum of terms was also evident in Thomas Dixson's second example in the section, which required him to find the sum of all integers from 1 through 19. This was found by multiplying 19 by 10.

On the next page, Thomas tackled two problems concerned with the sum of successive odd numbers. The first problem was concerned with 1 + 3 + 5 + 7 + 9 + 11 + 13, and Thomas answered this by calculating the product of 7 and 7. The second problem required him to find the sum of the odd numbers from 1 through 31, and he determined that by multiplying 16 by 16. Although he did not explain why he solved the problems using the methods he did, it is unlikely he used the result that the sum of the first n odd numbers is n^2. Rather, it seems likely, that he looked for patterns. He then found 2 + 4 + 6 + 8 + 10 + 12 by calculating the product of 6 and 7, which could have been reached by multiplying 6 times 14 and dividing by 2. Then 2 + 4 + 6 + ... + 34 + 36 was found by multiplying 18 by 19 (or 18 times 38 divided by 2).

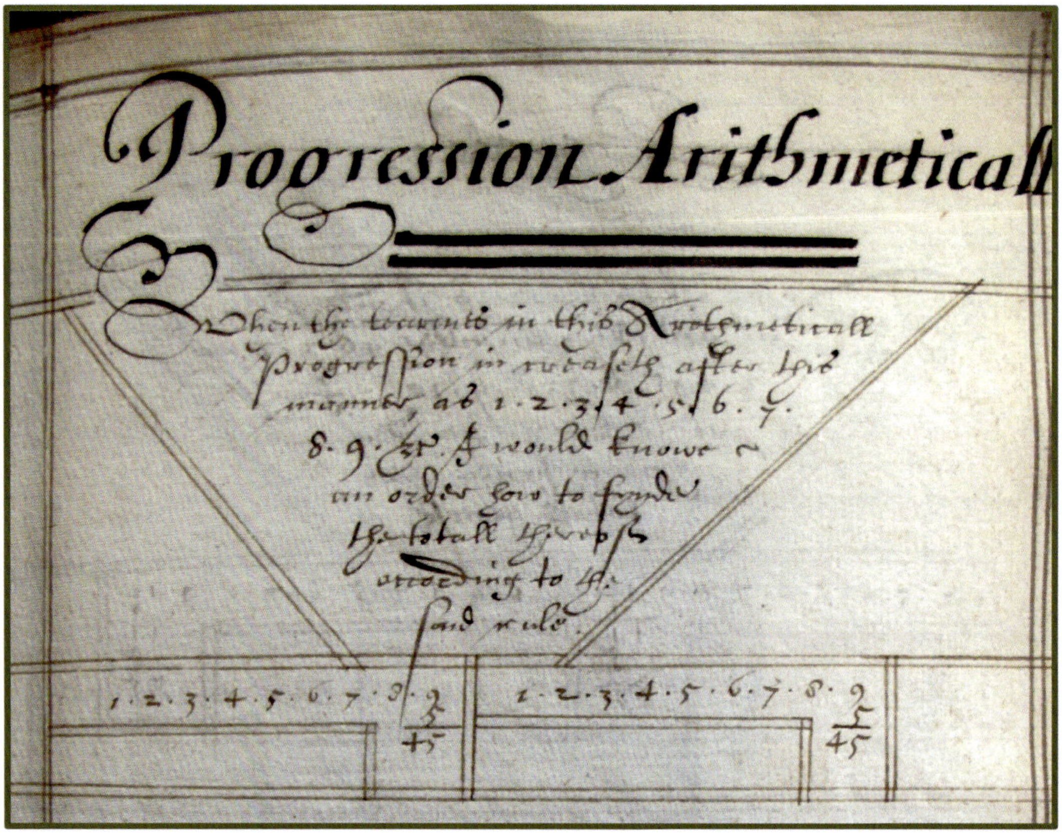

Figure 7.8. "Progression Arithmeticall" in Thomas Dixson's cyphering book (George A. Plimpton Collection, Rare Book & Manuscript Library, Columbia University in the City of New York).

Thomas did not give the impression that he was merely copying arithmetic from someone else's printed page. He seemed to be looking actively for patterns and generalizations. This was also the impression given from studying his work on what he labeled "Progression Geometricall." Figure 7.9 shows the very first problem he tackled under the heading "Progression Geometricall in double proportion." This involved finding the sum of 1 + 2 + 4 + 8 + 16 + 32 + 64, and from the outset this was found by subtracting 1 from the product of 64 and 2. Although no explanation was given as to why he felt he should do that, there can be no doubt that that is what he did. Similarly, with the next task, which required him to find the sum of 1 + 2 + 4 + 8 + 16 + 32 + ... + 524288 (each term to be added is the double of the previous term), he multiplied 524288 by 2 and subtracted 1. He then checked his calculation by using galley division. The pattern was extended in a later problem when he reasoned that 2 times (3 + 9 + 27 + 81 + 243 + 729 + 2187 + 6561) was equal to 3 less than the product of 6561 and 3. In a similar way, he stated that 2 times (4 + 16 + 64 + 256 + 1024 + 4096 + 16384 + 65536) was equal to 4 less than the product of 65536 and 4. These results were checked using galley division (see Figure 7.9).

The sections on progressions were followed by a section on the rule of three. That order was inconsistent with Leonardo of Pisa's *Liber Abbaci,* but consistent with Sacrobosco's order of topics, and with the order presented by the German, Adam Riese (Jackson, 1906). Unfortunately, we have not been able to find other British cyphering books from the first half of the 17th century to check if Sacrobosco's order was commonly assumed by those offering instruction in arithmetic, but the evidence from Thomas Dixson's (1630–1632) suggests that that could have been the case.

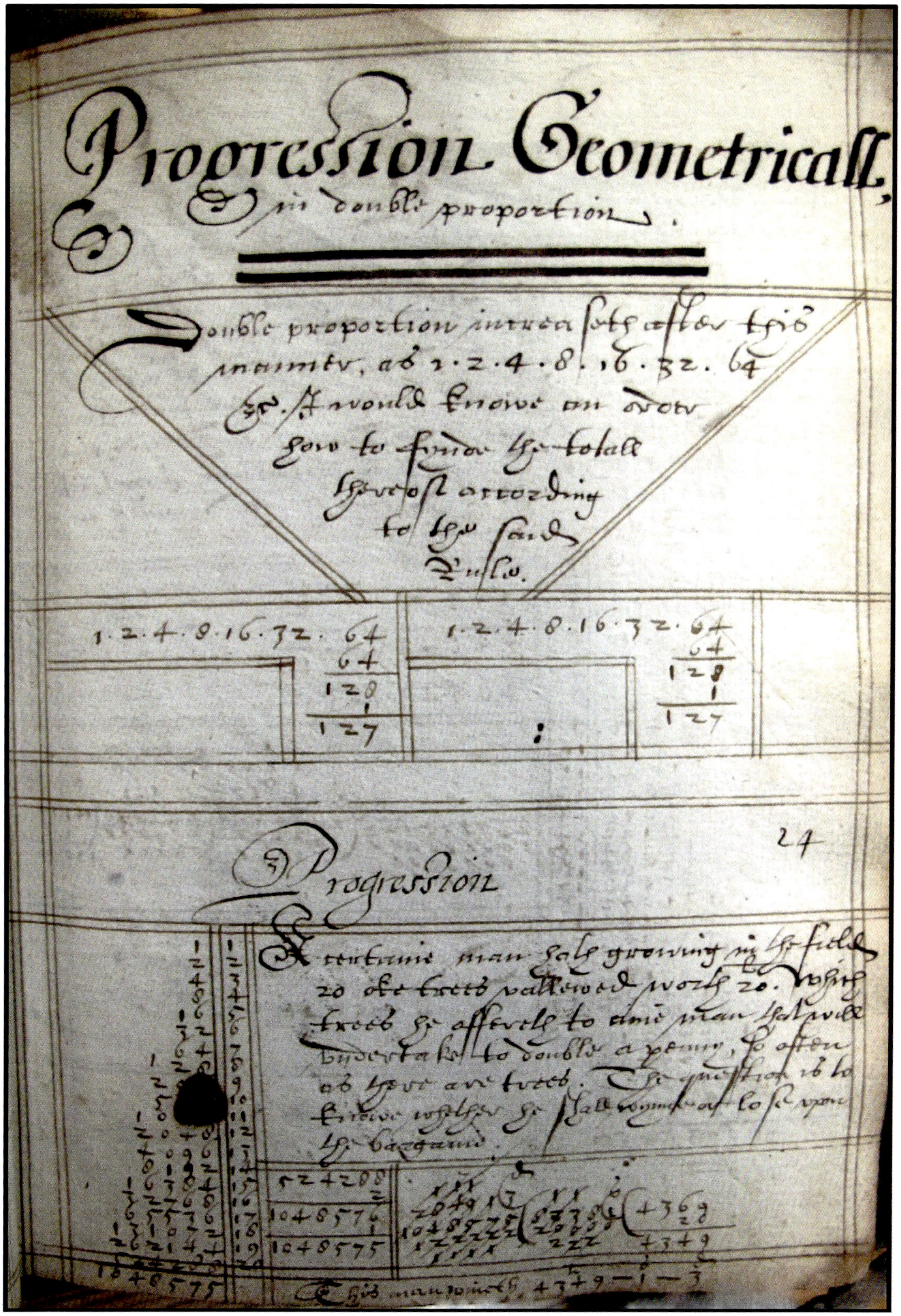

Figure 7.9. Thomas Dixson's patterns in "Progression Geometricall" tasks (George A. Plimpton Collection, Rare Book & Manuscript Library, Columbia University in the City of New York).

On this matter, Augustus de Morgan (1847) stated:

> That Leonard(o) of Pisa, as he is called, was the first who wrote in the new system is pretty generally affirmed by the Italians of the fifteenth and sixteenth century: and there can be no doubt of it. But because we trace the first formal user or expounder to Italy, it by no means follows that the other parts of Europe owe the system to that country in the first instance, however certain it may be that they owe much assistance in the course of the general establishment. ... Come how they might, however, we find ourselves, at the invention of printing, in possession of the knowledge that two distinct systems of arithmetic were current. The stream which had united in one bed had not mingled their waters; and for nearly a century, two distinct classes of writings present themselves. Their only common point is the use of Arabic or Indian numerals and method of notation. (p. ix)

It seems that Thomas Dixson's arithmetic had more of the Sacrobosco variety in it, and less of the Leonardo variety. Dixson's cyphering book raises the question of how much the Sacrobosco system was being used in the first half of the 17th century, in England. Jackson (1906), who did not refer to the influence of Sacrobosco on curricula, provided evidence that in sixteenth-century Continental Europe, arithmetical and geometrical progressions appeared in a variety of positions within textbooks. On this matter, more research is called for.

Concluding Comments

This chapter represents a beginning to a much wider and deeper investigation that needs to be carried out with respect to mathematics education in Great Britain in the seventeenth century. Although Thomas Dixson was certainly not the only person to prepare a cyphering book in England in the seventeenth century, there is not much known about the structure of pre-college mathematics education in England at the time. Isaac Newton's early mathematical training was probably based on a cyphering book owned by Henry Stokes, a teacher at Grantham Grammar School (which Newton attended in the mid-1650s). Stokes's cyphering book dealt with the extraction of cube roots, surveying, mensuration, trigonometry, and geometrical constructions (Rickey, 1987).

There was much to like about the quality of mathematics education revealed by our brief study of Thomas Dixson's cyphering book. Throughout the book a problem-solving approach is evident. Thomas did not take notes on how to do different types of problems—rather, he solved carefully sequenced sets of problems. Some of the problems were of the type that to solve them one needed to identify patterns and make generalizations. That was particularly the case with the problems on arithmetical and geometrical progressions.

One of the most intriguing aspects of our analysis was the revelation that the arithmetic content that Thomas Dixson dealt with was more in line with the curriculum offered in Sacrobosco's *Algorismus Vulgaris* than with the *abbaco* curriculum associated with Leonardo of Pisa's *Liber Abbaci.* Yet, during the next 200 years, the *abbaco* curriculum would become dominant, both in Great Britain and in North America. The main indicator that Sacrobosco's *Algorismus Vulgaris* provided the basic curriculum framework for Thomas Dixson's manuscript was that arithmetical and geometrical progressions came relatively early, and so did common fractions. Further research aimed at studying whether early in the 17th century, the Sacrobosco curriculum held sway in pre-college mathematics education, is needed. If, indeed that was the case then why and when did the move to *adopt* the *abbaco* curriculum occur? Such research could also throw light on the kind of mathematics education that the early settlers in colonial New England wanted for their children.

References

Baker, H. (1568/1591). *The well-spring of sciences, which teacheth the perfect worke and practice of arithmeticke, both in whole numbers and in fractions*. London, United Kingdom: Thomas Purfoot.

Baker, H. (1687). *Baker's arithmetick teaching the perfect worke and practice of arithmeticke, both in whole numbers and in fractions, whereunto are added many rules and tables of interest, rebate and purchases, &c, also the art of decimal fractions intermixed with common fractions for the better understanding thereof, newly corrected and contracted and made more plain and ease by Henry Phillippes.* London, United Kingdom: J. Richardson and William Thackery.

Bjarnadóttir, K. (2009, October 15). *Arithmetica—Arithmetic art, mathematics for all.* Paper presented at the Nordic Research Network on Special Needs Education in Mathematics (NORSMA) Conference, held at the University of Iceland.

Burton, W. (1833). *The district school as it was, by one who went to it.* Boston, MA: Carter, Hendee and Company.

Cocker, E. (1678/1725). *Cocker's arithmetick: A plain and familiar method ...* (42nd ed.). London, United Kingdom: H. Tracy.

Cubberley, E. P. (1920). *The history of education.* Boston, MA: Houghton Mifflin Company.

Davis, N. Z. (1960). Sixteenth-century French arithmetics on the business life. *Journal of History of Ideas*, *21*(1), 18–48.

De Morgan, A. (1847). *Arithmetical books from the invention of printing to the present time.* London, United Kingdom: Taylor and Walton.

Denniss, J. (2012). *Figuring it out: Children's arithmetical manuscripts 1680–1880.* Oxford, United Kingdom: Huxley Scientific Press.

Dilworth, T. (1784). *The young book-keeper's assistant; shewing him, in the most plain and easy manner, the Italian way of stating debtor and creditor* (9th ed.). London, United Kingdom: Richard and Henry Causton.

Ellerton, N. F., & Clements, M. A. (2012). *Rewriting the history of school mathematics in North America 1607–1861*. New York, NY: Springer.

Franci, R. (2009). Mathematics in the manuscript of Michael of Rhodes. In P. O. Long, D. McGee, & A. M. Stahl (Eds.), *The book of Michael of Rhodes: A 15th century maritime manuscript* (Vol. 3, pp. 115–146). Cambridge, MA: MIT Press.

Howson, G. (1982). *A history of mathematics education in England.* Cambridge, United Kingdom: Cambridge University Press.

Ifrah, G. (2000). *The universal history of numbers from prehistory to the invention of the computer.* New York, NY: John Wiley & Sons. Inc.

Jackson, L. L. (1906). *The educational significance of sixteenth century arithmetic from the point of view of the present time.* New York, NY: Teachers College Press.

Jones, H. S. (1954). Foreword by the Astronomer Royal. In E. G. R. Taylor, *The mathematical practitioners of Tudor & Stuart England 1485–1714* (pp. *ix-x*). Cambridge, United Kingdom: Cambridge University Press.

Karpinski, L. C. (1925). *The history of arithmetic.* Chicago, IL: Rand McNally & Company.

Knox, V. (1788). *Liberal education: Or a practical treatise on the methods of acquiring useful and polite learning*. London, United Kingdom: Charles Dilly.

Lam, L-Y (1966). On the Chinese origin of the galley method of arithmetical division. *The British Journal for the History of Science, 3*(1), 66–69.

Long, P. O., McGee, D., & Stahl, A. M. (Eds.). (2009). *The book of Michael of Rhodes: A 15th century maritime manuscript.* Cambridge, MA: MIT Press.

Money, J. (1993). Teaching in the market place, or "*Caesar adsum jam forte: Pompey aderat*": The retaining of knowledge in provincial England during the 18th century. In J. Brewer & R. Porter (Eds.), *Consumption and the world of goods* (pp. 335–377). London, United Kingdom: Routledge.

Monroe, W. S. (1917). *Development of arithmetic as a school subject.* Washington, D.C.: Government Printing Office.

Pedersen, O. (1985). In quest of Sacrobosco. *Journal for the History of Astronomy, 16,* 175–221.

Recorde, R. (1549). *Grounde of the arts: Teaching the worke and practise, of arithmeticke.* London, United Kingdom: R. Wolff.

Reynolds, B. E. (1993). The algorists vs the abacists: An ancient controversy on the use of calculators. *The College Mathematics Journal, 24*(3), 218–223.

Rickey, V. F. (1987). Isaac Newton: Man, myth, and mathematics. *College Mathematics Journal, 18,* 362–389.

Smith, D. E. (1908). *Rara arithmetica.* Boston, MA: Ginn and Company.

Stedall, J. (2012). *The history of mathematics: A very short introduction.* Oxford, United Kingdom: Oxford University Press.

Swetz, F. (1987). *Capitalism and arithmetic: The new math of the 15th century.* La Salle, IL: Open Court.

Taylor, E. G. R. (1954). *The mathematical practitioners of Tudor & Stuart England 1485–1714.* Cambridge, United Kingdom: Cambridge University Press.

Van Egmond, W. (1976). *The commercial revolution and the beginnings of Western mathematics in Renaissance Florence, 1300–1500.* PhD dissertation, Indiana University.

Van Egmond, W. (1980). *Practical mathematics in the Italian Renaissance: A catalog of Italian abbacus manuscripts and printed books to 1600.* Firenze, Italy: Istituto E Museo di Storia Della Scienza.

Walkingame, F. (1785). *The tutor's assistant being a compendium of arithmetic and a complete question book* (21st ed.). London, United Kingdom: J. Scratcherd & I. Whitaker.

Wingate, E. (1630). *Arithmetique made easie.* London, United Kingdom: Stephens and Meredith.

Yeldham, F. A. (1926). *The story of reckoning in the Middle Ages.* London, United Kingdom: George A. Harrap.

Yeldham, F. A. (1936). *The teaching of arithmetic through four hundred years (1535–1935).* London, United Kingdom: Harrap.

Chapter 8

Writing as if Arithmeticke: George Bickham's Cyphering Book, Circa 1740

Abstract: Of the 11 extraordinary cyphering books considered in this volume, George Bickham's was the only one that was specifically prepared for sale to others. We do not know how many copies of this book were printed, but we do know that the copy which we now hold is one of very few which remain. Bickham was a writing master who was regarded, in his time, as one of world's best penman. As was usually the case with writing masters, he paid much attention to the kind of writing and calligraphy needed to prepare a strikingly beautiful cyphering book, and this book provides testimony to that fact. However, there are hardly any model examples, or exercises for the student to do. The emphasis seemed to be on form rather than mathematical substance, and the language of commercial arithmetic is strongly emphasized. As would have been expected, topics are presented according to the normal *abbaco* sequence.

Key Words: *Abbaco* curriculum sequence; arithmetic and writing; calligraphy; ciphering book; ciphering tradition; cyphering book; cyphering tradition; George Bickham the elder; penmanship; writing master; writing mathematics.

Figure 8.1 shows the discolored front page of a copy of George Bickham the Elder's (c. 1740) cyphering book that the authors own. Bickham was one of England's greatest-ever writing masters and, sometime around 1740, this particular book was prepared and engraved "for the use of schools." Bickham's cyphering book was extraordinary because of the excellence it obviously displayed so far as penmanship and calligraphy were concerned. The book draws attention to the tradition that existed in Great Britain, Continental Europe, and North America in the seventeenth and eighteenth centuries, that a student who was preparing a cyphering book should strive for excellence in writing, and that headings should be in fine calligraphy. This tradition held that a person could not be a good teacher of arithmetic unless his own penmanship and calligraphy were at a high level (Bickham, c. 1740; Monaghan, 2007).

The emphasis on beautiful penmanship and high calligraphy for cyphering books was part of a long tradition that a well-educated person should be able to write well. Usually the first page of a cyphering book featured elaborate calligraphic headings. The name of the student preparing the book was given, and often the name of the town in which the student lived, and the year in which entries were first made in the book, were also given. Because the quality of penmanship and calligraphic headings was expected to be high, those preparing cyphering books often had next to them penmanship manuals, such as those prepared by George Bickham, in order that models of excellent writing, calligraphy, and decorative scrolls could be seen and emulated.

At a time when most adults could not write, the achievement of having prepared a beautiful and largely correct cyphering book was well-regarded by potential employers and college-entrance authorities.

Nerida F. Ellerton and M. A. (Ken) Clements, *Abraham Lincoln's Cyphering Book and Ten other Extraordinary Cyphering Books*, DOI 10.1007/978-3-319-02502-5_8,

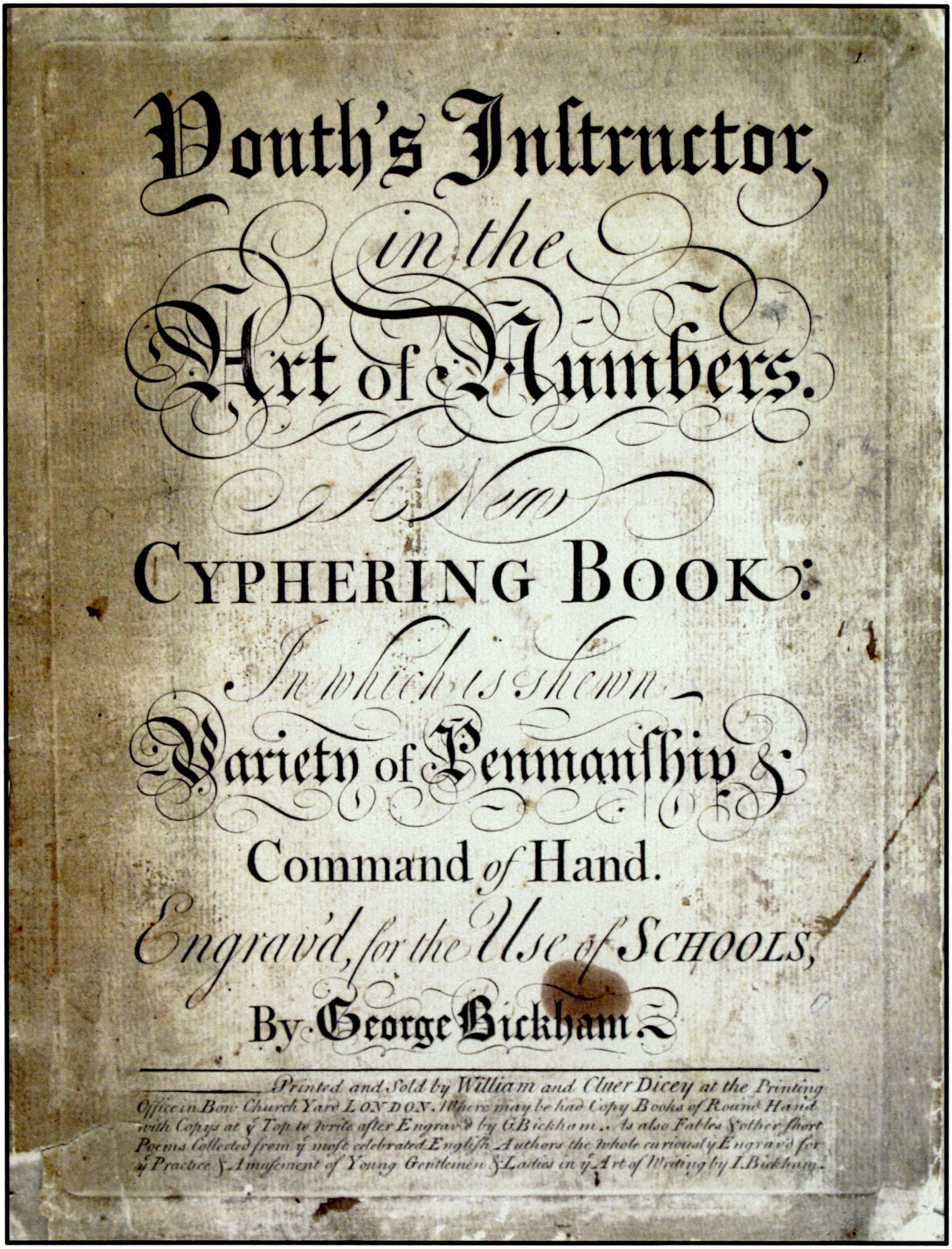

Figure 8.1. The front cover of George Bickham the Elder's (c. 1740) cyphering book.

On the next two pages we reproduce two engravings from Bickham's (1743) *The Universal Penman.* Bickham placed the first engraving (see Figure 8.2), which was by J. Bancks, opposite his title page. It demonstrates how exquisite the penmanship and calligraphy prepared using a quill, with home-made ink, could be. Sometimes the work was done at night, under the light of a candle or an oil lamp. In the second engraving (Figure 8.3), Bickham described such a person as a "writing master" and stated that the art of writing was "useful to the gentleman and scholar as well as the man of business" (p. 2).

Before reproducing, and analyzing, pages from George Bickham the Elder's (c. 1740) cyphering book, historical links between writing education and arithmetic education will be explored.

Linking Writing Education and Mathematics Education

The Cyphering Tradition

Throughout the period 1300–1800, and possibly even before that, many children in Southern and Western Europe prepared cyphering books in reckoning schools and in community and grammar schools (Beaujouan, 1988; Benoit, 1988). For example, Kool (1999) studied 36 fifteenth- and sixteenth-century Dutch-language cyphering books and found that these not only dealt with standard arithmetic themes but also offered solutions to practical problems. The preparation of these books was typically supervised by "writing masters," and "arithmetic" was regarded as an indispensable and important component of writing education. A fine cyphering book was one which not only followed the traditional *abbaco* sequence of arithmetic topics—in which concepts and rules were summarized, model examples were shown, and exercises solved—but one which also featured beautiful penmanship and calligraphy.

In the public mind, writing and arithmetic went together, largely because many of the best teachers of writing were also teachers of arithmetic. This link was so strong that in the German city of Nuremberg there was a guild of writing and arithmetic instructors which separated them from all other teachers (Brown & Coffman, 1914; Cubberley, 1920). During the period 1200–1850 these writing masters went by various names across Western Europe—in Italy they were called *maestri d'abbaco,* in France, *maistres d'algorisme*, in Germanic regions, *rechenmeisters*, and in England, *scriveners* (Howson, 1982; Ifrah, 2000; Swetz, 1987; Unger, 1888). It is not an overstatement to say that, together with specialist schools such as Christ's Hospital in London, they controlled the preparation of skilled personnel for accounting, computation, navigation, surveying, etc. (Struik, 1987). Indeed, during the period 1600–1800 traditional European Latin schools and universities increasingly felt the need to compete for students with the *rechenmeisters* and with those grammar schools which specialized in preparing students in mercantile arithmetic, navigation, surveying, and other mathematics-related subjects. The implemented curriculum—often referred to as the *abbaco* curriculum—in the reckoning schools relied heavily on the preparation of cyphering books, and the language used in these books was the local vernacular, and *not* Latin.

During the seventeenth century in Great Britain the link between writing and arithmetic increasingly became well established, and that link continued throughout the eighteenth century and into the nineteenth century (Carlo, 2005; Freeman, 1913). Students who studied arithmetic in British schools usually did so through a cyphering approach in which writing was emphasized (Heal, 1931; Monroe, 1917; Watson, 1902).

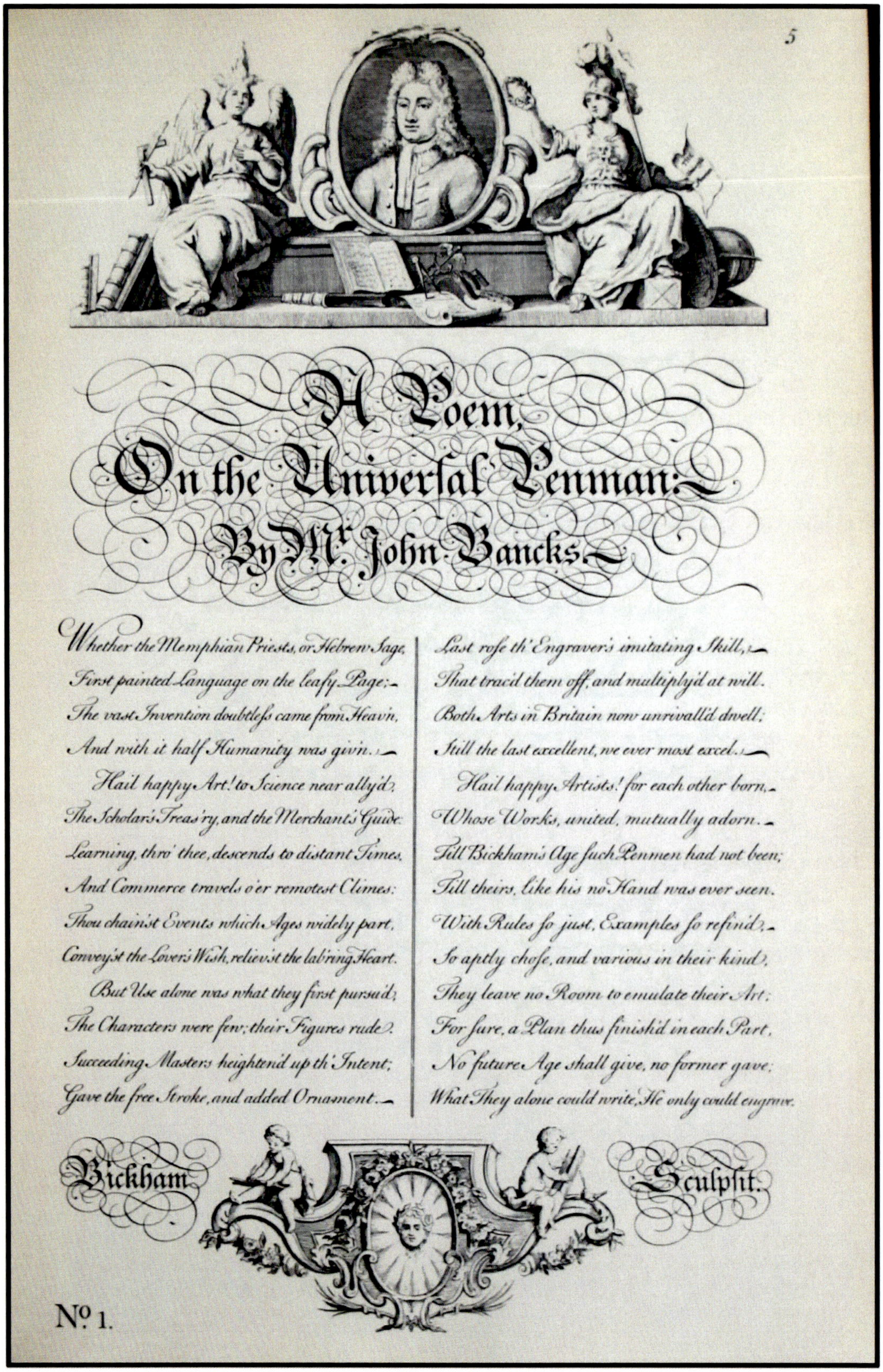

5

A Poem, On the Univerſal Penman: By Mr. John Bancks.

Whether the Memphian Priests, or Hebrew Sage,
First painted Language on the leafy Page;
The vast Invention doubtleſs came from Heav'n,
And with it half Humanity was giv'n.
Hail happy Art! to Science near ally'd,
The Scholar's Treas'ry, and the Merchant's Guide:
Learning, thro' thee, descends to distant Times,
And Commerce travels o'er remotest Climes:
Thou chain'st Events which Ages widely part,
Convey'st the Lover's Wish, reliev'st the lab'ring Heart.
But Use alone was what they first pursu'd;
The Characters were few; their Figures rude.
Succeeding Masters heighten'd up th' Intent;
Gave the free Stroke, and added Ornament.

Last roſe th' Engraver's imitating Skill,
That trac'd them off, and multiply'd at will.
Both Arts in Britain now unrivall'd dwell;
Still the last excellent, we ever most excel.
Hail happy Artists! for each other born,
Whose Works, united, mutually adorn.
Till Bickham's Age ſuch Penmen had not been;
Till theirs, like his no Hand was ever seen.
With Rules ſo just, Examples ſo refin'd,
So aptly choſe, and various in their kind,
They leave no Room to emulate their Art:
For ſure, a Plan thus finiſh'd in each Part,
No future Age shall give, no former gave;
What They alone could write, He only could engrave.

Bickham Sculpſit.

No. 1.

Figure 8.2. The writing master (by J. Bancks, in Bickham, 1743, p. ii).

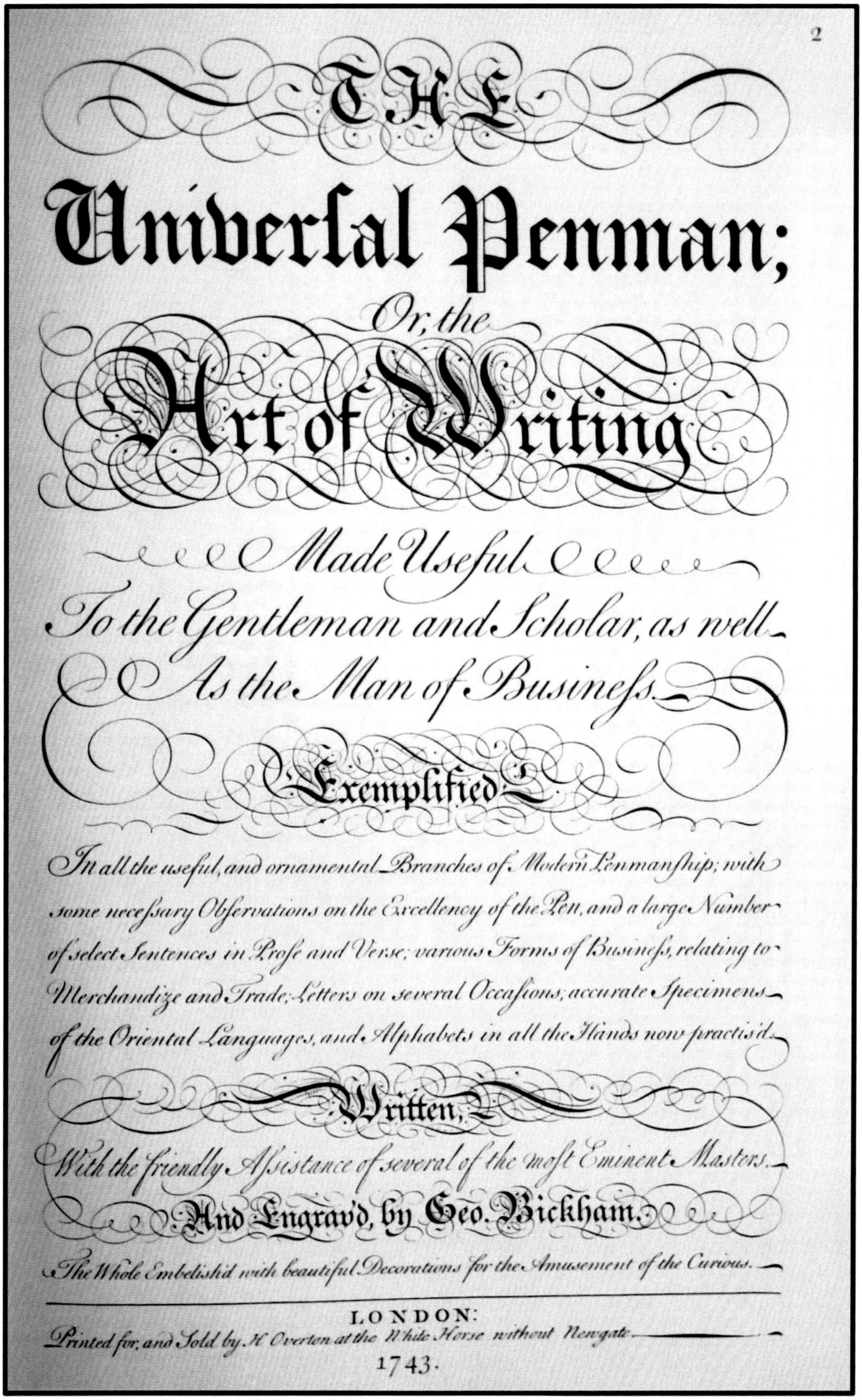

2

THE

Universal Penman;

Or, the

Art of Writing

Made Useful

To the Gentleman and Scholar, as well

As the Man of Business.

Exemplified

In all the useful, and ornamental Branches of Modern Penmanship; with some necessary Observations on the Excellency of the Pen, and a large Number of select Sentences in Prose and Verse; various Forms of Business, relating to Merchandize and Trade; Letters on several Occasions; accurate Specimens of the Oriental Languages, and Alphabets in all the Hands now practis'd.

Written,

With the friendly Assistance of several of the most Eminent Masters.

And Engrav'd, by Geo. Bickham.

The Whole Embelish'd with beautiful Decorations for the Amusement of the Curious.

LONDON:

Printed for, and Sold by H. Overton at the White Horse without Newgate.

1743.

Figure 8.3. The title page of Bickham's (1743) *The Universal Penman* (p. 2).

Schoolchildren were encouraged to practice their penmanship and calligraphy by making elegant entries in their cyphering books. In 1577, Christ's Hospital, a school located next to St Paul's in the heart of London, established an endowed writing school (Freeman, 1913), and immediately it included cyphering as a key component of its curriculum (Christ's Hospital, 1595). When, in 1673, the Royal Mathematical School was established and located within Christ's Hospital, school authorities immediately demanded high excellence in the scholars' penmanship (Howson, 1982; Trollope, 1834). Thus, "writing as also arithmeticke," already an important component of the European *abbaco* tradition, became an essential ingredient of the Royal Mathematical School's program (see Chapter 10 of this book).

As Heal's (1931) major study of writing masters revealed, John Ayres, George Bickham, William Butler, Edward Cocker, James Hodder, Humphry Johnson, and William Webster were among a large number of respected British writing masters who authored school arithmetics (Cajori, 1907; Carpenter, 1963; Cremin, 1970; Freeman, 1913; Heal, 1931; Monaghan, 2007; Reynolds, 1993; Smith, 2005). In Cocker's (1664) 56-page book, which was aimed at those who would tutor in writing and arithmetic, only 7 pages were dedicated to writing, but 49 to arithmetic. In 1671, *The Clark's Tutor for Arithmetick and Writing, or a Plaine and Easie way of Arithmetick* was published and attributed to Edmund Wingate, even though Wingate had died in 1656. In 1760 Charles Hutton, who would become a well-regarded British mathematician, opened a writing and mathematical school in Newcastle, England (Howson, 1982). Soon after that, Hutton (1764) described himself as a "writing master and teacher of mathematics" (p. iii).

One consequence of this link between writing and arithmetic instruction was that when a school official stated that "writing" was part of the curriculum (Ayres, 1682; Bickham, c. 1740; Champion, 1747; Colson, 1736; Jenkins, 1813; Johnson, 1719; Massey, 1763), this usually implied that students were expected to prepare cyphering books. The strength of that association has been lost in time, and that has resulted in the number of 17th- and 18th-century schools in which cyphering was an integral part of the curriculum being underestimated by historians (Small, 1914; Watson, 1902).

Translating "Writing as also Arithmeticke" to the North American Colonies

The strong connection between writing and arithmetic was translated into colonial America, and that explains the attention given to fine penmanship and calligraphy in many North American cyphering books (Ellerton & Clements, 2012; Thornton, 1996). Thus, for example, Thomas Pratt, after graduating from the Academy and College of Philadelphia in 1757, was appointed tutor, then writing master and usher in the English School of that College. In 1759 he took up the position of Professor of Mathematics at the College, but in 1760 he resumed as writing master. Then, in 1763, he was re-appointed Professor of Mathematics, which he retained until 1766 when he resumed his position as writing master (University of Pennsylvania Archives, 1749–1768).

Emigrants to North America from Great Britain were not the only colonial pioneers to value the writing component of arithmetic. In and around New York, for example, Dutch and Huguenot immigrants, placed much emphasis on "writing as also arithmeticke" (Ellerton & Clements, 2012). In New York in 1730 a Dutch-language textbook, *Arithmetica of Cyffer-Konst,* authored by a certain Pieter Venema, formerly a teacher in Holland, was published. A first edition of this text had been published in Holland in 1714 but, for the New York edition, Venema undertook "to make a clear and succinct cyphering book upon that excellent science that flourishes in the city [i.e., New York] and country" (quoted in Carpenter, 1963, p. 126). Venema (1730) was another to refer to himself as a writing master. The content in Venema's text followed the *abbaco* tradition, dealing first with whole number operations, and then proceeding to "trade, partnership and the handling of merchandise" (Carpenter, 1963, p. 126). Translated from the Dutch, the title page of the book was "Arithmetic or the art of cyphering, according to the coins, measurements and weights of New York" (Simons, 1924, p. 59).

Pages from George Bickham's (c. 1740) Cyphering Book

The cyphering book prepared by George Bickham, which will be the main focus of attention in this chapter, is rare. We have not been able to find any acknowledgement of its existence on the Internet, but we believe Bickham caused it to be published around 1740. Hence we have cited it as Bickham (c. 1740). When we purchased the book we were not aware how rare it was.

It will be in order here to make a few background comments on Bickham and on the cyphering book. Bickham led a movement in England in the early eighteenth century to popularize a form of round-hand penmanship and calligraphy initially introduced in Italy, France and other parts of Europe. *The Universal Penman*, Bickham's (1743) magnificent treasury of English roundhand calligraphy, included 125 pictorial scenes, over 200 script pictures, 19 complete animals, 275 lettered specimens, more than 100 panels, frames, cartouches, and more. These were contributed by a number of master calligraphers. Thus, what we have in this chapter is a book on how arithmetic cyphering books should be presented, prepared by someone who was recognized as one of the most outstanding writing masters in Great Britain—and indeed in the world.

Figure 8.4. Portrait of *George Bickham the Elder* (1684–1758). This was drawn in 1741 by Bickham's son, George Bickham the Younger.

George Bickham the Elder was an author, writing master and engraver of works on penmanship. His reputation within London flourished during the period 1712–1758. His most famous work was *The Universal Penman* series (1730–1750), containing examples by 25 contemporary writing-masters, on 212 folio copperplates.

.

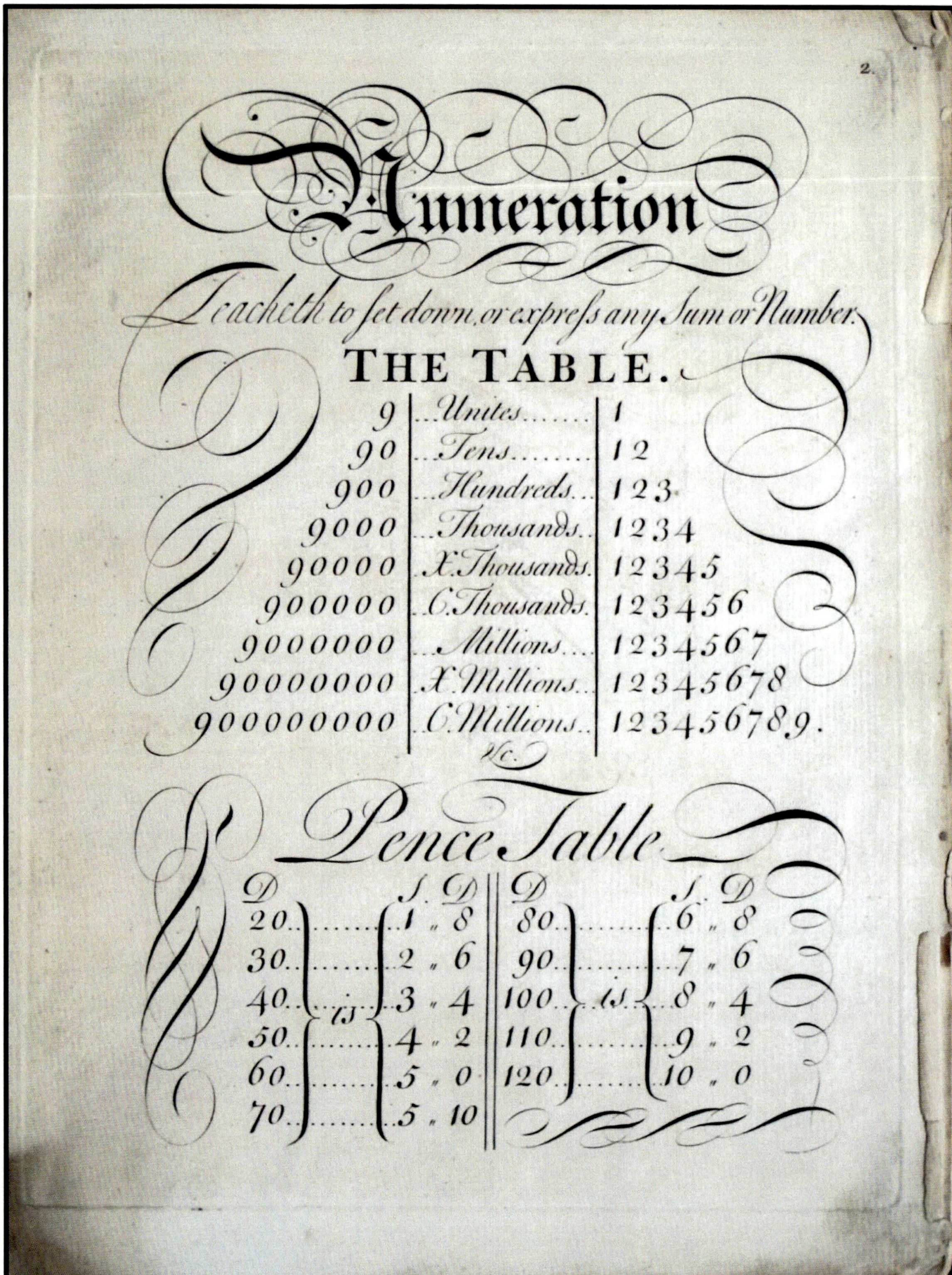

2.

Numeration

Teacheth to ſet down, or expreſs any Sum or Number.

THE TABLE.

9	Unites	1
90	Tens	12
900	Hundreds	123
9000	Thousands	1234
90000	X. Thousands	12345
900000	C. Thousands	123456
9000000	Millions	1234567
90000000	X. Millions	12345678
900000000	C. Millions	123456789.

&c.

Pence Table

D		S. D	D		S. D
20	is	1 . 8	80	is	6 . 8
30	is	2 . 6	90	is	7 . 6
40	is	3 . 4	100	is	8 . 4
50	is	4 . 2	110	is	9 . 2
60	is	5 . 0	120	is	10 . 0
70	is	5 . 10			

Figure 8.5. A "numeration table" and a "pence table," from George Bickham the Elder's (c. 1740) cyphering book (p. 2).

Figure 8.5 shows Bickham's presentation of a "numeration table" and a "pence table." Bickham advised that the numeration table "teacheth to set down, or express any Sum or Number." The table shows numbers from 1 to 100 million. The abbreviations *x.Thousands, x.Millions, C.Thousands, and C.Millions* were used to denote *ten thousands, ten millions, one-hundred thousands, and one-hundred millions,* respectively.

Although modern-day educators might shudder at the prospect of 10-year-old children being confronted with numbers reaching to hundreds of millions, students who began to cypher had usually learned to count and carry out simple arithmetical operations before they began to cypher. This earlier, pre-cyphering, education occurred in dame schools, in subscription schools, in area schools and sometimes in the private homes of the middle- and upper-classes—in which private governesses or tutors were employed.

Numeration was the first topic in the *abbaco* sequence, and was therefore often the first topic dealt with in an arithmetic textbook or a cyphering book. Figure 8.6, from Hodder's (1714) "520th edition" of *Arithmetick or that Necessary Art made most Easie*, a text which was published in London but widely used in North America as well as in Great Britain, shows a typical numeration table.

Relatively few numeration tables included at the beginning of cyphering books showed decimal fractions, because these were rarely taught in schools in the eighteenth century. After the introduction of decimal currency into the United States of America in 1792, decimal fractions were gradually introduced into the curriculum, but even then many students had not studied decimal fractions before they began to study topics like simple interest, compound interest, discount, and profit and loss (Ellerton & Clements, 2013).

Around 1740, when Bickham prepared his cyphering book, the British currency was based on the relationships 4 farthings make 1 penny, 12 pence make 1 shilling, and 20 shillings make one pound (and 21 shillings make one guinea). Some believed that with the ratios involved in that system of currency it was not necessary for children to learn decimal fractions. So, "Pence Tables," like the one that Bickham prepared (see Figure 8.5) were printed.

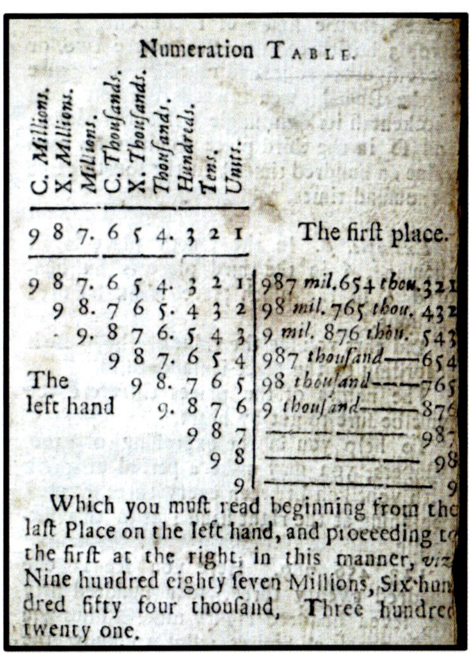

Numeration Table.

C. Millions. X. Millions. Millions. C. Thousands. X. Thousands. Thousands. Hundreds. Tens. Units.

9 8 7. 6 5 4. 3 2 1 The firft place.

9 8 7. 6 5 4. 3 2 1 | 987 mil. 654 thou. 321
9 8. 7 6 5. 4 3 2 | 98 mil. 765 thou. 432
9. 8 7 6. 5 4 3 | 9 mil. 876 thou. 543
9 8 7. 6 5 4 | 987 thoufand—654
The 9 8. 7 6 5 | 98 thoufand—765
left hand 9. 8 7 6 | 9 thoufand—876
9 8 7 | —987
9 8 | —98
9 | —9

Which you muft read beginning from the laft Place on the left hand, and proceeding to the firft at the right, in this manner, *viz.* Nine hundred eighty feven Millions, Six hundred fifty four thoufand, Three hundred twenty one.

Figure 8.6. The numeration table on page 3 of a widely-used arithmetic (Hodder, 1714).

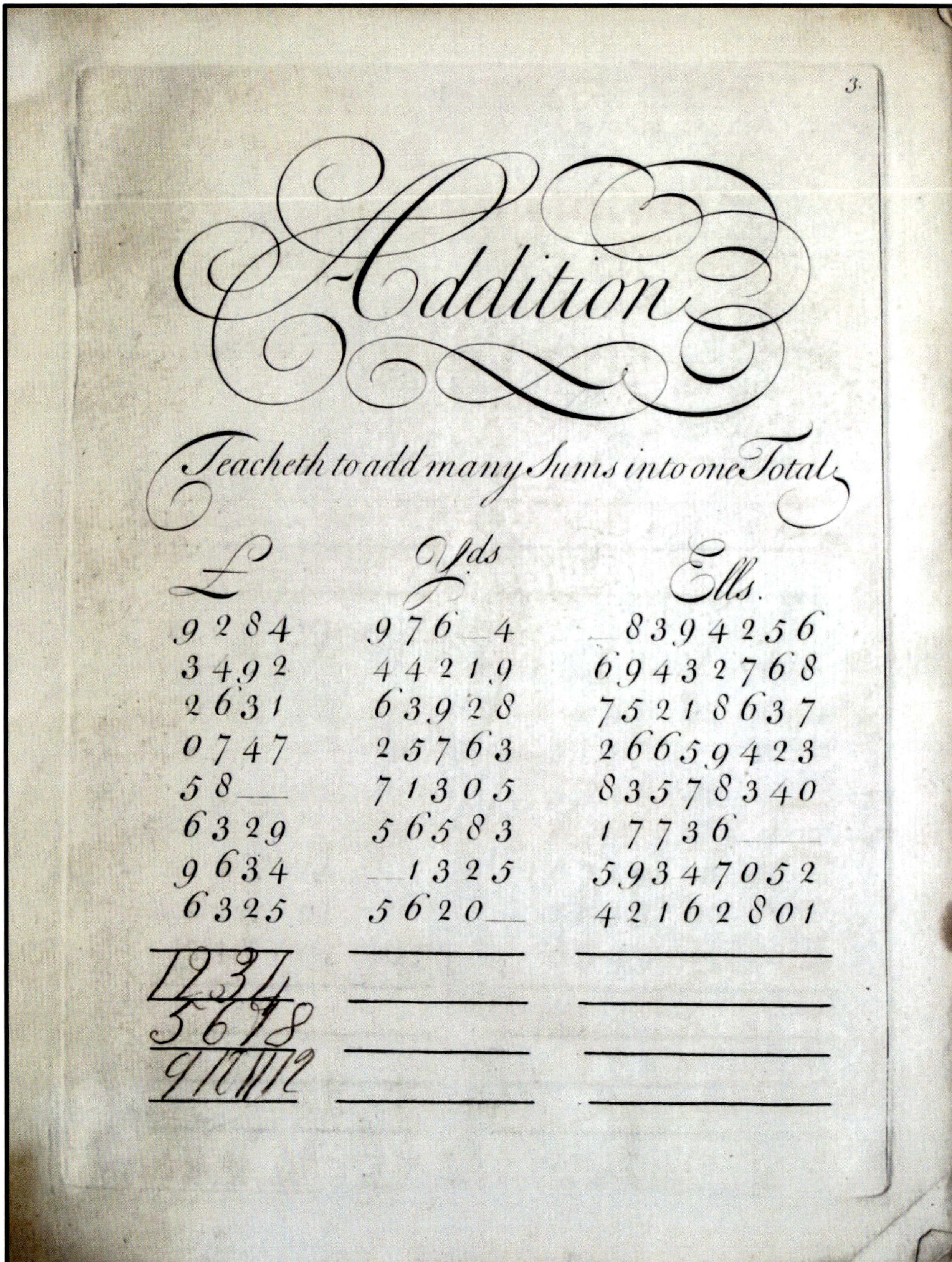

Figure 8.7. Addition tasks in George Bickham's (c. 1740) cyphering book (p. 3).

The addition tasks shown in Figure 8.7 were examples of "compound addition"—that is to say, quantities, and not "pure numbers," were being added. Sometimes a dash was used to indicate 0 (or zero). At the foot of each group of eight quantities to be "added" there were three spaces for answers, In the first space the sum of the eight quantities was to be inserted; in the second, the sum of seven of the eight quantities would be inserted; The answer in the second space would be subtracted from the answer in the first space, and the result should be the number not included when seven numbers were added. On the page shown, a previous owner has written 1234 5678 9101112 in the three spaces at the bottom of the first addition—of course, that was not what Bickham intended to be written in those spaces.

Notice the colloquial use of the word "sums" in "Addition Teacheth to add many sums into one Total."

In Figure 8.8, Bickham asked students to add the four lists of money (each list had six amounts of money to be added). It is interesting that farthings were indicated as common (or vulgar) fractions with denominators 2 and 4. In two of the examples shown it was assumed that students could cope with adding several fractions (like $^1/_4$, $^2/_4$, and $^3/_4$). Probably the students thought in terms of adding farthings.

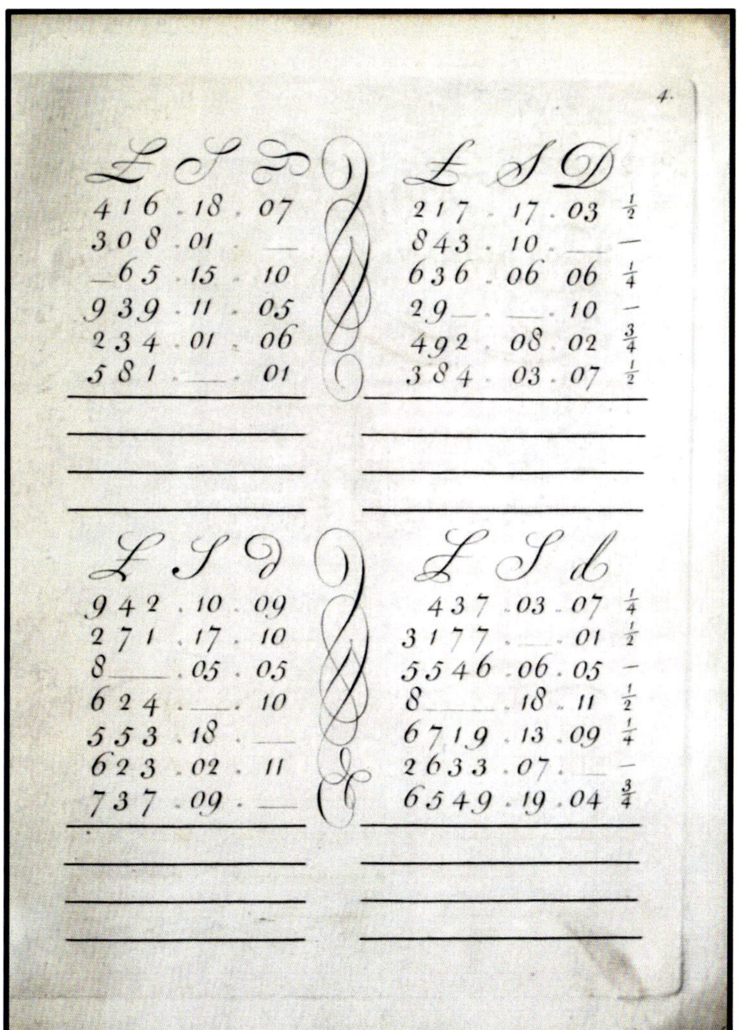
4.

£	S	D
416	18	07
308	01	—
65	15	10
939	11	05
234	01	06
581	—	01

£	S	D	
217	17	03	½
843	10	—	—
636	06	06	¼
29	—	10	—
492	08	02	¾
384	03	07	½

£	S	D
942	10	09
271	17	10
8	05	05
624	—	10
553	18	—
623	02	11
737	09	—

£	S	d	
437	03	07	¼
3177	—	01	½
5546	06	05	—
8	18	11	½
6719	13	09	¼
2633	07	—	—
6549	19	04	¾

Figure 8.8. Four "addition-of-money" tasks in Bickham's (c. 1740) cyphering book (p. 4).

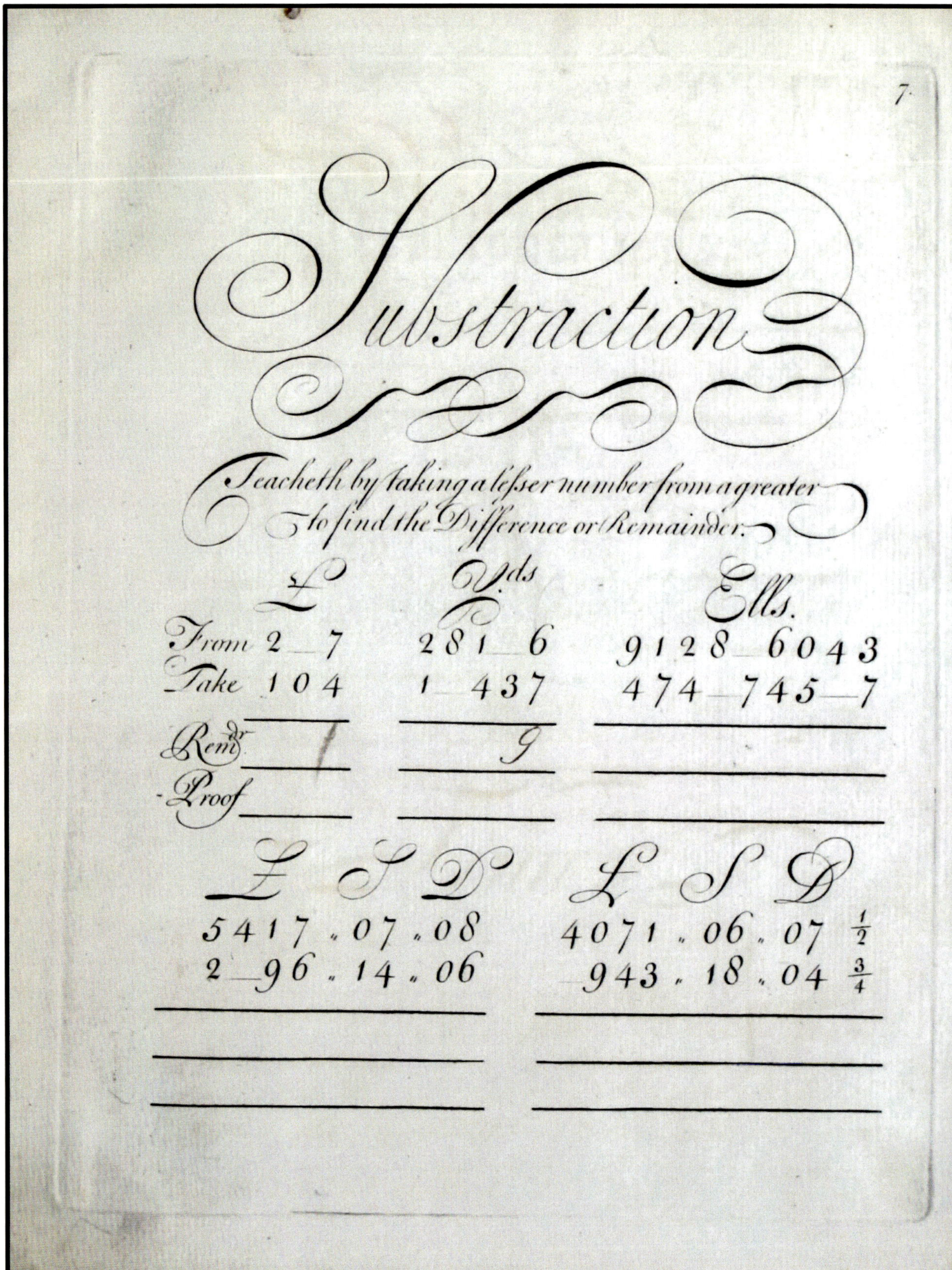

Figure 8.9. "Substraction" tasks in George Bickham's (c. 1740) cyphering book (p. 7).

Comments on Page 7 from Bickham's (c. 1740) Cyphering Book: Compound Operations

According to Daniel Adams (1822), an ell was a unit for measuring lengths of cloth but its equivalent length in inches varied from nation to nation. Thus, for example, in England, 45 inches made an ell; in Scotland it was 37 inches and one-fifth of an inch; a Flemish ell was 27 inches; but a French ell was 54 inches (which was twice a Flemish ell). This confusion of units was addressed by scholars during the French Revolution in the early 1790s, when a standardized metric system was introduced into France. In North America, the number of inches in an ell varied according to the European background of the majority of people in a region.

Notice that "substractions" were to be "proved" by additions. In the fourth example, the problem arising from having to "substract" ¾ from ½ would probably have been dealt with by considering ¾ and ½ as 3 and 2 farthings, respectively. Almost all students attempting these "substraction" tasks would never have formally studied common (or "vulgar") fractions.

Bickham (c. 1740) did not recommend an algorithmic procedure for subtraction. There were at least three different algorithms commonly used for subtraction in the eighteenth century, the most common being the "equal additions" method (Ellerton & Clements, 2012; Jackson, 1906). Consider, for a moment, the second of Bickham's subtraction tasks, which required the value of 28106 – 10437 to be found. By the equal additions algorithm, the student would start with "7 from 6 you can't do; add 10 to the 6, and say 7 from 16 is 9; then add the 10 back, to the 3, to get 4; then say, 4 from 0 you can't do, so add 10 to the 0; 4 from 10 is 6; then add the 10 back and say 5 from 11 is 6; and 1 from 8 is 7; finally, 1 from 2 is 1. Therefore, the answer is 17669 (yards). This can be 'proved' by adding 17669 to 10437, which would give 28106, as it should."

However, if the same task were to be attempted using the so-called "complementary" method, the language would be as follows: "7 from 6 you can't do; but 7 from 10 is 3 and 3 plus 6 is 9; then pay back the one 10, and say, 4 from 0 you can't do, so 4 from 10 is 6, and 6 add 0 is 6; then pay back the one hundred, and say 5 from 10 is 5 add 1 is 6; then pay back the one thousand, and say 1 from 8 is 7 and 1 from 2 is 1. So the answer is 17669 (yards)."

If the same task were to be attempted using the so-called "decomposition" method, the language would be as follows: "7 from 6 you can't do; so take 10 ones from the 100 (two places to the left), giving 7 from 16 is 9; then 3 from 9 is 6; and, since 4 from 0 you can't do, say 4 from 10, which is 6; then 0 from 7 is 7, and 1 from 2 is 1. So the answer is 17669 (yards)."

There are other algorithms for subtraction of whole numbers. According to Lambert Jackson (1906) and David Eugene Smith (1925), each of the three algorithms described above had histories stretching back to at least the 16th century. Ellerton and Clements (2012) analyzed the subtraction algorithms that could be unambiguously identified in the 212 North American cyphering books in their collection at that time, and found that the "equal additions" method was the most commonly used, with the "complementary" method being the second most commonly used. Not one student seemed to use the "decomposition" algorithm, despite the fact several leading eighteenth- and nineteenth-century textbook authors recommended that method. The "decomposition" algorithm is the most widely-used algorithm for subtraction in the United States of America today.

Any of the three methods could be used for any of the subtraction tasks shown in Figure 8.9. The reader is urged to try using the different algorithms for the last subtraction task—involving pounds, shillings and pence.

Units of measurement commonly employed in the eighteenth century are shown in Figure 8.10 and Figure 8.11. There were three different types of weights—"troy" (for jewelry and precious metals, as well as for corn and liquors), "avoirdupoise" (in England, used for weighing butter, cheese, groceries, etc.), and "apothecaries" (used for medicines). So far as wine, beer and ale measures were concerned, it should be noted that children often drank these beverages—especially in regions whether the quality of the water was inadequate. Learners were expected to memorize the numerical relationships between the different units, and be able to apply these in compound operations tasks (involving addition, subtraction, multiplication and division of quantities).

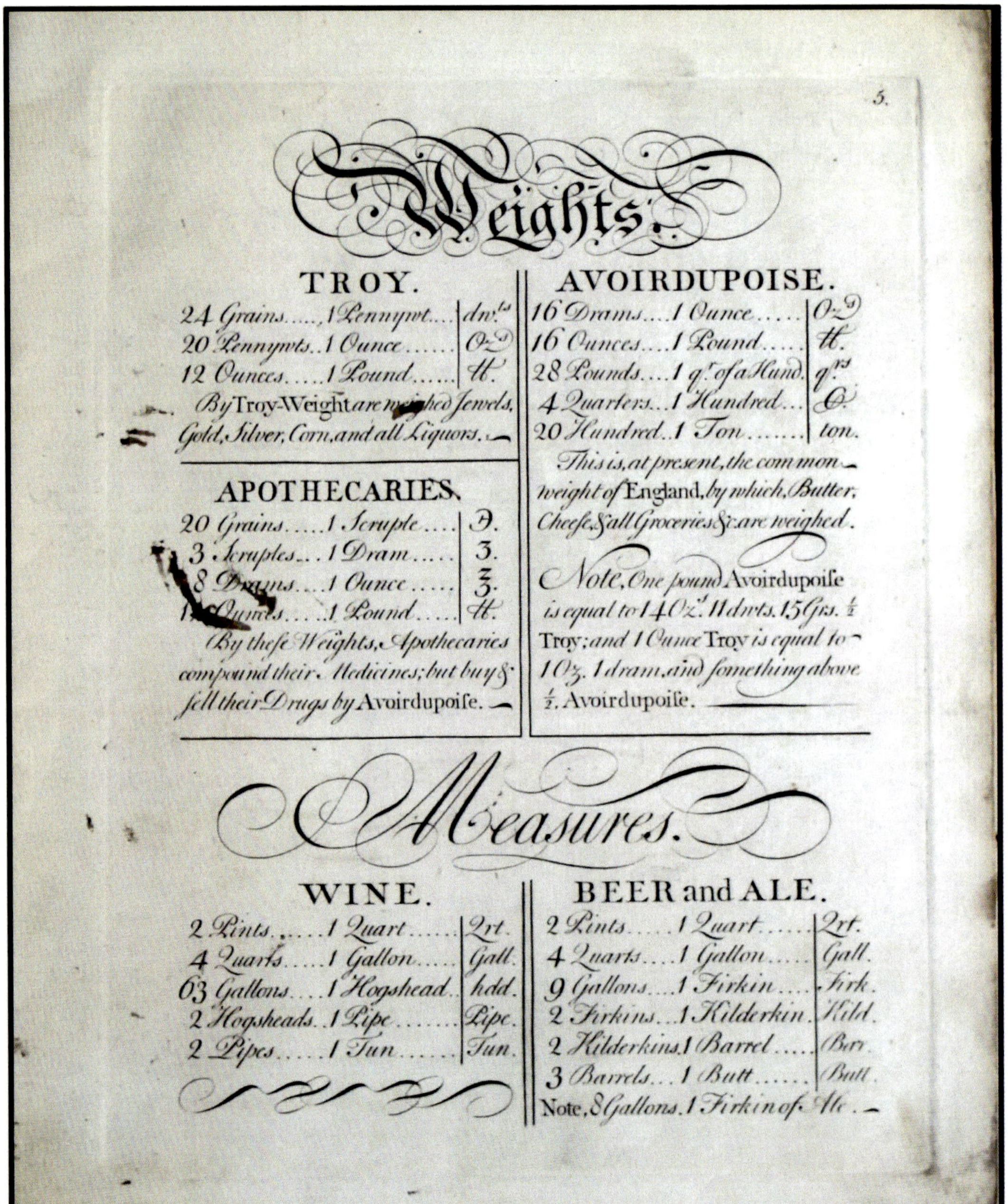
5.

Weights.

TROY.
24 Grains.....1 Pennywt....dwt.
20 Pennywts..1 Ounce......Oz.
12 Ounces.....1 Pound......lb.
By Troy-Weight are weighed Jewels, Gold, Silver, Corn, and all Liquors.

APOTHECARIES.
20 Grains.....1 Scruple.... ℈.
3 Scruples...1 Dram..... ʒ.
8 Drams....1 Ounce..... ℥.
12 Ounces....1 Pound.....lb.
By these Weights, Apothecaries compound their Medicines; but buy & sell their Drugs by Avoirdupoise.

AVOIRDUPOISE.
16 Drams....1 Ounce.....Oz.
16 Ounces....1 Pound.....lb.
28 Pounds....1 qr. of a Hund. qrs.
4 Quarters...1 Hundred...C.
20 Hundred..1 Ton......ton.
This is, at present, the common weight of England, by which, Butter, Cheese, & all Groceries &c. are weighed.
Note, One pound Avoirdupoise is equal to 14 Oz. 11 dwts. 15 Grs. ½ Troy; and 1 Ounce Troy is equal to 1 Oz. 1 dram, and something above ½. Avoirdupoise.

Measures.

WINE.
2 Pints.....1 Quart.....Qrt.
4 Quarts....1 Gallon.....Gall.
63 Gallons....1 Hogshead..hdd.
2 Hogsheads..1 Pipe.......Pipe.
2 Pipes.....1 Tun......Tun.

BEER and ALE.
2 Pints.....1 Quart......Qrt.
4 Quarts....1 Gallon.....Gall.
9 Gallons....1 Firkin.....Firk.
2 Firkins...1 Kilderkin. Kild.
2 Kilderkins 1 Barrel.....Barr.
3 Barrels...1 Butt......Butt.
Note, 8 Gallons 1 Firkin of Ale.

Figure 8.10. Measurement units in George Bickham's (c. 1740) cyphering book (p. 5).

Figure 8.11 displays further relationships between units which students were expected to memorize. The table showing Time units indicated that there were 4 weeks in a month, and 13 months 1 day and 6 hours in a Julian year (or 52 weeks & 6 hours, or 365 days & 6 hours, or 8766 hours). They were also expected to learn that exactly "365 days, 5 hours 49 minutes 4 seconds and 1/3" constituted one solar year.

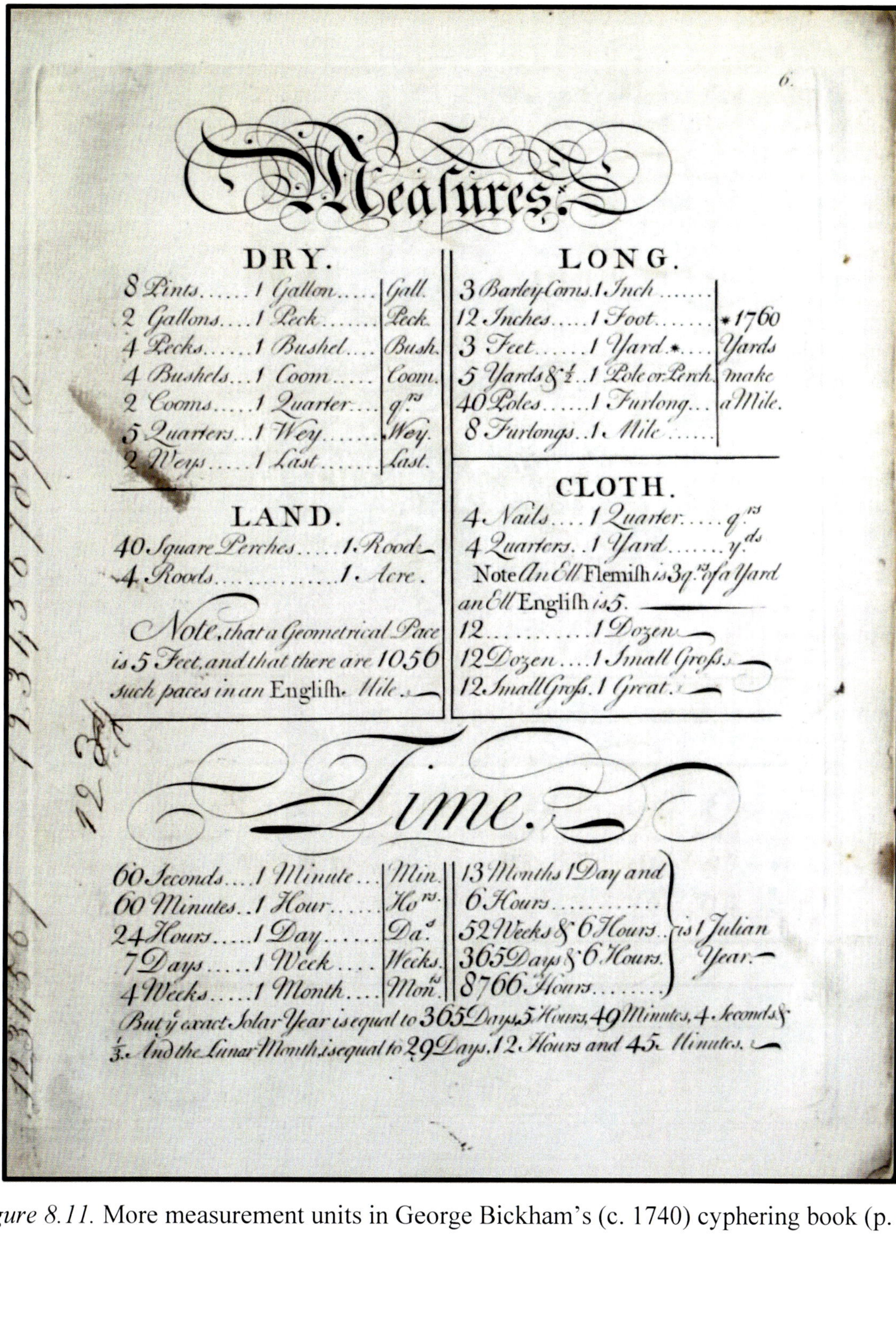
6.

Measures.

DRY.

8 Pints......	1 Gallon......	Gall.
2 Gallons.....	1 Peck.......	Peck.
4 Pecks......	1 Bushel....	Bush.
4 Bushels...	1 Coom.....	Coom.
2 Cooms.....	1 Quarter....	q.rs
5 Quarters...	1 Wey.......	Wey.
2 Weys.....	1 Last.......	Last.

LONG.

3 Barley Corns.	1 Inch	
12 Inches.....	1 Foot.......	*1760
3 Feet......	1 Yard.*....	Yards
5 Yards & ½..	1 Pole or Perch.	make
40 Poles......	1 Furlong...	a Mile.
8 Furlongs..	1 Mile......	

LAND.

40 Square Perches..... 1 Rood.
4 Roods............. 1 Acre.

Note, that a Geometrical Pace is 5 Feet, and that there are 1056 such paces in an English Mile.

CLOTH.

4 Nails.... 1 Quarter..... q.rs
4 Quarters.. 1 Yard....... y.ds
Note An Ell Flemish is 3 q.rs of a Yard an Ell English is 5.
12.......... 1 Dozen
12 Dozen.... 1 Small Gross.
12 Small Gross. 1 Great.

Time.

60 Seconds....	1 Minute...	Min.
60 Minutes..	1 Hour......	Ho.rs
24 Hours.....	1 Day......	Da.s
7 Days.....	1 Week....	Weeks.
4 Weeks.....	1 Month....	Mon.ts

13 Months 1 Day and 6 Hours............ }
52 Weeks & 6 Hours.. } is 1 Julian Year.
365 Days & 6 Hours. }
8766 Hours....... }

But y.e exact Solar Year is equal to 365 Days, 5 Hours, 49 Minutes, 4 Seconds & ⅓. And the Lunar Month is equal to 29 Days, 12 Hours and 45 Minutes.

Figure 8.11. More measurement units in George Bickham's (c. 1740) cyphering book (p. 6).

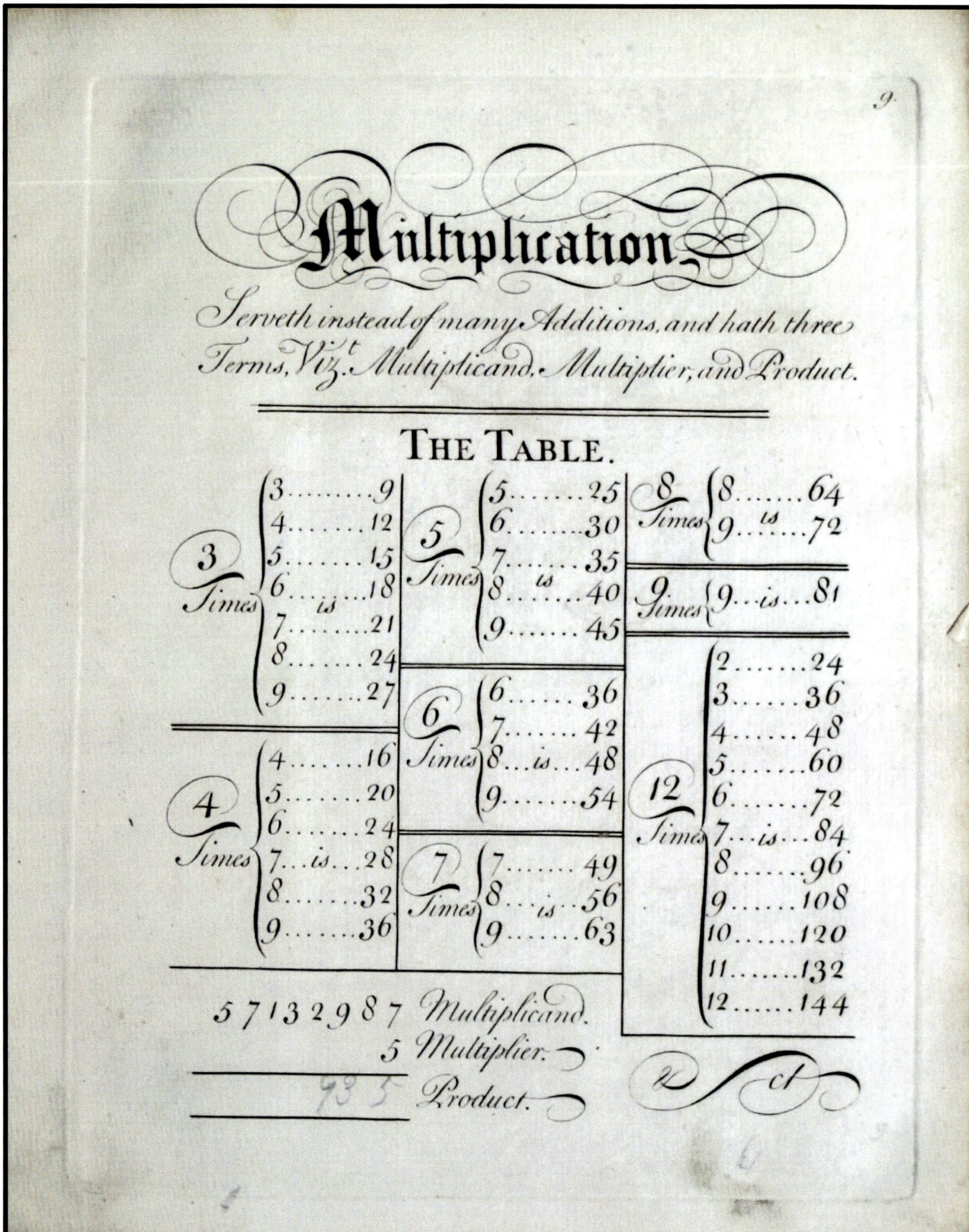

9.

Multiplication

Serveth instead of many Additions, and hath three Terms, Viz.t Multiplicand, Multiplier, and Product.

THE TABLE.

3 Times		4 Times	
3 9		4 16	
4 12		5 20	
5 15		6 24	
6 ... is ... 18		7 ... is ... 28	
7 21		8 32	
8 24		9 36	
9 27			

5 Times	6 Times	7 Times
5 25	6 36	7 49
6 30	7 42	8 ... is ... 56
7 35	8 ... is ... 48	9 63
8 ... is ... 40	9 54	
9 45		

8 Times	9 Times	12 Times
8 64	9 ... is ... 81	2 24
9 ... is ... 72		3 36
		4 48
		5 60
		6 72
		7 ... is ... 84
		8 96
		9 108
		10 120
		11 132
		12 144

5 7 1 3 2 9 8 7 Multiplicand.
5 Multiplier.
935 Product.

&c.

Figure 8.12. Multiplication tables in Bickham's (c. 1740) cyphering book (p. 10).

Comments on Page 9 from Bickham's (c. 1740) Cyphering Book: Multiplication

Multiplication tables (usually up to 12 times 12) were required to be remembered, and the terminology "multiplicand," "multiplier," and "product" was also to be known. The 12 times table was especially important for reduction tasks because there were 12 pence in a shilling and 12 inches in a foot. Figure 8.12 shows only one actual multiplication task, 57132987 by 5, and the terminology "multiplicand," multiplier" and "product" was introduced. This particular page is especially elementary, considering that children were not expected to prepare cyphering books until they had reached 10 years of age. One might have expected 10-year-olds already to have committed to memory the multiplication facts shown in Figure 8.12.

It should be observed that in Bickham's mind a multiplication table such as that shown in Figure 8.12 came after students had already carried out additions and subtractions on quantities involving various kinds of measures—concerned with troy weight, avoirdupoir weight, apothecaries' weight, wine, beer and ale and dry capacity measures, length, area, and time. In the world of school arithmetic, the application of the four operations to quantities was known as "compound operations." The authors of some textbooks dealt with a particular operation such as addition, and then immediately applied that operation in compound addition tasks before proceeding to the next operation. With such authors, the order of the early chapters in their arithmetic would be numeration, addition, compound addition, subtraction, compound subtraction, multiplication, compound multiplication, division, and compound division. With other authors, however, all of the operations would be dealt with in their pure number forms before any of the compound operations would be introduced. In those cases, the order for the early chapters would be numeration, addition, subtraction, multiplication, division, compound addition, compound subtraction, compound multiplication, and compound division. With Bickham's text, compound operation tasks immediately followed upon each particular operation.

Notice, however, that except for the case of farthings, Bickham (c. 1740) did not apply the four operations to tasks requiring the use of common fractions or decimal fractions. In early eighteenth-century schools, common fractions and decimal fractions often came much later in the implemented curriculum—in fact, sometimes well after the rule of three. Algorithms involving so-called "reduction" had been developed which enabled students to by-pass decimals, but often these were clumsy, and students who used them were unlikely to understand why they worked.

Figure 8.12 did not indicate how the multiplication operation could be linked to addition, or to division. Such matters could be dealt with by a teacher, probably during recitation sessions. So far as multiplication was concerned, many believed that students simply needed to memorize their tables. It was received wisdom that the two most important components of cyphering were the practice it gave for penmanship and calligraphy, and the opportunities it provided for recording important facts that needed to be committed to memory. It was someone else's task to devise ways and means by which students would come to understand how key ideas were related.

Many of the cyphering books in the writers' collection do not have pages in which multiplication tables were shown. As stated above, it was often assumed that students had already memorized such tables before they began cyphering.

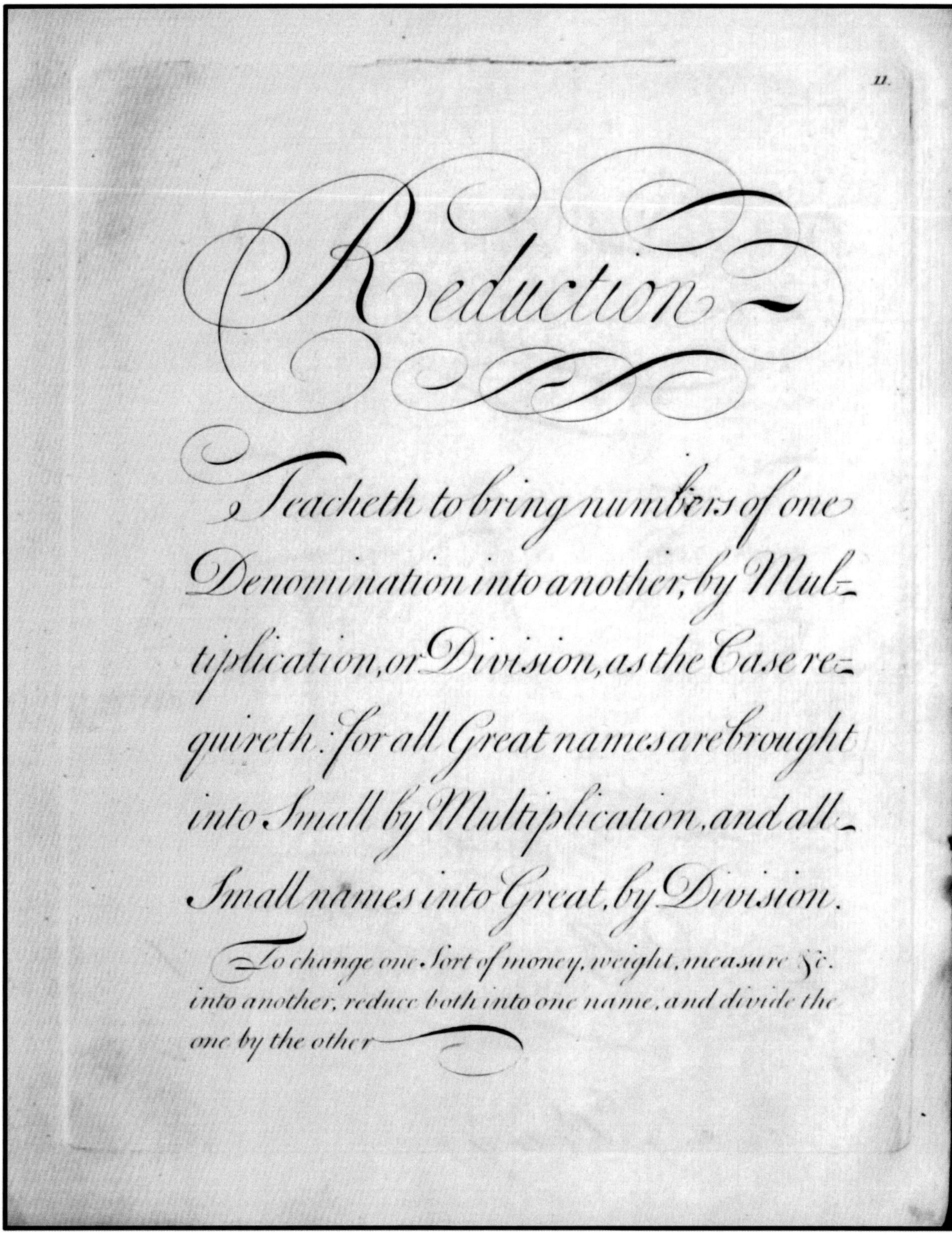

11.

Reduction

Teacheth to bring numbers of one Denomination into another, by Multiplication, or Division, as the Case requireth: for all Great names are brought into Small by Multiplication, and all Small names into Great, by Division.

To change one Sort of money, weight, measure &c. into another, reduce both into one name, and divide the one by the other

Figure 8.13. Notes on "reduction" in Bickham's (c. 1740) cyphering book (p. 11).

Comments on Page 11 from Bickham's (c. 1740) Cyphering Book: Reduction

How many farthings are there in 7 pounds 12 shillings and $3\frac{1}{2}$ pence? That type of problem was part of the *abbaco* curriculum studied by children aged between 10 and 13 years in Europe and North America in the seventeenth, eighteenth and early nineteenth centuries. The standard procedure was to multiply 7 by 20, then add 12; then multiply 152 by 12, and add 3; and then multiply the 1827 by 4 and add 2, to obtain 7310 farthings. But that was not the procedure that Bickham used in Figure 8.13. Bickham converted the 7 pounds to farthings, then the 12 shillings to farthings, then the 3-and-a-half pence to farthings. Then, the results of all the sub-calculations were added, to obtain 6720 + 576 + 14, or 7310, farthings.

Of course, the reverse kind of problem was also common—convert 7310 farthings to pounds, shillings and pence. Division was required—first by 4, then by 12 and then by 20—and remainders at each stage needed to be interpreted. Thus, for example, 7310 divided by 4 gave 1827 and 2 remainder, which needed to be interpreted as 1827 pence and 2 farthings; then, 1827 divided by 12 gave 152 and 3 remainder, which needed to be interpreted as 152 shillings and 3 pence; then, 152 divided by 20 gave 7 and 12 remainder, which was to be interpreted as 7 pounds and 12 shillings. So, bringing the parts together meant that 7310 farthings equaled 7 pounds 12 shillings and $3\frac{1}{2}$ pence. Notice that the 2 farthings were written as a fraction (this was done even before common—i.e., vulgar—fractions were studied).

Such reduction calculations could have been useful in everyday life (particularly for a clerk working in a merchant's office). Consider for example, the problem of finding the simple interest on 7 pounds 12 shillings and $3\frac{1}{2}$ pence at 6 percent per annum, for two years. This type of problem was usually found in the *abbaco* curriculum after the direct rule of three had been dealt with, but *before* decimal fractions, common fractions, and indeed percentage, had been formally studied. One method of dealing with it was to convert the 7 pounds 12 shillings and $3\frac{1}{2}$ pence to farthings, then use a rule-of-three direct proportions approach.

These "reduction" processes were widely used in societies in which there were many different units for weights and other measures. In Great Britain, decimal fractions were often part of the intended arithmetic curriculum for schools, but less often part of the implemented curriculum. Although after 1792 Federal (decimal) currency was gradually introduced into school mathematics programs in the United States of America, English currency, and the currencies of other European nations, were still widely used. In Salem, Massachusetts, for example, boys helped load or unload, measure, and tax, the precious cargoes on ships setting out for, or returning from, India and other Far-Eastern countries. There were tare and tret calculations to be made, taxes to be applied, and all of this had to be done, and recorded manually. Those who served as "supercargoes" on merchant ships needed to be able to barter, on the spot, with foreign merchants and make a profit. Not surprisingly, the topic "barter" was part of the *abbaco* curriculum (Ellerton & Clements, 2012).

Reduction was not only useful for problems involving money, but also for tasks concerned with weights and measures of many kinds. It was probably unrealistic to expect young learners to memorize all the relationships between units, and that was why it was common for important relationships of this kind to be recorded in cyphering books.

18.

Interest

Is the Advantage made by Lending of Money at a certain Stipulated Rate ℘ C.t ℘ Ann. which the Law, at present, fixes at 5 £ ℘ 100 £.

Simple Interest is that which arises only from the Principal it's self; but

Compound Interest arises both from Principal and Interest, that is, if at the Years end ye Interest is not pay'd, 'tis added to ye Principal, and the Sum accounted as a new Principal for the following Year.

Figure 8.14. "Simple and compound interest," in Bickham's (c. 1740) cyphering book (p. 18).

Comments on Page 18 from Bickham's (c. 1740) Cyphering Book: Simple and Compound Interest

Simple and compound interest are important real-life concepts and have been part of school arithmetic for centuries (Jackson, 1906). George Bickham's page on "interest" (see Figure 8.14) did not draw attention to any educational issues, but in fact there were several very challenging issues that might have been mentioned. One major issue, mentioned several times previously in this volume, arose from the fact that the traditional *abbaco* curriculum introduced simple and compound interest concepts before common and decimal fractions, and percentage, had been dealt with.

Although decimal fractions began to be used more widely in Europe in the seventeenth century, following the pioneering work of Simon Stevin and others (Clark, 2013; Ellerton & Clements, 2013), this was not reflected in most British school arithmetic curricula until well into the nineteenth century. Writers like John Kersey (1683), Edward Cocker (1684) and, in the United States, Thomas Jefferson (1785) advocated large-scale use of decimal-fraction concepts in many real-life applications, but the traditional *abbaco* curriculum stood firm in its resistance to that development. It placed the formal treatment of decimal fractions, common fractions, and percentage after simple and compound interest, discount, etc., and that tended to decide the order in which arithmetical topics were studied in both British and North American schools. Thus, long after official government introduction of the decimalized Federal currency in the United States of America in 1792, North American children were calculating simple and compound interest on amounts of money specified in pounds, shillings and pence, and they were not using decimal fractions.

Consider, for example, the typical task of finding the simple interest on 550 pounds 6 shillings and 3 pence over 5 years at $5\frac{1}{2}$ percent per annum. Percentage was dealt with by the idea that £100 grew to £105-10-0 after 1 year at $5\frac{1}{2}$ percent, and to £127-10-0 after 5 years. But without recourse to decimals, or even to common fractions, what could be done to find the simple interest on 550 pounds 6 shillings and 3 pence over 5 years at $5\frac{1}{2}$ percent per annum? One way to proceed was to resort to reduction procedures and to reduce each of £100-0-0, £127-10-0 and £550-6-3 to pence, and then to apply the direct rule of three. Once the answer was obtained in pence this then had to be converted back to pounds, shillings and pence. Although conceptually simple enough, the calculations which had to be carried out were messy, and likely to generate errors.

In other words, from a mathematics education perspective, Bickham's (c. 1740) cyphering book provided a superficial treatment of simple interest. It was not surprising that writing masters, like Bickham, regarded the calligraphy as the most important part of arithmetic. They were not concerned with any conceptual difficulties arising from the need for learners to sequence and then carry out messy calculations.

But who was there to change the state of affairs? Certainly, Thomas Jefferson (1785) tried to do that, in his brilliantly conceived notes on the need to introduce decimalized currency and weights and measures. Although he failed to win the day completely, his success in getting a decimal currency approved did lay the groundwork for a more coherent, better sequenced, school arithmetic curriculum in the United States.

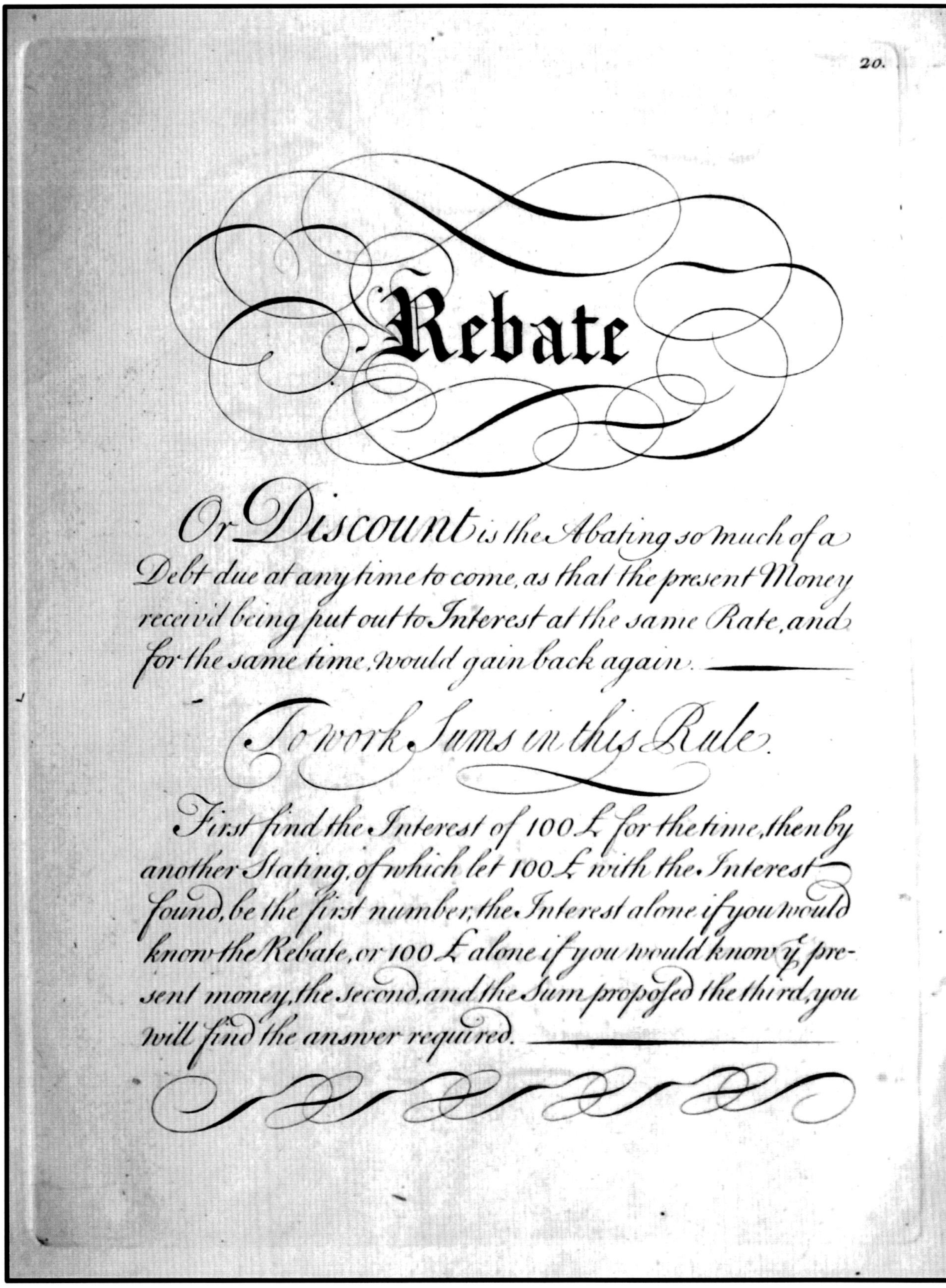

20.

Rebate

Or Discount is the Abating so much of a Debt due at any time to come, as that the present Money receiv'd being put out to Interest at the same Rate, and for the same time, would gain back again.

To work Sums in this Rule.

First find the Interest of 100£ for the time, then by another Stating, of which let 100£ with the Interest found, be the first number, the Interest alone if you would know the Rebate, or 100£ alone if you would know ye present money, the second, and the Sum proposed the third, you will find the answer required.

Figure 8.15. Notes on "rebate" in Bickham's (c. 1740) cyphering book (p. 20).

Comments on Page 20 from Bickham's (c. 1740) Cyphering Book: Rebate or Discount

"Rebate" (often called "discount") was another "beyond-the-rule-of-three" topic. Because "rebate" (or "discount") was placed in most arithmetics before decimal fractions or common fractions or percentage, traditionally the method for solving rebate problems was done by a method which avoided those concepts.

Bickham's explanation (see Figure 8.15) of how rebate problems should be tackled was complex to the point of being almost incomprehensible for a beginner:

> First find the interest of 100£ for the time, then by another stating, of which let 100£ with the interest found be the first number, the interest alone if you would know the rebate, or 100£ alone if you would know the present money the second, and the sum proposed the third, you will find the answer required.

What did this mean, and how did this work for actual situations?

Suppose, for example, John had borrowed money, but he received an inheritance so that he could pay out immediately the 500 pounds he still had to pay, but less an amount due to his paying 5 years earlier than was required. How much rebate should he receive if the interest originally payable was 6 per cent per annum? According to Bickham, John would have had to pay 30 pounds in (simple) interest over the 5 years, and so if x pounds was his rebate then:

$$130 : 30 :: 500 : x$$

Then, according to the rule of three:

$$x = \frac{30 \times 500}{130},$$

which would mean that John should receive a rebate of £115-7-8$\frac{1}{4}$. Of course, Bickham would not have used x, because most students had never been introduced to algebra.

Alternatively, Bickham's (c. 1740) rule (Figure 8.16) indicated that the amount John had to pay back (y pounds, say, with his rebate incorporated) was given by:

$$130 : 100 :: 500 : y$$

And this would give the amount to be paid as £384-12-3$\frac{3}{4}$ (which is £115-7-8$\frac{1}{4}$ less than £500).

It might be recalled that in the final extant leaf of Abraham Lincoln's cyphering book—the leaf was concerned with "discount"—Abraham attempted the problem: "What is the present worth and what the discount of £500 payable in 10 months at 5 per annum?" The "present worth," by Bickham's (c. 1740) rule, would be given by:

$$104\frac{1}{6} : 100 :: 500 : y$$

Thus, $y = 480$. So, the "present worth: would have been £480, and the "discount" £20. This calculation was straightforward if done with common fractions of a pound, but Abraham had not studied common fractions—so he reduced the pounds, shillings and pence to farthings, which made his calculations more tedious.

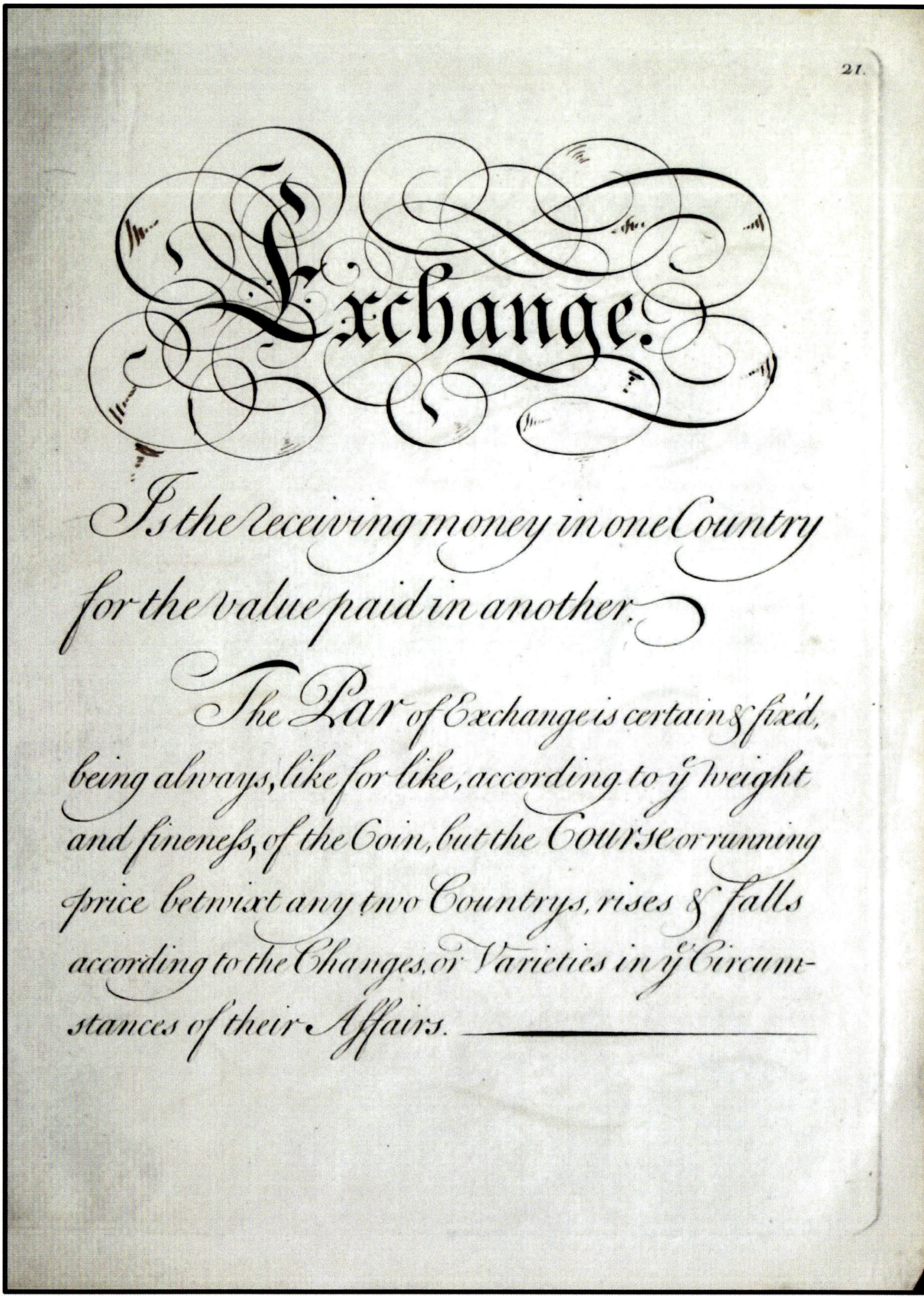
21.

Exchange.

Is the receiving money in one Country for the value paid in another.

The Par of Exchange is certain & fix'd, being always, like for like, according to ye weight and fineneſs, of the Coin, but the Course or running price betwixt any two Countrys, rises & falls according to the Changes, or Varieties in ye Circum-stances of their Affairs.

Figure 8.16. Notes on "exchange" in Bickham's (c. 1740) cyphering book (p. 21).

Comments on Page 21 from Bickham's (c. 1740) Cyphering Book

"Exchange" was a vitally important topic in the arithmetic curricula of eighteenth-century European and North American schools. In Europe, when one crossed the border from one nation to another, one had to deal with another currency. Many of the free immigrants in North America were from Europe, and they still thought in terms of the currencies with which they had grown up. When the fledgling United States of America was created in the late eighteenth century, a decimalized currency was introduced, but the different states strenuously retained their own versions of English currency (pounds, shillings, pence and farthings). To make matters worse, with these state currencies, a pound in one state did not necessarily carry the same value as a pound in another state.

Merchants, of course, traded with merchants operating from other states and nations. They quickly recognized that, in the absence of mechanical devices, other than the out-of-favor abacus, for calculation, it was important that they employed good reckoners.

In his notes on exchange, Bickham (c. 1740) pointed out that the value of a coin usually not only reflected "its weight and fineness" but also the "circumstances of their affairs" (see Figure 8.16). This same premise was in the forefront of Thomas Jefferson's (1784) mind when he set out his brilliant "Notes on Coinage" (held in the Library of Congress at Washington, DC). However, Jefferson (1785) went further—he framed his argument for the need for decimal currency around educational reasons. In 1792, after having returned from a period of government service in revolutionary France, Jefferson successfully shepherded through the administrational aspects of legislation on decimal currency making United States of America the first nation in the world to have a completely decimalized currency.

Bickham (c. 1740) wrote at a time when there was no fully decimalized currency, anywhere. He included entries on details of the currencies of Holland, Flanders and Germany, Spain, and Portugal. Figure 8.17 shows a copy of his statement on "Spain."

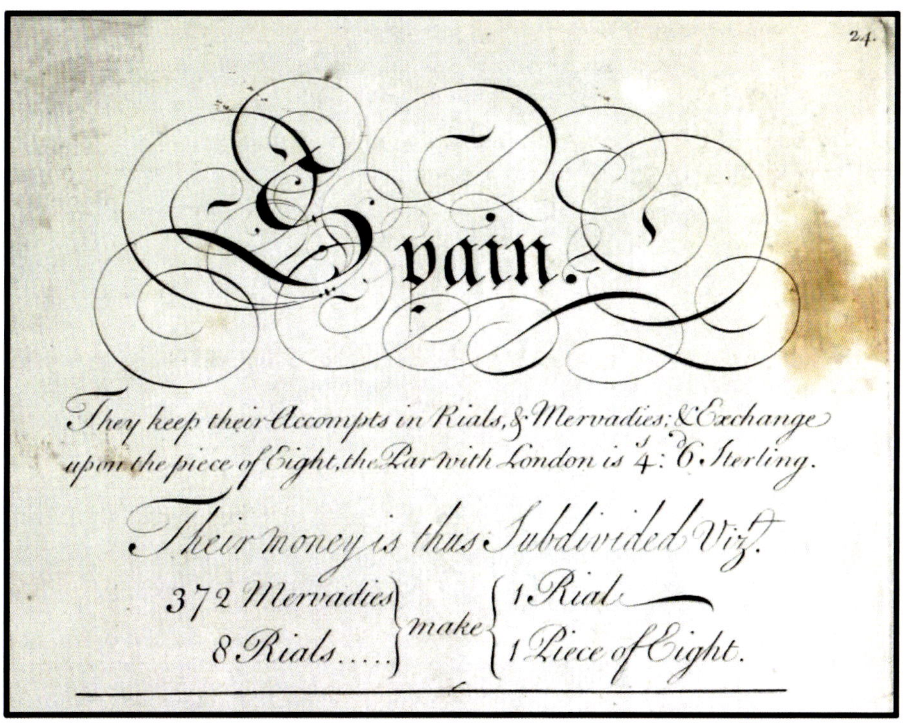
24.

Spain.

They keep their Accompts in Rials, & Mervadies; & Exchange upon the piece of Eight, the Par with London is 4: 6. Sterling.

Their money is thus Subdivided Viz.

372 Mervadies } make { 1 Rial
8 Rials.... } make { 1 Piece of Eight.

Figure 8.17. Charles Bickham's (c. 1740) notes on Spanish currency (p. 24).

Bickham's (c. 1740) Cyphering Book was Extraordinary, but ...

The cyphering book that George Bickham prepared and caused to be made available to the general public was extraordinary in the sense that it represented the best effort of one of the world's top writing masters to demonstrate the level of penmanship and calligraphy that students should strive to achieve in *their* cyphering books. Yet, it is obvious that George Bickham did not intend his cyphering book to be like a cyphering book that a student typically prepared. Bickham's cyphering book did not include any model examples, and he set relatively few exercises.

Indeed, this eighteenth-century cyphering book by one of the world's leading writing masters pointed to the superficiality of the writing-as-also-arithmeticke side of cyphering. A critic might say its approach to arithmetic education "elevated form above substance." To such a charge, Bickham would have defended his cyphering book on the following grounds:

1. On the cover of his book he emphasized that his book was concerned with showing "varieties of penmanship," and demonstrating "command by hand." His cyphering book did achieve those ends.
2. Teachers needed to have access to one or more "normal" cyphering books, or textbooks, to provide model examples, and exercises. Bickham's cyphering book was merely concerned with presenting models of penmanship and calligraphic excellence.
3. In eighteenth-century British mercantile environments, being able to write well was extremely important, as was being able to perform arithmetic operations confidently and competently. From Bickham's perspective, students needed to be trained to communicate the results of their arithmetical deliberations in an attractive and accurate way. In his book he tried to show how the writing side of that expectation could be achieved.

However, the possibility that someone using Bickham's cyphering book as a model would prepare a beautiful cyphering book, featuring high penmanship and calligraphy, but not understand the associated arithmetical principles and skills, could not be denied. In a mercantile world where, so often, time was to be equated with money, the wisdom of asking students to invest large amounts of time in the preparation of beautiful cyphering books was questionable. Around 1740, however, there were few who would have questioned the need for a cyphering book like Bickham's. But, a century later, around 1840, that would no longer have been the case. In the first half of the nineteenth century the assessment of mathematical learning increasingly began to be based on pencil-and-paper tests. In 1840, not all merchants regarded beautiful handwriting and calligraphy as vitally important, and the demise of the cyphering approach would come quickly (Ellerton & Clements, 2012).

At the time when Charles Bickham was preparing his cyphering book, written examinations were unknown, and the quality of cyphering books—as seen by parents, teachers, students, and others—was related more directly to appearance than to anything else. Charles Bickham knew that his book would serve as a guide to those who needed to prepare cyphering books in which the penmanship and calligraphy were of a high standard. Around 1740, many believed that so far as the preparation of cyphering books was concerned, form *was* indeed more important than substance.

References

Adams, D. (1822). *The scholar's arithmetic—Or Federal accountant.* Keene, NH: John Prentiss.
Ayres, J. (1682). *Arithmetick and writing.* London, United Kingdom: Author.

Beaujouan, G. (1988). The place of Nicolas Chuquet in a typology of 15th century French arithmetics. In C. Hay (Ed.), *Mathematics from manuscript to print 1300–1600* (pp. 73–88). Oxford, United Kingdom: Clarendon Press.

Benoit, P. (1988). The commercial arithmetic of Nicolas Chuquet. In C. Hay (Ed.), *Mathematics from manuscript to print 1300–1600* (pp. 96–116). Oxford, United Kingdom: Clarendon Press.

Bickham, G. (The Elder) (c. 1740). *Youth's instructor in the art of numbers. Cyphering book: In which is shown variety of penmanship & command of hand. Engrav'd for the use of schools.* London, United Kingdom: William and Cluer Dicey.

Bickham, G. (Ed.). (1743). *The universal penman.* London, United Kingdom: Henry Overton.

Brown, J. C., & Coffman, L. D. (1914). *How to teach arithmetic: A manual for teachers and a text-book for normal schools.* Chicago, IL: Row Peterson and Company.

Cajori, F. (1907). *A history of elementary mathematics with hints on methods of teaching.* New York, NY: The Macmillan Company.

Carlo, P. W. (2005). *Huguenot refugees in colonial New York: Becoming American in the Hudson Valley.* Brighton, United Kingdom: Sussex Academic Press.

Carpenter, C. (1963). *History of American schoolbooks.* Philadelphia, PA: University of Pennsylvania Press.

Champion, R. J. (1747). *The tutor's assistant in teaching arithmetic*. London, United Kingdom: Author.

Christ's Hospital (1595). *Dame Mary Ramsey's gift: Deeds relating to the maintenance by the Governors of Christ's Hospital of a grammar school in Halstead, Essex.* Reference Code CLC/210/G/BRB/041/MS13583. (Document held in LMA, London, United Kingdom).

Clark, K. M. (2013). "In these numbers we use no fractions": A Classroom Module on Stevin's Decimal Fractions. MathDL (The MAA Mathematical Sciences Digital Library).

Cocker, E. (1664). *The tutor to writing and arithmetic.* London, United Kingdom. H. Tracy.

Cocker, E. (1684/1720). *Decimal arithmetic, wherein is shewed the nature and use of decimal fractions in the usual rules of arithmetic, ...* (5th ed.). London, United Kingdom: J. Darby for M. Wellington.

Colson, J. (1736). *An arithmetical copy-book.* London, United Kingdom: Author.

Cremin, L. A. (1970). *American education: The colonial experience 1607–1783*. New York, NY: Harper & Row.

Cubberley, E. P. (1920). *The history of education.* Boston, MA: Houghton Mifflin Company.

Ellerton, N. F., & Clements, M. A. (2012). *Rewriting the history of school mathematics in North America 1607–1861.* New York, NY: Springer.

Ellerton, N. F., & Clements, M. A. (2013, September). *The mathematics of decimal fractions and their introduction into British and North America schools.* Paper presented to the Third International Conference on the History of Mathematics Education held at Uppsala University Sweden, September 25–28, 2013.

Freeman, F. N. (1913). Writing. In P. Monroe (Ed.), *A cyclopedia of education* (Vol. 5, pp. 819–827). New York, NY: The Macmillan Company.

Heal, A. (1931). *The English writing-masters and their copy-books 1570–1800.* Cambridge, United Kingdom: Cambridge University Press.

Hodder, J. (1714). *Hodder's arithmetic or that necessary art made most easie* (520th ed.). London, United Kingdom: M. Boddington.

Howson, G. (1982). *A history of mathematics education in England.* Cambridge, United Kingdom: C.U.P.

Hutton, C. (1764). *Schoolmaster's guide: Or a complete system of practical arithmetic adapted to the use of schools.* Newcastle Upon Tyne, United Kingdom: J. White and T. Saint.

Ifrah, G. (2000). *The universal history of numbers from prehistory to the invention of the computer.* New York, NY: John Wiley & Sons. Inc.
Jackson, L. L. (1906). *The educational significance of sixteenth century arithmetic.* New York, NY: Teachers College Press.
Jefferson, T. (1784). *Notes on the establishment of a money unit and of a coinage for the United States.* Handwritten manuscript held in the Library of Congress, Washington, DC.
Jefferson, T. (1785). *Notes on the establishment of a money unit and of a coinage for the United States.* Paris, France: Author.
Jenkins, J. (1813). *The art of writing.* Cambridge, United Kingdom: Author.
Johnson, H. (1719). *New treatise of practical* arithmetic (4th ed.). London, United Kingdom: Thomas Wood.
Karpinski, L. C. (1925). *The history of arithmetic.* Chicago, IL: Rand McNally & Company.
Kersey, J. (1683). *Mr Wingate's arithmetick, ...* (8th ed.). London, United Kingdom: J. Williams.
Kool, M. (1999). *Die conste vanden getale. Een studie van Nederlandstalige rekenboeken uit de vijftiende en zestiende eeuw, met een glossarium van rekenkundige termen* [*The art of numbers. A study of Dutch arithmetic books of the 15th and 16th century, with a glossary of arithmetical terms*]. Hilversum, The Netherlands: University College.
Massey, W. (1763). *The origin and progress of letters.* London, United Kingdom: Johnson.
Monaghan, E. J. (2007). *Learning to read and write in colonial America.* Amherst, MA: University of Massachusetts Press.
Monroe, W. S. (1917). *Development of arithmetic as a school subject.* Washington, DC: Government Printing Office.
Reynolds, B. E. (1993). The algorists vs the abacists: An ancient controversy on the use of calculators. *The College Mathematics Journal, 24*(3), 218–223.
Simons, L. G. (1924). *Introduction of algebra into American schools in the 18th century.* Washington, DC: Department of the Interior Bureau of Education.
Small, W. H. (1914). *Early New England schools.* Boston, MA: Ginn.
Smith, D. E. (1925). *History of mathematics: Volume II, Special topics of elementary mathematics.* Boston, MA: Ginn and Company.
Smith, E. B. (2005). "Let them compleately learn": Manuscript clues about early women's educational practices. In *A manuscript miscellany,* Summer 2005 Institute. Manchester, United Kingdom: Manchester University Press.
Struik, D. J. (1987). *A concise history of mathematics* (4th ed.). New York, NY: Dover.
Swetz, F. (1987). *Capitalism and arithmetic: The new math of the 15th century.* La Salle, IL: Open Court.
Thornton, T. P. (1996). *Handwriting in America: A cultural history.* New Haven, CT: Yale U.P.
Trollope, W. (1834). *A history of the royal foundation of Christ's Hospital.* London, United Kingdom: Pickering.
Unger, F. (1888). *Die methodik der praktischen arithmetick.* Leipsic, Germany: Teubner.
University of Pennsylvania (1749–1768). *Trustees of the University of Pennsylvania 1740–1900, Vol. 1 (1749–1768).* Philadelphia, PA: University of Pennsylvania Archives.
Venema, P. (1730). *Arithmetica of Cyffer-Konst, volgens de Munten Maten en Gewigten, te Nieu-York, gebruykelyk als mede een kort Ontwerp van de Algebra.* New York, NY: Jacob Goelet.
Watson, F. (1902). The curriculum and text-books of English schools in the first half of the 17th century. *Transactions of the Bibliographical Society VI* (Oct 1900–March 1902) (pp. 159–179). London, United Kingdom: Blades, East & Blades.
Wingate, E. (1671). *The clark's tutor for arithmetick and writing, or, a plain and easie way of arithmetick.* London, United Kingdom: Stephens and Meredith.

Chapter 9

Bound for Botany Bay? Circa 1791

Abstract: The 240-page navigation cyphering book featured in this chapter includes 80 artistic creations, many of them exquisite water-color miniatures. These miniatures are remarkable in the sense that they depict buildings, towers, hills, streams, etc., on which are labeled heights, widths, distances, angles, etc., relevant to mathematical tasks which are given in the text. Some of the labels are so small that it is difficult to read them with the naked eye. An important part of this chapter relates the beautiful artwork in the manuscript to the 18th- and 19th-century push to include serious drawing education in curricula designed for middle-class students and artisans. Another intriguing aspect of the chapter is that it draws attention to mathematical compromises, by mathematicians, with respect to trigonometric functions and logarithms. The title of the chapter, "Bound for Botany Bay? Circa 1791" was chosen because of the possibility that the diagrams and other artwork in the chapter depict actual scenes that the person who created the navigation cyphering book saw from a ship (or ships) that was (or were) part of the First Fleet from England to Botany Bay (New Holland) in 1787 and 1788.

Key Words: Botany Bay, Christ's Hospital, ciphering, cyphering, decimal fractions, drawing education, history of mathematics education, logarithms, navigation education, trigonometry.

A Navigation Cyphering Book

There is much that we do not know about the manuscript which will be examined in this chapter. We do not know the name of the person who prepared it, where it was prepared, or the year when it was prepared. Given that we know so little about its origins, why do we consider it to be an *extraordinary* cyphering book? Before addressing that question, it will be useful to summarize what we do know about the manuscript.

The 240-page manuscript has unusually small dimensions for a cyphering book—18.5 cm by 11 cm, that is to say, 7.5" by 4.5." It has full-leather binding, with beautiful gilt patterning on its spine cover. From the outside it looks like a high-quality, commercially-published academic treatise, but the interior pages feature copperplate handwriting, impressive calligraphic headings, and exquisite artwork.

Toward the end of the manuscript there is a handwritten log of a journey which, according to the log, took place in May 1791. However, there can be no guarantee that the journey summarized in the log actually occurred, because many navigation cyphering books included logs of fictitious journeys (Ellerton & Clements, 2012).

The manuscript has paper which is watermarked "C TAYLOR." Clement Taylor was a famous British paper maker based in Maidstone, Kent. Paper with the C TAYLOR watermark has been dated as early as 1787; two of Thomas Jefferson's letters held by the Library of Congress in the United States of America, dated 1792 and 1793, carry this same watermark (Clark & Murfin, 1995). From this evidence it seems reasonable to conclude that this manuscript was prepared some time during the period 1787–1795.

The writing on many of the pages often extends to the middle fold of the book, which suggests that the writing and illustrations on the pages, and the paintings on the fold-outs, were done *before*

Nerida F. Ellerton and M. A. (Ken) Clements, *Abraham Lincoln's Cyphering Book and Ten other Extraordinary Cyphering Books*,
DOI 10.1007/978-3-319-02502-5_9, © Springer International Publishing Switzerland 2014

the manuscript was finally bound. In this chapter it will be assumed that the manuscript upon which we are focusing was prepared sometime around 1791.

In the navigation cyphering book there are 80 illustrations altogether, including 51 miniature water-color paintings. Among the illustrations are:

- Two finished compass roses, one of which is particularly striking—it is reproduced as Figure 9.1;
- Three fold-out fortification water-color illustrations (two are shown in Figure 9.6);
- A large fold-out water-color painting showing a maritime image (see Figure 9.9).

But, why do we regard this manuscript as a *navigation cyphering* book?

Figure 9.1. The mariner's compass (from p. 36 of the navigation cyphering book).

What is a Navigation Cyphering Book?

Some cyphering books were prepared by persons training to be navigators, and we believe that that was the case with this manuscript. Elsewhere we have referred to handwritten texts of this variety as "navigation cyphering books" (Ellerton & Clements, 2012, p. 63).

Under What Circumstances was this Navigation Cyphering Book Prepared?

In the 18th and early-19th centuries many cyphering books were prepared by officers or midshipmen during voyages to distant locations—like, for example, when ships sailed between Western European nations and the East Indies, or between North America and the East Indies (Durkin, 1942; Ellerton & Clements, 2012; Rawley, 1981; Taylor, 1966). The Phillips Library, in Salem, Massachusetts, for example, holds up to 100 cyphering books prepared during merchant ventures to Africa, Europe and the East Indies (Gaydos & Kampas, 2010; Rawley, 1981). Some of these were arithmetic cyphering books, and others were navigation cyphering books.

Harvard University's Houghton Library holds two arithmetic cyphering books prepared by a certain William F. Allen in 1827. According to a note on a front end paper in one of these, the notebooks were sold for use on a voyage "from Salem towards India on the *Barque Pompey*." The order of the arithmetical content was largely, although not entirely, consistent with a standard middle-level *abbaco* sequence—the order was reduction, Federal currency, compound operations, decimal fractions, rules of three, inverse and compound proportion, vulgar fractions. commission, brokerage, discount, insurance, barter, loss and gain, alligation, single position, American customs duties. After 1702, in England, those who wanted to be naval schoolmasters needed a "naval schoolmaster's certificate," and a certain Nathan Prince, a Harvard College graduate and an experienced teacher of mathematics in Boston, went to London in the 1740s for the purpose of gaining that certificate (see, Taylor, 1966, pp. 139–140). A book published in London in 1801 was specifically written for instructors of "sea youth" including "schoolmasters of the Royal Navy" (Morrice, 1801).

There are, in fact, numerous extant examples of navigation cyphering books prepared by officers or midshipmen (Cavell, 2012) on voyages that took place in the eighteenth and nineteenth centuries (Ellerton & Clements, 2012). British naval authorities and merchants recognized that each ship on the high seas should have at least one person, and preferably two or more persons, who could use mathematical instruments to locate the ship's approximate position, and could recommend the best way forward given the prevailing conditions (direction of the wind, currents, etc.). Ship-owners typically employed "naval schoolmasters" who, when needed, assisted the navigators but mainly instructed midshipmen, and others, in reading, writing, arithmetic and, for those acquainted with trigonometry and logarithms, in navigation (Durkin, 1942; Taylor, 1966).

The Phillips Library in Salem, Massachusetts, holds a magnificent navigation cyphering manuscript which was completed in the 1760s and 1770s by a midshipman named Henry Tiffin. From notes within the manuscript it is clear that Tiffin prepared his manuscript while on board various sloops. The mathematical content of Tiffin's cyphering book was concerned with straight-edge compass constructions, trigonometry, logarithms, profiles of coastlines, and various kinds of sailing (plane, oblique, Mercator, middle latitude). The manuscript included several high-quality water-color paintings which Tiffin completed during journeys (Gaydos & Kampas, 2010). Another navigation cyphering book held in the Phillips Library, prepared by Ebenezer Bowditch between 1749 and 1751, includes a log for what was, probably, an actual journey between Salem and Charlestown, in South Carolina (Ellerton & Clements, 2012).

When we first examined the manuscript on which we are focusing in this chapter we were drawn to the possibility that it was prepared by a midshipman who was on board one of the 11 ships of the so-called "First Fleet" which sailed from Portsmouth, England, to Botany Bay, Australia, between May 1787 and January 1788. This venture was precipitated by the recommendation of Captain James Cook who, in 1770, had noted that a harbor, which he named "Botany Bay," was a likely base for settlement on the east coast of the vast, largely-unexplored Southern Continent known, at that time, as "New Holland." The First Fleet, which carried more than 1000 people altogether—including convicts, crewmen, military personnel, and other passengers—also carried animals, equipment, and provisions (Phillip, 1789). We conjectured that some years after the time of the First Fleet the person who prepared the manuscript added a log of a journey from Dundee in Scotland to Bergen in Norway, and that journey could have taken place in May 1791 (see pages 213–240 of the manuscript). The question whether that journey was real or fictitious is not relevant to our conjecture regarding the First Fleet because almost all navigation cyphering books included logs of journeys. Usually, the journeys were fictitious, but occasionally they were real. The reason for including a log was to have on record, in the navigation cyphering book, an example of how a log should be constructed.

Immediately before the log of the journey from Dundee to Bergen the manuscript under examination provided details of several other journeys—from Hull to Hamburg, from Stockholm to Petersburg, from Liverpool to Dublin, from Copenhagen to Berwick Upon Tweed, from Aberdeen to Hamburg, and from Berwick Upon Tweed to Yarmouth (see pages 186–233). No other navigation cyphering book that we have examined mentioned so many journeys. The entries of the unknown author left us with the impression that all of the logs had been created during actual journeys.

In relation to the possibility that the manuscript was prepared by someone attached to the First Fleet, we noted:

- There are several water-color miniatures in the manuscript depicting European brick structures with crosses (suggesting Roman Catholic churches). These appear early in the manuscript (e.g., on pages 22, 24 and 30), and are consistent with scenery that might have been seen from the First Fleet during the first weeks of its voyage (Phillip, 1789).
- On page 42 of the manuscript there is a reference to the Pike of Teneriffe, and a problem is presented and solved related to geographical features of that Pike. The First Fleet arrived at Santa Cruz, Teneriffe, in the Canary Islands, on June 3, 1787, and at that time Captain Arthur Phillip took on board fresh water, vegetables and meat (Phillip, 1789).
- On page 53 of the manuscript a roughly-drawn representation of what seems to be North and South America is shown. This is consistent with the fact that after leaving Teneriffe, the First Fleet proceeded to Rio de Janeiro in Brazil, South America (Hunter, 1793; Phillip, 1789).
- On page 60 there is a miniature showing a European-style building with a windmill next to it—consistent with the fact that after leaving Rio de Janeiro the First Fleet sailed to the Dutch-influenced Cape of Good Hope (South Africa), and remained there between October 13 and November 12, 1787 (Phillip, 1789).
- On page 43 of the manuscript a roughly-drawn picture of the earth's globe is shown, and on that New Holland (i.e., Australia) is pictured, but only partially—the west coast is shown, but the east coast is not. This suggests that the unknown person who prepared the manuscript had New Holland in mind.
- Throughout the manuscript there are numerous pictures showing two, or more than two, sailing ships—which is consistent with the artist being on a vessel sailing within a fleet rather than on a ship sailing alone.
- A large 13.5" by 7.5" fold-out water-color painting in this manuscript—between pages 122 and 123—shows two ships with French flags, surrounded by three ships carrying British flags. Small row boats seem to be emanating from all of the ships. The painting suggests that a cooperative spirit existed between the parties involved. It is possible that this painting was done on January 24, 1788, in Botany Bay (near what is now Sydney, Australia) by

> someone witnessing the interaction between two French ships that formed a "scientific expedition," led by Jean-François de La Pérouse, and British ships already at Botany Bay (Novaresio, 1996). The French expedition ships arrived just days after Governor Arthur Phillip's First Fleet had arrived, and apparently the French had expected to find a thriving colony at Botany Bay where they could repair ships and restock supplies. Hunter (1793) recorded that although the English fleet was in no position to offer much hospitality, the French crews were well received, and the French ships remained at Botany Bay until March 10, 1788.

This manuscript is unusual, and would be historically significant if it could be established that the main large fold-out painting was done in Botany Bay in January 1788, and that the other miniatures were created by a midshipman on the First Fleet. Unfortunately, although the data are consistent with our conjecture of a First Fleet connection, there is not enough direct evidence to confirm the conjecture, or even to support a claim that there is a high probability that the conjecture is true. Nowhere is the name of person who prepared the manuscript given, and nowhere is there any indication that any of the miniature paintings was conceived and painted at a particular place, on a particular day. Certainly, the watermark evidence suggested that the cyphering book was probably prepared between 1787 and 1795, and the "feel" of the book suggests that it was prepared by a midshipman on a voyage. And, although the large fold-out painting shows French and British ships together, this could have happened in many ports around the world. Yet, although there can be no guarantee that the picture was intended to depict events at Botany Bay in January 1788, that possibility remains.

Mathematical Considerations

From a mathematical perspective, the manuscript has three sections—dealing with mensuration, fortification, and navigation. In each, an applied problem-solving approach was adopted, with tasks presented as word problems. At all times the wording is succinct. From the outset it was assumed that decimal fractions, elementary trigonometric functions (sine, cosine, tangent, etc.), and logarithms (as calculational aids), could be used, and it seemed that a table of logarithms was available to the person who prepared the manuscript.

Mensuration

The manuscript began with a substantial section (pp. 1–66) on the mensuration of heights and distances. In Figure 9.2 we have reproduced eight exquisite, miniature water-color paintings from this section of the manuscript. Each miniature is, approximately, "actual size" (that is to say, the same size as it is in the manuscript), and each has its original elliptical (or oval) border.

What makes the miniatures especially interesting from a mathematical point of view is that each has dimensions (usually length or angle measurements) inscribed in it. These inscriptions are so small that they can hardly be seen by the naked eye. Figure 9.3 is an enlargement of Figure 9.2(d), and a circular construction used to create the outline of the elliptical border can still be seen. The miniature reproduced in Figure 9.3, from page 37 of the manuscript, is linked to the task: "From a ship at sea I observed a point of land to bear *ESE* and after sailing *NEbN* 18 miles I found its bearing to be *SSE*. How far was I from the point of land at each observation?"

Figure 9.2(a)
(page 7)

Figure 9.2(b)
(page 18)

Figure 9.2(c)
(page 31)

Figure 9.2(d)
(page 37)

Figure 9.2. Water-color miniatures from the navigation cyphering book.

Figure 9.2(e)
(page 56)

Figure 9.2(f)
(page 59)

Figure 9.2(g)
(page 60)

Figure 9.2 (h)
(page 64)

Figure 9.2 (Cont'd). Water-color miniatures from the navigation cyphering book.

Figure 9.3. An enlargement of the miniature shown in Figure 9.2(d).

In this section of the book it appears that our unknown author was creating mathematical problems that related to actual scenery that he could see from the ship, and then making miniature paintings of what he saw. If that was indeed the case, then it is unlikely that the measurements shown in small print on the miniatures were accurate—rather, they would have been estimates. Nevertheless, if the problems were created by our unknown author then his problem-posing achievements were amazing, and one would have to be impressed with his determination to create, and his success in creating, real-life examples.

Sloan (2000), in comparing the training in drawing offered at Christ's Hospital and at the Royal Naval Academy (hereafter "RNA") at Portsmouth, in England, made the point that RNA students were trained to draw figures as part of their "exercises in mathematics and mensuration, with labelled lines of trajectory and other calculations" (p. 105), while those at Christ's Hospital merely copied from drawings by their drawing master. The Royal Military Academy, at Woolwich, also trained its students "to accommodate the more technical and mathematical aspects of drawing necessary for preparing young men for trades and naval or military careers, as well as less technical aspects of landscape appropriate to young gentlemen" (Sloan, 2000, p. 106).

By contrast, the drawings of a Christ's Hospital student, James Slater Elly, in the mid-1750s, "contain neither lines of measurement nor calculation, and bear no relation to the written and mathematical exercises found on the facing pages" (Sloan, 2000, p. 106). According to Sloan, in the eighteenth century, British academies increasingly offered instruction in technical drawing. Thus, it would appear to be the case that the manuscript on which we are focusing in this chapter was consistent with the kind of technical training in drawing given to RNA and Royal Military Academy students. That same kind of training was also available to many students in private and dissenting academies which included technical aspects of drawing in their curricula (McLachlan, 1931; Sloan, 1986). We do not see the emphasis on technical drawing at Christ's Hospital, RNA, the Royal Military Academy academies as inconsistent with the possibility that our unknown author was a midshipman on the First Fleet to Botany Bay—for graduates of Christ's Hospital, RNA and the Royal Military Academy were apprenticed to ship masters, and to the Royal Navy, and could easily have been members of the First Fleet.

Despite the impressive determination to create real-life examples, examination of the mathematical writing in this mensuration section of the manuscript raises interesting questions about the extent to which the person who prepared the manuscript understood the mathematics with which he was concerned. Consider, for example, the mathematics recorded with respect to the following task which is introduced on pages 23 and 24 of the manuscript:

> Wanting to know the perpendicular breadth of a river, I took two stations on the bank where I was, and measured the angles subtended by the station line which was 80 yards, and a tree closs (sic.) by the other bank viz., 60°4' and 84°17'.What was the breadth?

The statement of the problem was illustrated by a miniature water-color painting (p. 24), an enlargement of which is shown in Figure 9.4. The Figure identifies two pertinent triangles, *ABC* and the right-angled *ADC*. The length of *CD* is required.

Figure 9.4. What is the breadth of the river?

Immediately after the statement of the problem the following "solution" appeared in the manuscript (pp. 24–25):

To find *AC*—

As Sine ∠ *C*	= 35°39' =	9.76554
To *AB* =	80 =	1.90309
So is Sine ∠ *B* =	84°17' =	9.99783
To *AC* =		11.90092
		2.13538

To find CB—

As radius	= 90°	=	10..........
To AC	= 136.6	=	2.13545
So is Sine ∠ A = 60°4'		=	<u>9.93782</u>
To CD			= 12.07327
			<u>2.07327</u>

The underscored 2.13538 and 2.07327 were 5-decimal-place logarithms of distance measures, and the answer to the problem (i.e., the breadth of the river), 118.4 yards—which was not specifically given—is the number whose base 10 logarithm, to five decimal places, is 2.07327.

Nowhere in the navigation cyphering book did the person who prepared the manuscript attempt to explain his reasoning for this working. In the first line, when he was finding AC, he wrote: "sine ∠C = 35°39' = 9.76554"; but, strictly speaking, that was nonsensical. To begin, ∠C might equal 35°39', but the statement sine ∠C = 35°39' is categorically wrong. Even allowing for the implications of trigonometrical theory accepted among mathematicians at the time, sine 35°39' was not equal to 9.76554, to 5 decimal places—what was true, according to the theoretical position, was that sine 35°39' was equal to 0.58283×10^{10}, and hence $\log_{10}$ sin 35°39' was equal to 9.76554 (because, $\log_{10}(5.8283 \times 10^{9})$ is equal to 9.76554, to five decimal places). All numbers shown on the right of the above setting out were logarithms, to base 10, to five decimal places. Thus, in the second line, the logarithm, base 10, of 80, was equal to 1.90309.

From both mathematical and educational perspectives, the setting out was inadequate. On the fourth line of the calculations, one finds "To AC =" and then 11.90092 appears. The fact was, the logarithm, base 10, of AC was *not* equal to 11.90092. On the next line 2.13538 is given, which is in fact the logarithm, base 10, of 136.6, to five decimal places. In these calculations, the 11.90092 was obtained by adding 1.90309 and 9.99783; then, the 2.13538 was obtained by subtracting 9.76554 from 11.90092. Then, the 136.6 was probably found by reading the logarithm tables "backwards," to find the antilogarithm of 2.13538. The answer to the original question was never formally given, there being no mention of the fact that the 2.07327 was the logarithm of the breadth (in yards).

Most mathematics students of the 21st century will wonder how sin 35°39' could be equal to 0.58283×10^{10}, because it has long been accepted that for any angle, θ^o say, the value of $\sin\theta^o$ must lie between -1 and 1 (or perhaps be equal to -1 or 1). In order to understand the trigonometrical concept of sine, as that concept was commonly defined in the 18th century, one needs to examine Figure 9.5, which is taken from Moore (1791, p. 23), but is identical to figures in earlier editions (e.g., Moore, 1779, p. 11). The radius of the circle shown in Figure 9.5 was assumed to be 10^{10}, and hence the sine of ∠A (in Figure 9.4) would be a large number whose logarithm (base 10) was 9.76554. But, nowhere in Moore's (1791) widely-used text is there any direct statement that it was being assumed that the radius measure of the circle shown in Figure 5 was 10^{10}. What Moore (1791) did say was:

> In a Right-angled Triangle ... to find an Angle, one of the given Sides must be made Radius; then say, as the Side made Radius is to Radius, so is the other given Side to the Sine, Tangent, or Secant, by it represented; which fourth Proportion must be looked for in the Tables of Sines, Tangents, &c. ... The same must be observed in stating all Cases in Trigonometry, as of the Rule of Three in Arithmetic, where we are taught to multiply the

second and third Terms together, to divide their Product by the first Term, and the Quotient will be the fourth Term, or the number Sought, and of the same Denomination as the second Term. (p. 25)

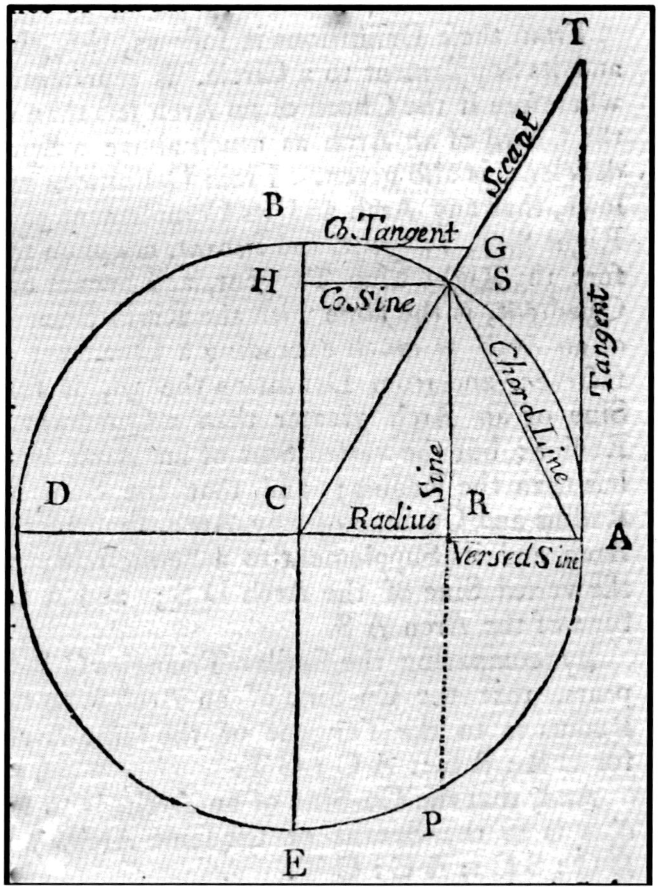

Figure 9.5. A line-segment approach to defining trigonometric functions (from Moore, 1791, p. 23).

Our unknown author might have referred to the 1791 edition, or to an earlier edition, of Moore's widely-used textbook, for Moore set his calculations out in the same way—except that Moore would have correctly placed the 136.6 to the left of 2.13538 and not to the left of 11.90092.

Most 21st-century students of trigonometry would think that the unknown author made use of the so-called "Sine Rule" in order to determine that the length measure of *AC* (in yards) was equal to the value of 80 × sin 84°19' divided by sin 35°39'—the 35°39' having been obtained by subtracting the sum of 84°17' and 60°4' from 180°. However, as Moore (1791) made clear, the logic of the calculation was related to the direct rule of three (see also Loomis, 1868).

Many 17th-, 18th- and 19th-century mathematicians who needed to carry out complicated calculations used tables of logarithms, and in so doing struggled to avoid having negative characteristics that arose for numbers between 0 and 1 (see, e.g., Hodgson, 1723; Hopkins, 1833; Moore, 1791). Take, for example, the following statement which appeared in an 1822 edition of Robert Gibson's well-known *A Treatise on Practical Surveying*, an edition which John D. Craig had "revised, corrected, and adapted to the use of schools, American surveyors, &c" (p. iii).

EXAMPLE 1

> Required, the sine of 36 degrees 40 minutes.
>
> Look at the head of the page for 36 degrees, and in the side column on the left hand, for 40 minutes; then below the word sine, on the same line with 40, we find 9.77609; which is that required. (Gibson & Craig, 1822, p. 86)

Thus, Gibson and Craig told their readers that the sine of 36 degrees 40 minutes was 9.77609. That was simply incorrect—probably what they meant to say was that, correct to five decimal places, the logarithm, base 10, of the sine of 36 degrees 40 minutes was 9.77609.

A consequence of this approach to trigonometry was that in the right-angled triangle *ACD,* shown earlier in Figure 9.4, it was *not* true that $AC \times \text{sine} \angle CAD$ was equal to *CD;* nor was it true that sine $\angle CAD$ was equal to the length measure of *CD* divided by the length measure of *AC.* Well into the 19th century the respected Elias Loomis (1868), Professor of Natural Philosophy and Astronomy at Yale College and author of a widely-used series of mathematics textbooks for schools and colleges, would write:

> The natural sine of 14°30' is 0.250380
>
> Its logarithm ... is (–1) + .398600.
>
> The characteristic of the logarithm is *negative*, as must be the case with all the sines, since they are less than unity. To avoid the introduction of negative numbers in the table, we increase the characteristic by 10, making 9.398600 and this is the number found for the logarithmic sine of 14°30'. The radius of the table of logarithmic sines is therefore, properly, 10,000,000,000, whose logarithm is 10. (p. 27)

It is a matter for speculation what Loomis meant by: "The radius of the table of logarithmic sines is therefore, properly, 10,000,000,000." Although such an approach was adopted by many applied mathematicians, it was never *universally* accepted. Thus, for example, John Robertson (1767), Fellow of the Royal Society and Master of the Royal Naval Academy at Portsmouth (and formerly Master in the Royal Mathematical School at Christ's Hospital), maintained that entries in tables of natural sines showed the decimal parts of the radius, supposed to be unity. Thus, for Robertson (1767), sine 84°17' would have been equal to 0.99503, and not 9.99783. Also, the logarithm (base 10) of sine 84°17' would have been, to 5 decimal places, –1 + 0.99783.

We have examined many eighteenth- and nineteenth century-textbooks on surveying and navigation, and have found that there is much confusion with respect to values of sine, cosine, etc. of angles, and the logarithms of these values. In an 1853 edition of Nathaniel Bowditch's *The New American Practical Navigator,* the reader was informed that log sine 28°37' was equal to 9.68029 (p. 33), and the value of log secant 126°20' was equal to 10.22732 (p. 33). Clearly, Bowditch was assuming the definitions deriving from a circle with radius 10^{10}, but he did not seem to worry that his approach meant that the value of the sine of an angle depended on the radius of the "standard" circle. As time went by, greater attention was given to tightening the ideas associated with trigonometric functions, and once that happened it was no longer possible to find a real-number value for log secant 126°20' because secant 126°20' is negative.

Thus, during the 17th, 18th and early 19th centuries, many applied mathematicians used double talk like this when referring to the logarithms of sines and cosines. Little wonder, then, that students like our unknown author appeared to be confused. Differences between what appeared in our unknown author's navigation cyphering book and in Moore's (1791) text might appear to be minor, but what Moore wrote at least made some mathematical sense—if one allows the radius measure of the standard circle to be 10^{10}. What our unknown author wrote on the matter, however, made little sense. His version of what appears to have been Moore's setting out meant that it was unlikely that he would have been able to explain the logic behind what he wrote down. That comment would apply to the working shown for numerous problems in the unknown author's text.

It is impossible to tell whether our unknown author copied from a textbook, or from another navigation cyphering book. Even mathematicians holding senior academic appointments were likely to be careless in some of their statements regarding logarithms and trigonometry. Thus, for example, the Rev. S. Vince (1810), Plumian Professor of Astronomy and Experimental Philosophy at the University of Cambridge, in England, included the assertion "tan 56°18' = $\angle$ A" on page 72 of his third "corrected" edition of *A Treatise on Plane and Spherical Trigonometry*.

Fortification

Following the section on heights and distances was a lengthy section on fortification (pages 69–125), and then followed a section on navigation (pages 126–240).

The section on fortification began with the statement:

> Fortification is that which teaches how to build walls and dig ditches round any place so as to enable the people within it to defend themselves in the easiest manner possible against assaults of an enemy from without. (p. 69)

Fortresses were built to last, and the remains of many can still be found in many parts of the world. Thus, for example, the authors have visited Fort Cornwallis, which was built by convict labor in Penang, Malaysia, in 1810, at a time when Penang was under British colonial rule.

Commercially-printed treatises on fortification usually featured characteristic diagrams such as those shown in Figure 9.6, which is taken from page p. 86 of the manuscript. Although designs of many ancient fortresses were based on mathematical considerations only rarely did one find a section on fortification in a cyphering book.

Figure 9.7 shows two of three water-color fold-outs showing typical fortification structures, and Figure 9.8 shows calculations intended to relate to the diagram on the right side in Figure 9.6. These calculations were only part of a long series of calculations which continued for more than four pages—and revealed how complex the designs of fortresses could be. Once again, however, the calculations were, from a strictly mathematical point of view, faulty, because no distinction was made between length measures and the logarithms (base 10) of those length measures.

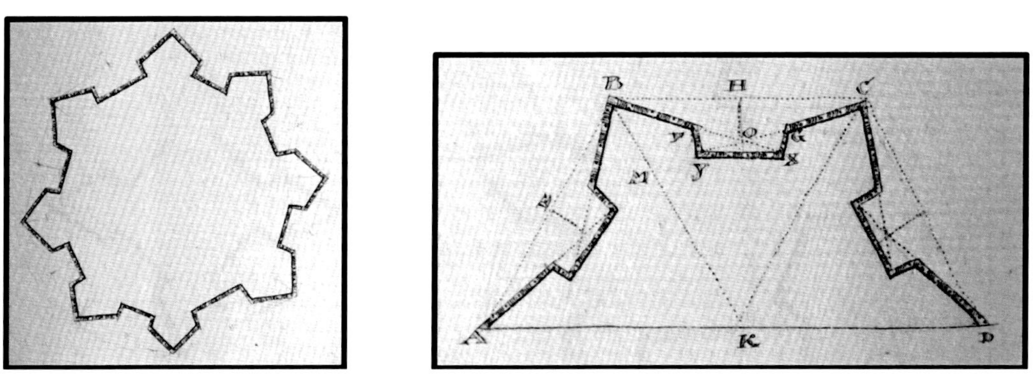

Figure 9.6. Designing a fortress (p. 86 of the navigation cyphering book).

Figure 9.7. Two fortification fold-outs (p. 105 and p. 107), showing four fortification designs: "Crown Work" (top), and "Ravelin," "Swallow's Tail," and Contre Queue D'Aronde" (bottom).

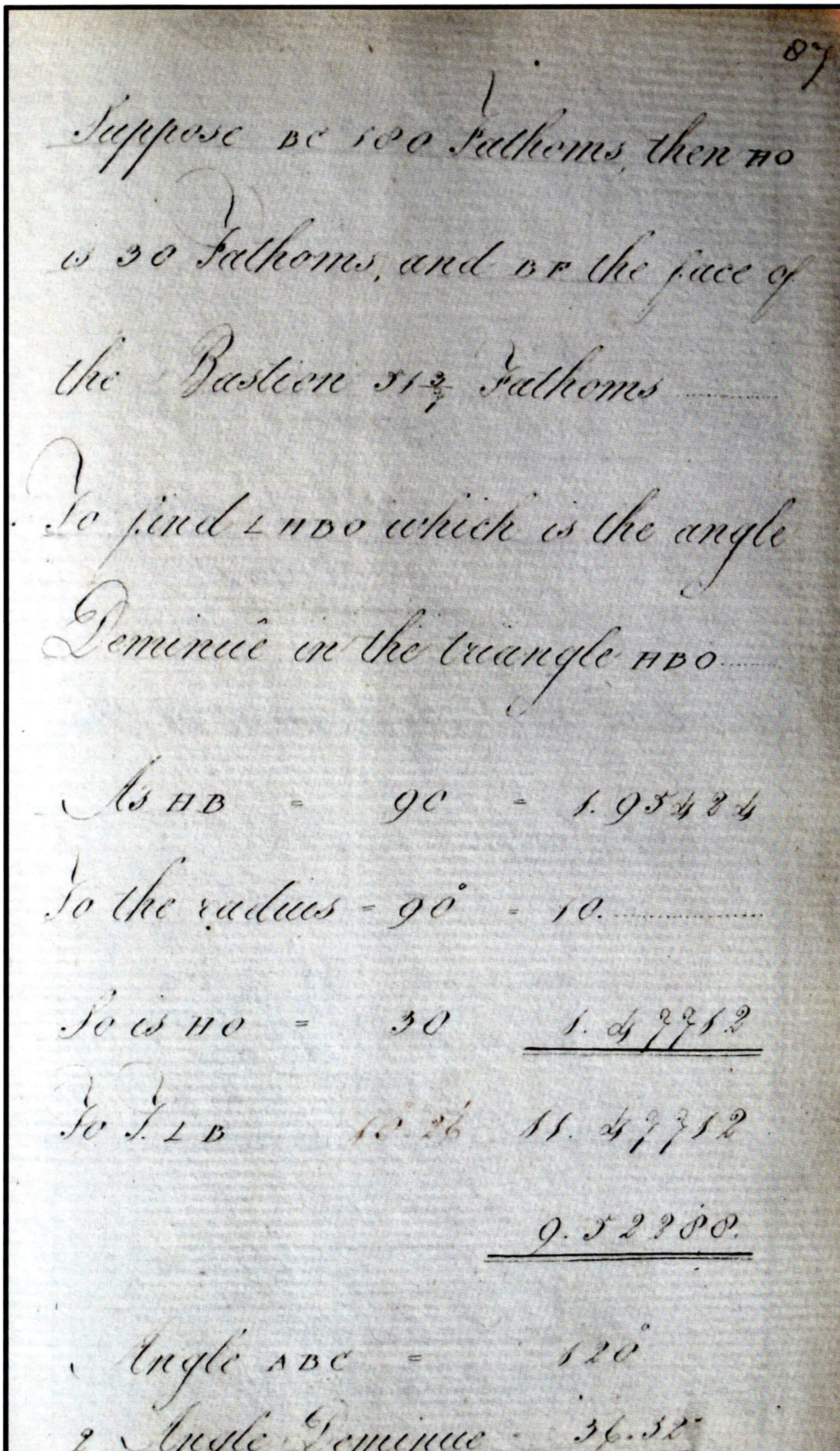

87

Suppose BC 180 Fathoms, then HO is 30 Fathoms, and BF the face of the Bastion $51\frac{3}{7}$ Fathoms

To find ∠ HBO which is the angle Deminuê in the triangle HBO

As HB = 90 = 1.95424

To the radius = 90° = 10.

So is HO = 30 1.47712

To S. ∠ B 18.26 11.47712

9.52288

Angle ABC = 120°

2 Angle Deminue = 36.52

Figure 9.8. A page of calculations related to the design of a fortress (p. 87).

Navigation

This was the longest section in the manuscript, from page 118 through to page 240. It dealt especially with the various types of sailing—plane sailing (pp. 122–160), transverse sailing (pp. 162–171), current sailing (pp. 172–175), parallel sailing (pp. 176–193), middle latitude sailing (pp. 194–221), and Mercator's sailing (pp. 222–240). Although there were no water-color miniatures in this section, there were numerous line drawings (all in ink). A typical page is shown in Figure 9.9 (which is p. 233 in the manuscript). The task was: "A ship from Lat 39°N Long 48°52'W sails NNW 112 miles. Required her present latitude and longitude."

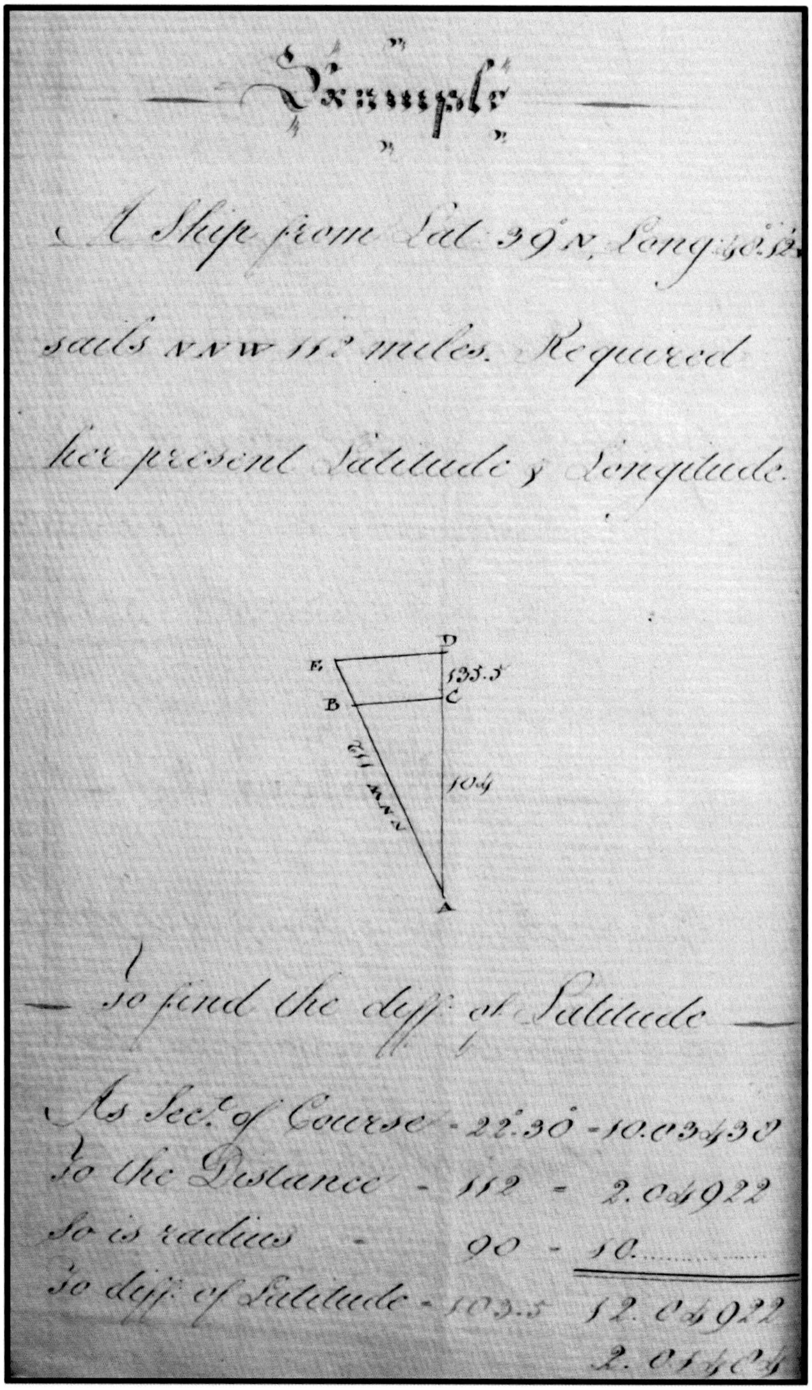

Figure 9.9. Response to a Mercator's sailing task (p. 233 of the navigation cyphering book).

An interesting aside is that the terminology "plane sailing" was used, rather than "plain sailing." Both spellings ("plane" and "plain") are often found in navigation cyphering books. The British geographer, Eva Taylor (1956) claimed that John Robertson, who worked at both Christ's Hospital and at the Royal Naval Academy, led an eighteenth movement that changed the spelling from "plain" (which was associated with what was known as a "plain chart") to "plane" (based on the idea that the earth could be regarded as an extended plane).

Why is this Cyphering Book Extraordinary?

Despite our reluctant setting aside of our First Fleet conjecture, we intuitively believed that the manuscript under examination provided a superb example of a navigation cyphering book prepared by a midshipman or officer, with the artwork and diagrams depicting scenes on a particular journey or on a combination of various journeys. When we reflected on why we thought the manuscript deserved the adjective "extraordinary" we were drawn to the fact that although we have a collection of about 500 cyphering books, prepared in a range of nations, the one which is the main focus of this chapter displays the highest quality artwork of any of them. The very fine calligraphic headings and copperplate writing, which were typical of what would be found on every page in the manuscript, were *not* extraordinary—in the sense that high quality penmanship and calligraphy can be found in quite a few of the cyphering books in our collection—but the quality of the artwork in the manuscript prepared by our unknown author, circa 1791, is unsurpassed. Furthermore, the artistic creations were generally not merely art for art's sake—the miniatures, with their very finely labelled measures of side lengths and angles—are amazing.

The impressive artwork in the manuscript deserves further comment. When preparing her University of London PhD dissertation, Kimberly Sloan (1986) studied, in some detail, the push to make the subject of drawing part of the curriculum of all "respectable" eighteenth-century British schools. Sloan's conclusions were:

1. In Great Britain, during the period from 1750 to 1900, drawing increasingly became part of the curriculum in many schools, private academies, "public" schools (such as, Eton and Harrow), and even at Oxford and Cambridge Universities. The subject was also often introduced, for serious study and application, into apprenticeship training schools.
2. An element of the middle classes in eighteenth-century Europe came to believe that in order to cope with the demands of ordinary living one needed not only to be able to read and write well, but also to draw well. Thus, for example, in 1768 the Drawing Society of Dublin, agreed to publish a series of volumes related to technical aspects of drawing, and these volumes included some very sophisticated treatments of mathematics, including higher algebra, geometry and navigation (De Morgan, 1860; Fenn, 1769).
3. Some influential persons asserted that learning to draw well would have benefits for learning other subjects. Thus, for example, the idea that training in how to draw well could improve one's appreciation and understanding of geometry was often expressed. That view was put forward by the philosopher John Locke, by the leading government adviser Samuel Pepys, and by the architect, Christopher Wren—all of whom maintained that drawing should be an integral part of the curriculum of all middle-class schools, and especially of those responsible for training surveyors, navigators, and architects. It was not a matter of surprise that at the government-subsidized Christ's Hospital school, located in central London, the curriculum included drawing. In the late seventeenth century, Pepys, Wren, Isaac Newton, the mathematician John Wallis, and the astronomer David Gregory, were all

closely connected with the development of the curriculum of the Royal Mathematical School within Christ's Hospital, and all students in that School were given expert instruction in drawing.

4. In the eighteenth century, outstanding British drawing masters such as John Ayres, Alexander Cozens, George Bickham (see Chapter 7 in this book) and Bernard Lens III, worked hard to educate the public that the kind of training in technical drawing offered at schools like Christ's Hospital, the Royal Naval Academy at Portsmouth, and the Royal Military Academy at Woolwich, produced better surveyors and navigators.
5. Drawing became a particularly popular past-time with women, and some claimed that no young lady deserved "the honourable character of an accomplished woman without a competent knowledge of drawing," which could "improve the imagination and strengthen the judgement by obliging you to examine the object you copy with greater care" (quoted in Sloan, 1986, pp. 238–239).

Probably, the following depiction of the fashionable damsel by Hannah More (as reported in Manwaring, 1925) represented a viewpoint held by many influential persons:

> Then comes my drawing master; he teaches me to paint flowers and shells, and to draw ruins and buildings, and to take pictures, and half a dozen fire screens which I began for mama ... I learn varnishing, and gilding, and japanning, and next winter I shall learn modelling, and etching, and engraving in mezzotinto, and aquatinto, for Lady Di Dash learns etching, and mama says as I shall have a better fortune than Lady Di, she vows I shall learn everything she does. (Quoted in Sloan, 1986, pp. 237–238)

Pride of Ownership

Despite the lack of information regarding the provenance of the navigation cyphering book under examination in this chapter, it is impossible to examine the manuscript carefully and not recognize pride of ownership. This is a beautiful manuscript, one which features exquisite artwork and fine penmanship and calligraphy. Although the mathematics lacks technical accuracy, it seems that the person who prepared the manuscript fully realized that the artwork and the associated calculations related to real situations.

Figure 9.10 is a reduced-size reproduction of the 29 cm by 23 cm (11.75" by 9.25") water-color fold-out which was inserted into the manuscript after page 122—at the end of the section on Fortification. Because the fold-out is more than 220 years old, the folds are clearly visible and detract from the visual effect of the painting. Nevertheless, the attention to detail, and the impressive depiction of the clouds, makes this a very worthwhile piece of art. Altogether, there are 51 paintings in the manuscript, and this is the largest. It is the only painting that does not directly illustrate a problem or situation described in the text.

The painting reproduced in Figure 9.10 attests to the pride of ownership felt by our unknown author. We expect that the three British and two French boats that are shown were painted from a real-life experience. We would like to think that the location was Botany Bay, close to what is now Sydney, Australia; and we would like to think that the date was January 24, 1788 (Hunter, 1793; Phillip, 1789). However, since the name of the painter is not known and no identifiable landmarks are shown in the painting, our conjectures probably represent little more than wishful thinking. It seems safe to assume, though, that the painting was included in the navigation cyphering book because our unknown author was not only very proud of his book, but also of his painting.

Figure 9.10. A reproduction of the large fold-out in the navigation cyphering book (p. 122).

Altogether, there were 80 artistic creations in the book and each would have required dedicated time for its creation. In addition, there were many mathematical calculations related to details depicted in almost all of the water-color paintings and line drawings. Undoubtedly, the author dedicated a large amount of time, effort, and thought to creating *his* book. After he had completed the pages, he had it bound, full-leather, and arranged for beautiful gilt decorations to be inscribed on the spine cover.

Indeed, this *was his* book, and he probably never prepared another like it. It would have been carefully looked after, and ultimately passed on, probably within the family. It survives, intact, and is, potentially at least, a document of considerable interest from a historical point of view.

From the point of view of artistic excellence this is, then, an extraordinary cyphering book. It could be argued that from a mathematical point of view our unknown author was neither accurate nor creative. He relied heavily on the kind of mathematical calculations found in navigation texts like Moore (1791). That said, throughout the text he continually used decimal fractions, logarithms, and elementary trigonometry, and that meant he was dealing with forms of mathematics that, in 1791 (or whenever he prepared his cyphering book), 99 percent of the British population did not know anything about. Furthermore, if it is true that he was constantly creating mathematics problems, and estimating measurements, associated with what he could see from his ship, then the quality of his problem posing, and the creativity that he displayed in the problems he created and illustrated, were remarkable.

Concluding Comments

It is the high quality of the artwork in this manuscript which makes it extraordinary. However, our analysis has revealed other historically interesting matters with respect to the manuscript. The first of these "other matters" is the way problems seem to have been fitted to scenes which our unknown author viewed from the ships on which he was serving as a midshipman. We cannot be certain that the author created the problems—perhaps, they were created by our unknown author's teacher(s); perhaps they were copied from another, older cyphering book. But if they were created by the author, then they represent an amazing feat in relevant problem posing.

Our analysis has revealed that the unknown author did not really understand the mathematics associated with using logarithm tables to expedite calculation involving trigonometrical relationships. The idea that the sine of an angle, for example, might have a value of almost 10^{10} seemed to lead to educational difficulties. Certainly, many textbook authors in the eighteenth century developed trigonometry from a line segment approach based on a circle, with the radius of the circle being assumed to be 10^{10}. From that definition, it had to follow that the sine (or cosine, etc.) of an angle could be anything between (or including) 0 and 10^{10}. But that conclusion meant that in a right triangle for which one of the interior angles was 30°, sine 30° was *not* equal to the length measure of the side "opposite" the 30° angle divided by the length measure of the hypotenuse. Future mathematicians would pay more attention to the definitional issues involved, and the apparent dilemma would be solved. In the eighteenth century, however, the use of logarithms was important with respect to calculations in surveying, navigation, and astronomy, and the circle-with-radius-10^{10} definitions were adopted to avoid negative values of characteristics when logarithms of sines, cosines, etc., of angles were needed. It is unusual for mathematicians to allow "sloppy" definitions for pragmatic reasons, but here is a case where that seems to have happened.

One consequence of educationally-deficient descriptions of the line-segment approach to trigonometry offered by some mathematicians was that many of the students who grappled with mensuration tasks for which they were required to find angle and side measurements of triangles came to use tables of logarithms in mechanical ways. They learned to follow a magical routine which, if they were careful, would enable them to arrive at correct answers (see, for example, working shown in 1790 by a student called "F. Wilcox," reproduced in Denniss (2012), p. 33).

The attitude of students seemed to be that some clever people "out there" knew how to use these calculational devices called logarithms, and one only had to follow the procedures they recommended in order to get correct answers. This way of thinking complemented the forms of education found in arithmetic cyphering books in which students used algorithms, developed over centuries by "algorists," to get answers without really knowing *why* they were doing what they doing. Richard Skemp (1976) called this "rules without reason."

References

Bowditch, N. (1853). *The new American practical navigator.* New York, NY: E. G. W. Blunt.

Cavell, S. A. (2012). *Midshipmen and quarterdeck boys on the British Navy 1793–1815*. Woodbridge, United Kingdom: Boydell Press.

Clark, P., & Murfin, L. (1995). *The history of Maidstone—The making of a modern county town.* Stroud, United Kingdom: Alan Sutton Publishing.

De Morgan, A. (1860). Drawing Society of Dublin. *Notes and Queries, 2*(9), 444.

Denniss, J. (2012). *Figuring it out: Children's arithmetical manuscripts 1680–1880*. Oxford, United Kingdom: Huxley Scientific Press.

Durkin, J. J. (1942). Journal of the Revd. Adam Marshall, schoolmaster, *U.S.S. North Carolina,* 1824–1825. *Records of the American Catholic Historical Society of Philadelphia, 53*(4), 152–168.

Ellerton, N. F., & Clements, M. A. (2012). *Rewriting the history of school mathematics in North America 1607–1861*. New York, NY: Springer.

Fenn, J. (1769). *First volume of the instructions given in the drawing school established by the Dublin-Society, pursuant to their resolution of the fourth of February, 1768; ... Under the direction of Joseph Fenn.* Dublin, Ireland: Alex M'Culloh.

Gaydos, T., & Kampas, B. (2010). *American and Canadian ciphering books, n.d., 1727–1864.* Salem, MA: Phillips Library at the Peabody Essex Museum.

Gibson, R., & Craig, J. D. (1822). *A treatise on practical surveying* (3rd ed.). Baltimore, MD: F. Lucas Jr., and Cushing & Jewett.

Hodgson, J. (1723). *A system of mathematics containing the Euclidean geometry, plane and spherical trigonometry, the projection of the sphere, both orthographic and stereographic, astronomy, the use of globes and navigation.* London, United Kingdom: Thomas Page, William and Fisher Mount.

Hopkins, W. (1833). *Elements of trigonometry.* London, United Kingdom: Baldwin and Craddock.

Hunter, J. (1793). *An historical journal of the transactions at Port Jackson and Norfolk Island with the discoveries that have been made in New South Wales and in the Southern Ocean since the publication of Phillip's voyage, compiled from the official papers, including the journals of Governors Phillip and King, and of Lieut Ball.* London, United Kingdom: John Stockdale.

Loomis, E. (1868). *Elements of plane and spherical trigonometry, with their applications to mensuration, surveying, and navigation.* New York, NY: Harper & Brothers.

Manwaring, E. (1925). *Italian landscape in eighteenth century England.* New York, NY: Oxford University Press.

McLachlan, H. (1931). *English education under the Test Acts: Being the history of the nonconformist academies, 1662–1820*; Manchester, United Kingdom: Manchester University Press.

Moore, J. H. (1779). *The practical navigator and seaman's new daily assistant, being a complete system of practical navigation, improved and rendered easie to any common capacity* (4th ed.). London, United Kingdom: B. Law.

Moore, J. H. (1791). *The practical navigator and seaman's new daily assistant, being an epitome of navigation, including the different methods of working the lunar observations, with every particular requisite for keeping a complete journal at sea.* London, United Kingdom: B. Law.

Morrice, D. (1801). *The young midshipman's instructor (designed to be a companion to Hamilton Moore's Navigation): With useful hints to parents of sea youth, and to captains and schoolmasters of the Royal Navy.* London, United Kingdom: Knight and Compton.

Novaresio, P. (1996). *The explorers*. New York, NY: Stewart, Tabori & Chang.

Phillip, A. (1789). *The voyage of Governor Phillip to Botany Bay, with an account of the establishment of the colonies of Port Jackson & Norfolk Island, to which are added the journals of Lieuts. Shortland, Watts, Ball & Capt. Marshall.* London, United Kingdom: John Stockdale.

Rawley, J. A. (1981). *The trans-Atlantic slave trade*. New York, NY: W. W. Norton.

Robertson, J. (1767). *A general treatise of mensuration* (3rd ed.). London, United Kingdom: J. Nourse and J. Wilcox.

Skemp, R. (1976). Instrumental understanding and relational understanding. *Mathematics Teaching, 77,* 20–26.

Sloan, K. M. (1986). *The teaching of non-professional artists in eighteenth-century England.* PhD thesis, Westfield College, University of London.

Sloan, K. (2000). *"A noble art": Amateur artists and drawing masters c.1600–1800.* London, United Kingdom: British Museum Press.

Taylor, E. G. R. (1956). All plain sailing. *Journal of Navigation, 9*(2), 230–232.

Taylor, E. G. R. (1966). *The mathematical practitioners of Hanoverian England 1714–1840.* Cambridge, United Kingdom: Cambridge University Press.

Vince, S. (1810). *A treatise on plane and spherical trigonometry; with an introduction, explaining the use of logarithms adapted to the use of students in philosophy.* Cambridge, United Kingdom: J. Smith.

Chapter 10

From the Royal Mathematical School: Charles Page, 1825

Abstract: Charles Page was born in 1810, and between 1818 and 1825 he was a student at the Royal Mathematical School within Christ's Hospital, London. Over his final two years as a "mathemat" (i.e., student in the Royal Mathematical School) he prepared the 671-page navigation cyphering book which is the subject of analysis in this chapter. Throughout the manuscript the standard of penmanship, calligraphy, artwork, and mathematics is at a very high level—indeed, much of the mathematics in this cyphering book is at a higher level than the mathematics in any of the other extraordinary cyphering books considered in this volume. Careful examination of the entries in Charles Page's manuscript suggests, however, that many of them were taken down from dictation. In the section on quadratic equations there are unchecked errors, and we were left with the impression that although the appearance of pages was invariably impressive, the quality of Charles' learning about mathematics and about navigation may have been less than impressive. Details in this chapter provide a glimpse of the expectations on mathemats in the 1820s.

Key Words: Christ's Hospital; ciphering book; cyphering book; Isaac Newton; James Hodgson; logarithms; navigation cyphering book; navigation education; Royal Mathematical School; Royal Military Academy; Royal Naval Academy; sailing; Samuel Pepys; trigonometry.

Background: Christ's Hospital, Newgate Street, London

In 1552, a new school was opened in the extensive buildings of the dissolved Grey Friars Monastery in Newgate Street, London. This school, named Christ's Hospital, would be largely financed by the government (Pearce, 1901). More than a century later, in 1673, King Charles II, of England, and his advisers, recognizing that sea-power was becoming crucially important—not only for national defense and status, but also for colonial expansion and for mercantile and industrial development (Howson, 1982; Howson, Keitel & Kilpatrick, 1981)—started a mathematical section at Christ's Hospital for 40 scholarship "poore boyes" aged from about 10 years. Initially, these students were supported out of royal finances, and their program supervised by a board which included the nation's top scientists (Christ's Hospital, 1953; Howson, 1982; Pearce, 1901; Plumley, 1976; Sloan, 1986, 2000; Wilson, 1820), but later financing came mainly from benefactions. Education was free until 1892. Each boy who graduated would be required to become an apprentice in some sea-faring capacity.

In time, Christ's Hospital would become one of the world's best training schools for navigators. According to Jones (1954), Christ's Hospital was the first English school in which mathematics became an important part of the curriculum. When Alexander Dallas Bache (1839), Benjamin Franklin's grand-son, visited Christ's Hospital in the 1830s, he declared, that its grand scale was "so much beyond any other in Great Britain, and, indeed, in the world" (pp. 65–66). Hans (1951a) described the Royal Mathematical School at Christ's Hospital as the first "modern school" established anywhere in the world (p. 532).

The government believed that many boys whose fathers had died at sea, especially while serving in the Royal Navy, deserved serious consideration for admission to the Royal Mathematical School. Because of the frequency of wrecks at sea, piracy, accidents, etc., there were many young boys who

Nerida F. Ellerton and M. A. (Ken) Clements, *Abraham Lincoln's Cyphering Book and Ten other Extraordinary Cyphering Books*, DOI 10.1007/978-3-319-02502-5_10, © Springer International Publishing Switzerland 2014

had lost sea-faring fathers, and there was a demand for boys from such families to be admitted to Christ's Hospital.

By the beginning of the 19th century, Great Britain was unquestionably the leading maritime power in the world (Mahan, 1898; Padfield, 2005; Waters, 1958). As early as 1697, Peter the Great, Czar of Russia, showed interest in the Royal Mathematical School at Christ's Hospital and two of its graduates (one was 15 years old, the other 17) were appointed to lead a training school for navigators that opened in 1701 in Russia (Allan, 1984; Cross, 2007; Hans, 1951a; Howson, 1982). Graduates of the Moscow School of Mathematics and Navigation would subsequently become teachers in the famous cyphering schools that Peter the Great established in 1716 to teach arithmetic and geometry to children drawn from all levels of Russian society (Cracraft, 1971; Okenfuss, 1973).

The decision to establish the Royal Mathematical School was motivated by a desire to assist the British government to provide the kind of technical education needed for Great Britain to keep up with the military might of the Dutch, French and Spanish nations. In 1600, Maurits, Prince of Orange and Stadholder of Holland and Zeeland, founded a school for military engineers in Leiden (Waters, 1958). The statutes and the curriculum for this school had been drawn up by Simon Stevin, a mathematician and engineer, and lessons were given in the vernacular, rather than in Latin. The school was intended for "illiterate" masons, navigators, carpenters, surveyors and engineers, and after attending "theoretical" classes in the winter, selected students were conscripted to the Dutch army to "apply" what they had learned. Although the school was not a university, during the 17th century it profoundly influenced mathematics curricula in Dutch universities (Van Berkel, 1988).

From the outset, the Royal Mathematical School at Christ's Hospital enhanced the theoretical and practical backgrounds of likely recruits for the British mercantile service and for the British Navy (Hans, 1951a, 1951b). It was not until 1709 that the Christ's Hospital board found someone—James Hodgson—who was suitably competent both mathematically and pedagogically to direct the program in a satisfactory way. Hodgson would be Master of the Royal Mathematical School for 46 years (Christ's Hospital, 1953; Plumley, 1976; Stewart, 1999).

Charles Page's Navigation Cyphering Book

Who Was Charles Page?

Charles Page was born in Ramsgate, in Kent, England, on April 17, 1810, to Captain Charles Page and Harriet Page. The eldest of 13 children, he was presented for admission to Christ's Hospital by Sir William Curtis, President of Christ's Hospital, and entered the Royal Mathematical School in London on June 10, 1818. School records indicate that he left Christ's Hospital on October 20, 1825, having completed the Royal Mathematical School's program, and was bound, for seven years, as apprentice to George Stephens, commander of the 403-ton ship *Boyne* (Christ's Hospital, 1857; Ken Mansell to Nerida Ellerton, personal communication, January 18, 2014). He never married, and emigrated to Australia in 1832. Three of his brothers (John, Joseph, and Henry) would also emigrate to Australia (www.queenslandfamilytrees.com and www.ancestry.com).

Charles Page's Navigation Cyphering Book

During his time at Christ's Hospital, from 1818 to 1825, Page did what was expected of all mathemats in the Royal Mathematical School—he prepared a navigation cyphering book, setting out the mathematics and navigation that he had studied during his final years at the school. Like all cyphering books prepared by mathemats, Charles's book had magnificent leather binding with large gilt crests on the front and back covers. The beautifully-inscribed title page (see Figure 10.1) indicated that he had acquired "the elements of navigation whilst educating in the royal Mathematical School, Christ's Hospital, founded by his late Majesty, King Charles 2nd."

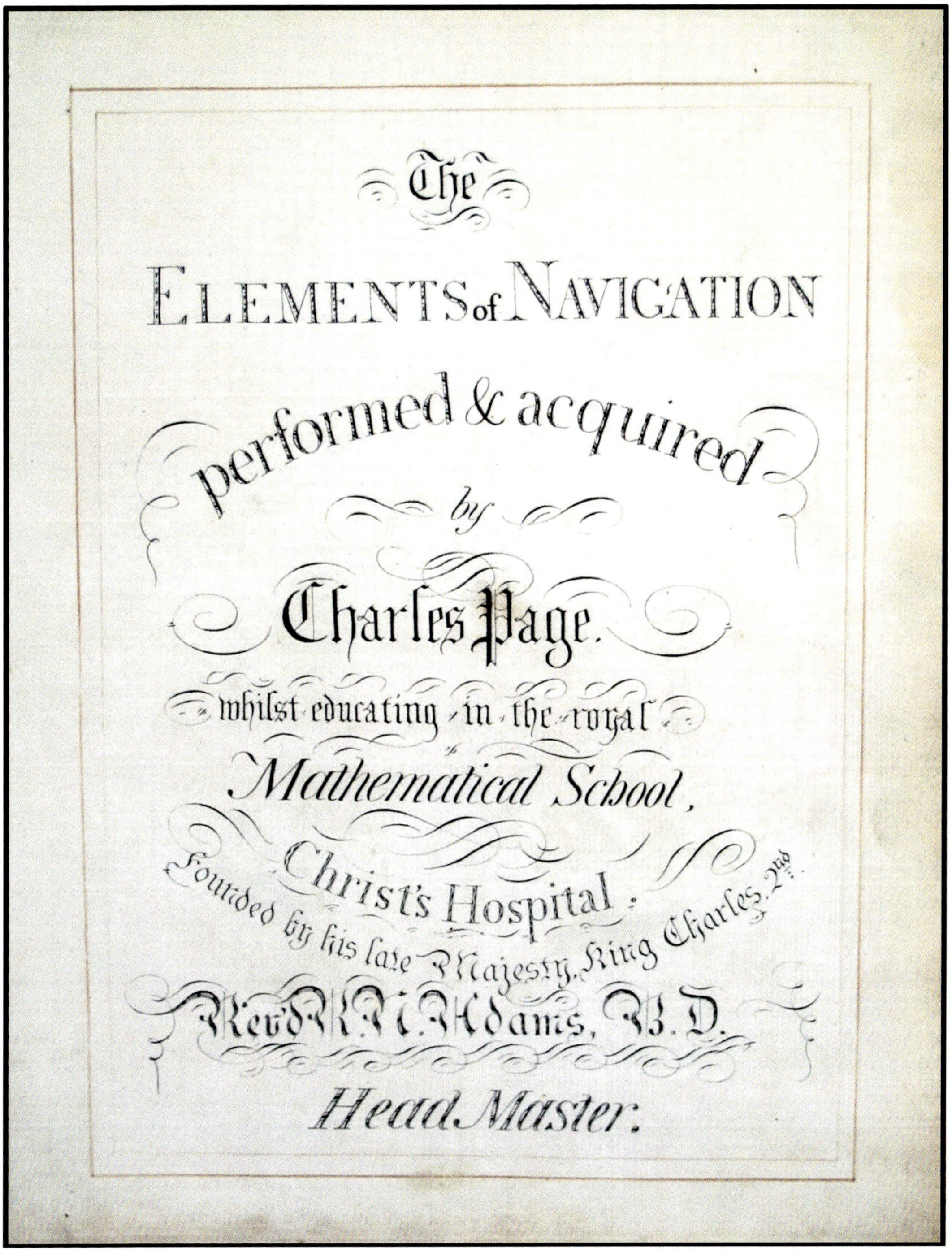

Figure 10.1. The first page in Charles Page's cyphering book.

Page's book had 671 handwritten pages, each with dimensions 11.5" by 9.5." It was divided into 14 sections, which dealt with arithmetic, geography, the *Elements of Euclid*, plane trigonometry, chronology, algebra, navigation, oblique sailing, windward sailing, globular sailing, spherics, astronomy, the use of globes, day's works, and "journal of a voyage from England towards Madeira." The last section set out a log of a possible but nevertheless fictitious journey.

In this chapter at least one page from each of the sections will be shown as a figure. Usually, but not always, figures will be placed on the left side of a pair of pages, and on the right side the significance for mathematics education or navigation education of what was shown in the figure (on the left side) will be discussed. The chapter will conclude with an analysis of the contribution of Christ's Hospital to the development of the cyphering tradition, as that tradition operated in Great Britain and in North America, from 1673 through about 1850.

The Royal Mathematical School's curriculum was periodically changed by the Governors during the first 200 years of the School's existence, especially when a new Master was appointed. Charles Page's cyphering book included a section on algebra, a topic which had not always received separate attention in early versions of the School's curriculum. Those who were admitted were supposed to be poore boyes, but the definition of poore boye sometimes allowed students from well-to-do families to be included. Ten new students were admitted every year, and it was intended that ten students would leave every year. Careful reading of the Christ's Hospital minute books, now held in the London Metropolitan Archives, makes it clear that in the nineteenth century students in the Royal Mathematical School who did not make good progress were ejected from the program. The course was academically rigorous and the daily cyphering regime very demanding. Students needed to work diligently if they were to survive. That said, the Royal Mathematical School was well supplied with excellent navigation instruments, and the "mathemats" daily could use the beautiful, very long, and very flat, cyphering desks which had been included in the special classroom that, in 1710, Hodgson had persuaded the school to build for the mathemats (see Figures 10.2 and 10.3, which are taken from "Blues" and the Royal Navy, 1900, p. 247).

Figure 10.2. The classroom of the "mathemats" in the 1890s—the old desks had belonged to an earlier building erected in 1710 (from "Blues" and the Royal Navy, 1900, p. 247).

Figure 10.3. The Christ's Hospital Museum in the 1890s (from "Blues" and the Royal Navy, 1900, p. 247).

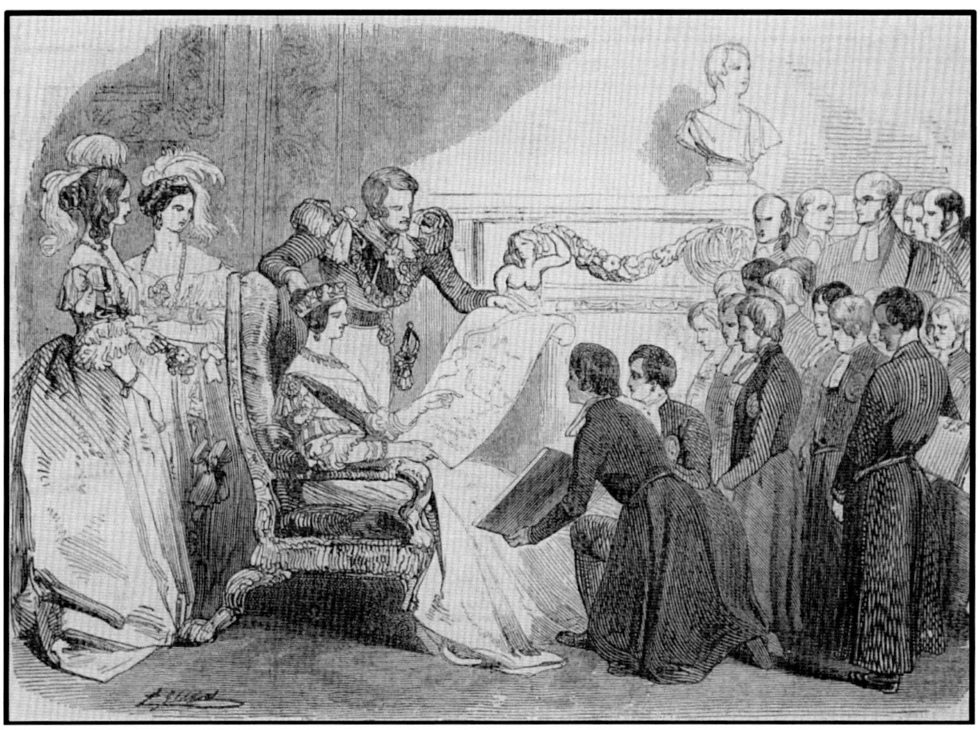

Figure 10.4. Queen Victoria inspecting a mathemat's cyphering book (reproduced from *The Illustrated London News*, June 8, 1843, p. 19).

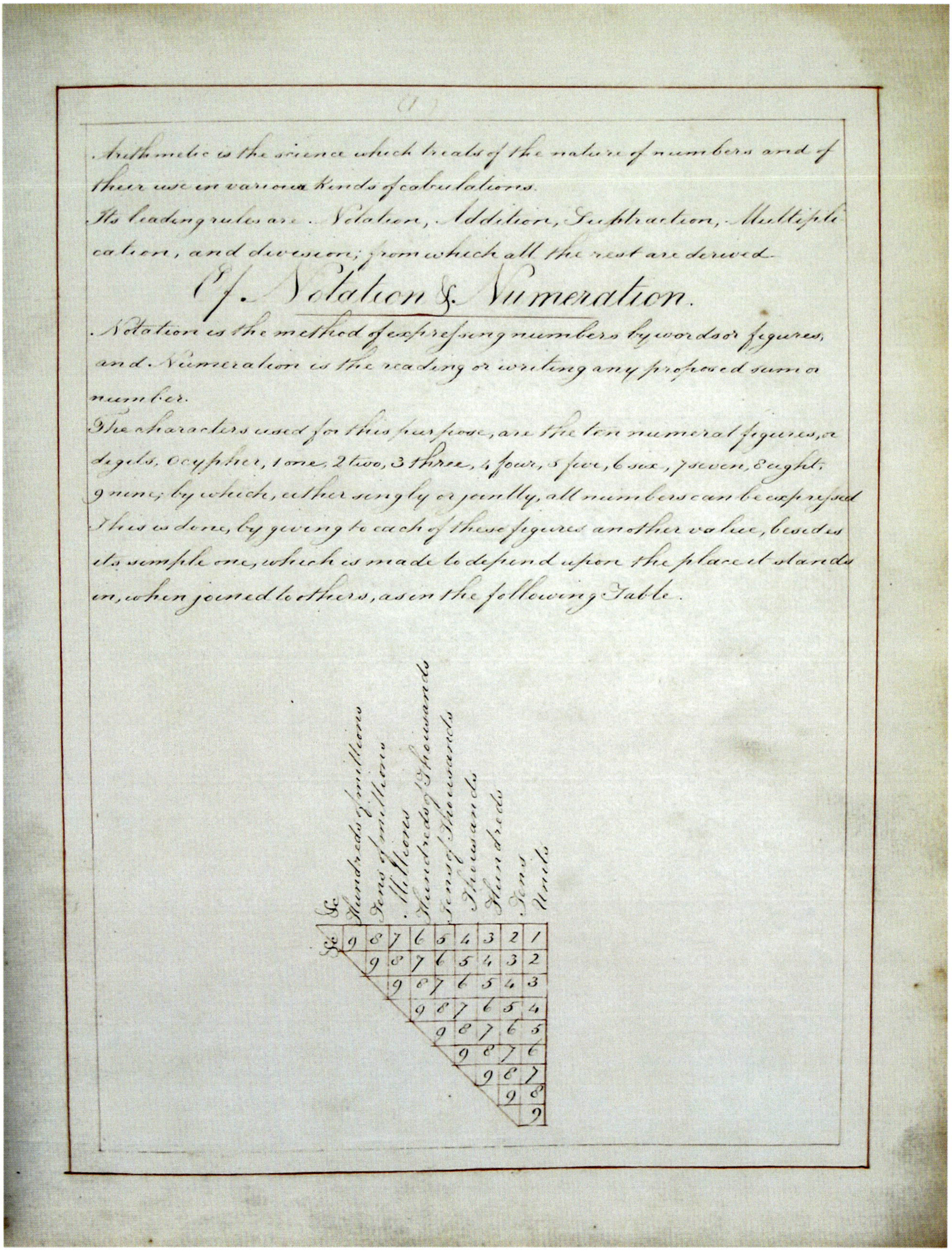

(1)

Arithmetic is the science which treats of the nature of numbers and of their use in various kinds of calculations.

Its leading rules are Notation, Addition, Subtraction, Multiplication, and division; from which all the rest are derived.

Of Notation & Numeration.

Notation is the method of expressing numbers by words or figures, and Numeration is the reading or writing any proposed sum or number.

The characters used for this purpose, are the ten numeral figures, or digits, 0 cypher, 1 one, 2 two, 3 three, 4 four, 5 five, 6 six, 7 seven, 8 eight, 9 nine, by which, either singly or jointly, all numbers can be expressed This is done, by giving to each of these figures another value, besides its simple one, which is made to depend upon the place it stands in, when joined to others, as in the following Table.

&c	Hundreds of millions	Tens of millions	Millions	Hundreds of Thousands	Tens of Thousands	Thousands	Hundreds	Tens	Units
&c	9	8	7	6	5	4	3	2	1
		9	8	7	6	5	4	3	2
			9	8	7	6	5	4	3
				9	8	7	6	5	4
					9	8	7	6	5
						9	8	7	6
							9	8	7
								9	8
									9

Figure 10.5. Numeration, in Charles Page's cyphering book (p. 1).

The Arithmetic Section of Charles Page's Cyphering Book

The first of the 14 mathematical sections in the cyphering book was simply titled "Arithmetic," and, from a mathematical point of view, it was the most elementary of all the sections. There were 59 numbered pages altogether in this section, and the topics dealt with were numeration, simple addition, simple subtraction, simple multiplication, simple division, compound operations (for addition, subtraction, multiplication and division), reduction, simple proportion, vulgar fractions, decimal fractions, involution, evolution, simple interest, and logarithms.

This section on arithmetic basically followed an extended *abbaco* arithmetic curriculum (Ellerton & Clements, 2012). However, the topics were dealt with more quickly than usual, and, the last topic, logarithms, was not part of the mathematics curriculum of most other schools.

Figure 10.5 shows the first page in the arithmetic section. It was concerned with "notation and numeration," which was traditionally the first topic in a course dealing with *abbaco* arithmetic. There are several points worth noting about this page:

- The superbly crafted handwriting, on a neatly ruled page, was characteristic of what might be expected from a Royal Mathematical School student. Charles Page was probably about 15 years of age when he prepared the first page. It would have initially been prepared on a separate, loose, page, and later it would have been united with other pages that he would prepare while completing his final two years at the School. In keeping with school tradition, the level of penmanship displayed in Figure 10.5 (and indeed throughout the whole cyphering book) was so high that it was difficult to imagine a 15- or 16-year-old boy being able to write like that.
- The numeration table shown at the bottom of Figure 10.5 displayed the powers of 10 from "units" to "hundreds of millions." Almost all commercially-published school arithmetics intended for students who had learned to count, started with a numeration table, and then proceeded to the four operations on whole numbers. Then followed the so-called "compound operations" (that is to say, applying the four operations to quantities such as money, time, length, area, capacity, etc.), and ascending and descending "reduction" (expressing the amount a quantity given in a unit as the same quantity, but with respect to a different unit). Finally, the end of the elementary *abbaco* curriculum would be reached by studying the golden rule, or "rule of three" (which was concerned with proportions).
- By definition, *abbaco* arithmetic employed the Hindu-Arabic numeration system which, with its place-value qualities as well as its use of zero as a place-holder, was the preferred system of European merchants. This system, which had been translated from Arab nations into Europe during the twelfth century, had quickly revolutionized European commercial transactions. Curiously, university departments of mathematics had been loath to take it up, but by the beginning of the nineteenth century its usefulness, both in commerce and in the development of number theory, was well recognized.

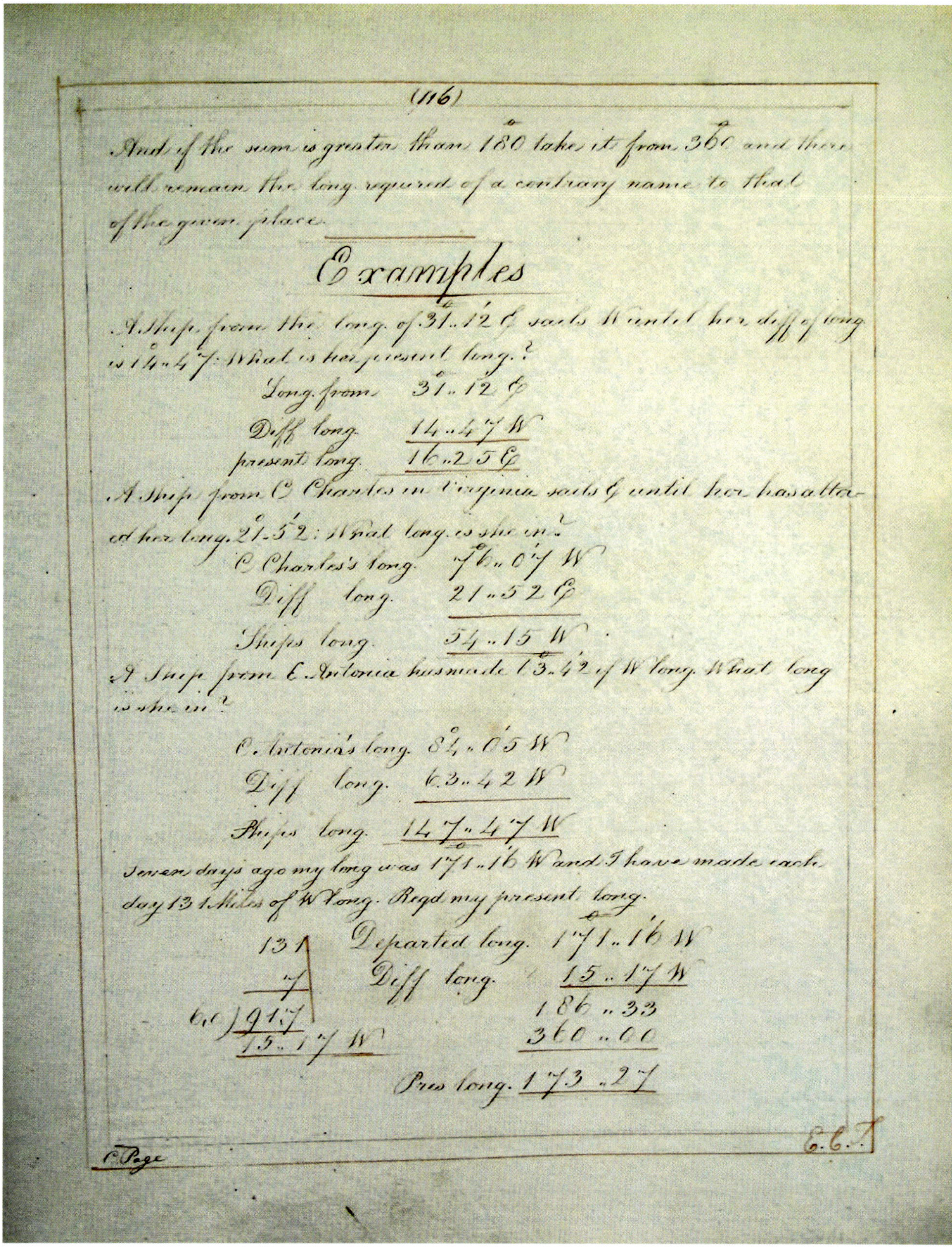

Figure 10.6. Some "Geography" problems in Charles Page's cyphering book (p. 116).

The Geography Section in Charles Page’s Cyphering Book

This was the second of 14 major sections in the cyphering book. It was natural that the study of geography be thought of as related to the study of mathematics, because international commercial transactions relied so heavily on calculations concerned with latitude and longitude, and time variations. There were 50 numbered pages altogether in this section.

There are four examples shown in Figure 10.6, and each was concerned with longitude.

1. A ship from the longitude of 31° 12’ east sails west until her difference of longitude is 14° 47’. What is her present longitude?

   The solution was set out as follows:

Long. from	31° 12’ E
Diff. long.	14° 47’ W
Present long.	16° 25’ E

Although this was a relatively simple problem, the solution did not show that the 47’ needed to be subtracted from 72’ to get 25’, and then 1° needed to be “paid” back, so that next calculation needed was 31° less 15°, which was 16°. Alternatively, the original longitude could be expressed as 30° 72’ E, and if 14° 47’ were to be subtracted from that then the answer would be 16° 25’ E. A third method would be to subtract 47 from 60, to get 13, and then add 12 to the 13 to get 25; then “pay back” the 1 to arrive at 15 from 31 is 16. These three approaches corresponded to the three most commonly-used subtraction algorithms of the day (Ellerton & Clements, 2012; Jackson, 1906).

2. A ship from C. Charles in Virginia sails east sails west until her longitude has altered 21° 52’. What longitude is she in?

   The solution was set out as follows:

C. Charles long.	76° 07’ W
Diff. long.	21° 52’ E
Ship’s long.	54° 15’ W

Once again, any of the three main subtraction algorithms could have been used.

3. A ship from E. Antonia has made 63° 42’ of W longitude. What longitude is she in?

   The solution was set out as follows:

E. Antonia’s long.	84° 05’ W
Diff. long.	63° 42’ W
Ship’s long.	147° 47’ W

Unlike the first two problems, this one required addition.

4. Seven days ago my longitude was 171° 16’ W. and I have made each day 131 miles of W longitude. Required my present longitude.

   The solution was set out as follows:

131	Departed long.	171° 16’ W
7	Diff. long.	15°17’ W
60/ 917	Ship’s long.	186° 33’
15.17 W		360° 00’
		173° 27’

This problem requires more knowledge than that needed to solve the first three problems.

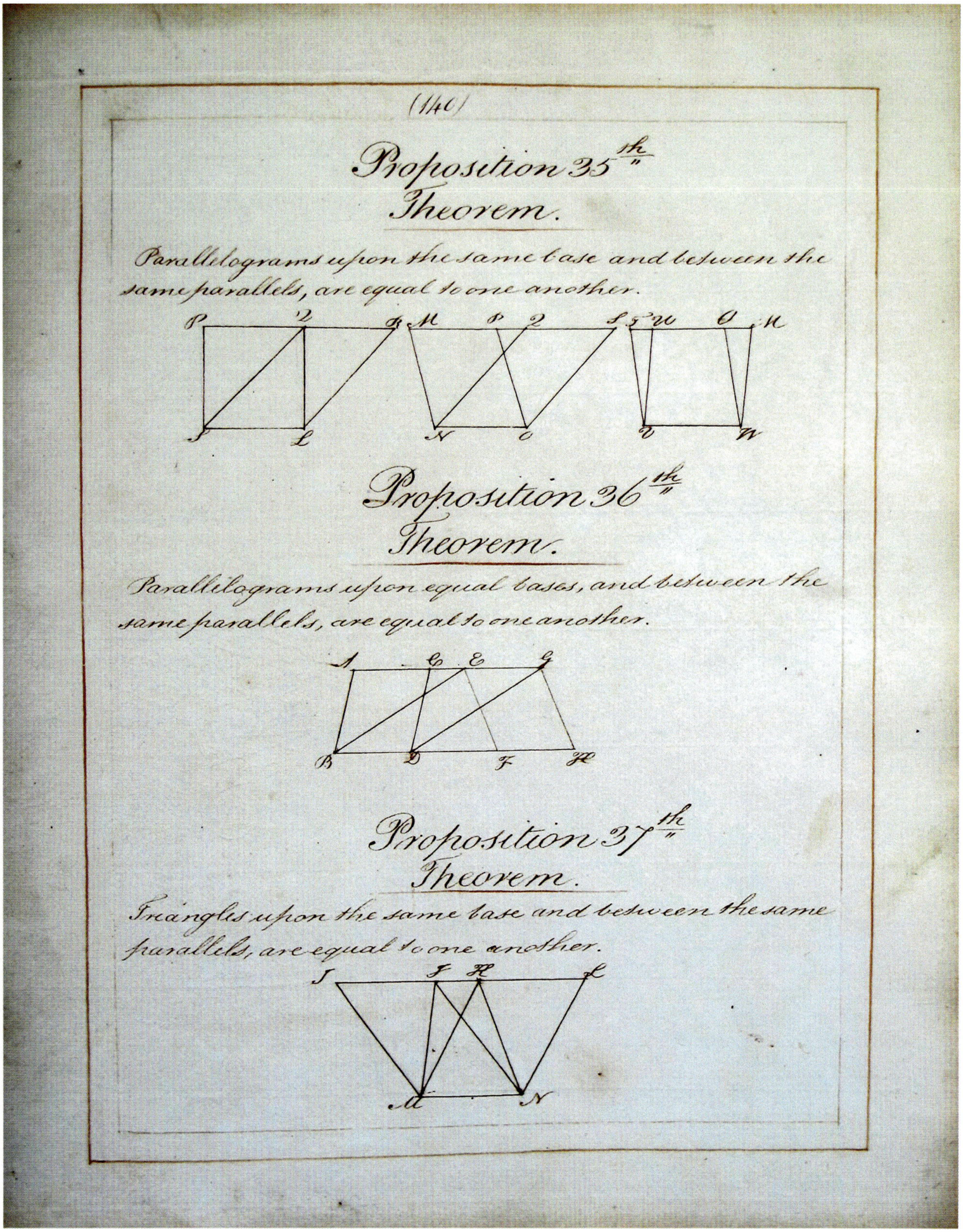

Figure 10.7. Propositions 35, 36 and 37, Euclid's "*Elements*" Book I (p. 140 in Charles Page's cyphering book).

The Euclidean Geometry Section in Charles Page's Cyphering Book

This was the third of 14 major sections in the cyphering book. The first six books of Euclid's *Elements*—the most famous text ever written on geometry—were summarized, and the summary occupied 73 pages (i.e., more than 10 percent of all the pages in the cyphering book). Figure 10.7 shows Charles Page's entries for Propositions 35, 36 and 37. This section is replete with definitions, axioms and theorems, all beautifully illustrated with diagrams drawn in ink. No proofs of theorems were given, and one wonders whether Charles and his fellow Christ's Hospital students ever had to set out proofs (which, of course, were included in Euclid's *Elements*).

Around the time Charles was preparing his cyphering book there was much controversy in the worlds of mathematics and philosophy with respect to the status of Euclid's *Elements.* Three outstanding mathematicians of the day—the Hungarian Janos Bolyai, the German Carl Friedrich Gauss, and the Russian Nicolai Ivanovitch Lobachevsky, were about to put forward geometries that did not share all the axioms that Euclid had taken for granted. These three are jointly regarded as the fathers of what is known today as non-Euclidean geometry. But Immanuel Kant (1724–1824), the great German philosopher, argued in his *Critique of Pure Reason* that Euclidean Geometry represented a form of pure reasoning, something which was not of empirical origin, and an absolute necessity for logical thought.

In England, the old universities tended to cling to Euclidean geometry. Students were expected to learn Euclidean proofs by heart, word for word, comma for comma, diagram for diagram. In America, it is said that Abraham Lincoln memorized the first six books of Euclid, and structured some of his greatest speeches—his Gettysburg address, for example—around Euclidean form (undefined terms, definitions, theorems, proofs) (Dorfman, 1966; Hirsch & Van Haften, 2010). Euclidean geometry had been part of the Royal Mathematical School's curriculum from its beginnings (see, e.g., Hodgson, 1723; Stewart, 1999), and would remain so throughout the 18th and 19th centuries. One can never know whether being required to copy Euclid's *Elements* assisted the Royal Mathematical School's graduates to become better navigators than they would otherwise have been. But to study the *Elements* is to be confronted with an all-embracing logical structure that more or less compels the learner to think about the foundations for sound reasoning.

The purpose of the Royal Mathematical School was to provide the kind of academic training that young men would need if they were to be as well prepared as possible to take up apprenticeships in navigation and, ultimately, higher positions such as ship navigators and captains. These young men needed to have an understanding of concepts like angle, congruence, similarity, two-and three-dimensional polygons and polyhedra, spheres, conic sections, etc. They also needed to be able to reason soundly within spatial contexts—for example, what would be the shortest distance via the surface of the earth from where my ship is now to Port *X,* and why am I confident that what I'm thinking now would in fact give the shortest distance? It would be hard to imagine a better, more comprehensive training than that offered by Euclid's *Elements*, provided the instructional approach resulted in more than mere memorization of words.

The Royal Mathematical School's curriculum included sections on plane trigonometry, spherical trigonometry, surveying, and mensuration of superficies, and each of those involved non-trivial thinking about, and application of, geometrical concepts. But once students left Christ's Hospital the practical side of geometry would assume greater importance than the theoretical aspects.

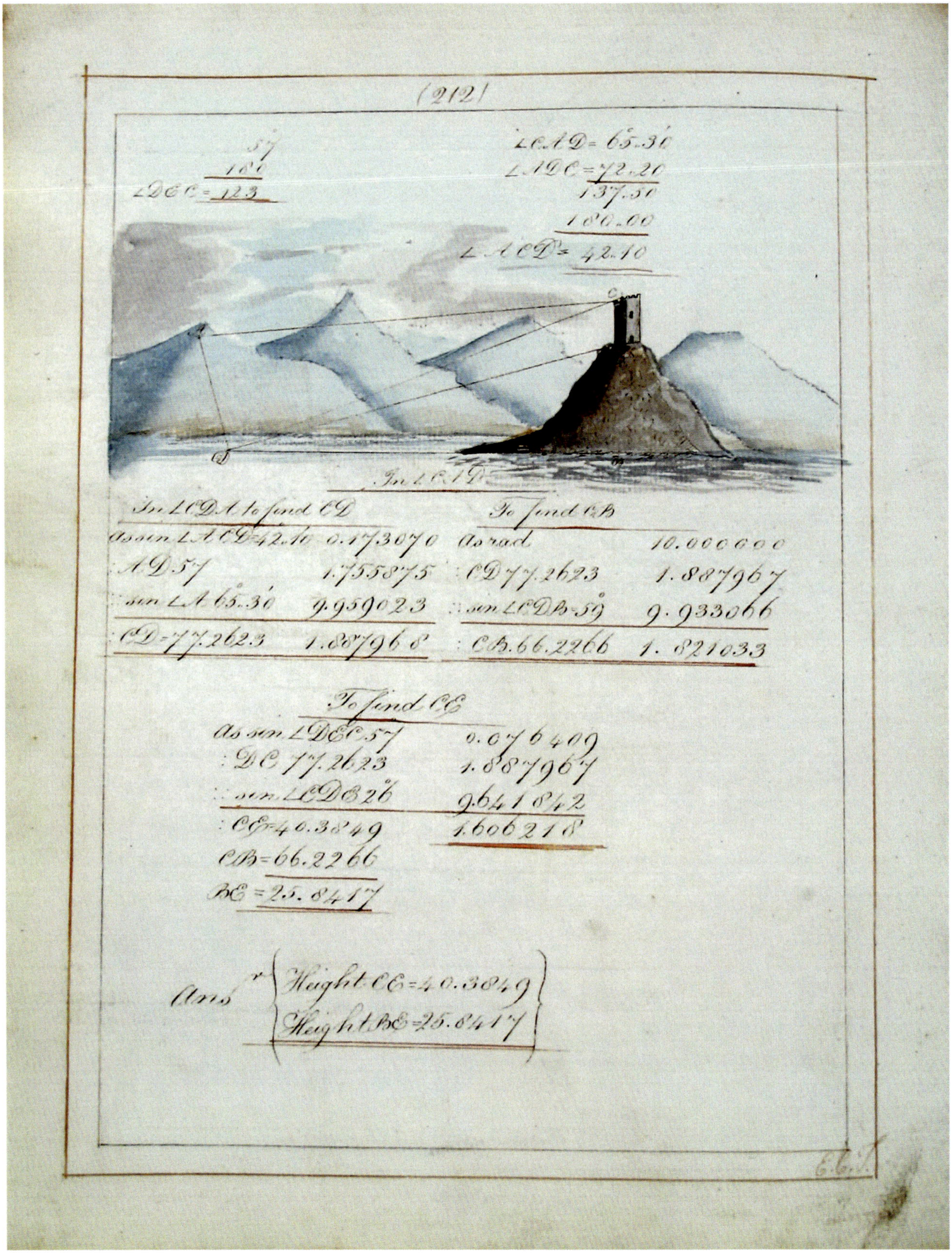

Figure 10.8. A plane trigonometry task in Charles Page's cyphering book (p. 212).

The Plane Trigonometry Section of Charles Page's Cyphering Book

This was the fourth of the 14 major sections in the cyphering book. The notes on plane trigonometry occupied 22 pages in Charles Page's cyphering book, and Figure 10.8 shows part of the solution to a problem which involved finding the height of a castle from an external point. It was recognized that knowledge of how to find solutions to such problems could be helpful in real-life, navigational situations. The solution included the beautiful, and fully labelled, water-colored diagram shown. Learning to draw and color such diagrams had long been an important feature of the Royal Mathematical School's curriculum—although Sloan (2000) has presented evidence that, in the early years of the School, the diagrams did not always obviously match the mathematics demanded by the mensuration tasks shown in the navigation cyphering books.

The first page of the section on plane trigonometry adopted what was known as the line-segment definition for the six trigonometric functions sine, cosine, tangent, cosecant, secant, and cotangent. The diagram shown in Table 10.8—another version of it was given in Chapter 9 of this volume—could be found in almost all elementary treatments of plane trigonometry, and had always been used in the Royal Mathematical School (see, Allen, 1970; Hodgson, 1723). It was widely used wherever trigonometry was studied in the 18th and early 19th centuries (Van Sickle, 2011).

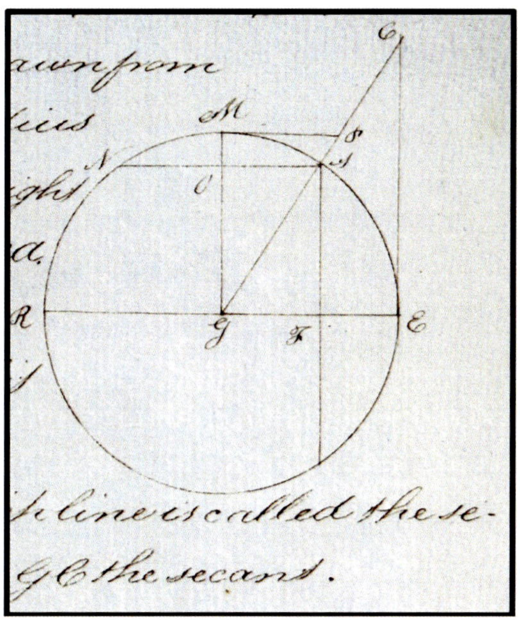

Figure 10.9. The standard line-segment approach to plane trigonometry in Charles Page's cyphering book, p. 195).

Notice that in notes associated with Figure 10.9, the definition was for the "right sine of an arc," "the right tangent of an arc," etc., and not the "sine of an angle," the "tangent of an angle," etc.

Much of the section in Charles Page's cyphering book shows how plane trigonometry, linked with logarithms, could be used to find angles and side lengths for triangles. This was applied to real-life problems involving the calculation of heights and distances (see, for example, the situation depicted in Figure 10.8). The challenge for students was to identify appropriate triangles, with known angles and side lengths.

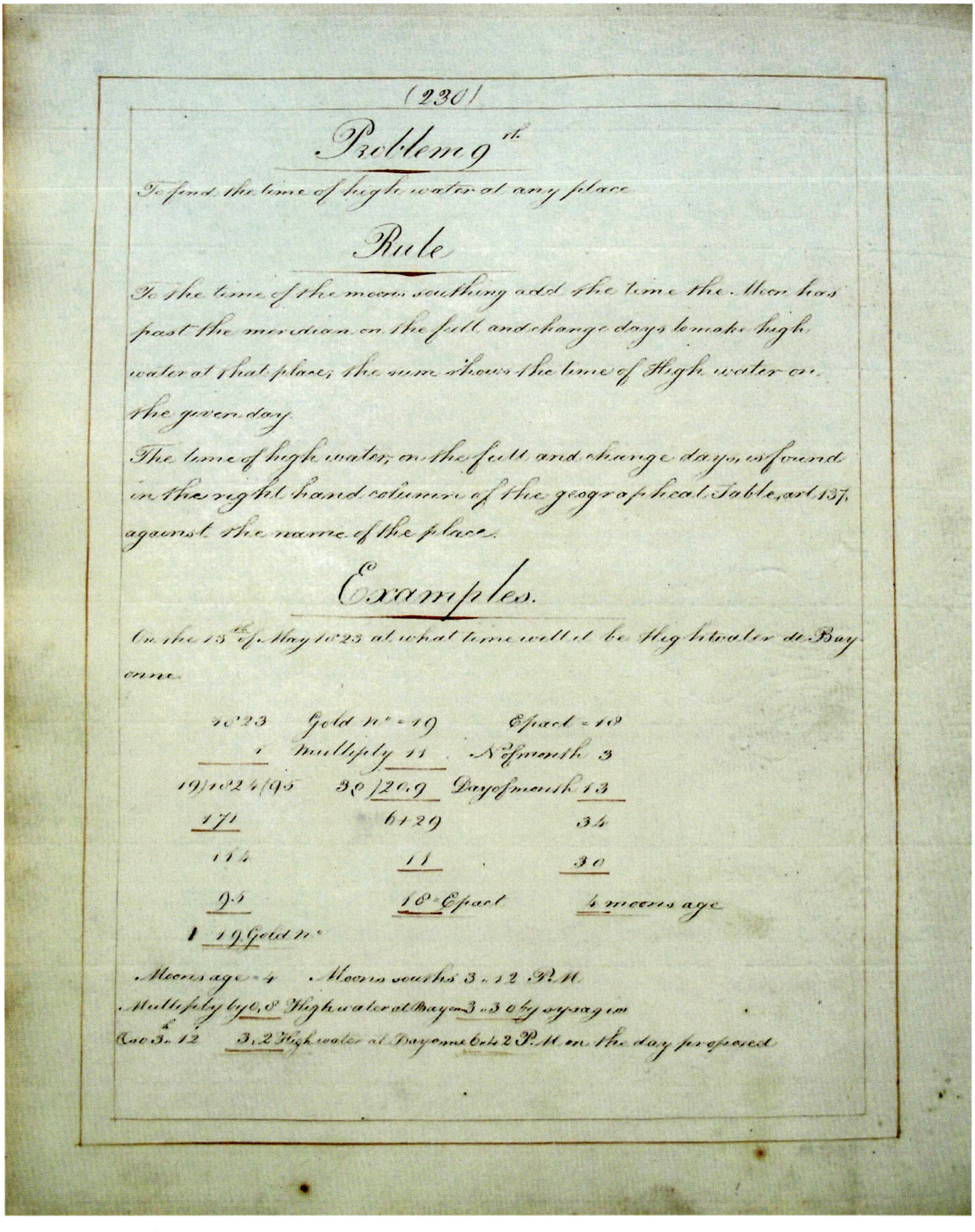

(230)

Problem 9th

To find the time of high water at any place

Rule

To the time of the moon's southing add the time the Moon has past the meridian on the full and change days to make high water at that place; the sum shows the time of High water on the given day.

The time of high water, on the full and change days, is found in the right hand column of the geographical Table, art 137, against the name of the place.

Examples.

On the 13th of May 1823 at what time will it be High water at Bayonne

1823 Gold No = 19 Epact = 18
1 multiply 11 No of month 3
19)1824(95 30)209 Days of month 13
171 6–29 34
114 11 30
95 18=Epact 4 moons age
19 Gold No

Moons age = 4 Moons southe 3 . 12 P.M.
Multiply by 0.8 High water at Bayonne 3 . 30 by voyage
Ans 3 . 12 3 . 2 High water at Bayonne 6 . 42 P.M. on the day proposed

Figure 10.10. Page 230, in the "Chronology" section of Charles Page's cyphering book.

Chronology in Charles Page's Cyphering Book

The fifth section in the cyphering book was devoted to the mathematics of "chronology," that is to say, the mathematics associated with the ordering of events in time. Young navigators needed to take into account differences in time arising from differences in latitude and longitude. It was important that they be able to calculate how long a journey was likely to take, and the expected time of arrival. From the outset, the Christ's Hospital curriculum included a significant component on chronology (see, for example, Section 19 in Hodgson, 1723).

Figure 10.10 shows page 230 from Charles Page's manuscript, which dealt with "Problem 9th," on how to find the time of high water at any place. The rule that was given stated: "To the time of the moon's southing add the time the moon has past (sic.) the meridian on the full and change days to make high water at that place; the sum shows the time of high water on the given day" (p. 230). Page added that "the sum of high water on the full and change days is found in the right hand column of the Geographical Table art 137 against the name of the place."

It was obviously important that someone on a large ship should be able to find the time of high water for a harbor in which the ship had entered or would enter. It was equally obvious that it would be no easy matter to do that, and that the person who would have been expected to know how to do it was the navigator. Students in the Royal Mathematical School at Christ's Hospital were training to be navigators. It was reasonable to think that often in the future, in distant harbors, theirs would be the task of calculating times of high water.

Following the statement of the rule, an example was given: "On the 13th of May, 1823, at what time will it be high water?" Quite complex calculations were then made, and clearly only persons with the requisite training in terminology and the associated mathematics would have known what to do.

Did Charles Page Understand what he Wrote in his Navigation Cyphering Book?

It is appropriate here to point out the huge difference between the kind of *abbaco* mathematics recorded in a typical arithmetic cyphering book and the mathematics recorded in navigation cyphering books prepared within the Royal Mathematical School at Christ's Hospital. To become a navigator a student needed to learn to use logarithms and trigonometry, which were not topics usually covered in the elementary and mid-level *abbaco* curriculum.

But did Charles understand the principles that were behind what he wrote in his cyphering book? With the thought in mind that perhaps he merely copied much of the difficult work from a textbook, and that he either copied working for the problems and exercises, or was helped to do them by the master or by an usher, we consulted the book *Elements of Navigation; Containing the Theory and Practice,* by John Robertson and William Wales (1780)—a text made available to Christ Hospital mathemats. In Volume 1 of the fourth edition of this book, published in London in 1780, we found a section on Chronology (pages 364–374), and compared what was there with the notes in Charles Page's cyphering book. The wording in the cyphering book was almost identical with that in the textbook. The examples solved in Charles' book were similar to, or identical with, examples in Robertson and Wales' (p. 191) text. That finding led us to check whether the same was true for other sections. We shall report our findings on this matter from time to time in this chapter.

North American ship-owners struggled to find competent navigators. In the 1790s, in Salem, Massachusetts, for example, Elias Hasket Derby, one of the wealthiest men in the United States, operated a navigation school to train navigators for the barges which he required for his trading ventures with India and the Far East. Throughout the period 1780–1830 many Harvard College students took courses in which they prepared navigation cyphering books which reflected, in their content and sequencing, the Christ's Hospital model (Ellerton & Clements, 2012).

The conceptual leaps that students needed to make in order to progress from the traditional *abbaco* curriculum to a navigation curriculum like that followed at Christ's Hospital, and the requirement that students should graduate from the Royal Mathematical School when they were about 16 years old, made it difficult for the students to learn thoroughly the mathematics which was required if they were to be able to do the work expected of professional navigators. Although the Royal Mathematical School at Christ's Hospital attempted to solve that problem, it was one thing to establish a tough curriculum suitable for prospective navigators, another thing to find persons capable of teaching that curriculum, and a third thing to find students capable of understanding what they would be taught, given the time and age restrictions.

Figure 10.11 shows page 551, from the section on the "Use of Globes" in Charles Page's cyphering book. The solution of the set problem continued for the whole of the next page. The point of showing Figure 10.11 here is to help readers to appreciate the points being made early in the last paragraph. On the page shown, the diagram is extremely complex—it is a wonder that any 15- or 16-year-old boy could actually draw it. The problem involved spherical triangles, and in the history of schooling considerably less than one percent of 15- or 16-year olds would be expected to be able to cope with the mathematics needed to solve such a problem. Throughout the solution, plane trigonometrical functions and logarithms were used.

In Charles Page's book there were many pages with mathematics pitched at the same difficulty as that shown in Figure 10.11. And, of course, as they were grappling with the high level of mathematics, the mathemats knew that nothing but their very best handwriting, calligraphy, and illustrations would suffice. It is hardly surprising that some mathemats compromised by choosing form rather than substance—probably reasoning that so long as they copied from Robertson and Wales's text, and made sure that the entries in their book *looked* beautiful, then all was likely to be well. There would be an end-of-course verbal examination by an external authority, but it was unlikely that complex explanations for difficult problems would be required.

When Charles Page was at Christ's Hospital, in the 1820s, Christ's Hospital was not the only place where future navigators were being trained. In 1733, the Royal Naval Academy (RNA) had been established, as a shore-side facility, and a comprehensive curriculum, not unlike that at Christ's Hospital, put in place. However, according to Dickinson (2007), the Academy did not achieve the objective of becoming the preferred path for training naval officers. Although the traditional route of a sea-going "apprenticeship" remained important, many officers were appointed as a result of family ties, and patronage. In 1806 the Academy was reconstituted as the Royal Navy College, and in 1816 that College was amalgamated with the School of Naval Architecture, which had been established in 1811 (Dickinson, 2007). At that time, the future role of Christ's Hospital for training prospective navigators was unclear, because there were political pressures seeking to make it easier for boys from well-connected families to achieve the ranks of navigator, and captain, without having to complete rigorous theoretical training programs, with associated apprenticeships. The "sailor King," William IV, the British monarch between 1830 and 1837, would comment that "there was no place superior to the quarterdeck of a British man of war for the education of a gentleman" (quoted in Dickinson, 2007, p. 32). More will be said about King William IV in the next chapter of this book, but the main point, here, is to note that in the 1820s, at the very moment when Charles Page was preparing his navigation cyphering book at Christ's Hospital, the Royal Mathematical School was under enormous pressure to produce graduates whose theoretical knowledge was recognized as at least equal to that of graduates of the School of Naval Architecture.

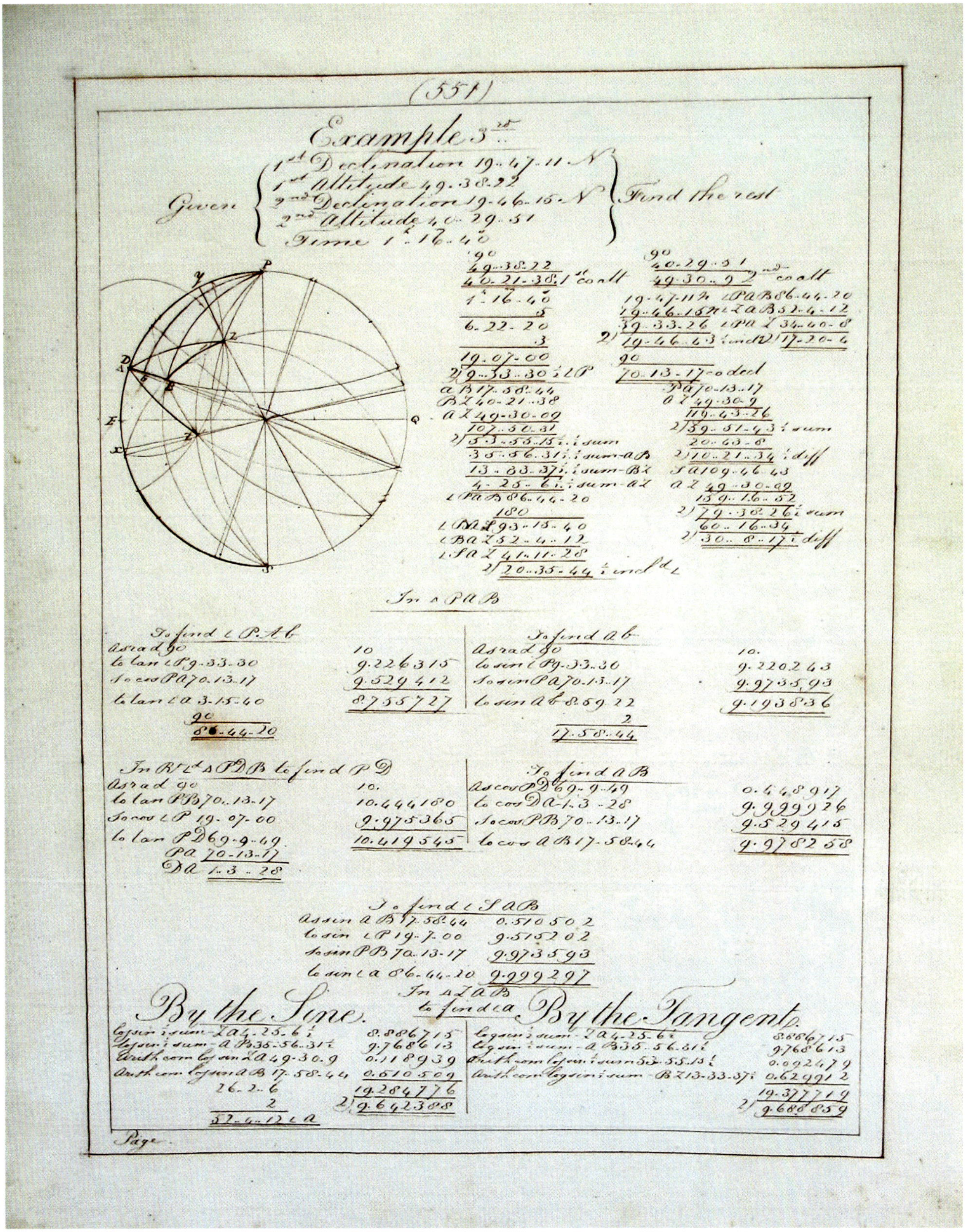

Figure 10.11. Part of Charles Page's solutions to a complex task (p. 551).

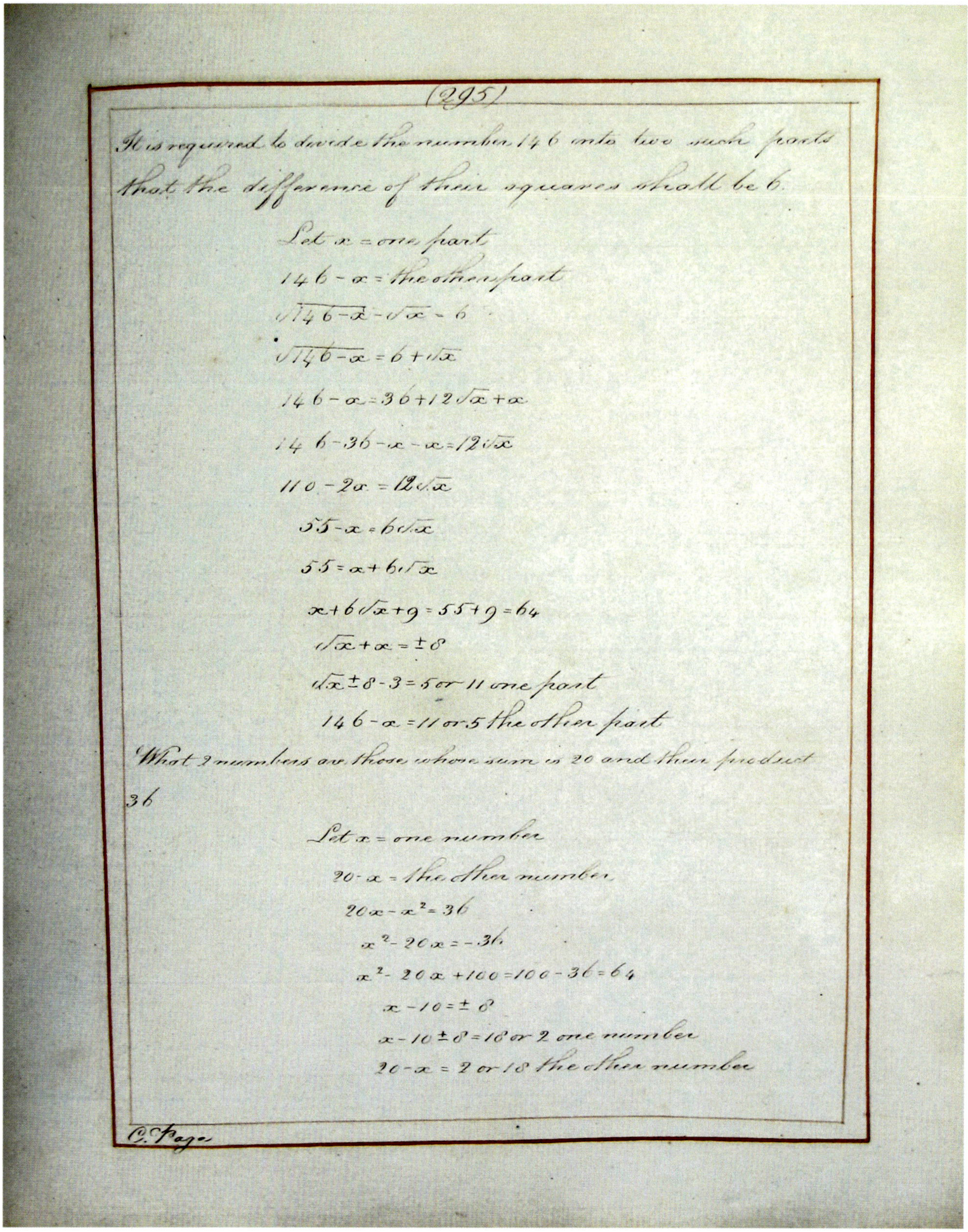

(295)

It is required to divide the number 146 into two such parts that the difference of their squares shall be 6.

Let x = one part

$146 - x$ = the other part

$\sqrt{146-x} - \sqrt{x} = 6$

$\sqrt{146-x} = 6 + \sqrt{x}$

$146 - x = 36 + 12\sqrt{x} + x$

$146 - 36 - x - x = 12\sqrt{x}$

$110 - 2x = 12\sqrt{x}$

$55 - x = 6\sqrt{x}$

$55 = x + 6\sqrt{x}$

$x + 6\sqrt{x} + 9 = 55 + 9 = 64$

$\sqrt{x} + x = \pm 8$

$\sqrt{x} \pm 8 - 3 = 5$ or 11 one part

$146 - x = 11$ or 5 the other part

What 2 numbers are those whose sum is 20 and their product 36

Let x = one number

$20 - x$ = the other number

$20x - x^2 = 36$

$x^2 - 20x = -36$

$x^2 - 20x + 100 = 100 - 36 = 64$

$x - 10 = \pm 8$

$x - 10 \pm 8 = 18$ or 2 one number

$20 - x = 2$ or 18 the other number

C. Page

Figure 10.12. Page 295 from the "algebra" section of Charles Page's cyphering book.

Algebra in Charles Page's Cyphering Book

The sixth section in the cyphering book was devoted to algebra, and it occupied 60 pages. Algebra had not been part of the original Christ's Hospital curriculum in the late 17th century, but that changed after 1709, when Christ's Hospital published a complex work on algebra by the Swiss mathematician Johannes Alexander. This text had been translated into English by Samuel Cobb, a master at Christ's Hospital, and 750 copies of the Latin edition together with 1000 copies of the English translation were printed. A 128-page appendix, written by Humfry Ditton (1709), a master at Christ's Hospital, was added to Alexander's (1709) original text.

Figure 10.12 shows solutions to two tasks which generated quadratic equations. In 1825, algebra was not commonly studied in British or North American schools by students aged between 10 and 16 years. Although quadratic equations are not far along the algebra curriculum continuum, in 1825 it would not have been expected that a school student would be able to solve word problems of the kind shown in Figure 10.12.

In Figure 10.12 a logical connective for "therefore" (e.g., ∴) was not used to connect any two lines logically. It seems to have been assumed that the reader would be able to follow the logic from what was written. Elsewhere (Ellerton & Clements, 2012), we have referred to the genre in which solutions to mathematics problems were presented without the associated logic being explained as PCA ("**P**roblem, **C**alculation, **A**nswer"). In *abbaco* arithmetic cyphering books this tendency had become almost universal in the early 1800s, but here it was translated into school algebra.

The question at the top of Figure 10.12 was, "It is required to divide the number 146 into two such parts that the difference of their squares shall be 6." Depending on how x was defined, either $x^2 - (146 - x)^2 = 6$, or $(146 - x)^2 - x^2 = 6$, could have been appropriate; but Charles Page got the incorrect $\sqrt{(146 - x)} - \sqrt{x} = 6$. His equation could have been correct if the word "squares" in the question had been "square roots," and x had been defined to be the smaller of the two numbers. For the problem stated in Figure 10.12, the two required numbers would be $73\frac{3}{146}$ and $72\frac{141}{146}$. Given that these are "clumsy," it can be presumed that the intended problem was the one involving square roots. When trying to solve this intended problem, Charles wrote:

$$\text{Let } x = \text{one part and } 146 - x \text{ the other part.}$$
$$\sqrt{(146 - x)} - \sqrt{x} = 6.$$

After reaching the statement $x + 6\sqrt{x} + 9 = 64$, he incorrectly maintained that this was equivalent to $\sqrt{x} + x = \pm 8$. That should have been $\sqrt{x} + 3 = \pm 8$. According to Charles this implied that $\sqrt{x} \pm 8 - 3 = 5$ or 11 for one part and $146 - x = -11$ or 5, for the other part.

What Charles wrote was wrong, and suggested he did not understand what he was doing. He had not thought much about the logic of the problem, and he had not checked his answer for correctness. All of the foregoing suggests that he copied his "solution" from another source. If that was indeed the case then that was unfortunate, for an important part of the cyphering tradition was that only correct statements should be entered into cyphering books. A cyphering book was prepared so that it would become a reference book for life.

Thus, a serious error was made and, because no check was attempted, the entry in the book remained as incorrect. On the next page of Charles' cyphering book he "solved" two other word problems, one that reduced to a quadratic equation, and the other to a cubic equation. With the first he did not interpret whether his solutions responded to the original question, and with the second, he obtained only one of three possible solutions.

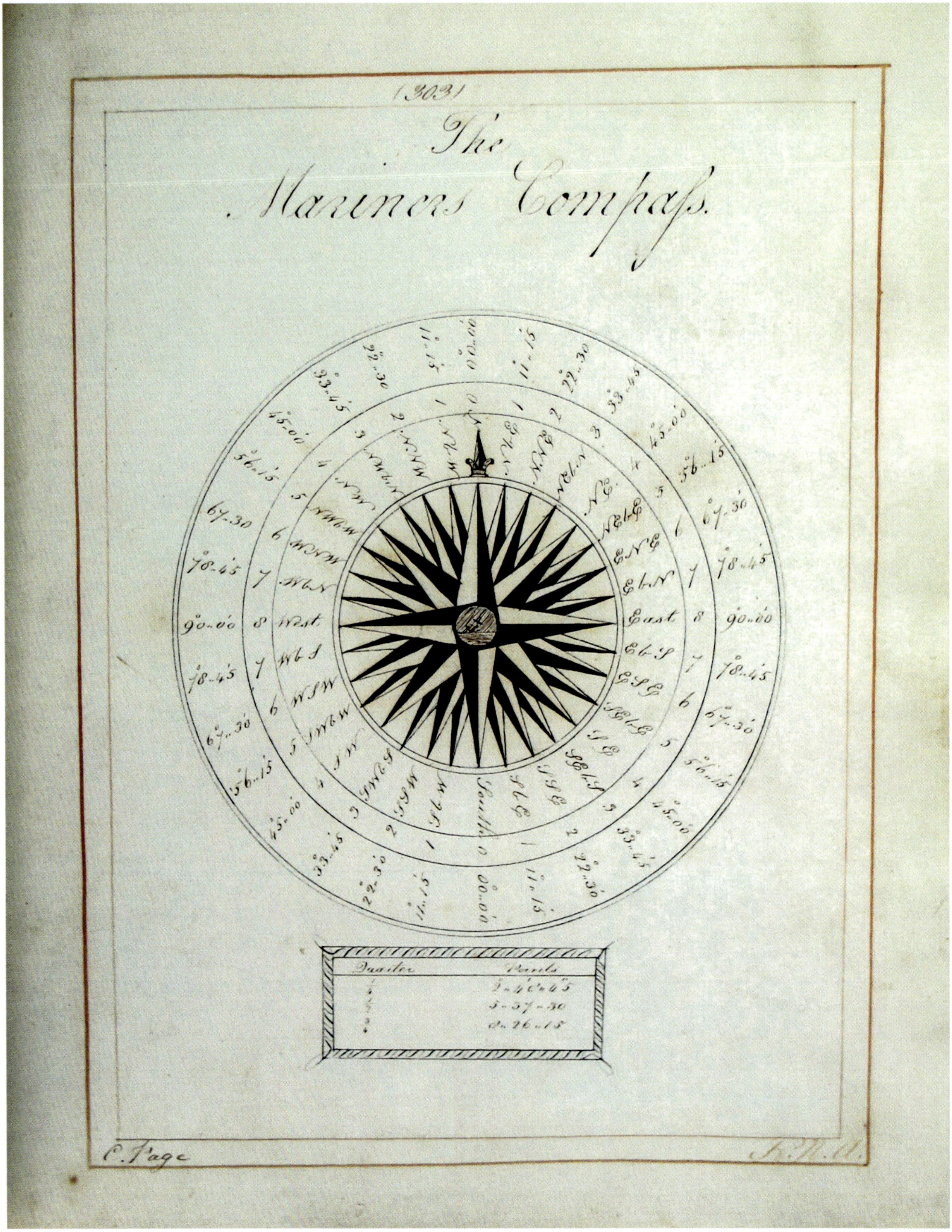

Figure 10.13. The mariner's compass—from Charles Page's cyphering book (p. 303).

The Section on Navigation in Charles Page's Cyphering Book

The seventh section was concerned with definitions and principles of navigation. It provided an introduction to the terminology and principal ideas of navigation—the elaboration of which would be the main object for the rest of the book. This introduction occupied 25 pages.

On the first page of this section a "mariner's compass" was defined as "an artificial representation of the horizon of every place by means of a circular piece of paper called a card divided like the horizon into degrees and points which are called rhums." That verbal definition made less sense than the beautiful black and white illustration shown in Figure 10.13.

The supposed clarification of terminology continued—the following statement is unedited:

- A rhumb line is a right line drawn from the centre of the compass to the horizon. And is named from the point on the horizon it falls in with;
- The course is an angle which the rhumb line the ship sails on makes with the meridian, and it is sometimes reckoned in degrees and sometimes in points of the compass;
- Distance is the number of leagues, miles, &c, intercepted between two places reckoned on a rhumb, or it is the way or length a ship has gone on a direct course in a given time;
- Difference of latitude is the distance a ship has made north or south of the place she set out from, and is reckoned in miles and parts of a mile on a meridian;
- Departure is the east or west distance a ship has made from the meridian of the place she departed from reckoning on the parallel of latitude she is arrived and in the plane she is arrived in and in the plane chart it is the same and difference of longitude;
- If a ship sails due north or south she sails on a meridian and makes no departure and difference of latitude is the same.

One wonders where the sequence of words used in these "definitions" came from. Although the definitions would appear to be loose, there is a certain ring of authority about them—as if they were dictated by an experienced mariner who was making up the definitions "on-the-spot."

There are numerous intricate scale diagrams in this section. Each diagram is carefully drawn in ink, obviously with the aid of a straight edge and a compass. Each diagram models a situation described in an example. Calculations, usually involving logarithms, are shown

Here is a typical example:

A ship in latitude $1^{o}3'6$ N is bound to a port lying to the S eastward distant 420 miles and 180 to the eastward and having plied on the following courses SEbS59, SSE$\frac{1}{2}$E84, S$\frac{1}{2}$W92, WSW4.3, WbN25, SbE$\frac{1}{2}$E62, ESE89, EbN37, ESE$\frac{1}{2}$E95.

Following this statement of the problem, numerous calculations were shown, and then an elaborate black-and-white diagram was drawn. Finally, the answer "3.48.42 Pres, Lat" was reached. Red ink was used to highlight important parts of the solution. The modern reader finds it difficult to imagine that all of this was achieved by a 15- or 16-year-old boy using a quill.

This kind of task was obviously integral to a boy's training to be a navigator. Charles Page was studying in the Royal Mathematical School, one of the world's foremost schools for the teaching and learning of mathematics. But the raison d'être of the School was to produce persons with the kind of knowledge needed to succeed as apprentice navigators, and ultimately as fully qualified, and highly practical navigators.

Traverse Sailing

Figure 10.14 shows an example from a section on "transverse sailing in Charles Page's cyphering book. A "traverse" was defined as "a compound course wherein several different successive courses and distances are known" (p. 313).

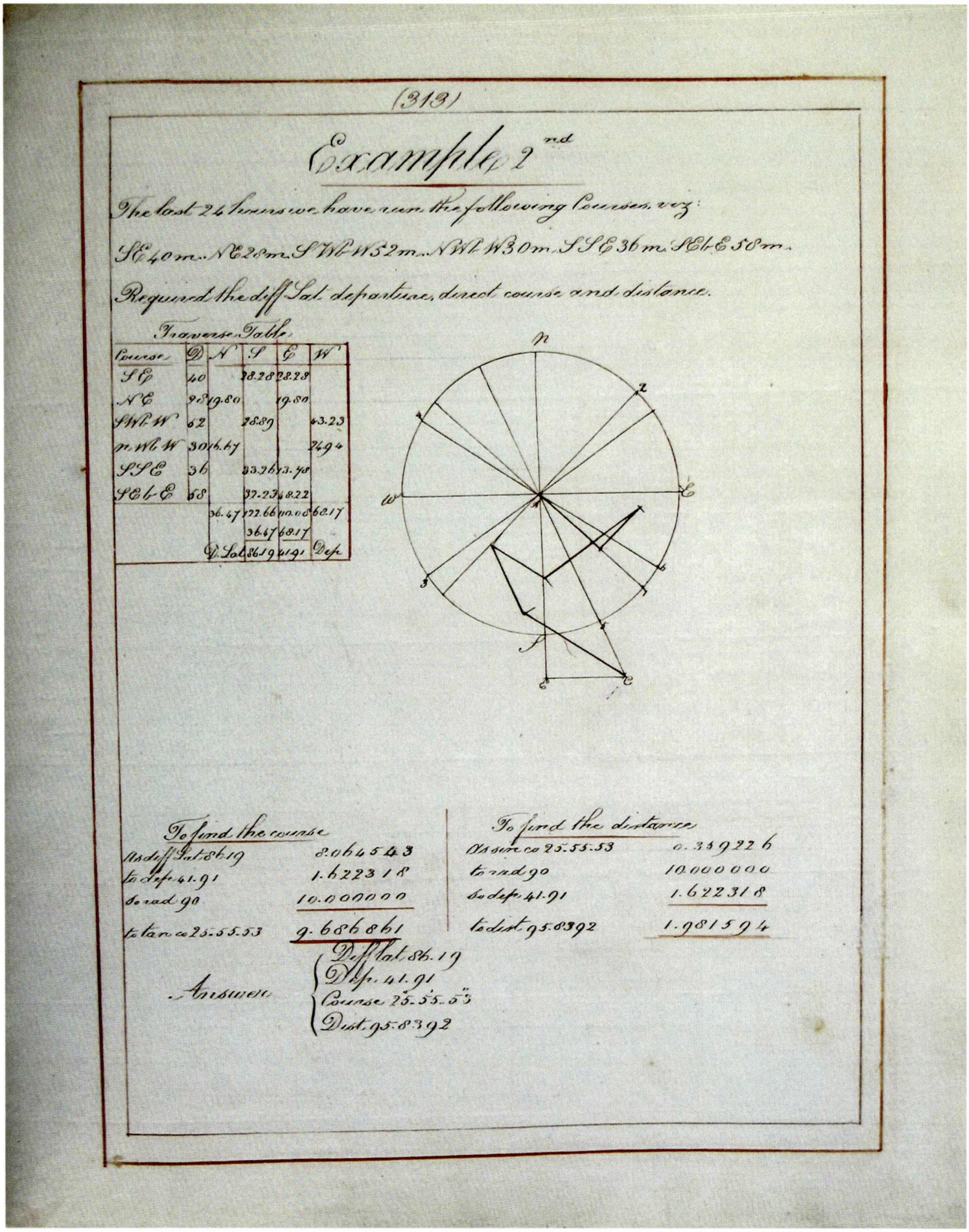

(313)

Example 2nd

The last 24 hours we have run the following Courses, viz:

SE 40 m. NE 28 m. SWbW 52 m. NWbW 30 m. SSE 36 m. SEbE 58 m.

Required the diff. Lat. departure, direct course and distance.

Traverse Table

Course	D	N	S	E	W
SE	40		28.28	28.28	
NE	28	19.80		19.80	
SWbW	52		28.89		43.23
NWbW	30	16.67			24.94
SSE	36		33.26	13.78	
SEbE	58		32.23	48.22	
		36.47	122.66	110.08	68.17
			36.47	68.17	
	D. Lat		86.19	41.91	Dep.

To find the course

As diff Lat. 86.19	8.064543
to dep. 41.91	1.622318
So rad. 90	10.000000
to tan. co 25.55.53	9.686861

To find the distance

As sine co 25.55.53	0.359226
to rad 90	10.000000
So dep. 41.91	1.622318
to dist. 95.8392	1.981594

Answer: Diff. lat. 86.19; Dep. 41.91; Course 25.55.53; Dist. 95.8392

Figure 10.14. An example of "transverse sailing" in Charles Page's cyphering book (p. 313).

Oblique Sailing, Windward Sailing, Current Sailing and Globular Sailing

The next four sections in Charles Page's cyphering book were concerned with oblique, windward, current and globular sailing. Altogether, these occupied 76 pages, and Figures 10.15, 10.16, 10.17 and 10.18 illustrate the kinds of entries which Charles made within the sections. The following textual descriptions of key terms appeared in the cyphering book:

Oblique Sailing

"In the previous sections the bearing and distances of places were determined by the solution of right-angled triangles, but at sea it is often the case that only oblique positions can be observed" (p. 331). Figure 10.15 shows an example from a section on "oblique sailing" (p. 335).

Windward Sailing (see Figure 10.16)

"When the wind is directly or partly against a ship's direct course to the place where she is bound to, she reaches her port by a kind of zigzag ... course which is made by sailing with the wind first of the ship and then on the other. ..." (p. 345).

Current Sailing (see Figure 10.17)

"A current is a progressive motion of the water which causes all floating bodies to move that way towards which the stream is directed. ... The setting or current is that part of the compass toward which the compass runs, and the drift of a current is the rate it runs per hour. ..." (p. 359).

Globular Sailing (see Figure 10.18)

"Globular sailing attempted to take account of the difference of latitude, difference of longitude, departure, course deduced from the spherical figure of the earth. ..." (p. 371).

For each of the above four types of sailing the set of words was only part of the overall description which appeared in Charles Page's notes. But, even when complete sets of words were given it would have been evident that the descriptions came from the voice of experience, and did not satisfy a mathematician's desire for tight definitions. The text for each of the four types was accompanied by suitably illustrated examples, and it was expected that this would fill in the gaps left by mathematically inadequate descriptions.

There were other types of sailing. "Parallel sailing" was "the art of finding out what distance a ship should run due east or due west in sailing from the meridian of one place to that of another place in any parallel of latitude" (p. 373). "Middle latitude sailing" was "an answer of solving the several cases of globular sailing by the principles of plane and parallel sailing" (p. 378). "Mercator's sailing" was the art of resolving the several cases by plane trigonometry with the assistance of a table of meridian parts" (p. 387).

The various types of sailing clearly needed to be experienced to have real meaning. Although some students in the Royal Mathematical School came from families with naval backgrounds, the School was intended for "poore boyes"—who had not necessarily had any experience at sea. The last entry in Charles Page's cyphering book was a log of a journey from England towards Madeira, but almost certainly this was a fictitious journey. Probably, Charles would have had to wait until he was an apprentice before he experienced the "real thing." The preceding notes on the types of sailing were similar to those in corresponding sections of Robertson and Wales' (1780) text.

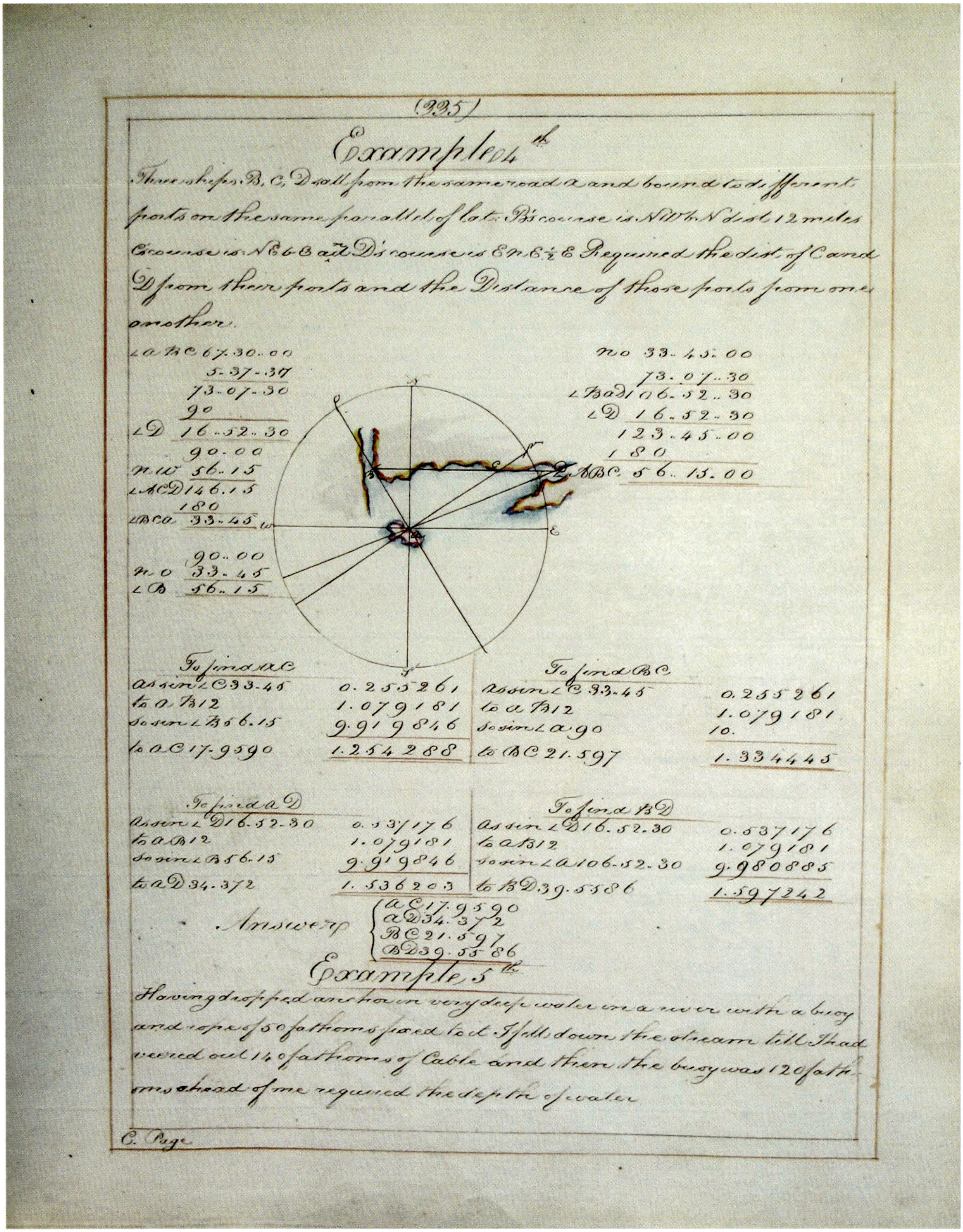

Figure 10.15. A page on "oblique sailing" in Charles Page's cyphering book (p. 335).

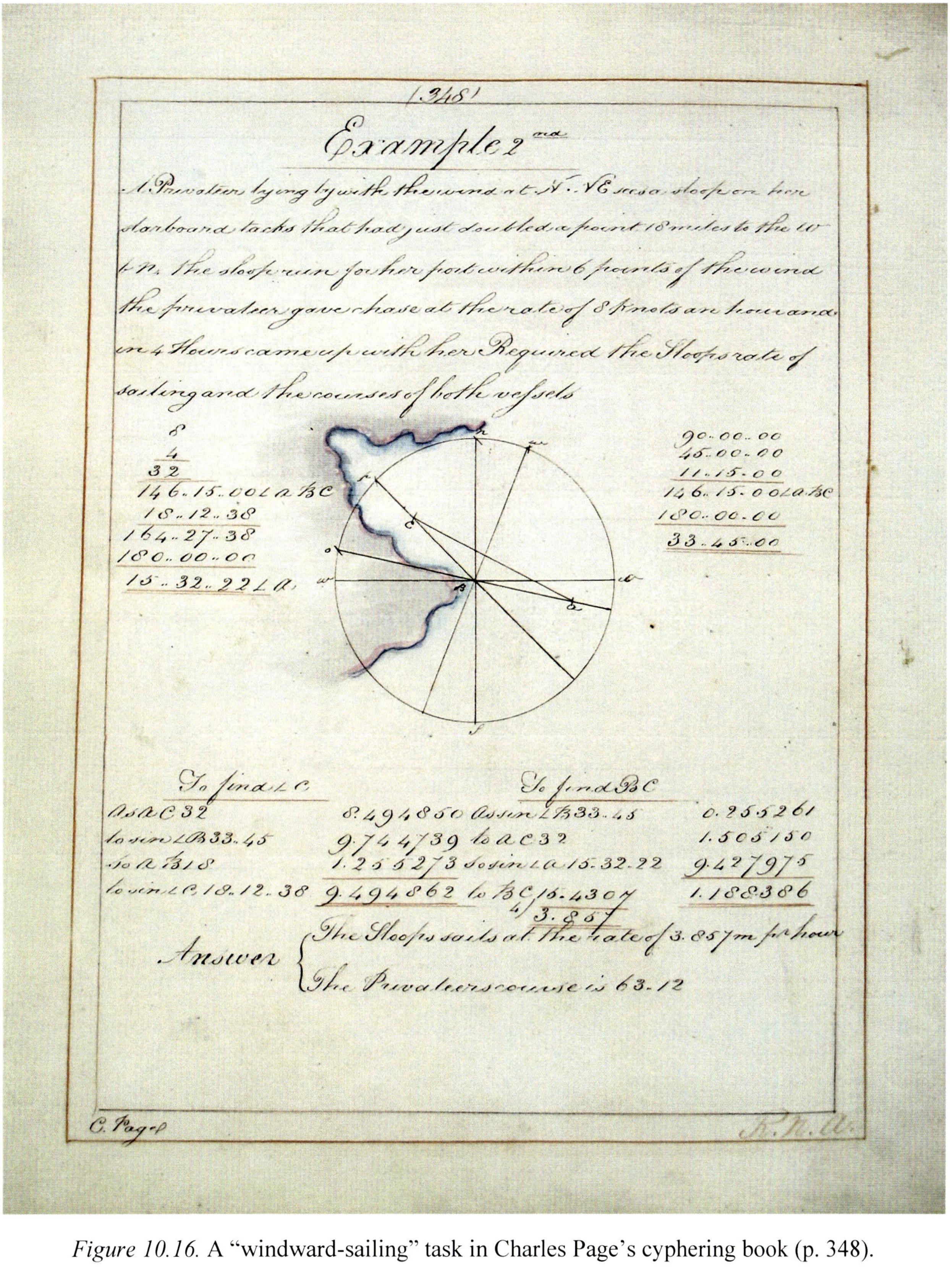

Figure 10.16. A "windward-sailing" task in Charles Page's cyphering book (p. 348).

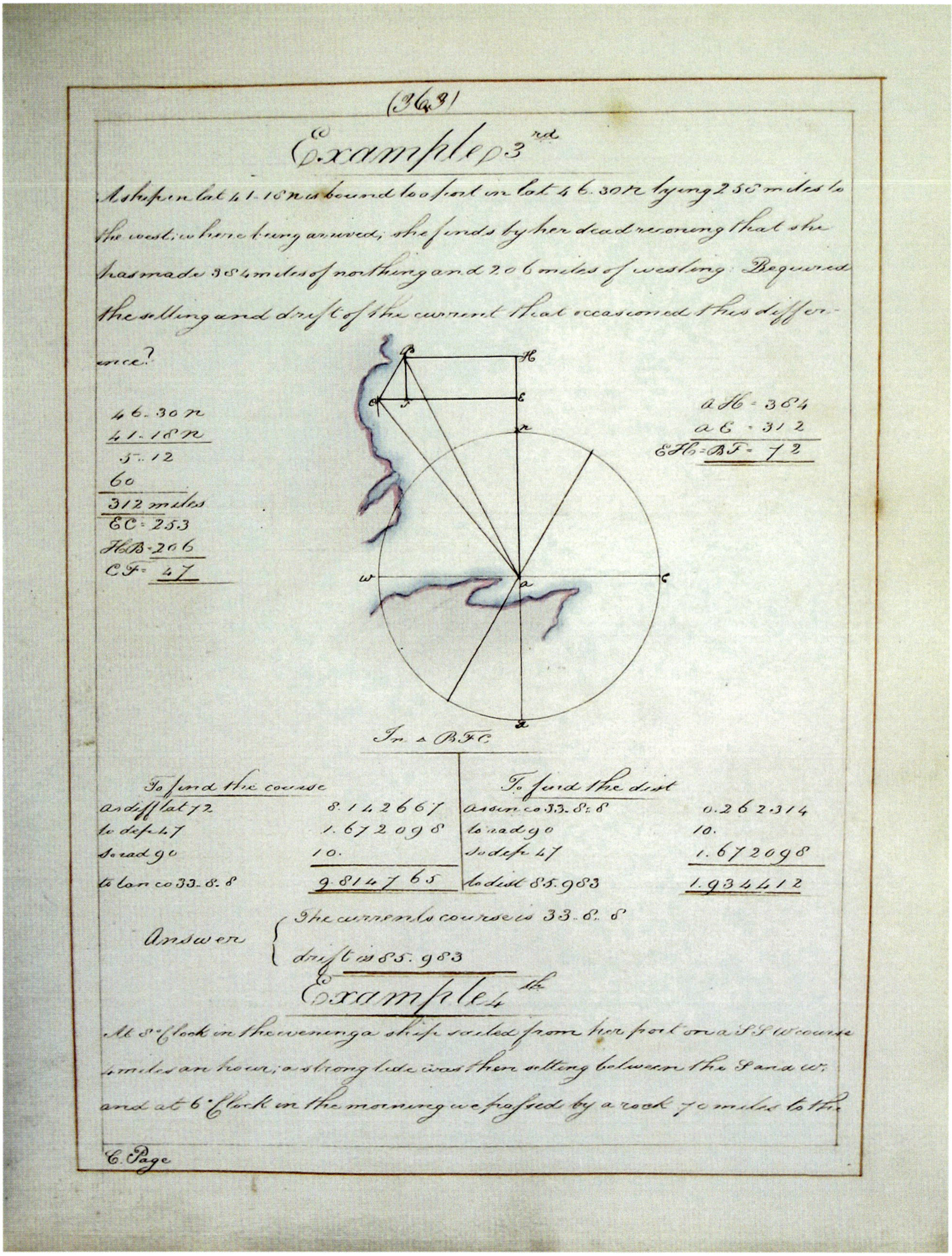
(363)

Example 3rd

A ship in lat 41-18 n is bound to a port in lat 46.30 n lying 255 miles to the west; where being arrived, she finds by her dead reconing that she has made 384 miles of northing and 206 miles of westing. Required the setting and drift of the current that occasioned this difference?

46-30 n
41-18 n
5-12
60
312 miles
EC = 253
HB = 206
CF = 47

aH = 384
aE = 312
EH = BF = 72

In △ BFC

To find the course

As diff lat 72	8.142667
to dep 47	1.672098
So rad 90	10.
to tan co 33-8-8	9.814765

To find the dist

As sin co 33-8-8	0.262314
to rad 90	10.
So dep 47	1.672098
to dist 85.983	1.934412

Answer { The current's course is 33-8-8, drift is 85.983

Example 4th

At 8 o'Clock in the evening a ship sailed from her port on a SSW course 5 miles an hour; a strong tide was then setting between the S and W. and at 6 o'Clock in the morning we passed by a rock 70 miles to the

C. Page

Figure 10.17. "Current sailing" in Charles Page's cyphering book (p. 363).

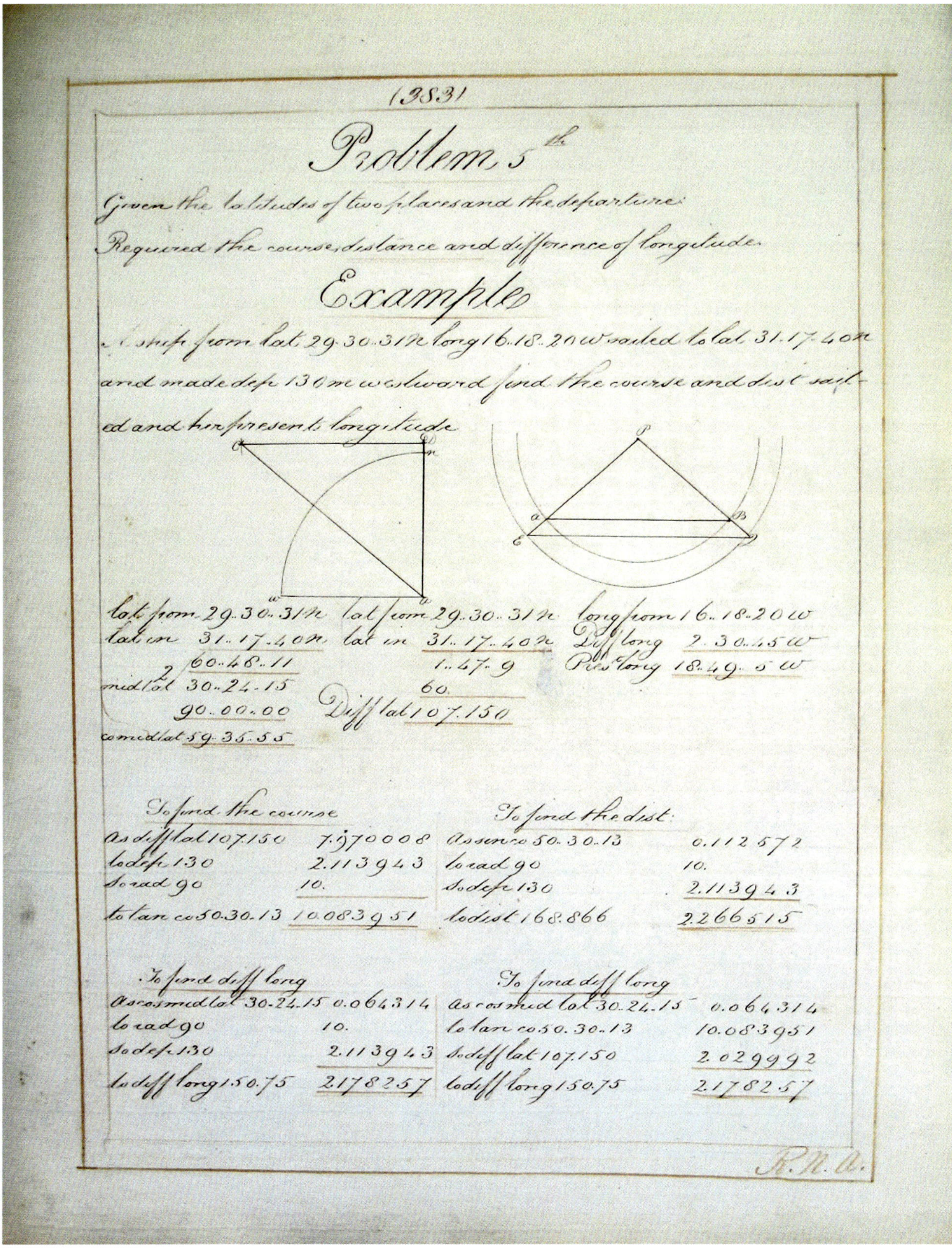

Figure 10.18. A "globular-sailing" task in Charles Page's cyphering book (p. 383).

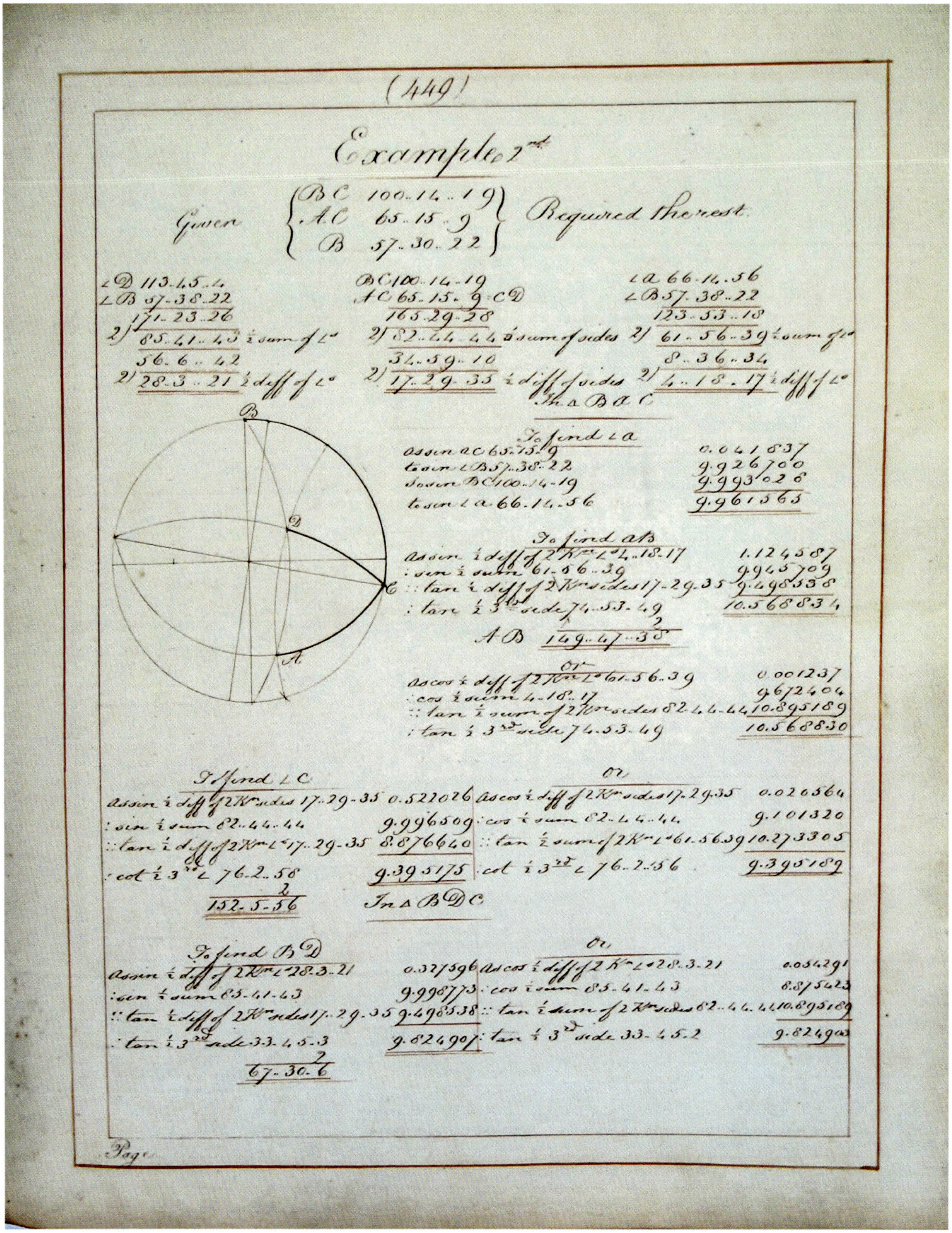

Figure 10.19. Charles Page's solution for an example on oblique spherics (p. 449).

The Sections on "Spherics," "Spheric Geometry," "Spheric Trigonometry," and "Oblique Spherics" in Charles Page's Cyphering Book

Using mathematics to model sailing on the earth's surface demands more than a knowledge of plane geometry and plane trigonometry, of course, because the earth's surface is not flat, but can be approximated by a sphere. Therefore, there were substantial sections on "spherics" (pp. 407–413), "spheric geometry" (pp. 415–429), "spheric trigonometry" (pp. 430–443), and "oblique spherics" (pp. 444–470) in Charles Page's cyphering book.

The relative difficulty of the mathematics in these sections can be appreciated, perhaps, by examining Figure 10.19, which shows an early example in the section on oblique spherics. In that Figure, *B and C* are located on a meridian of longitude and *A* and *C* are located on a parallel of latitude. The question simply stated: "Required the rest," by which was meant find all possible angles and distances for relevant spherical triangles. The handwriting is small, but it can be seen that Charles systematically went about completing the task.

There are many examples in these sections of Charles' book which resemble the one shown in Figure 10.19. Most 21st-century teachers of secondary school mathematics would believe that only a tiny proportion of 15- or 16-year-old children would have any chance of coping with the cognitive demands of the mathematics needed for this task. We do not know whether Charles merely copied the diagram, the calculations and the explanations for the task shown in Figure 10.19; nor do we know if the same is true for many other tasks in these sections. The question arises: When Charles became a navigator, on the open sea, facing real navigational problems, was he able to clarify in his own mind exactly what the problems were, and was he able to recognize which of the methods he had used (and were recorded in "his book") would fit the problems? In other words, did his Christ's Hospital training equip him to be able to do the modeling required when it was needed in real-life situations?

Of course, we will never know the answer to that question. We do know two pertinent things about Charles Page's Christ's Hospital experience: first, in November, 1824, Christ's Hospital awarded him a "Medal of 10 shillings" for the high quality of his work (the highest award was a "Superior Drawing Medal," followed by a "Medal of 15 shillings") (Christ's Hospital, 1823–1830). Second, on October 20, 1825, he was bound as an apprentice, for seven years, to George Stephens, Master commanding the 403-ton ship *Boyne* (that information is from page 13 of Christ's Hospital (1857) *Register of Christ's Hospital Royal Mathematical School Boys Apprenticed to Ships' Masters*, held by the London Metropolitan Archives).

We own two other Christ's Hospital navigation cyphering books one prepared in the mid-1830s by Henry James Calkin Andrew, and the other in the late 1850s by Richard Murdoch Robertson. Andrew's cyphering book covers the same mathematical and navigational themes as Page's, but the handwriting, calligraphy, and diagrams are not as neat. Robertson's was concerned only with navigation topics. In fact, in 1836, H. J. C. Andrew was awarded a lesser prize of 5 shillings for the quality of work. But, he did graduate successfully in 1837 from the Royal Mathematical School, and like Charles Page succeeded in gaining a position as an apprentice navigator (Christ's Hospital, 1836–1840). In 2009 we purchased H. J. C. Andrew's cyphering book in Queensland, Australia, and our investigations revealed that after he had completed his apprenticeship he took up a maritime position in New Zealand, where he became one of that nation's most respected maritime authorities.

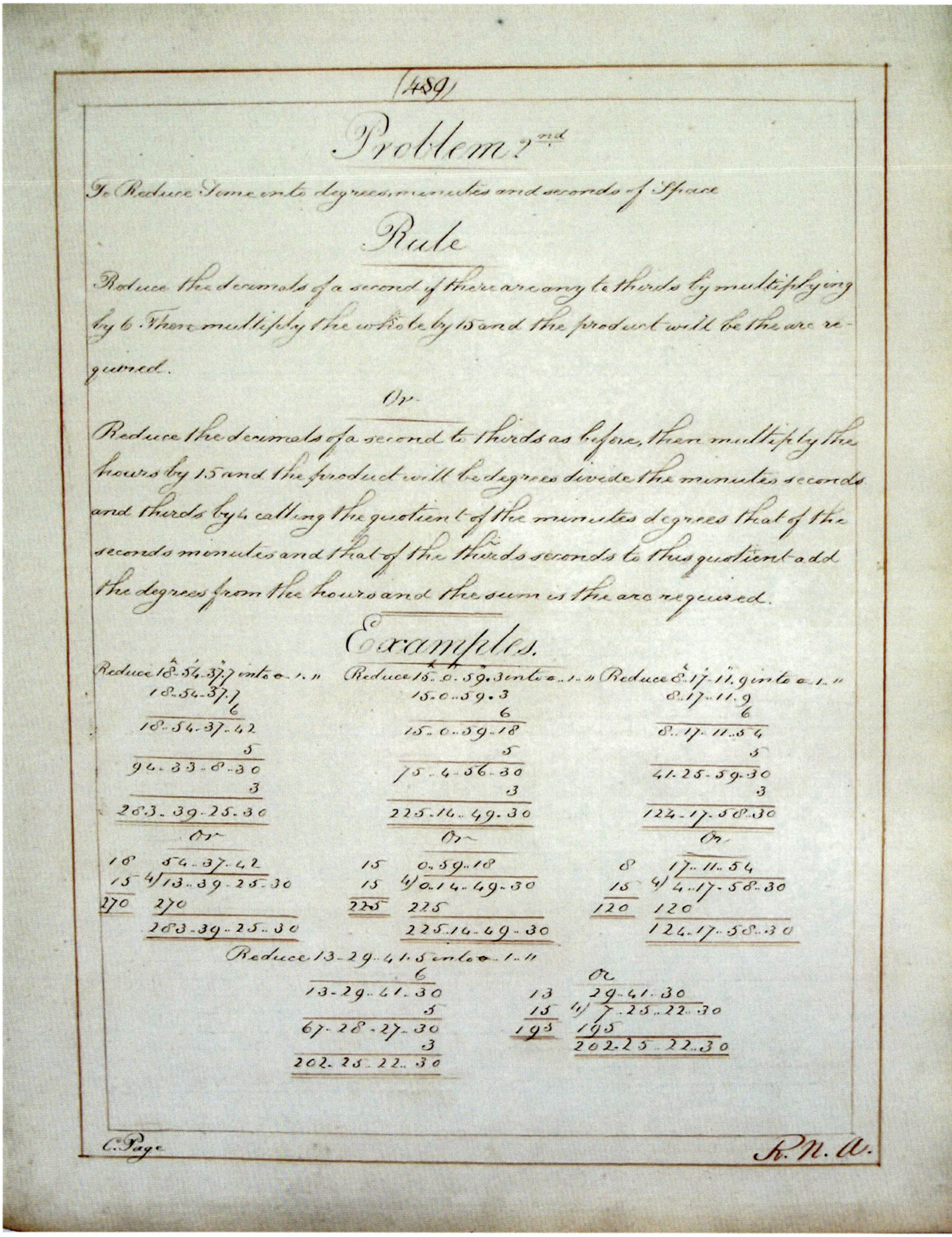

(489)

Problem 2nd

To Reduce Time into degrees, minutes and seconds of Space

Rule

Reduce the decimals of a second if there are any to thirds by multiplying by 6 Then multiply the whole by 15 and the product will be the arc required.

Or

Reduce the decimals of a second to thirds as before, then multiply the hours by 15 and the product will be degrees divide the minutes seconds and thirds by 4 calling the quotient of the minutes degrees that of the seconds minutes and that of the thirds seconds to this quotient add the degrees from the hours and the sum is the arc required.

Examples.

Reduce 18h.54′.37″.7 into ° ′ ″

18.54.37.7
6
18.54.37.42
5
94.33.8.30
3
283.39.25.30

Or

18 54.37.42
15 4)13.39.25.30
270 270
283.39.25.30

Reduce 15.0.59.3 into ° ′ ″

15.0.59.3
6
15.0.59.18
5
75.4.56.30
3
225.14.49.30

Or

15 0.59.18
15 4)0.14.49.30
225 225
225.14.49.30

Reduce 8.17.11.9 into ° ′ ″

8.17.11.9
6
8.17.11.54
5
41.25.59.30
3
124.17.58.30

Or

8 17.11.54
15 4)4.17.58.30
120 120
124.17.58.30

Reduce 13.29.41.5 into ° ′ ″

6
13.29.41.30
5
67.28.27.30
3
202.25.22.30

Or

13 29.41.30
15 4)7.25.22.30
195 195
202.25.22.30

C. Page R.N.A.

Figure 10.20. "Astronomy" problems in Charles Page's cyphering book (p. 489).

The Section on "Astronomy" in Charles Page's Cyphering Book

The next section in Charles Page's cyphering book occupied just 15 pages, and was concerned with astronomy. The section began: "Astronomy is the science which treats of the motions and distances of the heavenly bodies, and of the appearances thence arising" (p. 477).

Figure 10.20 shows a page of exercises from this section. The problem statement at the top of the page was intriguing: "To reduce time into degrees, minutes and seconds of space." The rule given was equally intriguing: "Reduce the decimals of a second if there are any by multiplying by 6. Then multiply the whole by 15 and the product will be the arc required." An alternative to this rule was also given: "Reduce the decimals of a second to thirds as before, then multiply the hours by 15 and the product will be degrees. Divide the minutes, seconds and thirds by 4, calling the quotient of the minutes degrees, that of the seconds minutes, and that of the thirds seconds. To this quotient add the degrees from the hours and the sum is the arc required" (p. 489).

Several examples illustrating how the rule could be applied are shown in Figure 10.20.

Throughout history many great mathematicians concerned themselves with astronomy. In the 1690s, Isaac Newton, John Wallis, and David Gregory played important roles in the definition of the Christ's Hospital mathematics curriculum (Allan, 1984; Howson, 1982; Sloan, 1986; Turnbull, 2008), and it was hardly surprising, therefore, that the curriculum had a section devoted to astronomy. But its presence in the curriculum was due to more than the influence of those involved in the curriculum-framing process. For many centuries navigators relied on the positions of heavenly bodies when attempting to determine the position of their ships—that was especially the case before the so-called "longitude problem" was solved (Dash, 2000; Sobel, 1995).

From the beginning, the Royal Mathematical School's curriculum included a section on astronomy. James Hodgson (1723), in the volumes he specifically wrote for the School's program, included a detailed discussion of the relationship between astronomy and navigation (Section XVI). Well-trained, experienced navigators were able to use the sun and the stars, when they were visible, to locate the north and south celestial poles. A strong understanding of the night sky enabled a navigator to determine where the invisible south celestial pole was. It was known, too that the angle that Polaris was above the horizon was almost exactly equal to the observer's latitude. And, since objects on the celestial equator rise due east and set due west, it was worth a navigator's time to learn how to find the equator in the night sky. But, whereas astronomers were concerned with understanding when things would be visible, navigators were more interested in the horizon.

In the examples shown in Figure 10.21, and the associated discussion, the impression is given that any detailed notes on links between astronomy and navigation would be inappropriate, because there would be too much to be learned given the time available. As a result, the notes merely indicated rules for doing something, and the examples showed how these rules might be applied. In such circumstances, there was a danger that rules would be merely copied, with little understanding, from a textbook. We compared the chapter on astronomy in Robertson and Wales's (1780) *Elements of Navigation; Containing the Theory and Practice* with Charles Page's entries on the subject in the 1820s, and found that sections in the two were often identical.

On the ocean, in a storm, the night sky might not be visible, and much knowledge and skill was required of a navigator charged with the responsibility of keeping track of the location of the ship, and the direction in which it was going. Being lost at sea, for even a short time, was dangerous, and if a destination was missed because of poor navigation skills, then the food supply might not be able to be replenished. Diseases such as scurvy represented a major threat, as did piracy. It was absolutely important that navigators kept ships on time and on tried and trusted routes. Their efficiency could be enhanced if they knew how to take advantage of knowledge gained through the night sky, if and when that sky was visible.

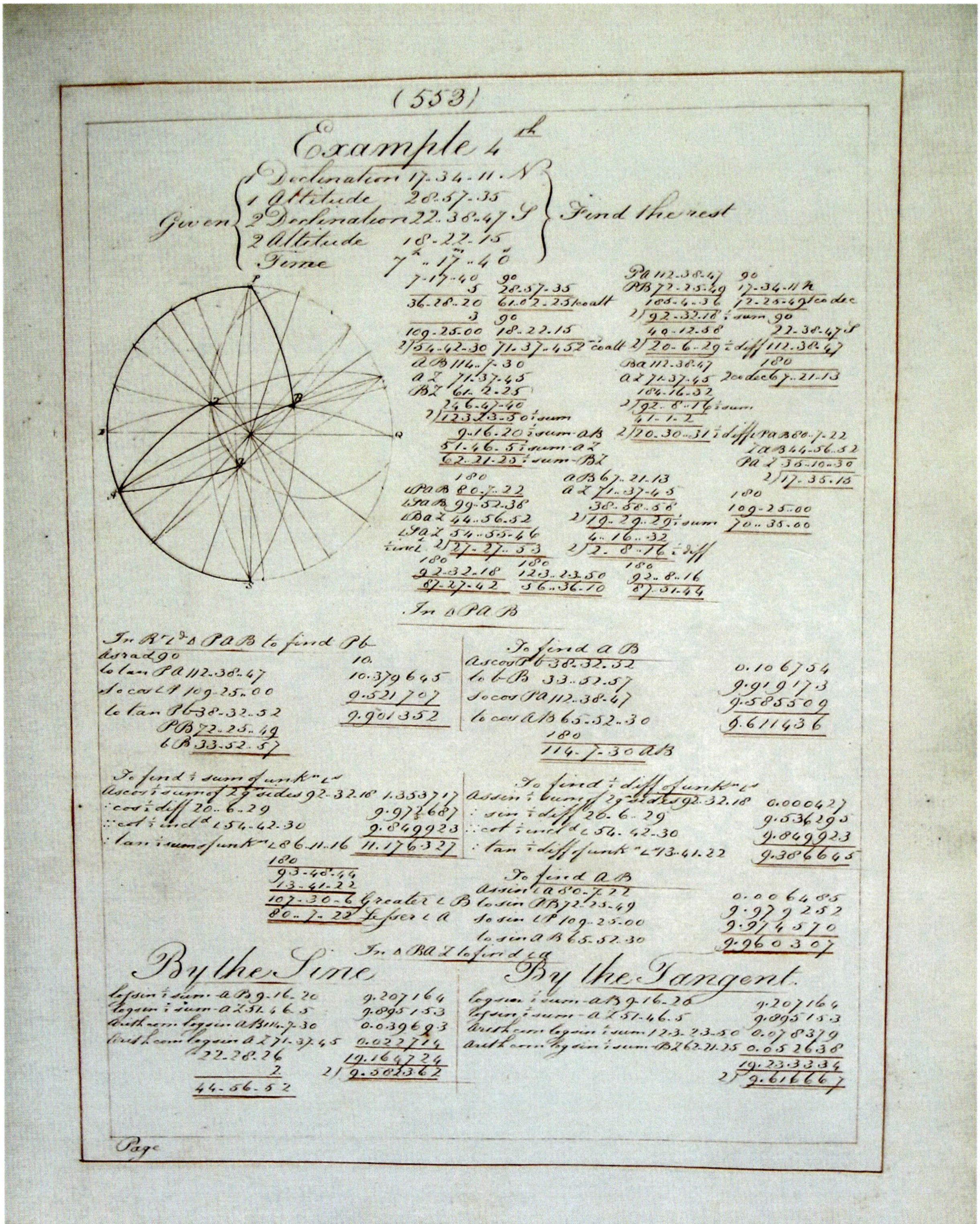

Figure 10.21. "Given 1 declination, 1 altitude, 2 delineation, 2 altitude, time. Find the rest" (from Charles Page's cyphering book, p. 553).

The Use of the Globes

The next section in Page's cyphering was headed "The Use of the Globes," which referred to "two spherical bodies called the terrestrial and celestial globes, the convex surfaces of which are supposed to give a true representation of the earth and heavens" (p. 497). The section occupied 60 pages, and most of the pages showed solutions to stated problems. The solutions often featured incredibly intricate diagrams, and the detailed calculations were beautifully presented in very small handwriting. Two ink colors, black and red, were used. Even though Christ's Hospital students received careful training in technical drawing (Sloan, 1986), it is difficult to imagine how 15-year-olds could have coped with the cognitive and artistic demands of the tasks.

The statements of the first 21 problems in the sections are given below:

1. Find the latitude and longitude of any place on the terrestrial globe.
2. To find any place on the globe, the latitude and longitude of which are given.
3. To find the distance of any two places on the globe.
4. To find the sun's place, right ascension, and declination on any day.
5. To rectify the globe for the latitude, zenith and noon.
6. At any given hour at a given place, to find what hour it is at any other place.
7. The latitude of a place and day of the month being given, to find the sun's declination, meridian altitude, right ascension, amplitude, oblique ascension, ascensional difference, and thence the time of rising, setting, and length of day and night.
8. Given the latitude of the place and day of the month, to find the sun's altitude and azimuth, either when he is due east or west, at 6 o'clock or at any other hour while he is above the horizon.
9. Given the latitude, day of the month, and sun's altitude, to find the azimuth and hour of the day.
10. To find the latitude and longitude of any star,
11. To find the right ascension and declination of a star.
12. On any day and in any given place to find when a proposed star rises, sets, souths, and its meridian altitude.
13. To represent the appearance of the heavens at any time in any given place.
14. Given the sun's longitude and the obliquity of the ecliptic, required the sun's right ascension and declination.
15. Given the obliquity of the ecliptic and the sun's declination, required the sun's longitude and right ascension.
16. Given the obliquity of the ecliptic and the sun's right ascension, required the sun's longitude and declination.
17. Given the latitude of a place and the sun's declination, required the time and place of setting, the meridian altitude, with the length of the day and night.
18. Given the latitude of a place and the sun's declination, required the sun's altitude and azimuth and hour when the sun is due east of west.
19. Given the latitude of a place and the sun's declination and altitude, required the hour from noon and the sun's azimuth.
20. Given the obliquity of the ecliptic and a star's right ascension and declination, required its latitude and longitude.
21. Given the latitude and longitude of a fixed star, and also the obliquity of the ecliptic, required the right ascension and declination of the star. ...

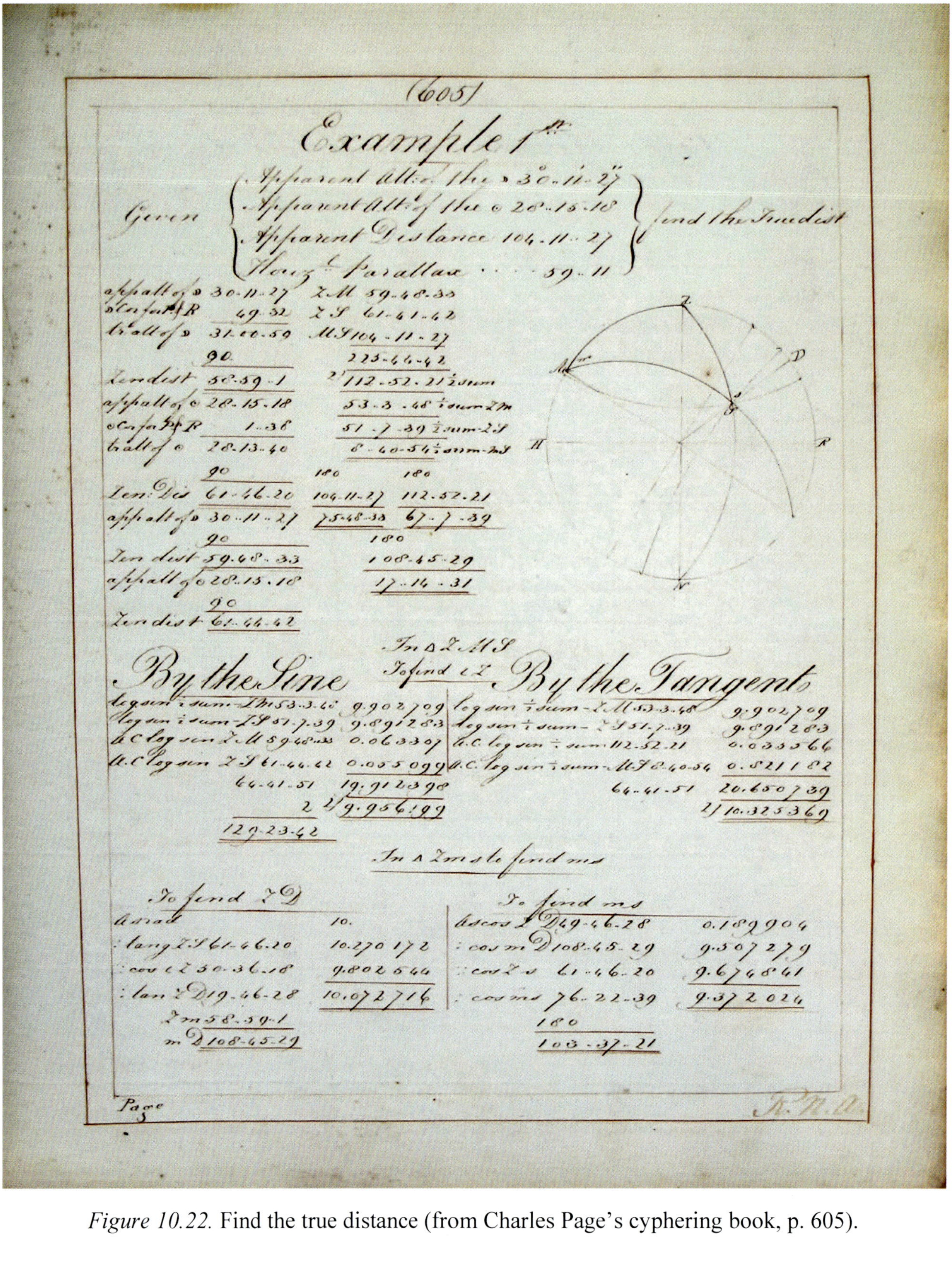

Figure 10.22. Find the true distance (from Charles Page's cyphering book, p. 605).

All in a Day's Work

This next section, titled "Day's Work," was lengthy—it occupied 90 pages. Its stated purpose was to provide details of "the practical methods of finding a ship's place every day at noon and setting the course and distance she has to sail" (p. 559). It began by stating that for navigators the following matters were all in a day's work:

- The means of measuring and correcting the run of the ship;
- The nature and use of the sea and azimuth compasses;
- How to find the sun's amplitude and azimuth;
- How to find the variation of the compass and to correct the courses the ship has sailed on that account;
- How to find the ship's lee-way and thence to correct the courses;
- To determine the nature and use of instruments proper for observing the altitudes and distances of celestial objects;
- How to find the latitude and longitude by celestial observations;
- How to correct the daily account of latitude and longitude.

As if to prove that the task of a navigator on a ship was demandingly complex, the section then provided details about what should be done, including what calculations needed to be made, in order that the tasks associated with each of these eight points could be achieved.

The expectations set out for what a navigator was supposed to achieve in a day's work seemed to be remarkable. Consider, for example, what a navigator might be expected to do to find the "true distance" for the situation shown in Figure 10.22.

On a large merchant ship, or a ship engaged in warfare, in the eighteenth and nineteenth centuries there was usually one experienced navigator. That person was privileged in the sense that his on-board living quarters were next to the Captain's cabin. He would be assisted by, and responsible for, at least one apprentice navigator, who would be expected to take control of more mundane navigational matters while the experienced navigator was asleep or otherwise occupied. The merchant ships looked to Christ's Hospital to provide well-trained navigation apprentices—and, in fact, the British government paid an allowance to merchant seamen who accepted Christ's Hospital graduates as apprentices (Hans, 1951a, 1951b; Pearce, 1901; Plumley, 1976). Guardians of those boys accepted into the Royal Mathematical School had to sign a statement that on graduation the boys would be bound as apprentices for seven years.

Navigators were expected to make on-board calculations. Theirs was the responsibility of guiding ships, safely, to expected destinations as quickly as possible. They had to deal with all the variations that could arise at sea. There would be times when they would not be able to see the sun, moon or stars. Taking into account wind patterns, currents, and weather conditions they were expected to steer the best possible courses. Sometimes their decisions were influenced by matters other than the weather—for example, they knew that they should try to avoid entering seas in which piracy was rife. They had with them the instruments that they would need in order to measure what had to be measured, and they also had tables of logarithms and trigonometric ratios to assist them in their calculations. But, most important of all, was the knowledge that they carried in their heads, and in their navigation cyphering books, about what to do, when to do it, and how to do it.

The section on "Day's Work" in Charles Page's cyphering book makes it very clear that the most important calculations needed specific mathematical knowledge, the kind of knowledge that required a form of training that went well beyond merely getting correct answers.

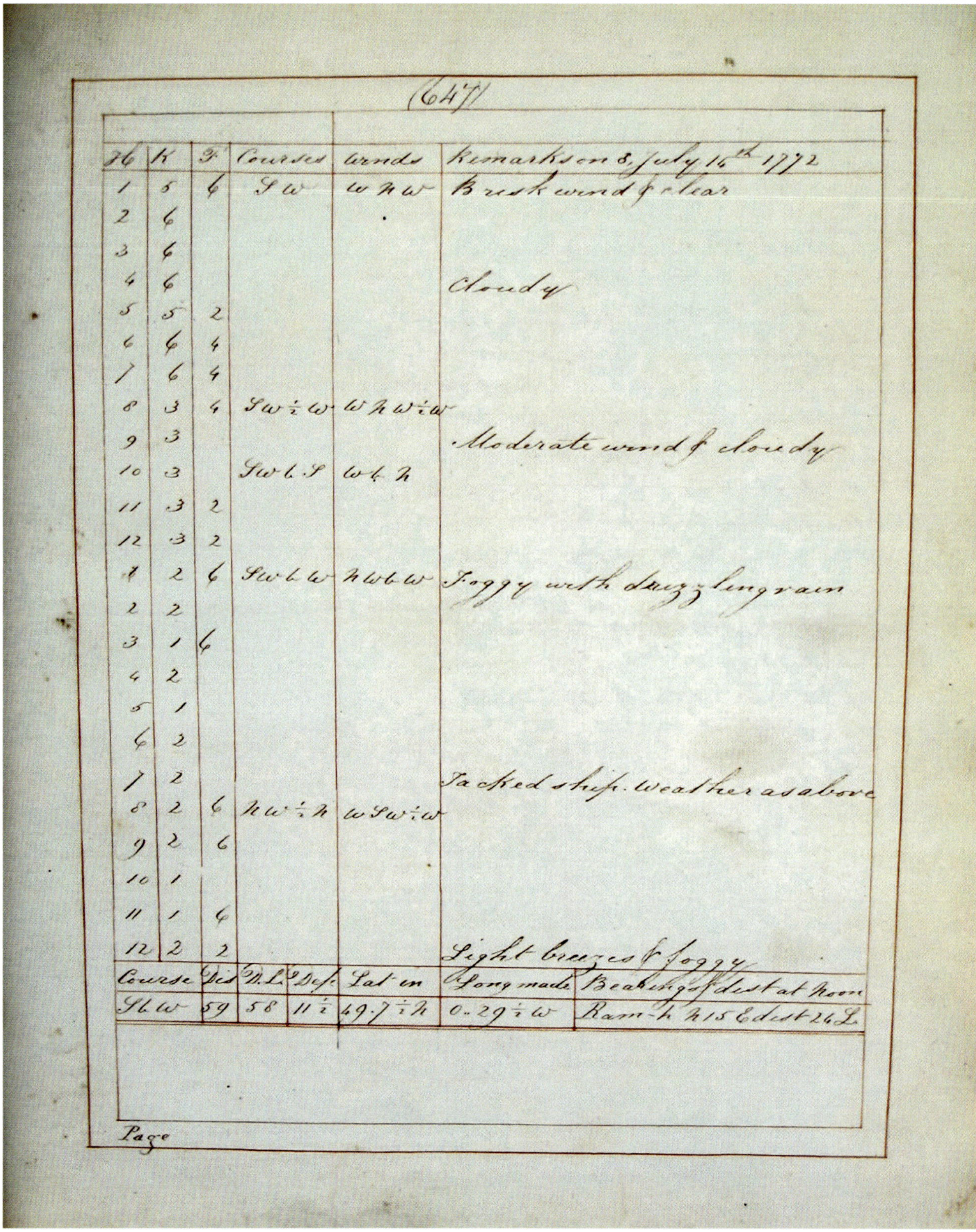

(647)

H	K	F	Courses	Winds	Remarks on 8, July 16th 1772
1	5	6	SW	WNW	Brisk wind & clear
2	6				
3	6				
4	6				Cloudy
5	5	2			
6	6	4			
7	6	4			
8	3	4	SW½W	WNW½W	
9	3				Moderate wind & cloudy
10	3		SWbS	WbN	
11	3	2			
12	3	2			
1	2	6	SWbW	NWbW	Foggy with drizzling rain
2	2				
3	1	6			
4	2				
5	1				
6	2				
7	2				Tacked ship. Weather as above
8	2	6	NW½N	WSW½W	
9	2	6			
10	1				
11	1	6			
12	2	2			Light breezes & foggy

Course	Dis	N.L.	Dep.	Lat in	Long made	Bearings & dist at Noon
SbW	59	58	11½	49.7½N	0.29½W	Ram-h N15E dist 26 L

Page

Figure 10.23. Part of a log of a journey from England towards Madeira (from Charles Page's cyphering book, p. 647).

A Journal of a Voyage from England Towards Madeira

This 31-page section was the last section in the 671-page cyphering book. It provided details of the kind of daily entries that a navigator was required to make in a journal during voyages. Keeping an up-to-date log was an important component of the day's work of all navigators, and almost all Christ's Hospital navigation cyphering books had, as a final section, an example of a log for a complete voyage (Ellerton & Clements, 2012). Although students who wrote the details of the voyages in their navigation cyphering books never (or rarely) actually made the journey being summarized, entries could have been based on a log that some navigator had made in the past.

The "voyage" described in Charles Page's cyphering book had, apparently, begun at 2 pm, July 2, 1772, at Deplford, near London. According to the log, at that time "the pilot came on board, ordered the ship to become loose from the shear hulk (wind NWbW)." At 5 pm the ship "made sail down the river (wind NE) until it made fast at the shear hulk at Woolwich, the wind not permitting us to proceed further." "At 4 am cast loose from the hulk (wind NNW) and at 10 am anchored in long reach, moored with a cable each way" (p. 643).

Then on July 3, "employed taking the gunners' stores on board. Read the Articles of War to the ship's company and punished William Barnette with 12 lashes for drunkenness and disobedience. Fine weather and moderate wind (NW) " (p. 643).

And so the entries continued—there were usually between three and five lines per day. Once the open sea was reached, however, the entries became much more detailed, with times, speeds, course directions, wind directions, and remarks (sometimes quite technical), being recorded by the hour—Figure 10.23 shows a page from Charles Page's log. At the bottom of a page a succinct summary for the past 24 hours of the journey was given. This included summary data for "latitude in" and "longitude made." In the summary section, comments such as "moderate wind & cloudy" were made, from time to time.

Between pages 417 and 433 of Volume 1 of James Hodgson's (1723) detailed text—which was specifically written for students at the Royal Mathematical School by the Master-in-charge of the School—a log of a journey, which purportedly took place in 1721, was shown. Every day, entries mentioned the present bearing of the ship, and its distance from "The Lizard" (from which the journey had begun). Hodgson (1723), who was a Fellow of the Royal Society," wrote in his Preface that the volumes set out the methods he used in teaching navigation, and the rules by which he himself operated. He continued:

> I have endeavoured thro'out the whole course of this work to deliver everything with as much plainness as possible. ... I can't recollect that there is one thing left undemonstrated, that is capable of it. ... the calculations are all done to the greatest degree of exactness, and have all gone through my own hands. (pp. vii-viii)

The publication of Hodgson's volumes continued a tradition that had begun when the Royal Mathematical School was established—masters wrote textbooks setting out the intended curriculum (see, e.g., Moore & Perkins, 1681; Robertson & Wales, 1780). The students' navigation cyphering books provided evidence of the implemented curriculum, and regular recitation sessions with the master-in-charge and with ushers, and annual "verbal" examinations of individual students by outside experts, checked the attained curriculum. Christ's Hospital graduates quickly became citizens of the world and after completing their apprenticeships many chose to live in North America and in British colonies like New South Wales and New Zealand. Around 1800, the Royal Mathematical School's curriculum was influential at Harvard University (Ellerton & Clements, 2012), and in many other parts of North America (Coldham, 1990).

Some Final Comments on the Royal Mathematical School at Christ's Hospital

When we first examined Charles Page's navigation cyphering book we felt stunned. The first noticeable features were the large gilt crests embossed on both the front and back leather covers. Then, on opening the book, we were confronted with the statement that this book had been produced by a child being educated at the Royal Mathematical School at Christ's Hospital, "founded by his late Majesty King Charles 2nd." Then, we were overwhelmed by the undeviating high quality of the penmanship, calligraphy, and artistic work. After carrying out investigations into the history and circumstances of the Royal Mathematical School, we found that Charles Page was probably 14, 15 or 16 years of age when he made most of the entries in his cyphering book. How could he have achieved such excellence at such a young age? Further investigation revealed that Christ's Hospital had probably the strongest writing education tradition of any school in the nation (Mann, 1865).

We were more than curious, too, about the mathematics we found on display in the cyphering book. Charles was just 16 when he left the school. As far as we know, he had not come from a family background with known brilliance with respect to mathematics. How could he apparently have learned so much about mathematics, so quickly? His cyphering book included work on Euclidean geometry, logarithms, plane trigonometry, and spherical trigonometry, and these were applied, at a high level, it seemed, to all aspects of navigation.

When we studied carefully the section on algebra in Charles' book we began to realize that he had not always obtained a relational understanding of what he had written in his book. He did not link parts of an argument together logically, and he did not always check his answers for correctness. We began to wonder whether our assessment of his mathematics education at Christ's Hospital had been colored by the beautiful penmanship, calligraphy, and illustrations. It seemed that some of what he wrote had been dictated to him.

Although we were not well qualified to assess degrees of correctness of his entries on navigational topics, we were not fully convinced that he understood what he was writing. We reflected that it would hardly be surprising if he had not, for he had written much, on a wide range of navigational themes, and he had certainly spent a lot of time making sure that his entries looked beautiful. We could not reject the possibility that although his cyphering book scored high in both its appearance and the level of mathematics, it could have been sub-standard so far as educational efficiency was concerned. After all, his book included 671 handwritten pages, all carefully crafted. Assuming that he prepared the book over a two-year period (and that the mathematics education he received during the earlier years at Christ's Hospital was concerned with elementary mathematical ideas which were not recorded among the 671 pages), then he prepared, on average, almost one page of his book each day, seven days a week, for two years. More research into what actually took place in the classrooms in the Royal Mathematical School is clearly needed.

It will be in order to close with some comments on the way history has treated the Royal Mathematical School within Christ's Hospital, and especially the way it has treated James Hodgson, the Master-in-Charge between 1709 and 1755. Almost all who have written about the early years of the School have pointed out that the early masters struggled to cope with the combination of mathematical, navigational, and pedagogical demands that it placed on teachers (Howson, 1982; Trollope, 1834). It is arguable that the person who saved the School was James Hodgson. Hodgson would become a council member of the Royal Society, and wrote numerous texts, including *Theory of Navigation* (1706) and the two-volume *System of Mathematics* (1723). Such was the success of the Royal Mathematical School under Hodgson's leadership that the Royal Navy Academy, at Portsmouth, established in 1733 for the training of naval officers, would model its curriculum on Hodgson's Christ's Hospital program. The Naval Academy was intended to train the officers for the

Royal Navy. Ultimately, Royal Naval authorities would be critical of the Royal Mathematical School ("Blues" and the Royal Navy, 1900).

When Master of the Royal Mathematical School, Hodgson, recognizing that the curriculum was too difficult for boys aged between 10 and 16, persuaded School authorities to allow boys to stay on to 18 years, or to even more than that. This was not liked by the classically-trained headmasters of Christ's Hospital, who accused the "mathemats" of being over-aged bullies who terrorized the younger boys in the school (Trollope, 1834; Wilson, 1820).

James Hodgson

- Was Master of the Royal Mathematical School from 1709 to 1755. During his time as Master, the School began to flourish, and its reputation as a fine school was established.
- Was elected Fellow of the Royal Society in 1703, and member of the Council of the Royal Society from 1733;
- Published numerous books on mathematics, especially the mathematical side of navigation, astronomy, fluxions, and perspective in drawing;
- Mastered Newton's theory of fluxions and incorporated that in his 2-volume *System of Mathematics* (1723), written especially for Christ's Hospital students.

Figure 10.24. James Hodgson (1672–1755), engraving by George White after Thomas Gibson.

Ultimately, the Royal Mathematical School was compelled to revert to the policy whereby boys should not remain at school beyond the age of 16. This reversion to the original policy was agreed to by the highly-regarded Master of the Mathematical School, William Wales, formerly navigator for James Cook. Wales revised a textbook earlier authored by John Robertson—who had been Master at Christ's Hospital but became Master-in-Charge of the Naval Academy—and thus the two institutions offered very similar curricula. Wales was well-liked, and his name is usually glorified in histories of the Royal Mathematical School (e.g., Trollope, 1834), whereas James Hodgson's work does not receive the credit which, perhaps, it deserves. It is noteworthy that in 1773, well before Wales was at Christ's Hospital, the Royal Mathematical School and Royal Marine Academy at Portsmouth were described, by an independent reviewer, as "the two most eminent mathematical schools" in Great Britain (A Society of Gentlemen, 1773, p. 130).

The reversion to the original policy of boys not proceeding beyond the age of 16 created the same sort of difficulty that Hodgson had set out to solve. According to an early manuscript from the Royal Library of King George II, under Robertson and under Wales the mathemats were reduced merely to copying text from Robertson's textbook—and under those circumstances the students' understanding of relevant principles and procedures may have been compromised. The continuation of the age policy, together with the maintenance of the original high-level curriculum, or something close to it, could explain why Charles Page seemed to lack genuine understanding of some of the more difficult topics that he studied while attending the Royal Mathematical School.

In 1843, Horace Mann, the great American educator, visited Christ's Hospital, and although he was not impressed with its very authoritarian, overly-religious, atmosphere, he thought the boys' handwriting was the best that he had seen anywhere in the world (Mann, 1865).

Charles Page was from a sea-faring family—his father was the master of a packet boat, and his youngest brother, Henry, became a master mariner. The first ship on which Charles served, the *Boyne*, sailed to the East Indies in 1825, and records show that in 1826 the *Boyne* was used to transport 201 convicts and a number of passengers from London via County Cork (Ireland) to Port Jackson (Sydney) in Australia. Charles Page's period of apprenticeship was almost completed when, on May 9, 1832, he and his brother John sailed from Portsmouth on the *Munro* to become "free settlers" in New South Wales There were 199 male convicts aboard, and these were accompanied by crew and by a few passengers.

Shipping records suggest that Charles never lost his affinity with the sea, for he took numerous trips, as a passenger, up and down the east coast of Australia, and also between Australia and New Zealand. His mother, Harriet, died on May 1, 1869, and the death notice published in the *Sydney Morning Herald* on July 24, 1869, made reference to her son "Mr C. Page, Rocky Mouth, Clarence River (New South Wales) and to Captain Henry Page of Newtown (Sydney)." In another death notice, published a few days later, also in the *Sydney Morning Herald*, another son—John Code Page of Woodfordia, Clarence River—and a grandson, John, living in the state of South Australia—were mentioned. In Sand's *Directories, Sydney and New South Wales* (1858–1893), Charles was listed as living in Surry Hills (Sydney) and his occupation was given as a stationer. Joseph Cope Page was listed as a stock and station agent, Henry as a master mariner, and John without a specified occupation—all in Sydney. Joseph died in Randwick (Sydney) in 1910, Henry in Brisbane (Queensland) in 1888, and records suggest that Charles died in 1882.

To what extent did Charles, in his post-Christ Hospital life, apply the mathematics he learned at the Royal Mathematical School? Certainly, he embarked on many sea voyages, firstly as an apprentice, but then as a private passenger. His fine penmanship and drawing skills may be what drew him to becoming a stationer in New South Wales. Perhaps he passed on some of the skills he had acquired at Christ's Hospital to his customers and to others in his family and the surrounding community. But, regardless of the influence that this one student may have had during his lifetime, it is safe to say that the training given to mathemats at Christ's Hospital was destined to make its mark in unlikely places around the world (see, e.g., Coldham, 1990; Mansell, 2011).

Forty-six Royal Mathematical School workbooks are known to exist, each of which is not unlike the one prepared by Charles Page (Clifford Jones, personal communication to Nerida Ellerton, February 9, 2014). The oldest of these surviving workbooks was prepared by John Cox in 1723, and the "youngest" (which we hold) by Richard Murdoch Robertson in 1858.

According to Jones, "the workbooks of the Royal Mathematical School pupils were strongly bound and taken to sea when they joined their ships." Most of the surviving handwritten manuscripts are between 500 and 700 pages long. Jones added that not all of the boys who prepared the manuscripts originally entered Christ's Hospital on the Royal Mathematical School Foundation—some transferred to the Royal Mathematical School later. Most of the 46 extant manuscripts are located in the United Kingdom, or Australia or the United States of America.

We have seen other similar handwritten manuscripts, prepared at schools like the Royal Naval Academy, at Portsmouth, and the Royal Hospital School at Greenwich. In the second half of the eighteenth century, Harvard College offered a navigation course in which students prepared manuscripts which resembled those prepared by students at the Royal Mathematical School. At that time there were direct links between Harvard College and Christ's Hospital. Clearly, the role which navigation education had on mathematics education, and *vice-versa,* in Great Britain and in North America, could be a fertile area of investigation for researchers.

Acknowledgment

We gratefully acknowledge the generous assistance of Ken Mansell and Clifford Jones, volunteers at the Christ's Hospital Museum, in the preparation of this chapter. Without Ken's help we would not have been able to clarify certain aspects associated with the admission of Charles Page's to Christ's Hospital, and with his life after graduating from the Royal Mathematical School. Clifford's comments on the early history of the Royal Mathematical School, and on the present locations of cyphering books prepared by former Royal Mathematical School students, were invaluable. Clifford Jones is currently writing a history of the Royal Mathematical School.

References

A Society of Gentlemen. (1773). Review of Robertson's *The Elements of Navigation. The Critical Review: Or Annals of Literature, 35*, 130–134.

Alexander, J. (1709). *A synopsis of algebra,* London, United Kingdom: J. Barber.

Allan, G. A. T. (1984). *Christ's Hospital.* London, United Kingdom: Town & Country Books.

Allen, J. B. L. (1970). *The English mathematical schools, 1670–1720.* PhD thesis, University of Reading.

Bache, A. D. (1839). *Report on education in Europe, to the trustees of the Girard College for Orphans.* Philadelphia, PA: Bailey.

"Blues" and the Royal Navy. (1900, November 24). *The Navy and Army Illustrated,* 245–248.

Christ's Hospital (1823–1830). *General Committee minutes*, Volume 17. Handwritten manuscript, London Metropolitan Archives Reference CLC/210/B/005/MS12811/017.

Christ's Hospital (1836–1840). *General Committee minutes*, Volume 19. Handwritten manuscript, London Metropolitan Archives Reference CLC/210/B/005/MS12811/019.

Christ's Hospital (1857). *Register of Christ's Hospital Royal Mathematical School boys apprenticed to ships' masters 1816–1857.* Handwritten manuscript, London Metropolitan Archives Reference CLC/526/MS30338.

Christ's Hospital (1953). *The Christ's Hospital book.* London, United Kingdom: Hamish Hamilton.

Coldham, P. W. (1990). *Child apprentices in America from Christ's Hospital, London, 1617–1778.* Baltimore, MD: Genealogical Publishers Co.

Cracraft, J. (1971). *The church reform of Peter the Great.* London, United Kingdom: Macmillan.

Cross, A. (2007). *By the banks of the Neva.* Cambridge, United Kingdom: Cambridge University Press.

Dash, J. (2000). *The longitude prize.* New York, NY: Frances Foster Books.

Dickinson, H. W. (2007). *Educating the Royal Navy*. London, United Kingdom: Routledge.

Ditton, H. (1709). *An appendix to the algebra of John Alexander.* London, United Kingdom: J. Barber.

Dorfman, M. (1966). Lincoln's arithmetic education: Influence on his life: Part II. *Lincoln Herald, 68*(3), 108–120.

Ellerton, N. F., & Clements, M. A. (2012). *Rewriting the history of school mathematics in North America 1607–1861.* New York, NY: Springer.

Hans, N. A. (1951a). The Moscow School of Mathematics and Navigation. *The Slavonic and East European Review, 29*(73), 532–536.

Hans, N. A. (1951b). *New trends in education in the 18th century.* London, United Kingdom: Routledge.

Hirsch, D., & Van Haften, D. (2010). *Abraham Lincoln and the structure of reason.* New York, NY: Savas Beatie LLC.

Hodgson, J. (1723). *A system of the mathematics, containing the Euclidean geometry, plain and spherical geometry, the projection of the sphere, both orthographic and stereographic, astronomy, the use of globes and navigation.* London, United Kingdom: Thomas Page.

Howson, G. (1982). *A history of mathematics education in England.* Cambridge, United Kingdom: Cambridge University Press.

Howson, G., Keitel, C., & Kilpatrick, J. (1981). *Curriculum development in mathematics.* Cambridge, United Kingdom: Cambridge University Press.

Jackson, L. L. (1906). *The educational significance of sixteenth century arithmetic.* New York, NY: Teachers College Press.

Jones, H. S. (1954). Foreword by the Astronomer Royal. In E. G. R. Taylor, *The mathematical practitioners of Tudor & Stuart England 1485-1714* (pp. *ix-x*). Cambridge, United Kingdom: Cambridge University Press.

Mahan, A. T. (1898). *The influence of sea power upon history 1660-1783.* Boston, MA: Little, Brown, and Company.

Mann, M. (1865). *Life of Horace Mann.* Boston, MA: Walker, Fuller, and Company.

Mansell, K. (2011). *Christ's Hospital in the Victorian era.* Whitton, United Kingdom: Ashwater Press.

Moore, J., & Perkins, P. (1681). *A new systeme of the mathematicks designed for the use of the Royal Foundation of the Mathematical School in Christ's Hospital.* London, United Kingdom: Robert Scott.

Okenfuss, M. J. (1973). Technical training in Russia under Peter the Great. *History of Education Quarterly*, *13*(4), 325–342.

Padfield, P. (2005). *Maritime power and the struggle for freedom: Naval campaigns that shaped the modern world.* Woodstock, NY: The Overlook Press.

Pearce, E. H. (1901). *Annals of Christ's Hospital*. London, United Kingdom: Hugh Reed.

Plumley, N. (1976). The Royal Mathematical School within Christ's Hospital. *Vistas in Astronomy*, *20*, 51–59.

Robertson, J., & Wales, W. (1780). *The elements of navigation containing the theory and practice, with the necessary tables and compendiums for finding the latitude and longitude at sea, to which is added a treatise on marine fortification* (4th ed.). London, United Kingdom: Christ's Hospital.

Singh, P., & Ellerton, N. F. (2013). International collaborative studies in mathematics education. In M. A. Clements, A. J. Bishop, C. Keitel, J. Kilpatrick, & F. Leung (Eds.), *Third international handbook of mathematics education* (pp. 827–860). New York, NY: Springer.

Sloan, K. M. (1986). *The teaching of non-professional artists in eighteenth-century England.* PhD thesis, Westfield College, University of London.

Sloan, K. (2000). *"A noble art": Amateur artists and drawing masters c.1600–1800*. London, United Kingdom: British Museum Press.

Sobel, D. (1995). *Longitude: The true story of a lone genius who solved the greatest scientific problem of his time*. New York, NY: Walker and Company.

Stewart, L. (1999). Other centres of calculation, or where the Royal Society didn't count: Commerce, coffee-houses and natural philosophy in early modern London. *British Journal of the History of Science, 32*, 132–153.

Trollope, W. (1834). *A history of the royal foundation of Christ's Hospital.* London, United Kingdom: Pickering.
Turnbull, H. W. (Ed.). (2008). *The correspondence of Isaac Newton 1676–1687* (Vol. II). Cambridge, United Kingdom: The Royal Society and Cambridge University Press.
Van Berkel, K. (1988). A note on Rudolf Snellius and the early history of mathematics in Leiden. In C. Hay (Ed.), *Mathematics from manuscript to print 1300–1600* (pp. 156–161). Oxford, United Kingdom: Clarendon Press.
Van Sickle, J. (2011). *A history of trigonometry education in the United States: 1776–1900.* PhD dissertation, Columbia University.
Waters, D. W. (1958). *The art of navigation in England in Elizabethan and early Stuart times*. New Haven, CT: Yale University Press.
Wilson, J. I. (1820). *A brief history of Christ's Hospital*. London, United Kingdom: Author.

Chapter 11

Fit for a King? William Beattie, Circa 1810 and 1835–1836

Abstract: William Beattie prepared both of the manuscripts that are analyzed in this chapter. The first has 466 pages and is concerned mainly with mensuration. Many of the notes and examples in that manuscript are from a 1748 London edition of William Hawney's classic text, *The Compleat Measurer.* The second manuscript has 309 pages and concerns navigation. It is based almost entirely on a 1796 edition of John Hamilton Moore's *The New Practical Navigator.* It is conjectured that although the title page of the measurement manuscript is dated 1835, in fact the manuscript was prepared by Beattie around 1810, when he was a student at Clarencefield Academy in Dumfriesshire, Scotland. Beattie became a medical doctor, and served as personal physician and then private secretary to the man who would become King William IV—and would be remembered as the "Sailor King." It is conjectured that the navigation cyphering book, dated 1836 on its title page, was specially prepared by Beattie for his ailing King (who died in 1837). Although the penmanship and calligraphy in both manuscripts are of a high standard, Beattie seemed to value the mensuration manuscript more highly than the navigation manuscript.

Key Words: Ciphering book; copybook; cyphering book; John Hamilton Moore; King William IV; mensuration; navigation; navigation cyphering book; navigation education; sailing; William Beattie; William Hawney.

This chapter is about two cyphering books, both of which were prepared by William Beattie. One is mainly concerned with mensuration—that is to say, with the principles and practices of measurement—and the other with navigation. They have 466 pages and 309 handwritten pages, respectively, and each has dimensions 33 cm by 20 cm (i.e., approximately 13" by 8").

The mensuration cyphering book's title page indicates that it was prepared at Clarencefield Academy in 1835, and was based on a 1748 edition of a textbook on measurement authored by William Hawney. The back board of this manuscript is slightly separated from the spine, and the front end papers and title pages are together but are no longer bound with the remainder of the manuscript. However, the quality of the penmanship, calligraphy, and illustrations is high throughout, and the half-bound leather manuscript has marbled hard covers. On the spine cover, "Hawney's Mensuration W. Beattie" is printed in gilt, and there are also elaborate gilt decorations. Throughout the handwritten text, titles and subtitles are routinely presented in water-colored block print. The main theme of the manuscript, mensuration, is consistent with the fact that William's father had been an architect, surveyor, and builder (Austin, 1885).

The second manuscript also has marbled hard covers with half-leather binding. On the spine cover, "Moore's Navigation W. Beattie" is printed in gilt, and there are beautiful surrounding gilt decorations. The interior title page indicates that the manuscript was based on "Moore's Navigation," and was prepared by William Beattie, of Clarencefield Academy. A date is given—February 29, 1836.

All of the above suggests that the provenance for these manuscripts might easily be established. However, details in this chapter will show that that was not the case.

Nerida F. Ellerton and M. A. (Ken) Clements, *Abraham Lincoln's Cyphering Book and Ten other Extraordinary Cyphering Books*, DOI 10.1007/978-3-319-02502-5_11, © Springer International Publishing Switzerland 2014

Who Was William Beattie?

William Beattie (1793–1875) came from a well-to-do family in Dumfriesshire, Scotland. He was a highly literate 19th-century medical doctor who was close to numerous leading European society figures, writers and poets of his day. Between 1806 and 1811 he was a boarder at Clarencefield Academy in Dumfriesshire, and during that time he gained a competent knowledge of Latin, Greek, and French. In 1812, he began medical studies at Edinburgh University, and in 1818 he graduated from that University with an M.D. degree (Austin, 1885). Between 1822 and 1826 he accompanied the Duke of Clarence, King George III's son, on three trips to Germany as the Duke's person physician (Beattie, 1831). When, in 1830, the Duke ascended to the British throne, as King William IV, Beattie continued in his service. He served the Duke (and King) for 14 years altogether, including three as private secretary (Austin, 1885).

King William IV was born in 1765 and died in 1837. He enlisted in the Royal Navy when he was 13 years old and in 1785 became a lieutenant. In 1786 he was Captain of HMS *Pegasus*, and later, in 1786, he was stationed in the West Indies under Horatio Nelson—with whom he shared a friendship. In 1788 he was Commander of the frigate HMS *Andromeda*, and in 1790 he was promoted to Rear-Admiral in command of HMS *Valiant*. Around 1790 he was made Duke of Clarence. Between 1830 and 1837, when King of England, he was known as "the Sailor King" (Somerset, 1980; Ziegler, 1973). The painting shown in Figure 11.1 was done around 1800, well before he was king.

Under What Circumstances Were the Manuscripts Prepared?

When Were the Manuscripts Prepared?

We conjecture that Beattie completed the volume that was concerned with Hawney's *Mensuration* sometime between 1807 and 1811, when he was a student at Clarencefield Academy. We further conjecture that the second manuscript—the one that focused on navigation—was completed in 1835 and 1836, after Beattie had been personal secretary to William IV. The Sailor King felt an affinity with all matters maritime, and we conjecture that William Beattie personally prepared the navigation manuscript as a very special tribute to his King.

At the foot of the title page of the mensuration manuscript, William Beattie printed "Clarencefield Academy April 16th, 1835," and on the title page for the navigation manuscript he printed Clarencefield Academy, February 29th, 1836" (see Figure 11.2). Although we accept that the front pages reproduced in Figure 11.2 were prepared in 1835 and 1836, we do not believe that Beattie prepared the mensuration manuscript in 1835. It is likely that the front title page was dated 1835 but the rest of the manuscript was completed about 25 years earlier when Beattie was a student at Clarencefield Academy. We would conjecture that the beautiful, high-class binding, and gilt labeling, of both manuscripts were completed in 1836, or after that.

The penmanship and calligraphy in the mensuration manuscript were less mature than in the navigation manuscript. From our experience of having examined many cyphering books, we would judge that the penmanship, calligraphy, and illustrations in the mensuration manuscript were appropriate to a youth of about 17 or 18 years, whereas the navigation manuscript seems to have been prepared by an older person. The first page of mathematical content in the mensuration manuscript is shown in Figure 11.3, and the first page of mathematical content in the navigation manuscript is shown in Figure 11.4.

Figure 11.1. Portrait of the Duke of Clarence, painted by Martin Archer Shee, c. 1800 (public domain, via Wikimedia Commons). In 1830, the Duke would become King William IV.

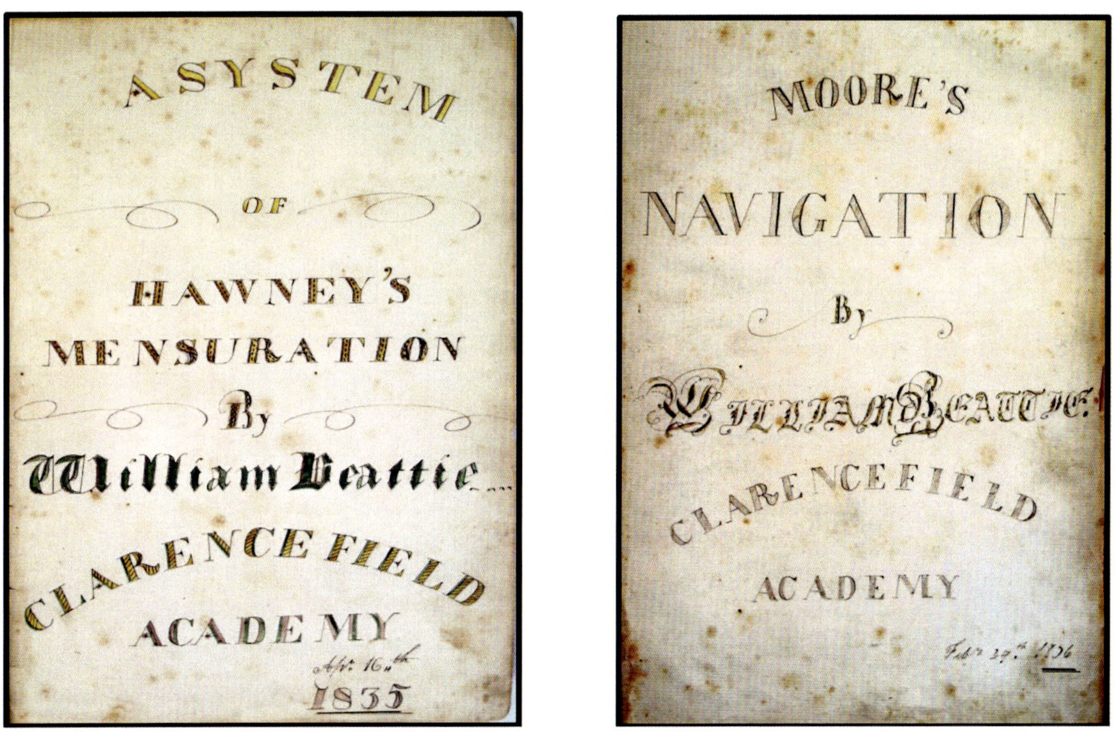

Figure 11.2. The front-page inscriptions to the two manuscripts.

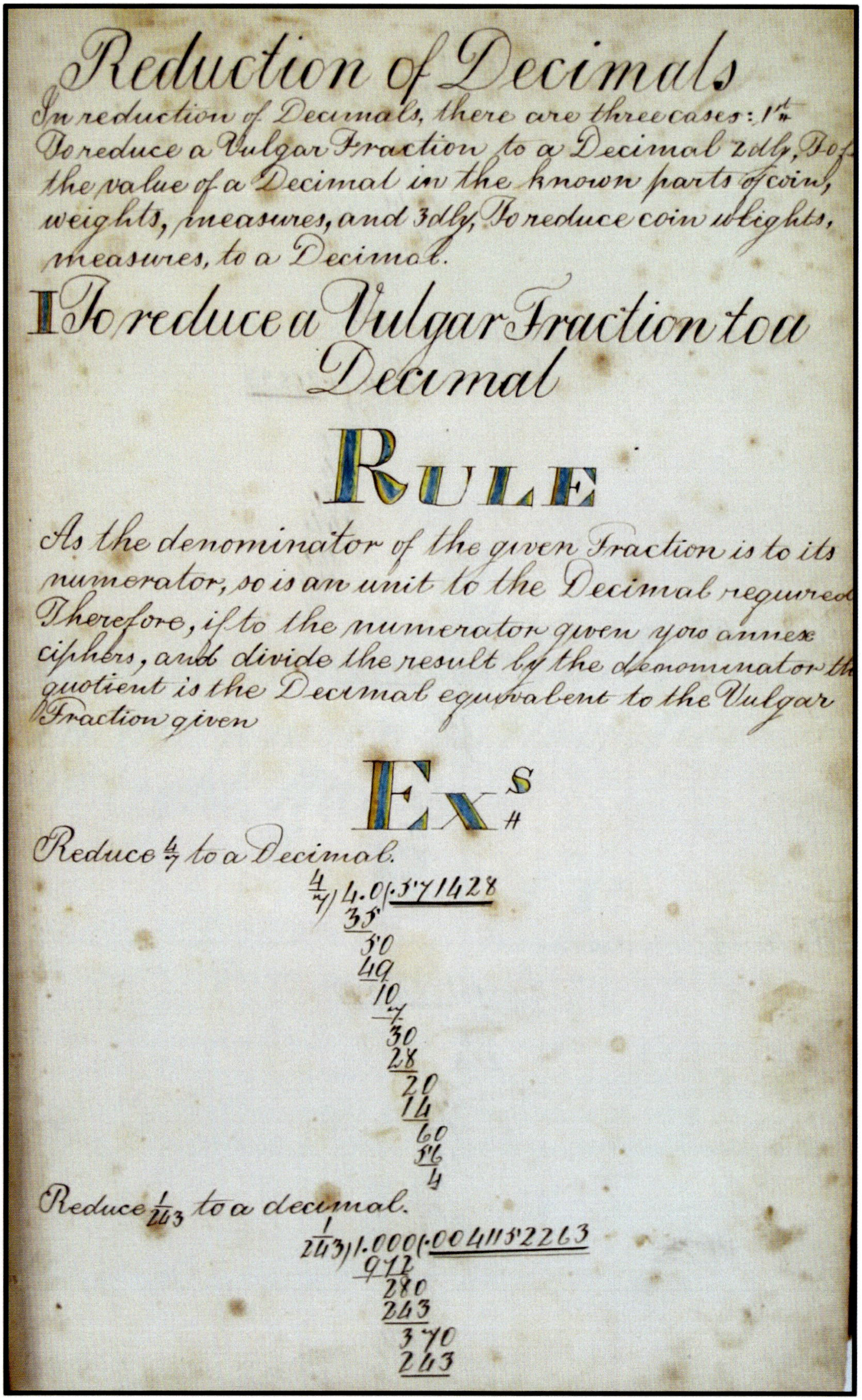
Reduction of Decimals

In reduction of Decimals, there are three cases: 1st To reduce a Vulgar Fraction to a Decimal 2dly, To find the value of a Decimal in the known parts of coin, weights, measures, and 3dly, To reduce coin weights, measures, to a Decimal.

I To reduce a Vulgar Fraction to a Decimal

RULE

As the denominator of the given Fraction is to its numerator, so is an unit to the Decimal required Therefore, if to the numerator given you annex ciphers, and divide the result by the denominator the quotient is the Decimal equivalent to the Vulgar Fraction given

Exs

Reduce $\frac{4}{7}$ to a Decimal.

7)4.0(.571428
35
50
49
10
7
30
28
20
14
60
56
4

Reduce $\frac{1}{243}$ to a decimal.

243)1.000(.0041152263
972
280
243
370
243

Figure 11.3. The first content page of Beattie's mensuration manuscript.

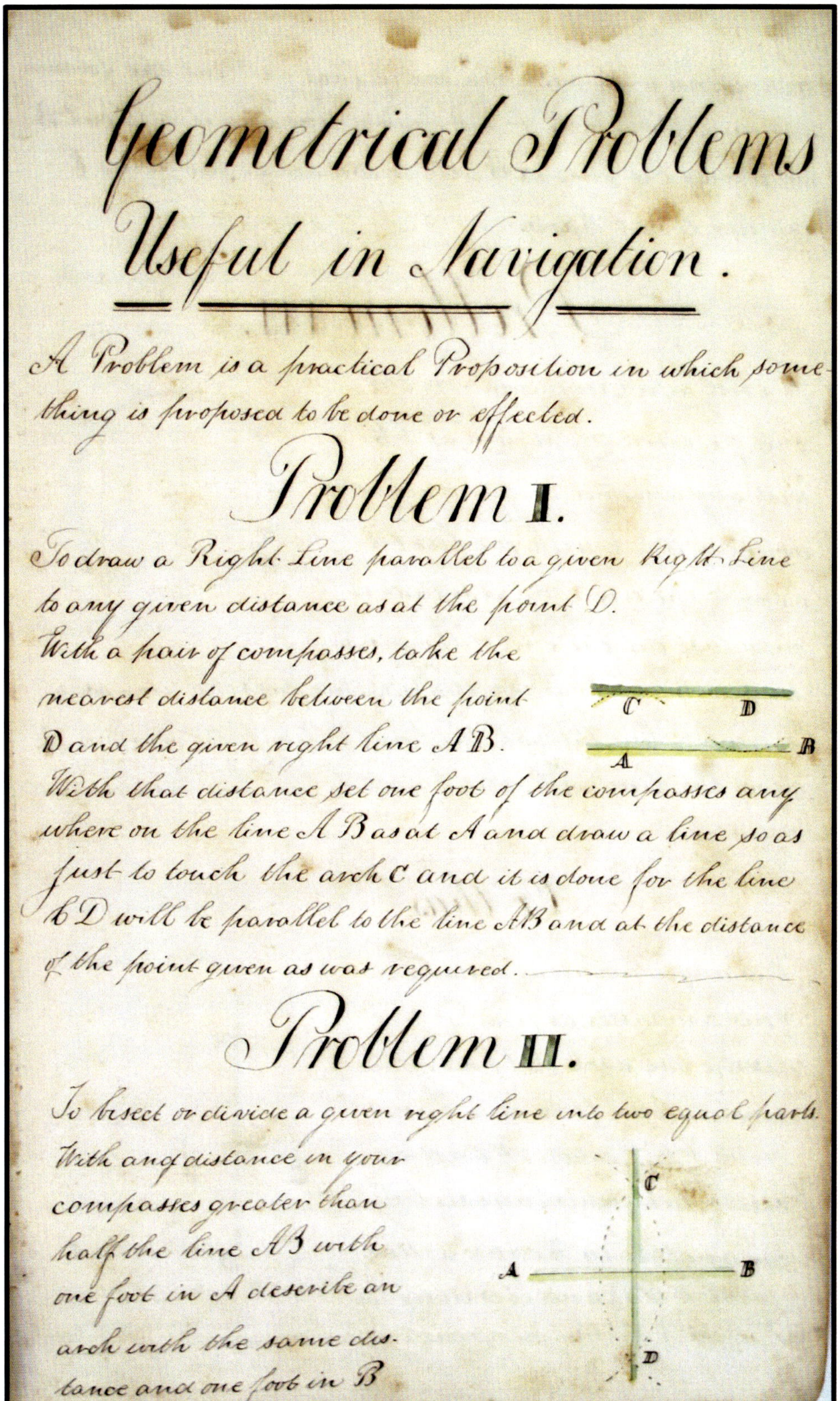

Geometrical Problems Useful in Navigation.

A Problem is a practical Proposition in which something is proposed to be done or effected.

Problem I.

To draw a Right Line parallel to a given Right Line to any given distance as at the point D.

With a pair of compasses, take the nearest distance between the point D and the given right line A B.

With that distance set one foot of the compasses any where on the line A B as at A and draw a line so as just to touch the arch C and it is done for the line C D will be parallel to the line A B and at the distance of the point given as was required.

Problem II.

To bisect or divide a given right line into two equal parts.

With any distance in your compasses greater than half the line A B with one foot in A describe an arch with the same distance and one foot in B

Figure 11.4. The first content page of Beattie's navigation manuscript.

Why might Beattie have dated the mensuration manuscript as April 16th, 1835, when he had originally prepared it at Clarencefield Academy, 25 years earlier? We conjecture that Beattie, as personal assistant to William IV, decided to prepare a navigation manuscript as a personal gift for his ailing "Sailor King." He also decided that it would be good to print Clarencefield Academy on the title page because, after all, William IV had been the Duke of Clarence. The theme closest to William IV's heart was navigation, and so Beattie chose to devote much time to preparing a beautiful handwritten manuscript on navigation, and to have the manuscript handsomely bound—in a manner fit for a king. Beattie had, on hand, the mensuration manuscript that he had prepared, probably around 1810, at Clarencefield Academy. Our hypothesis is that the mensuration manuscript served as a model for the navigation manuscript that Beattie prepared for the King. It made some sense for him to have his original mensuration manuscript lavishly bound, and the date "April 16th, 1835" indicated after the existing lettering for "Clarencefield Academy" at the very bottom of the title page. By contrast, the date, February 29th, 1836, on the title page of the navigation manuscript, was not crammed in at the foot of the page.

There is no evidence other than the date given at the foot of the title page that the mensuration manuscript was prepared in 1835. However, there is almost conclusive evidence that the navigation manuscript was prepared in 1835 and 1836. That evidence is in the form of "the log of a journal of a voyage from London to Madeira and Teneriffe in the Betsy of London, James B. Beck, Commander, Kept by William Beattie, Mate," which occupied 23 pages (pages 188–210) of the navigation manuscript. In that log it was clearly stated that the journey began on Jan 7th, 1836. Almost certainly, the journey summarized in the log never actually occurred, for it was common for navigation cyphering books to include details of fictitious journeys. Indeed, on the first page of the log (p. 188), there was an introduction stating that the purpose of the log was to exemplify how journals should be kept on journeys.

The navigation manuscript included 187 carefully-prepared pages *before* the log of the fictitious journey was begun, and so it can reasonably be assumed that the first entries in the manuscript were made well before January 7th, 1836. Thus it can be safely concluded that much of the navigation manuscript was prepared in 1835, or perhaps even before that.

William Beattie's *Mensuration* Manuscript as a Cyphering Book

From Figure 11.2 it can be seen that William Beattie linked his manuscript on mensuration to what he called a "system of Hawney's mensuration." The first edition of Hawney's *Complete Measurer: or, The Whole Art of Measuring* had been published in London in 1717. Hawney's book became a standard work, and its various editions were widely used for over a century.

We own a copy of the 1729 London edition of William Hawney's *The Compleat Measurer; or the Whole Art of Measuring,* and compared entries Beattie made in his manuscript with those in that edition of Hawney's textbook. We quickly found that although the rules as stated by Hawney were often repeated, word for word, in Beattie's manuscript, that was not always the case with the examples and exercises. For example, Hawney (1729) defined a task as "Let .7565 of a pound sterling be given to be reduc'd to shillings, pence and farthings" (p. 5), but Beattie asked, "What is the value of .7575 of a pound sterling?" The questions were similar but not identical, and Hawney and Beattie set out their solutions slightly differently:

Hawney (1729)

.7565
20
s ____
15.1300
12
d ____
1.5600
4
f ------
2.2400

Beattie

.7575
20
15.1500
12
1.8000
4
3.2000

Hawney (1729) showed, by multiplying by 20, that .7565 pounds was equal to 15.1300 shillings, and then without explaining what he was doing, he set aside the 15 and multiplied the .1300 by 12, to get 1.5600 pence; he then set aside the 1 penny and multiplied .5600 by 4 to get 2.2400 farthings. He did not state the answer to his problem, but it was: ".7565 pounds is 15 shillings, 1 penny and 2 farthings." He might have added that there would have been .24 of a farthing "left over."

Beattie adopted the same approach, but used an abbreviated form of presentation. Like Hawney, he did not specifically state his answer. Beattie's question ("What is the value of .7575 of a pound sterling?") was not as clearly worded as Hawney's question ("Let .7565 of a pound sterling be given to be reduc'd to shillings, pence, and farthings").

The pattern of copying definitions, rules and cases from Hawney's text—often word for word, but occasionally with minor modifications—and then changing the examples and exercises, continued throughout Beattie's manuscript on mensuration. Thus, for example, Beattie solved the problem of reducing 17.69 yards to the decimal of a mile in the following way:

1760 / 17.69
.010051136

That task could not be found in Hawney's (1729) text. Beattie offered no explanation for how he obtained his answer, or what his answer might mean. He did not say why he gave the answer to nine decimal places. Nor was there any comment on why it might be useful to write 17.69 yards as a decimal of a mile. Be that as it may, Beattie's response to the problem was neatly set out, and the level of penmanship in the writing of the numerals was high.

Figure 11.5 shows a typical page from Beattie's manuscript, and Figure 11.6 shows corresponding sections in Hawney's (1729) text (p. 14 and p. 18). Part of the cultural practice of preparing cyphering books was to devote time to making headings as beautiful as possible, and Beattie fulfilled that expectation with his heading for "Contracted Multiplication of Decimals." Although the rule statement was similar to the corresponding statement in Hawney (1729), it departed from it toward the end.

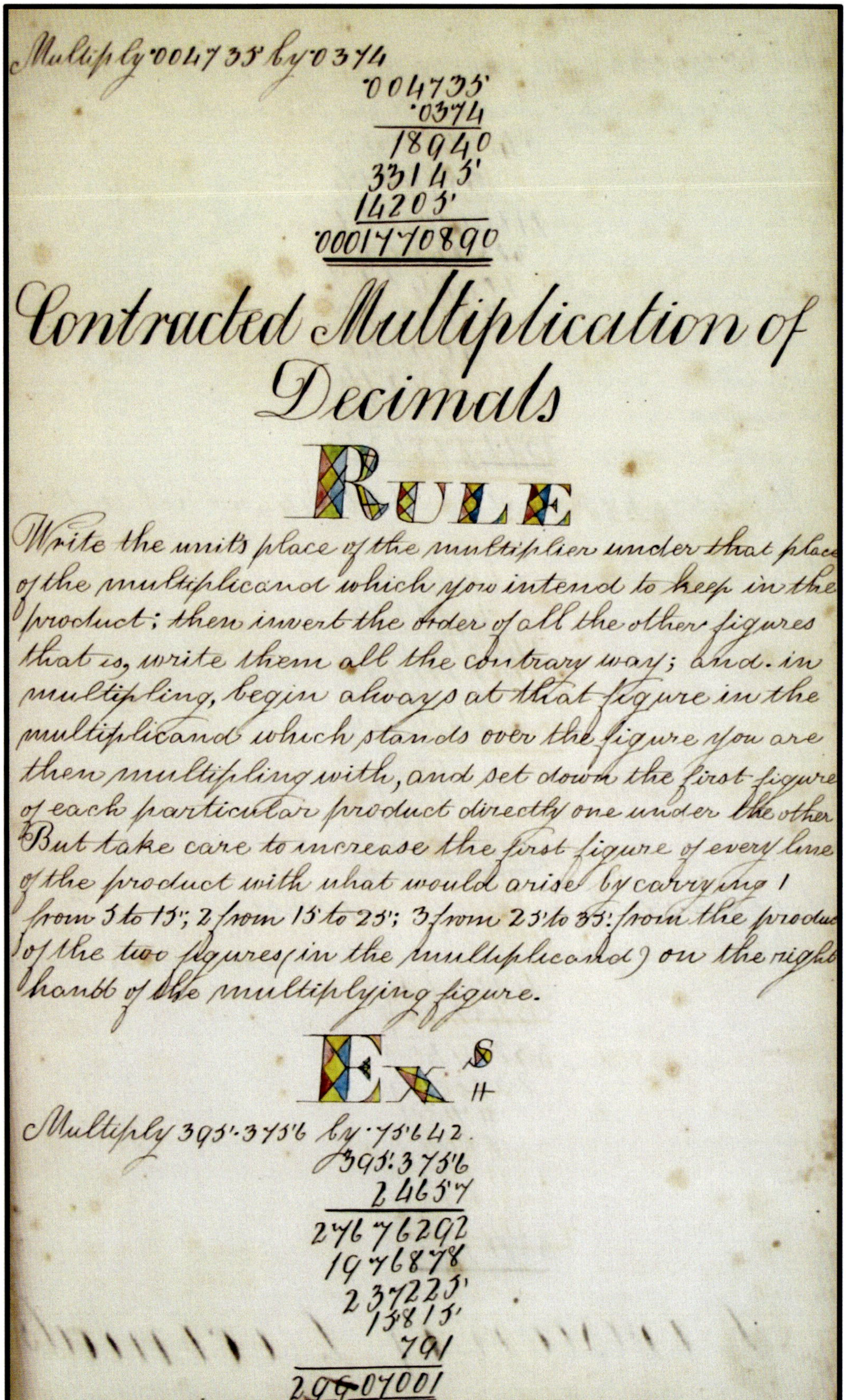

Multiply ·004735 by ·0374

·004735
·0374
18940
33145
14205
·0001770890

Contracted Multiplication of Decimals

Rule

Write the units place of the multiplier under that place of the multiplicand which you intend to keep in the product; then invert the order of all the other figures that is, write them all the contrary way; and in multipling, begin always at that figure in the multiplicand which stands over the figure you are then multipling with, and set down the first figure of each particular product directly one under the other But take care to increase the first figure of every line of the product with what would arise by carrying 1 from 5 to 15; 2 from 15 to 25; 3 from 25 to 35: from the product of the two figures (in the multiplicand) on the right hand of the multiplying figure.

Ex^s

Multiply 395·3756 by ·75642.

395·3756
24657
27676292
1976878
237225
13815
791
29907001

Figure 11.5. A typical page from Beattie's mensuration cyphering book.

14 *Contracted Multiplication.* Part I.

Contracted Multiplication of Decimals.

Becauſe in Multiplication of Decimal Parts and mix'd Numbers, there is no Need to expreſs all the Figures of the Product, but in moſt Caſes two, three, or four Places of Decimals will be ſufficient; therefore, to contract the Work, obſerve this following

RULE.

Write the Unit's Place of the Multiplyer under that Place of the Multiplicand, whoſe Place you intend to keep in the Product; then invert the Order of all the other Figures, that is, write them all the contrary Way. Then, in multiplying, always begin at that Figure in the Multiplicand which ſtands over the Figure you are then multiplying withal, and ſet down the firſt Figure of each particular Product directly one under the other; but yet a due Regard muſt be had to the Increaſe ariſing from the Figures on the right Hand of that Figure in the Multiplicand which you begin to multiply at. This will appear more plain by Examples.

18 *Contracted Multiplication.* Part I.

More Examples for Practice.

Multiply 395·3756 by ·75642, and prick off four Places in Decimals.

```
395·3756 the Multiplicand.
  24657. the Multiplyer revers'd.
-------
2767629 the Product by 7 increas'd with 7 × 6.
 197688 the Product by 5 increas'd with 5 × 5.
  23722 the Product by 6 increas'd with 6 × 7.
   1582 the Product by 4 increas'd with 4 × 3.
     79 the Product by 2 increas'd with 2 × 5.
-------
299·0699 the Product requir'd.
```

Figure 11.6. Parts of page 14 and page 18 from Hawney (1729).

The example that Beattie gave was exactly the same as one shown by Hawney. Interestingly, the task was to multiply 395.3756 by .75642, but both Hawney and Beattie seemed to multiply 395.3756 by .24657. A 21st-century reader would naturally think that both had made a mistake, but they had not—for they were applying the algorithm for "contracted multiplication of decimals." Between 1200 and 1850, many so-called "algorists" worked at developing methods for expediting calculations with Hindu-Arabic numerals (Ellerton & Clements, 2012; Swetz, 1987), and the now-long-forgotten contracted multiplication of numerals was one of their inventions.

Beattie prepared the main headings in attractive calligraphy, and then copied the corresponding rules, almost word for word, into his cyphering book. The worked examples and exercises that he included in his book (and there were many of them) were often originals, in the sense that they were not copied from Hawney's text.

Sometimes, even the examples were copied. For instance, calculations for the task of finding the area of an ellipse—which Hawney (1729) referred to as an "oval" or "ellipsis" (p. 124) and Beattie referred to as an "ellipsis"—whose longer diameter *AB* was 61.6 and shorter diameter *CD* was 44.4—were shown in Hawney's (1729) text (pp. 124–125), and these were repeated in Beattie's cyphering book. Even so, the statements of the rule for finding the area of an ellipsis were slightly different: Hawney (1729) stated, "Multiply the transverse diameter by the conjugate diameter, and that product by .7854" (p. 124); and Beattie wrote, "Multiply the transverse or longer diameter by the conjugate or shorter diameter, and that product by .7854." Neither Hawney nor Beattie gave any hint on where the .7854 came from.

Although Beattie copied much of the text from a later edition of Hawney's (1729) 384-page book, he sometimes used different worked examples and exercises, and he often modified Hawney's wording of rules and cases. He did not merely copy from the text, and in that sense his manuscript qualifies as a cyphering book (and not merely as a copybook). His impressive achievement, overall, was to produce a 466-page cyphering book, featuring above-average penmanship, attractive calligraphy, and colorful diagrams.

A brief section on surveying in the book included complex diagrams (see, for example, Figure 11.7, which shows a problem concerned with measuring the area of an irregular field).

A remarkable aspect of the mensuration cyphering book and of Hawney's (1729) text was that nowhere were logarithms used to support the often lengthy calculations that were needed. The calculations were carried out using standard four-operation algorithms for natural numbers and decimal fractions. By contrast, logarithms were used to expedite calculations throughout Beattie's navigation cyphering book.

William Beattie's Navigation Cyphering Book

We purchased the two cyphering books by William Beattie that we are examining in this chapter from an auction in 2011. As part of that same transaction we acquired, from the same seller, a copy of John Hamilton Moore's (1796) *The New Practical Navigator.* The seller assured us that the copy of Moore's famous text had been owned by William Beattie. According to its title page, Moore's (1796) text included "the substance of the examination every candidate for a commission in the Royal Navy, and officers in the honourable East India Company's service, must pass through, previous to being appointed." The book was "particularly recommended to the attention of the young gentlemen designed for, or belonging to, the sea" (p. iii).

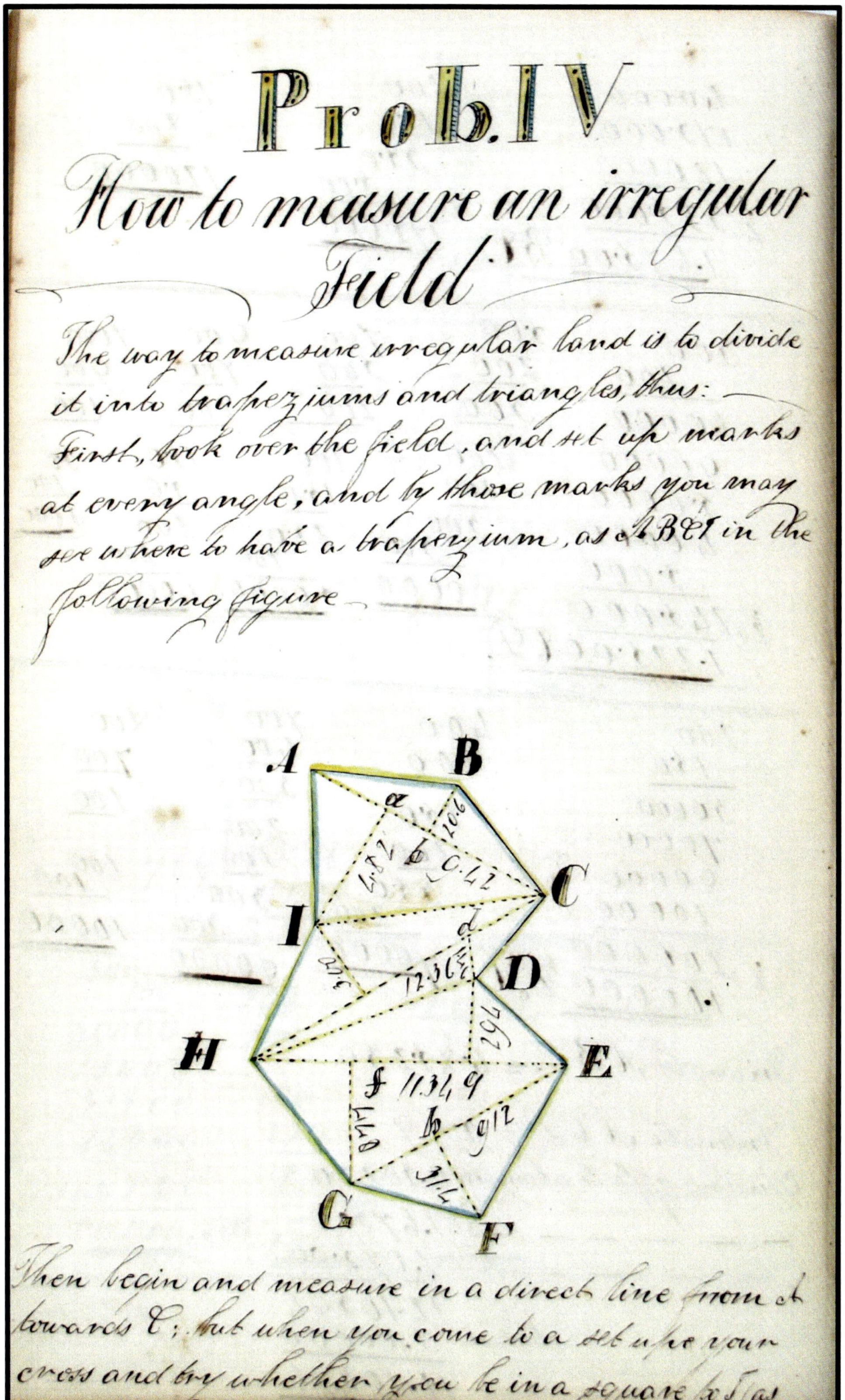

Prob. IV

How to measure an irregular Field

The way to measure irregular land is to divide it into trapeziums and triangles, thus: — First, look over the field, and set up marks at every angle, and by those marks you may see where to have a trapezium, as ABCI in the following figure —

Then begin and measure in a direct line from A towards C; but when you come to a set up your cross and try whether you be in a square to I as

Figure 11.7. How to measure an irregular field (from Beattie's cyphering book on mensuration).

Of special significance to this chapter, perhaps, was the description given of John Hamilton Moore, as "teacher of navigation, hydrographer and chart-seller to His Royal Highness, the Duke of Clarence" (p. iii). So, even in 1796 Moore seemed to have had links with the future King of England. That explains why, perhaps, 40 years later William Beattie was willing to spend so much time preparing a special cyphering book, based on Moore's text, for an ailing William IV.

The Section on Geometry in the Navigation Cyphering Book

Like many other navigation cyphering books, this cyphering book began with statements of "geometrical problems useful in navigation" (Ellerton & Clements, 2012). Most of these problems required elementary Euclidean constructions with a straight-edge and compass (e.g., "To bisect or divide a given right line into two equal parts," and "To erect a perpendicular on the end of a given right line"). In the cyphering book, the wording of these problems, the methods by which they could be solved, and the associated illustrations, were taken directly from Moore's (1796) text.

In her history of geometrical education in North America, Nathalie Sinclair (2008) argued that geometry curriculum in 18th- and early-19th-century North American colleges "*was* literally the *textbook* [original emphasis], with university students memorizing one proposition after another in the given sequence" (p. 15). Evidence from Beattie's navigation cyphering book suggests that the same statement might apply to geometrical education in Great Britain in the first half of the nineteenth century. The words in Beattie's navigation cyphering book are often the same as those in the corresponding paragraphs of Moore's (1796) text. But a careful examination of the associated diagrams shows clearly that all straight-edge and compass constructions were carefully carried out, presumably by Beattie himself. Arcs from compass constructions remain clearly visible in all of the tasks.

The Sections on Logarithms and Trigonometry

In Chapter 9 of this present volume (the chapter titled "Bound for Botany Bay?") we found that the midshipman who prepared the navigation cyphering book under examination in that chapter did not seem to understand the theory associated with logarithms and the application of that theory to trigonometry. In particular, the author of that manuscript did not appreciate that the radius of a circle had to be 10^{10} if it were to generate values of sines of angles that could be as high as 10^{10}. This lack of understanding of logarithms and how they related to 18th- and 19th-century trigonometry resulted in the unknown author of that chapter making mathematically incorrect statements. We surmised in that chapter that the unknown author had used an earlier edition of Moore's textbook as his reference for setting out numerical calculations based on logarithms.

When preparing his navigation cyphering book, William Beattie (and, perhaps, King William IV) seemed to have difficulty linking logarithms and trigonometry. That was probably because nowhere in Moore's (1796) text could we find a statement that the radius of the circle for which trigonometric functions were defined had to be 10^{10} length units. A procedure is given for using Moore's extensive set of mathematics tables to show, for example, that the logarithm (base 10) of sine $28^{\circ}17'$ was equal to 9.68029 (copied from Moore, 1796, p. 26), but there is no clear justification why that procedure was, from a mathematical point of view, soundly based.

It was not a matter of surprise, then, that we were able to find evidence that Beattie merely copied sections from Moore's text without understanding what he was writing about. Thus for example, Beattie wrote that "in any right-angled plane triangle, if the hypothenuse be made the radius of a circle the other two sides or legs will be the tangent of its opposite angles and the hypothenuse." This was, presumably, based on a statement in Moore (1796), but in that statement

Moore had correctly written "in any right-angled plane triangle, if the hypothenuse be made the radius of a circle, the other two sides, or legs, will be the sines of their opposite angles" (p. 34), rather than the incorrect "... the tangent of its opposite angles." If one does not understand what one is copying, it is easy to make a copying error, and in this case the introduced error prevented an understanding of the meaning of the text.

But, usually, Beattie did not merely copy Moore's text without thinking about what he was doing. On page 41 of Moore's book, for example, seven problems were set "to exercise the learner in trigonometry," and although Moore provided the answers to the problems he did not show how those answers were obtained. In his navigation cyphering book, however, Beattie showed complete working for all seven problems. The same is true of other exercises set by Moore (see, for example, exercises on plane sailing on pages 64–65, and exercises on Mercator's sailing on p. 111). Sometimes, too, Beattie took much time to make beautiful diagrams—more elaborate and obviously more colorful than the corresponding diagrams in Moore's text.

Sections on the Various Types of Sailing

Moore's textbook included sections on various types of sailing—plane, traverse, middle latitude, and Mercator's—and it was not surprising, therefore, that these types were dealt with in Beattie's navigation manuscript. Indeed, there are about 80 handwritten pages on the types of sailing in Beattie's book. Throughout these sections, the calligraphy is of a medium level, the handwriting neat, and mature, and diagrams clear but basic. There is a sense that although Beattie was dutifully copying much of the text from Moore's book, the desire for excellence in presentation had subsided.

However, some of the exercises that Moore set without offering solutions were dealt with in some detail in Beattie's manuscript. Figure 11.8 shows a typical page. The exercise which was being addressed in Figure 11.8 was recorded in Beattie's book as: "Requiring the bearing and distance of Hang. Cliff in Shetland, in Lat. 60.7 N and Long. 50 W. and the north cape of Lapland in Lat. 71°10 N. Long 26°1' E." On page 111 of his book, Moore set this task as an exercise, and although answers were given, the working showing how to arrive at those answers was not given. By contrast, Beattie devoted almost two pages to showing relevant working for the exercise. Although he managed to obtain Moore's answers, his explanations were difficult to follow.

The working shown in Figure 11.8 makes it clear that Beattie did not merely copy *all* of Moore's text. In that sense, his manuscript still qualified as a cyphering book, and was not purely a copybook. That said, it is likely that anyone reading what Beattie wrote would not follow the mathematics presented, because Beattie's transcription of the task was inaccurate. The two versions (Moore's statement and an unedited version of Beattie's transcription) are shown in Figure 11.9.

Notice that Beattie's statements of the latitude and longitude of Hang. Cliff were slightly different from those given by Moore. Also, the longitudes given for the North Cape of Lapland were not the same. Beattie corrected his transcription errors in his subsequent working, but confusion had been introduced by the difference between what he wrote in the statement of the problem and in his working toward finding solutions to it. Furthermore, in his use of logarithms, Beattie was cryptic. He did arrive at correct answers—but it would have been difficult for another person to follow what he had written.

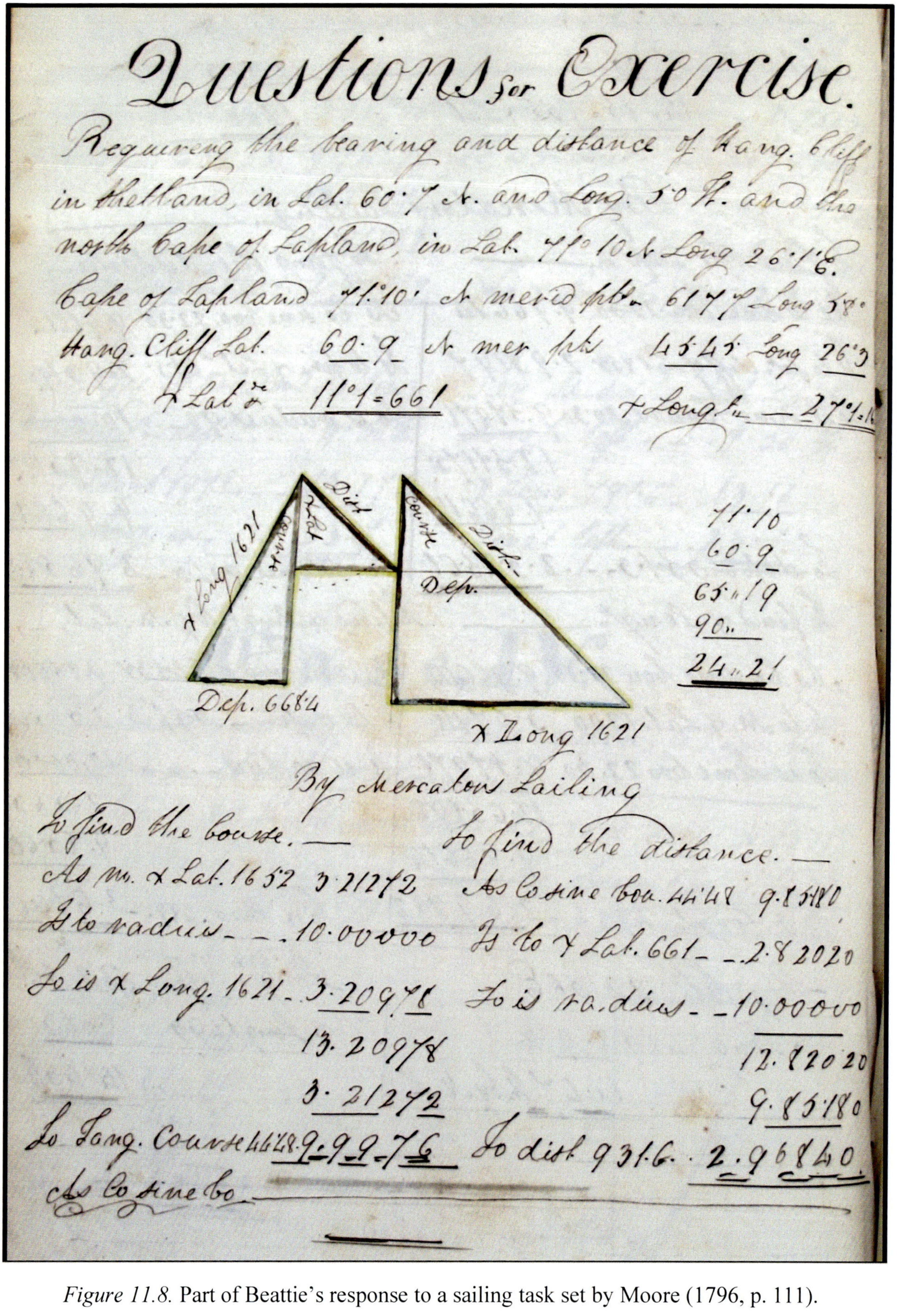

Questions for Excercise.

Requireing the bearing and distance of Hang. Cliff in Shetland, in Lat. 60°.7 N. and Long. 5°0 W. and the north Cape of Lapland, in Lat. 71°10 N Long 26°1 E.

Cape of Lapland 71°10' N merid pts 6177 Long 58°
Hang. Cliff Lat. 60.9 N mer pts 4545 Long 26°3
D Lat 11°1=661 D Long 27°4=1...

71°10
60.9
65"19
90"
24"21

By Mercators Sailing

To find the Course.—		To find the distance.—	
As m. D Lat. 1652	3.21272	As Cosine Cou. 44°48	9.85180
Is to radius	10.00000	Is to D Lat. 661	2.82020
So is D Long. 1621	3.20978	So is radius	10.00000
	13.20978		12.82020
	3.21272		9.85180
To Tang. Course 44°48	9.99706	To dist 931.6	2.96840

As Co sine Co

Figure 11.8. Part of Beattie's response to a sailing task set by Moore (1796, p. 111).

Moore's (1796) Statement of the Task:	*Beattie's (Unedited) Version of the Task:*
Required the bearing and distance of Hang. Cliff in Shetland, in latitude 60°9' N, and long. 58' W. and the North Cape of Lapland, in lat. 71°10' N. long. 26°3' E.?	Requireng the bearing and distance of Hang. Cliff in Shetland, in Lat. 60.7 N and Long. 50 W. and the north Cape of Lapland in Lat. 71°10 N. Long 26°1' E.

Figure 11.9. Two versions of what was intended to be the same task.

We gained the impression that toward the end of his navigation cyphering book Beattie had lost interest in the navigational issues with which he was dealing and that all he wanted to do was to fulfill a promise—either to himself or to his King—to complete the book in a satisfactory manner. Although a medical doctor, Beattie was, fundamentally, a literary scholar, with interests in the classics, literature, poetry, religion and art (Austin, 1885; Beattie, 1831). The only time we sensed that he made a large effort to make something really beautiful was when he set out to copy "a chart from England to the Cape Verd Island" from Moore (1796, p. 119). Moore's version of the chart was in black and white and had a basic map-like appearance; Beattie's water-colored version (see Figure 11.10) was something which must have taken many hours to complete.

The Journal

In the navigation cyphering book there are 22 handwritten pages providing commentary on how to keep a journal. Then there is a 23-page entry on "A Journal of a Voyage from London to Madeira and Teneriffe in the Betsy of London, James B. Beck Commander kept by William Beattie Mate."

Some preliminary notes in the manuscript, made immediately before the summary notes for the "journal," were almost identical to the text provided by Moore (1796) before his 18-page summary of a "Journal of a Voyage from London to Madeira and Teneriffe in the Frances of London, William Johnson, Commander, kept by Joseph Mills, Mate." The only real difference between the entries in the two journals for the journeys were the dates (Beattie's "journey" began on January 7th, 1836, and William Johnson's on April 8th, 1795). Undoubtedly, except for the dates, details in Beattie's journal were taken from Moore (1796).

It would be easy to be critical of the fact that Beattie more or less copied the journal entries from Moore (1796) into his own navigation manuscript. However, it should be recalled that on the title page of his manuscript Beattie acknowledged that in the process of preparing his navigation manuscript he would be relying heavily on Moore's textbook. Furthermore, the purpose for showing the "journal" was to provide a model of a journal for prospective navigators. Seen from that perspective, there was nothing wrong with Beattie copying out the journal entries provided by Moore. In any case, hopefully King William IV, a former sea captain, would have had experience with such journals—although these would have been done by his navigators, and not by himself.

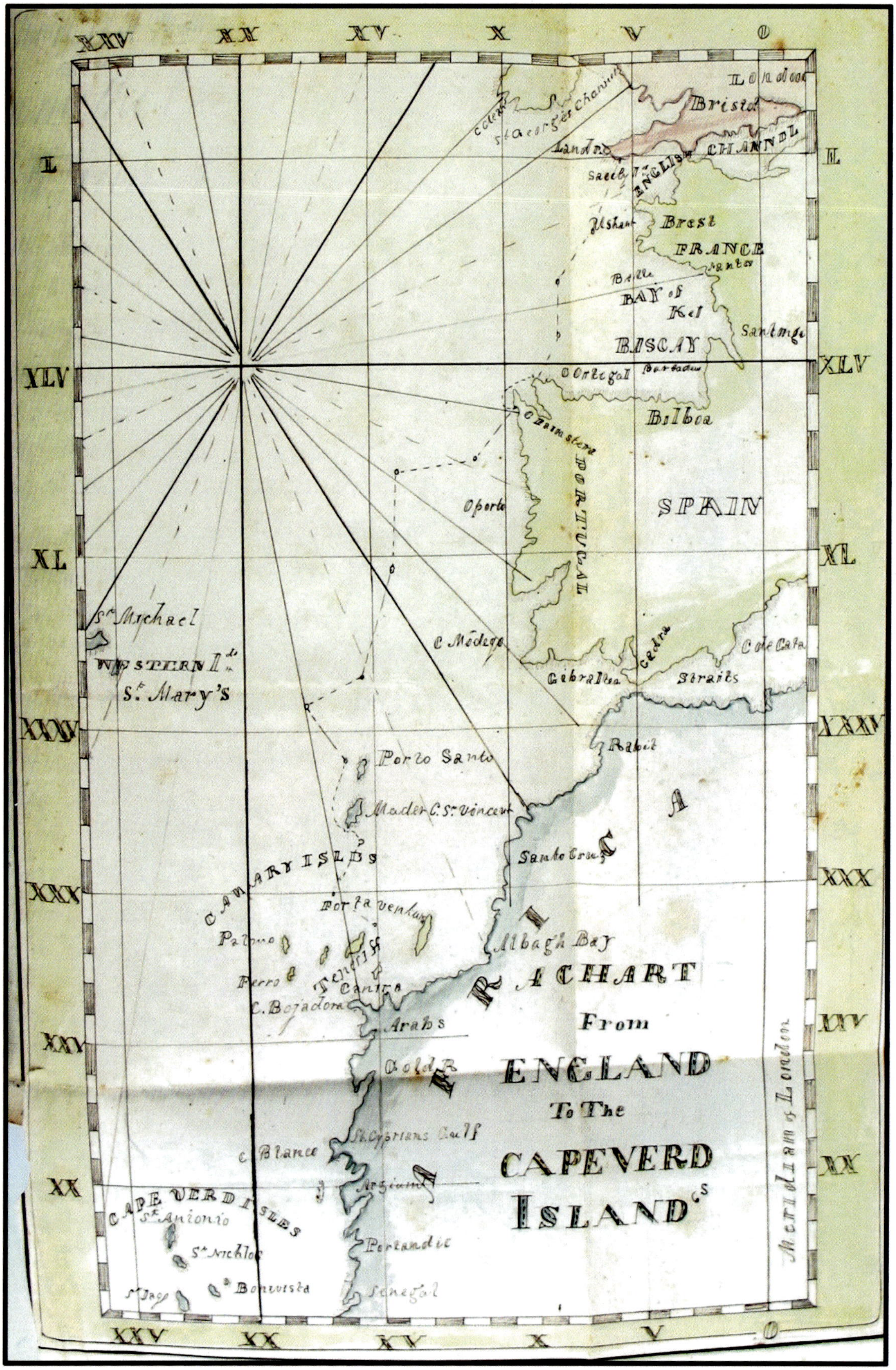

Figure 11.10. William Beattie's "A chart from England to the Cape Verd Island."

Concluding Comments

Of all the cyphering books on which we have focussed our attention, in the various chapters of this book, the Beattie manuscripts are closest to being pure copybooks—in the sense that large sections were copied from texts by William Hawney and John Hamilton Moore. Nevertheless, in both the mensuration and the navigation manuscripts Beattie did record solutions to numerous problems for which the solutions were not given by Hawney (1729) or Moore (1796). Because of that, we have been happy to regard the manuscripts as cyphering books.

The very existence of Beattie's manuscripts provide testimony to the power of the belief among those inclined towards higher education in the cyphering era that the process of using a quill to copy information from a known authority (such as an older cyphering book, or a well-regarded textbook like Moore's (1796) *The New Practical Navigator*) into a book had much educational merit (Ellerton & Clements, 2011). Thus, for example, we own an 80-page handwritten manuscript prepared by different members of the Terhune Family, of New Jersey, around the middle of the 18th century. The text in this manuscript was copied, in high calligraphic form, from John Love's (1688) *Geodaesia, or the Art of Surveying and Measuring of Land, Made Easie, ...* Love's book was commonly available in and around New Jersey at the time, but members of the Terhune family thought it worth the huge effort needed to copy it, by hand, with a quill (Ellerton & Clements, 2012). We conjecture that in his final years King William IV, the Sailor King, regretted that he had never prepared his own navigation cyphering book, and that his private secretary, and friend, William Beattie, decided to honor his King by preparing a navigation cyphering book for him.

Whether that conjecture is true will probably never be known. But there can be little doubt that the effort that Beattie put into preparing the navigation cyphering book was immense—especially if one takes into account that, fundamentally, Beattie was much more inclined toward literary and artistic creativity (e.g., Beattie, 1855) than toward matters concerned with mathematics, or science, or navigation. In the early 1800s, cyphering books were valued greatly as personal and educational artifacts, and somehow Beattie, in his early forties, found the will and dedication to prepare the navigation cyphering book.

Jeremy Kilpatrick (2012), in commenting on the cyphering approach to learning mathematics, stated that the approach "has both an extensive history and a complex rationale" (p. vi). According to Kilpatrick, in the approach, "the learner was to become an independent problem solver by learning to identify problems of various types, learning the rules for their solution, solving such problems, having each solution checked by a tutor or teacher and, only when the solution was correct, copying that solution into a cyphering book using exemplary calligraphy" (p. vi). Kilpatrick went on to say that "the handwriting process was not meant to be only copying; it was to be an accompaniment to thinking. The learner was not simply inscribing the solution onto paper; he or she was inscribing it into memory" (p. vi). In this chapter, the rich description that Kilpatrick offered of the cyphering approach most likely would not apply to the learning that might have been generated during the preparation of Beattie's navigation cyphering book—because, probably, that manuscript was prepared under extraordinary circumstances.

It is interesting to reflect on the possibility that William Beattie's mensuration cyphering book was prepared around 1810, when Beattie was attending the Clarencefield Academy, and that Beattie remembered that he had learned much in the course of preparing that manuscript. And, so, he persuaded himself to prepare a navigation cyphering book as his special gift to a dying friend.

Whatever the circumstances surrounding their preparation, it would be difficult to deny that these two cyphering books, one dedicated to mensuration and one to navigation, were extraordinary. Furthermore, they testified to the enormous educational and cultural values that even some members of the upper classes associated with the preparation of cyphering books.

Acknowledgement

We gratefully acknowledge the encouragement and assistance of Peter and Vivienne Murray, of Melbourne, Australia, in the preparation of this chapter.

References

Austin, P. B. (1885). William Beattie (1793–1875). In L. Stephen (Ed.), *Dictionary of national biography* (Vol. 4, pp. 25–27). London, United Kingdom: Smith, Elder & Co.

Beattie, W. (1831). *Journal of residence in Germany, written during a professional attendance on their Royal Highnesses the Duke and Duchess of Clarence ... during their visits ... 1822, 1825, and 1826.* London, United Kingdom: British Library.

Beattie, W. (Ed.). (1855). *Life and letters of Thomas Campbell.* New York, NY: Harper & Brothers, Publishers.

Ellerton, N. F., & Clements, M. A. (2011). Unique mathematics books from a lost tradition. *The Guild of Book Worker's Journal, 11,* 28–39.

Ellerton, N. F., & Clements, M. A. (2012). *Rewriting the history of school mathematics in North America 1607–1861.* New York, NY: Springer.

Hawney, W. (1729). T*he compleat measurer; or, the whole art of measuring* (3rd ed.). London, United Kingdom: J. and J. Knapton.

Kilpatrick, J. (2012). Foreword. In N. F. Ellerton & M. A. Clements, *Rewriting the history of school mathematics in North America 1607–1861* (pp. v–vi). New York, NY: Springer.

Love, J. (1688). *Geodaesia, or the art of surveying and measuring of land, made easie, showing by plain and practical rules how to survey, protract, cast up, reduce or divide any piece of land whatsoever; with new tables for the ease of the surveyor in reducing the measures of land. Moreover, a more facile and sure way of surveying by the chain than has hitherto been taught. As also, how to lay out new lands in America, or elsewhere: and how to make a perfect map of a river's mouth or harbour, with several other things never yet published in our language.* London, United Kingdom: John Taylor.

Moore, J. H. (1796). *The new practical navigator, being an epitome of navigation, explaining the different methods of working the lunar observations, and all the requisite tables used with the nautical almanac, in determining the latitude and longitude, and keeping a complete reckoning at sea; illustrated by proper rules and examples; the whole exemplified in a journal kept from England to the island of Teneriffe: also, the substance of the examination, every candidate for a commission in the Royal Navy, and Officer in the Honourable East India Company's service, must pass through previous to their being appointed; this with the sea terms, are particularly recommended to the attention of all young gentlemen designed for, or belonging to the sea.* (12th ed.). Tower Hill, United Kingdom: E. Law.

Sinclair, N. (2008). *The history of the geometry curriculum in the United States.* Charlotte, NC: Information Age Publishing, Inc.

Somerset, A. (1980). *The life and times of William IV.* London, United Kingdom: Weidenfeld Nicholson.

Swetz, F. (1987). *Capitalism and arithmetic: The new math of the 15th century.* La Salle, IL: Open Court.

Ziegler, P. (1973). *King William IV.* London, United Kingdom: Harper & Row.

Chapter 12

Lessons from Extraordinary Cyphering Books

Abstract: Six lessons about school mathematics, and its history, are drawn from the analyses of the 11 extraordinary cyphering books. These lessons relate to: (a) the intended curriculum that the *abbaco* and Sacrobosco traditions, and the British and North American textbook authors, defined for school mathematics during the period 1600 through 1880; (b) the implemented curriculum, as evidenced by entries in cyphering books; (c) the lack of hard data with respect to the attained curriculum—that is to say, with respect to what was actually learned by students during the cyphering era; (d) the remarkable value that many students placed upon their cyphering books; (e) the ethnomathematical circumstances which surrounded the preparation of the extraordinary cyphering books; and (f) the educational significance of qualitative differences between British and North American cyphering books.

Key Words: *Abbaco* curriculum; *attained* mathematics curriculum; end of the cyphering era; ethnomathematics; history of mathematics; history of school mathematics; implemented mathematics curriculum; intended mathematics curriculum; Irish ballads; Irish education; Irish poetry; lessons from history; Sacrobosco.

Differences between the 11 extraordinary cyphering books studied in Chapters 2 through 11 have been so great that one might wonder whether *any* general lessons for mathematics education might sensibly be gleaned from them. One might expect that one can only *generalize*, legitimately, about what *normally* happens by studying *typical* artifacts and scenarios arising out of *representative* circumstances. But such a point of view presupposes that there is some agreement on the meaning of words such as "normal," "typical," "generalize," and "representative." On the other hand, one could argue that one can learn much about normality by studying extremes.

Lessons from History?

Despite the challenge inherent in George Santayana's (1905) famous statement that "those who cannot remember the past are condemned to repeat it" (p. 284), there are many scholars who dispute the premise that it makes sense for historians to attempt to identify lessons from history. The following two quotations from a report in 2005 by James H. Sheehan, then President of the American Historical Association, are interesting with respect to that issue:

> Does the past provide lessons for the present, guidance for the future? In addition to telling us who we are, does history help us know what to do? I suspect that not many of us still share the confidence expressed by Lord Acton in his inaugural lecture of 1895 that "the science of politics is the one science that is deposited by the streams of history, like the grains of gold in the sand of a river; and the knowledge of the past, the record of truths revealed by experience, is eminently practical, as an instrument of action and a power that goes to making the future" (*Essays on Freedom and Power*, New York, 1957, pp. 25–26). Many of us doubt that the facts of the past are as discrete and quickly

Nerida F. Ellerton and M. A. (Ken) Clements, *Abraham Lincoln's Cyphering Book and Ten other Extraordinary Cyphering Books*, DOI 10.1007/978-3-319-02502-5_12,

> recognizable as grains of gold glittering in the sand; nor are we sure that the truth of past experience can so easily be deployed as an instrument for action. (p. 5)

Sheehan (2005) continued:

> However professionally skeptical we may be about learning from the past, there is no doubt that we try to do it all the time. We constantly tell stories about the past to our students, friends, children—and to ourselves—stories that are supposed to convey moral and practical lessons about how to behave. Physicians compile histories of their patients' diseases in order to make diagnoses and determine therapies. Military units write after action reports that provide the basis for assessing the reasons for success or failure. And, of course, historical lessons are part of every political discussion and debate. Again and again, our political leaders use the past to warn, admonish, and inspire the public; to criticize their opponents; and to justify their own policies. Historical analogies, comparisons, and metaphors are all around us; they are a source of collective wisdom on which we must rely. It is unlikely that we could live without them even if we wanted to. (p. 5)

Sheehan then went to say that any "alleged lessons of the past" should be subjected to "persistent critical scrutiny" (p. 5).

So, it is with some trepidation that we offer this final chapter as an initial attempt at identifying lessons from the 11 extraordinary cyphering books that we have examined. Our strategy will be to identify six "lessons" and *hope* that they *are* scrutinized. We contend that much greater attention than ever before needs to be given to the effects of the cyphering tradition and, in particular, to our claim that elements in present-day school mathematics have their origins in that tradition.

Six Lessons from the Extraordinary Cyphering Books

We claim that so far as the history of school mathematics is concerned there are important lessons to be gleaned from our study of extraordinary cyphering books. In this final chapter we identify six lessons from and about the history of school mathematics that we believe *we* learned while studying, and reflecting on, the 11 extraordinary cyphering books.

Stated succinctly, the six lessons are:

1. The *intended* mathematics curricula that were, somehow, "behind" individual cyphering books were *not* always the same as the *implemented* curricula and, indeed, it was often the case that the two kinds of curricula were quite different.
2. Although what appeared in the cyphering books did not provide us with full details with respect to the *implemented* curricula, analyses of those manuscripts offered strong evidence for what those preparing them were actually doing as they studied school mathematics.
3. It is very difficult to ascertain *attained* curricula merely by examining cyphering books. That is largely because it is impossible to determine the extent to which students *reflected* on what they entered in their cyphering books. It is also due to ambiguities associated with the meaning of words like "understanding."
4. The personal "sacredness"—for the want of a better term—of cyphering books was always impressive. Most of the extraordinary cyphering books came to be regarded as precious objects by those preparing them, with the students believing that the manuscripts would become their own personal guidebooks for life. It is difficult for 21st-century readers to

appreciate the fact that those preparing cyphering books often came to think of them as, next to the Bible, the most valuable of all the books they would ever own.

5. The contextual and ethnomathematical circumstances surrounding the preparation of cyphering books had a large impact on the students who prepared them, and especially on what they thought about and valued as they prepared the manuscripts.
6. During the nineteenth century the cyphering tradition in Great Britain evolved in noticeably different ways from the cyphering tradition in the United States of America. By the 1840s, British cyphering books tended to have a different physical appearance from cyphering books in the United States, and a significant part of the mathematics in the British cyphering books was presented in ways which were different from how the mathematics was presented in North American cyphering books. There was much more direct copying of material in Great Britain than in North America, and the calligraphy and penmanship in Great Britain appeared in more stereotypical forms.

We shall now elaborate on each of the above "lessons." Then, we will conclude the book with some final comments, and suggestions for further research.

Lesson 1: The Intended and Implemented Curricula were Different

Toward the end of the eighteenth century it was common for colleges in the United States of America to devise mathematics-entrance requirements for prospective students. Thus, for example, in 1786 Columbia College, in New York, required students to have completed arithmetic "to the rule of three," and in 1800 Yale College, in Connecticut, expected its beginning students to know "rules of vulgar fractions" (Broome, 1903).

The goal of reaching the rule of three was a powerful motivator to those who prepared all seven *arithmetic* cyphering books that we have examined in this volume. It was particularly evident in the early (c. 1667) North American manuscript (see Chapter 2), which dealt with very little mathematical content beyond definitions and terminology yet still, magically, and all of a sudden, reached the rule of three. Thomas Prust (Chapter 3) devoted much time to the rule of three, as did each of Martha and Elisabeth Ryan (Chapter 4), Mary Walters (Chapter 5), and Abraham Lincoln (Chapter 6). Three of the British cyphering books—by the unknown author (c. 1791) of the manuscript described in Chapter 9, by Charles Page (Chapter 10), and by William Beattie (Chapter 11)—were of the navigation variety, and although these students made use of the rule of three in their calculations with logarithms, the manuscripts were not working towards a treatment of the rule of three, presumably because navigation students were expected to know that rule from the outset. The other two British cyphering books studied in this book—those by Thomas Dixson (Chapter 3) and George Bickham (Chapter 8)—were principally concerned with arithmetic, and seemed to be locked into the standard elementary *abbaco* sequence which culminated in rule-of-three considerations.

Throughout the period 1200–1850, in Western Europe and subsequently in the colonies established in North America, anyone who had cyphered to the rule of three—sometimes referred to as the single rule of three, or the rule of three direct, or the golden rule—was regarded as an efficient reckoner who was qualified to be a teacher. Those few who studied *beyond* that rule, in topics like the rule of three inverse, or the double rule of three, or any of simple interest, compound interest, discount, loss and gain, barter, alligation, fellowship, false position, common fractions, decimal fractions, involution and evolution, mensuration, arithmetical and geometrical progressions, and permutations and combinations, could claim to be advanced reckoners who were ready for employment as applied problem solvers (Ellerton & Clements, 2012; Karpinski, 1925). At a time when mechanical calculation devices and tables of logarithms were not available to "ordinary" people, those who could demonstrate advanced written or mental calculation skills, and could apply these using the rule of three, were few and far between.

Our analyses of the various arithmetic cyphering books has made it clear that having cyphered to the rule of three was a highly ambiguous standard. That is to say, knowing that someone had cyphered to the rule of three did not mean much if one wanted to predict the kind of mathematics which that person had studied. Certainly, in North America many implemented curricula did not correlate well with what appeared in standard arithmetic textbooks, which is hardly surprising considering that most students of arithmetic did not own, or have ready access to, textbooks (Ellerton & Clements, 2012). Thus, for example, although the content of the circa 1667 cyphering book was totally different from that for Thomas Dixson's circa 1630 cyphering book, both of the students who prepared those books could have claimed to have cyphered to the rule of three. Unlike the other 10 extraordinary cyphering books, Mary Walters' book was delightfully concerned with introducing arithmetic through Irish ballads, poetry and song, yet she also cyphered to the rule of three. Abraham Lincoln, showing much determination, cyphered to the rule of three—for him it was very important that he not only got all of his calculations right but also that, as he cyphered, he definitely knew what he was doing.

A lesson for historians of education is that it is unwise to attempt to predict what actually happened when students studied arithmetic during the cyphering era merely by examining the contents of various arithmetic textbooks. Different students often approached the same topic in different ways. Often, the intended curriculum, as set out in a textbook, did not match the implemented curriculum in ways that might have been expected.

Lesson 2: Entries in Cyphering Books Offer the most Insight into the Implemented Curriculum

Lao Genevra Simons (1936) was one historian of mathematics education who recognized the central importance of cyphering books (which she called "student notebooks") for providing evidence with respect to implemented curricula. She wrote:

> A great deal has already been said about the custom of keeping student notebooks during this period of difficulty in obtaining books from England and of printing books in the colonies. If all the notebooks now hoarded by descendants of graduates of the early American colleges or lying neglected and forgotten in attics and closets, if all these notebooks could be presented to the several college libraries or historical societies, the history of early American education would be greatly enriched. In these notebooks, there is found the content and scope of the curriculum of the day in evidence that is unmistakable. (p. 588)

One can find many instances in which historians have examined various forms of data—like for example, textbooks, summaries of intended curricula, and reminiscences of students and teachers—and have then claimed that certain curricula were being followed. But, with respect to the teaching and learning of elementary forms of mathematics in British and North American schools up to about 1850, the most powerful evidence is likely to come from cyphering books. Entries in those books show clearly (or, as Simons stated, "in evidence that is unmistakable") what students actually spent their time on as they attempted to learn mathematics.

Logarithms and trigonometry in the implemented curriculum. The navigation cyphering books by an unknown author, circa 1791, by Charles Page at the Royal Mathematics School, and by William Beattie, attest to the forms of mathematics which those who aspired to become navigators were expected to study. But they attest to far more than that. For instance, our analyses of those manuscripts revealed that students were introduced to logarithms and trigonometry without becoming aware of the assumptions behind the five-figure tables of logarithms and of sines and cosines that they were asked to use.

Preference for form over substance in the implemented curriculum. Was the vast amount of time that students at Christ's Hospital, in London, spent on preparing pages of exquisite writing, and illustrations, time well spent? It seems that many of the pages in navigation cyphering books were copied from other written sources, or taken down as notes in dictation sessions—which raises the question: Did the students *understand* what they wrote in their books? Then there is the further question: What does the word "understand" mean in that context? We were impressed with the beautiful miniature water-color paintings in the unknown author's (c. 1791) manuscript, and reflected on whether those paintings were done by a midshipman as he observed scenes that were within his vision from his ship; we wondered how he could have labeled the side lengths and angles with such tiny writing. The possibility that William Beattie's navigation cyphering book was written to honor the King of England intrigued us—and made us think about how important cyphering books were deemed to be, and how their importance so rapidly diminished during the second half of the nineteenth century. But did Charles Page, our unknown author (c. 1791), and William Beattie (1836), *understand* the mathematics that they wrote in their cyphering books?

The magnificent calligraphy and artwork in Martha and Elisabeth Ryan's composite cyphering book undeniably represented a worthwhile artistic achievement, but the question still remains—how well did the sisters learn the mathematics that they recorded? That "attained curriculum" issue will be further discussed later in this chapter. But, there could be no doubt that the arithmetic in Martha and Elisabeth Ryan's cyphering book went well beyond the elementary, and from that perspective we have direct evidence that even around 1780 some females in North America were expected to deal with complex ideas in arithmetic. Why were Martha and Elisabeth able to spend so much of their time preparing such a beautiful cyphering book? Were they simply required to do what they did by a zealous tutor, or did they choose to create their cyphering books, voluntarily?

Abraham Lincoln doggedly stuck to his task of cyphering up to the rule of three. He then proceeded to deal with topics *beyond* the rule of three—like the double rule of three, loss and gain, discount, and simple and compound interest. Yet, later in life he would make a point of claiming that he cyphered to the rule of three. Pages in Lincoln's cyphering book testify to the implemented curriculum that he experienced. Although that implemented curriculum differed from what he later claimed he studied at school, it was reasonably consistent with the intended curriculum set out in arithmetic textbooks by authors like Thomas Dilworth (1806) and Stephen Pike (1822).

We have little doubt that the person who prepared the early (c. 1667) North American cyphering book was not very concerned about mastering the arithmetic that he or she was studying. By contrast, Thomas Prust *did* want to understand what he was doing. And, in keeping with her Irish upbringing, Mary Walters genuinely seemed to enjoy being able to learn arithmetic through interesting ballads, poetry, and songs. But George Bickham (c. 1740), for all his incredible calligraphy and penmanship, did *not* seem to be much concerned about the quality and extent of the calculational aspects of his arithmetic. Yet, Thomas Dixson seemed to revel in displaying detailed calculations in as succinct and beautiful form as possible. His cyphering book, and Thomas Prust's, drew attention to the importance of the galley method of division which was still being used in the seventeenth and early eighteenth centuries.

Sacrobosco and *abbaco* curricula influences? Analysis of cyphering books can also identify changes in curriculum sequence. For example, when we were preparing our recently-published book on the early history of school mathematics in North America (Ellerton & Clements, 2012), we were not aware of the influence of John Holywood—better known as Johannes de Sacrobosco—on curricular sequences for school arithmetic in Great Britain. We assumed that the *abbaco* sequence, deriving from Leonardo of Pisa, was standard in England and was transplanted into the early schools of New England, in North America.

In Thomas Dixson's cyphering book the summation of the terms of arithmetical and geometrical progressions was dealt with much earlier than was normal in the *abbaco* sequence. That early consideration of arithmetical and geometrical progressions was consistent with the Sacrobosco tradition. Furthermore, in the manuscript prepared by the unknown writer of the mid-1660s, in New England, the process of "halving" was considered, despite there being only a small number of pages. Halving related to the concept of "mediation," which was emphasized by Sacrobosco but was not often found in the *abbaco* sequence. So, it seemed that the Sacrobosco influence reached across to New England quite early during the colonization of North America. The Thomas Prust manuscript, which was started by Thomas Prust in England but was then brought to, and continued in, North America, also showed some Sacrobosco touches. The influence of Sacrobosco's ideas in North America is, clearly, a researchable theme.

Decimal fractions and lag time in the implemented curriculum. Despite decimal fractions having been formally defined by mathematicians before the end of the sixteenth century, the first mention of decimal fractions in our collection of 350 North American cyphering books was in 1766. However, we know that George Washington used decimals, logarithms and trigonometry in the cyphering book he prepared in the 1740s (Rickey, Crackel, & Silverberg, 2013), as did those who prepared several of the early 18th-century navigation cyphering books held in the Phillips Library in Salem, Massachusetts (see Gaydos & Kampas, 2010). Our research revealed that it was not until after the introduction of decimal currency in the United States that decimal fractions slowly but steadily became part of the implemented curriculum in most U.S. schools. By contrast, even well into the nineteenth century, it was uncommon to find sections on decimal fractions in British arithmetic cyphering books—despite the fact that some seventeenth-century British authors of commercially-printed arithmetics, like, for example, Edmund Wingate (1624, 1630) and Edward Cocker (1695), emphasized the calculational power of decimal arithmetic.

William Beattie made extensive use of decimals in his mensuration cyphering book, which was probably prepared at Clarencefield Academy, Scotland, around 1810. A major factor influencing Beattie's use of decimals was his heavy reliance on Hawney's *The Compleat Measurer; or the Whole Art of Measuring.* We examined carefully our 1729 third-edition copy of Hawney's text, and found that the 70 pages making up Part I of that book paid much attention to the arithmetic of decimal fractions.

When we examined the content on the extant pages in Abraham Lincoln's cyphering book we could not help but feel that he was severely hampered by not having studied decimal and vulgar fractions. But he was not aware that he was disadvantaged by his lack of knowledge concerning those topics, and he set about learning topics like discount, and simple and compound interest using the convoluted methods that had evolved as part of *abbaco* arithmetic.

Why should lag-times have been so long with respect to such apparently useful topics as decimal fractions and logarithms? The answer to that question depends on the vantage point taken by the person giving the answer. From the perspective of research mathematicians, the sixteenth-century applied mathematicians, the algorists, and the reckoning masters, hindered progress by teaching students according to an antiquated curriculum that did not take advantage of the latest research in pure mathematics. Some research mathematicians lamented the fact that Steven's decimals and Napier's logarithms came to be used purely as calculational devices. From the point of view of service mathematicians, however, the problem lay with technology—to take advantage of the calculational power of logarithms one needed not only a strong grasp of decimal fractions, but also ready access to tables of logarithms. This double requirement was only feasible for a small section of the population. For most school teachers, however, no problem existed because they did not understand decimals and logarithms and, in any case, had plenty to do keeping their students occupied with the "old curriculum." The lesson is this: the implemented curricula, as found in cyphering books, not only represented what the learners were asked to do, but also the knowledge and priorities of their teachers.

Figure 12.1 draws attention to how three different groups typically see mathematics from different perspectives. Some scholars write from a research mathematician's perspective; others from the perspective of applied mathematics; and others see mathematics from the point of view of learners, and think in terms of the teaching and learning of the subject.

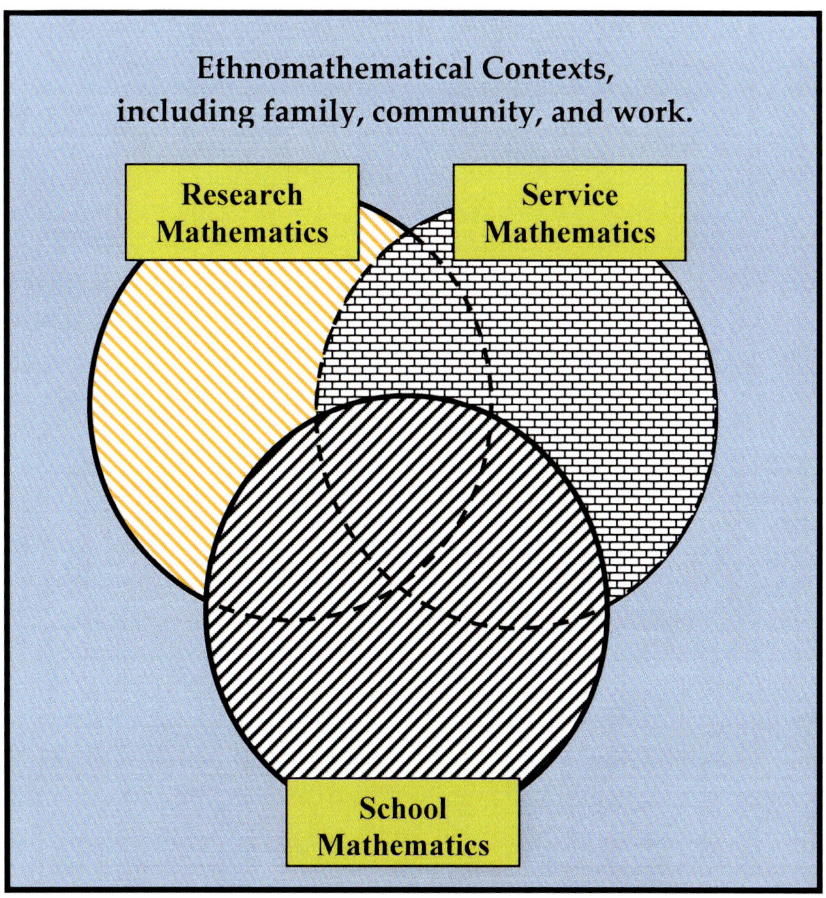

Figure 12.1. Different ways of "seeing" mathematics (from Ellerton & Clements, 2013, p. 10).

Take, for example, the following comment by the mathematician Carl B. Boyer (1991) in relation to attitudes to the introduction of logarithms during the period 1570–1700.

> Logarithms were hailed gladly by Kepler not as a contribution to thought, but because they vastly increased the computational power of the astronomer. Viète was not exactly "crying in the wilderness"; it is nevertheless true that most of his contemporaries were primarily concerned with the practical aspects of mathematics. Bürgi was a clockmaker, Galileo was a physicist and astronomer, and Stevin was an engineer. It was inevitable that these men should have preferred parts of mathematics that gave promise of applicability to their fields. Bürgi and Stevin, for example, aided in the development of decimal fractions, and Bürgi and Galileo were rivals in the manufacture and sale of a practical computing device known as the proportional compass. (pp. 315–316)

Boyer obviously valued the research aspect of mathematics so highly that he found it difficult to concede that service mathematics and school mathematics were legitimate forms of mathematics.

Another person to hold that viewpoint was the U.S. mathematician, Eric Temple Bell. Writing in 1945, Bell stated that Leonardo of Pisa (Fibonnaci) was indirectly responsible for "the deluge of practical manuals on elementary computation and the flood of commercial arithmetics which have poured from the printing presses of the world ever since the fifteenth century" (p. 214). He added that "in spite of their great practical utility, none of these indispensable works has contributed anything of importance to the development of mathematics" (p. 214).

In a similar vein, Bertrand Russell (1926), after describing mathematics as having a cold and austere form of beauty, went on to say that like musical ability, mathematical ability was a gift of the gods, something "quite rare, even to a moderate degree" (p. 177).

By contrast, Lambert Jackson (1906), at Columbia University early in the twentieth century, wrote from the perspective of a mathematics educator. He emphasized how the work of the European reckoning masters of the sixteenth century contributed to the popularization of mathematics, as people began to realize that the subject could be simultaneously theoretically interesting and useful. According to Jackson, although the sixteenth-century reckoning masters were responsible for putting too much commercial arithmetic into both the intended and implemented arithmetic curricula for schools, they nevertheless took "a step in the direction of making arithmetic concrete and paved the way for a better and more teachable system of subject matter" (p. 219). For the reckoning masters, mathematics was to be seen as "do-able," and important for *all* people—and not just a tiny, privileged few (Allanson, 1968).

Figure 12.1 places school mathematics in the forefront. Research mathematicians provide foundational support needed for the development of satisfactory, logically-cohesive, subject matter and service mathematicians show how mathematics can usefully be applied. It is only through the coordinated efforts of the three groups that mathematics can become a constructive force, accessible to the whole community.

In Western Europe, the time lag between the theoretical formulation of decimal fractions and their large-scale application to real-world problems in navigation, astronomy, surveying, and fortification was at least 50 years. Admittedly, it did not take long before forward-thinking educators, like Edmund Wingate (1630), began to include decimal fractions in the intended arithmetic curriculum for schools. However, cyphering-book evidence indicates that in Great Britain and North America it was not until well into the eighteenth century that decimal fractions began to be incorporated into the implemented curricula of some schools. With the introduction of decimal currency in the United States in the 1790s, decimal fractions steadily became part of the implemented curriculum in most schools in that nation. But, in Great Britain, despite decimal fractions receiving attention in a majority of British arithmetic textbooks, it was not until about 1850 that most schools included the topic in their implemented curricula.

In any particular nation the lag time was different in different parts of that nation. Thus, for example, when, deep in rural Indiana in the mid-1820s, a young Abraham Lincoln was preparing solutions to questions on simple and compound interest, and discount, he did not use decimal fractions, even though his problem-solving efforts could have been greatly facilitated if he had. The reason why Abraham did not use decimals was simple—he had never studied decimal fractions formally, and his teachers did not know enough about them to point him in the right direction. True, he might have had access to textbooks by Dilworth (1806) and Stephen Pike (1822), in which decimal fractions were considered, but his teachers asked him to follow the normal *abbaco* sequence—and sections on decimal fractions in textbooks, and in old cyphering books, came after simple interest and discount. In this case the sequence of topics in the intended curriculum, and lack of teacher knowledge, held back progress toward more modern learning.

The lesson here is easily stated: it is impossible to estimate the quality of a student's mathematics learning merely by examining the textbook(s) used by that student.

Lesson 3: Examining Intended and Implemented Curricula did not tell us much about how well Students Learned Mathematics

The cyphering process was supposed to have its own inbuilt mechanisms for assessing the quality of learning. There were four aspects of this assessment process, and the first three of these were associated with what was known as the *recitation* process, which occurred when individual students consulted with their teachers before and after making entries into their cyphering books (Thayer, 1928). The four aspects were:

1. When a student came to see the teacher for recitation—which happened at least once a day, and sometimes more than that—the teacher would check that entries which had been made in the student's cyphering book since the last recitation session had been done attractively and correctly. In many schools this was not an idle matter for either the students or their teachers, and students who did not make satisfactory entries into their cyphering books were likely to be punished.
2. Then, after the teacher had checked the quality of the entries that a student had made since the last recitation, the teacher would question the student to see if he, or she, had understood the mathematics that was under consideration. Sometimes, this meant little more than the student being asked to state what he or she had written, word for word, perhaps from memory. Some teachers, however, required evidence of far more than memorization. They wanted their students to be able to *explain* the *meanings* of what they had written. In such cases, a teacher would ask the student questions about what had been written, and the student would be expected to respond to the satisfaction of that teacher.
3. In the third phase of the recitation process the teacher would define what the student would need to do before the next recitation session and, in particular, what entries would need to be made in the cyphering book.
4. The final, summative, phase of the evaluation process would come in the last week of term when, with much fanfare, a public examination of a school's work would be held. On such an occasion, a school committee comprising a respected group of local professionals (e.g., a doctor, a lawyer, a clergyman, a retired army officer), would ask questions of all individual

> students. The students' cyphering books would be on display, and they were looked at not only by committee members but also by parents who were in attendance. A teacher's future employment could depend on the attractiveness of the cyphering books, and hence the emphasis on excellent penmanship and high calligraphy was not surprising.

Note that until about 1840, in both Great Britain and North America, assessment of the quality of learning was never based on written examinations (Henry, 1843; Roach, 1971; Rotherham, 1852; Watson & Kandel, 1911). We have argued elsewhere that the introduction of external written examinations would prove to be the death knell for the use of cyphering books in Great Britain and in the United States—but the cyphering tradition would linger longer in Great Britain than in North America (Ellerton & Clements, 2012). As evaluation by written examinations became standard, time spent on preparing a beautiful cyphering book became less defensible from a cost-benefit point of view. If a teacher wanted to keep his or her job it became increasingly important for that teacher to prepare students as well as possible for the end-of-term written examination. That meant there was no sense in requiring students to prepare beautiful cyphering books (Ellerton & Clements, 2012). Such was the power of the pencil-and-paper written examinations introduced after 1840 that the end of the 600-year-old cyphering era would come surprisingly quickly.

The navigation cyphering book prepared by William Beattie would not have been subjected to careful and critical examination by a teacher. It was prepared by a man in his forties, and therefore would not have been systematically checked by another. However, in the case of Charles Page, at Christ's Hospital, the cyphering book was inspected by a committee of navigation experts from Trinity House who were primarily concerned with whether students understood, and could apply, what they had learned. Thus, the students were not only expected to prepare beautiful cyphering books, but they were also expected to be able to understand and apply what they had written (Wilson, 1820).

We found it inspirational to read Mary Walters' cyphering book, for we gained the impression that she not only enjoyed preparing it, but she also learned the arithmetic which was embodied in the poems within it. But, at the end of the day, we do not have hard data to inform us on whether Mary understood relationally the concepts and principles she studied—and we do not know whether she could apply what she learned.

The oldest British cyphering book that we examined was by Thomas Dixson, and it seemed to us that he gained a strong grasp of elementary arithmetical operations. However, we do not know if Thomas could apply confidently and competently in everyday life what he learned. Likewise, we know nothing of the quality of learning resulting from George Bickham's cyphering approach. It seemed to us that Bickham was much more concerned with the quality of his students' penmanship than with the quality of their knowledge of arithmetical concepts. The oldest North American book we examined was the very brief 7-page manuscript which was probably prepared in New England in the late 1660s—our distinct impression was that the person who prepared that manuscript did not learn much arithmetic at all. The emphasis was on the language of operations, and more than anything else the manuscript seemed to be concerned with the author doing just enough to be able to claim that he, or she, had reached the rule of three.

Lesson 4: Cyphering Books Provide Data on Students' Attitudes and Values

Most of those who prepared cyphering books seemed to place enormous value on them. That is not surprising, given the effort and time they spent in doing the calligraphic headings, finding the correct solutions to tasks, and in writing those solutions with high-level penmanship. There would be very few, if any, other handwritten manuscripts that the students would prepare during their lifetimes into which they would inject so much of themselves.

Mary Walters was a young, Irish-background girl who made it clear that she very much enjoyed writing out poems which embodied mathematical tasks, and then solving them. Her 80-page cyphering book testified to her heritage, and to the way Irish educators thought that arithmetic was best learned. In that sense, her cyphering book was a cultural statement as much as anything else: "This is who I am," she declared—not in those exact words, but in the way that she did what she did. The beautiful rose that she painted on reaching the rule of three was a personal celebration of what was an important achievement for her.

Around 1630, in England, Thomas Dixson was prepared to devote large amounts of time to practicing the galley method of division, and other operations on whole numbers. Probably, he was under pressure from a teacher to present his calculations as neatly as possible, but that cannot hide the feeling of pride emanating from the pages of his cyphering book. One wonders how many other students in England around 1630 prepared similar cyphering books to Thomas's. John Denniss (2012) has looked at between 250 and 350 cyphering books, most of them of British origin, and the oldest one that he found was prepared around 1680. We expect future researchers to be able to show that the cyphering tradition was well established in Great Britain well before 1680, and that part of the tradition was that there was something almost sacred about one's own cyphering book.

We did not include the cyphering book prepared by Pierce Lacy Tanner in Maryland in 1809 within our collection of extraordinary cyphering books, but that book was prepared by a father for his son, James Tanner. On the inside of the front cover of the manuscript is a moving handwritten inscription: "SACRED; from James Tanner to his grandchild Edmund Bullock. This book was done by his great-grandfather, Pierce Lacy Tanner, nearly 50 years ago" (quoted in Ellerton & Clements, 2011, p. 28).

Perhaps, after he left school Abraham Lincoln did not continue to value the cyphering book that he had labored so hard to prepare in Spencer County, Indiana, in the early 1820s. But his step-mother, Sarah Bush Johnson Lincoln, who loved her step-son, certainly did value it. At the time when Abraham Lincoln was assassinated in 1865 she had retained just one of her step-son's handwritten books—and that was the cyphering book that he had prepared at school, 40 years earlier.

What moved the middle-aged William Beattie, someone who was not greatly interested in scientific or navigation matters, to prepare a 309-page navigation cyphering book with beautiful calligraphic headings, penmanship and illustrations? We cannot be sure, but we think he prepared it as a special gift for King William IV, the "Sailor King."

One cannot examine Martha and Elisabeth Ryan's composite cyphering book without being moved by the huge effort the sisters, with the assistance of at least one tutor, put into making it the most beautiful book that anyone could create. Fittingly, today, more than 230 years after it was prepared, it holds pride of place in the Southern Historical Collection in the Wilson Library at the University of North Carolina in Chapel Hill.

All of the cyphering books that we studied when preparing this volume had something special about them. Although a 21st-century reader might conclude that our circa 1667 manuscript was not greatly valued by the person who prepared it, the fact is that it was sufficiently valued that it was kept and has survived for almost 350 years. Thomas Prust's declaration, "Thomas Prust his booke Amen 1702," probably sums up what we are trying to say most clearly—Thomas was declaring that this book was his precious possession, and that was that!

The lesson is this: individuals who attempt to learn mathematics learn to attach values to what they doing—to the subject itself, and to the importance of their own efforts to learn mathematics. We believe that this is an important message from the cyphering era to modern mathematics educators.

Lesson 5: To Understand Cyphering Books Requires Knowledge of Contextual and Ethnomathematical Circumstances Surrounding the Preparation of the Books

The dominant approach to school mathematics in Western European nations in the seventeenth century was in line with the cyphering tradition, and it was inevitable that that approach would be translated into the early European settlements in North America. But, in the New World, none of the necessary ingredients for success—strong *mathematical knowledge* among teachers, well-honed *teaching power*, *student readiness* for learning arithmetic, and adequate *resources*—was usually present (Ellerton & Clements, 2012).

The oldest North American manuscript that we examined in this book was probably prepared in New England in the second half of the 1660s. One can only begin to imagine the physical and social contexts in which it was prepared. The settlers were struggling to survive, and for about eight months of the year all youngsters old enough to work in the fields were needed to help prepare for the all-important harvest. In most families there was no time for fancy things like learning arithmetic. But, during the winter months there was time, and so for those families able to get access to paper brought across from Europe, the idea that youngsters could learn to read, write and cypher was attractive. But who would teach them? What, precisely, should they be asked to learn? And, with respect to arithmetic, were the youngsters ready to learn what children of the same age were being asked to learn in Europe?

We will not answer those questions here for we do not know enough about the actual situation in which our unknown author, circa 1667, somehow prepared his, or her, 7-page manuscript. Nor will we attempt to describe the circumstances in which Thomas Dixson prepared his cyphering book in England in the 1630s, for we do not know much about his situation, either. But we will comment more on how local situations and values greatly influenced what happened when some youngsters attempted to learn mathematics—and that is, essentially, our fifth lesson.

Ethnomathematical Considerations

Modern mathematics education researchers use the term "ethnomathematics" when they examine mathematical aspects of everyday behaviors of community and family members, and how those everyday behaviors affect the mathematical development of learners immersed in particular cultures. Those who research ethnomathematics study the shared mathematical language, knowledge, skills and preferences of those who grow up within well-defined communities (D'Ambrosio, 1986; Emigh, 2002; Lave & Wenger, 1998). They recognize that those in one community, geographical region, trade, or profession, respond differently, mathematically speaking, to those in other communities, and know that that can be the case even when what is being studied in the different situations appears to be very similar. From a mathematics education perspective, the extraordinary manuscripts studied in this book testify to the overriding influence of special situations on student learning and on student valuing.

In this section we shall consider, necessarily briefly, just four of those special situations—those relating to Charles Page, at Christ's Hospital in inner-London in the 1820s; to Martha and Elisabeth Ryan, at a plantation in Bertie County, North Carolina around 1780; to Mary Walters in Baltimore, Maryland, around 1820; and to Abraham Lincoln, in rural Indiana in the early 1820s.

Charles Page's mathematical education from an ethnomathematical perspective. In Chapter 10 we indicated that Charles Page was the eldest of 13 children, and that he entered the Royal Mathematical School at Christ's Hospital School, in London, in 1818, when he was eight years of age. For the next seven to eight years he lived within the Christ's Hospital grounds, and in 1825 he graduated from Christ's Hospital and took on a 7-year apprenticeship.

As a mathemat at Christ's Hospital, Charles became part of a school which was located in the heart of London and had a remarkable, government-supported mathematics program aimed at preparing boys to become navigators. Charles' cyphering book was not very different from those prepared by other Christ's Hospital mathemats. Elsewhere, we have studied eight navigation cyphering books prepared by boys who attended Christ's Hospital at various times between 1746 and 1835 (Ellerton & Clements, 2012), and have observed that the manuscripts were very similar in their structure and in the mathematical and navigational content they embraced.

During his seven or eight years at Christ's Hospital, and especially during 1824 and 1825, when he was based in the Royal Mathematical School's special buildings, Charles would have had enormous pressure placed on him to make sure that he prepared the best quality navigation cyphering book—from the points of view of attractiveness and mathematical and navigational rigor—that he could possibly prepare. He was provided free of charge with the paper for his manuscript, the quills, and the ink. The large well-lit room in which the mathemats prepared their manuscripts had long, flat desks—especially designed to support the cyphering process. Textbooks, written by past Christ's Hospital masters, which covered all of the mathematical and navigational material in the carefully-defined Royal Mathematical School curriculum, were freely made available to all mathemats. Outstanding writing masters and drawing masters were on hand to train Charles to write and to draw beautifully, and the Master of the Mathematical School and his tutors were responsible for his mathematical and navigational development.

Being a mathemat meant that Charles lived and breathed mathematics almost every day. He knew that ultimately his level of understanding of what he was studying would be rigorously tested by experts from Trinity House. He could not avoid soaking up the atmosphere and the expectations, and he knew that, almost certainly, he would be apprenticed as a midshipman when, finally, he would leave Christ's Hospital. He imbibed a very special kind of mathematical culture, which would have affected the ways he thought for the rest of his life.

Martha and Elisabeth Ryan's mathematical education from an ethnomathematical perspective. Sisters Martha and Elisabeth Ryan were living on a plantation in Bertie County on the east coast of North Carolina when their remarkable and beautiful composite cyphering book was prepared. With the help of at least one tutor, Martha prepared the first 38 pages in 1776 and 1777, and the next 14 pages in January 1781. Also with the help of a tutor, Elisabeth prepared her 99 pages in 1780 and 1781. Each of the sisters was intensely interested in the progress of the Revolutionary War between the "rebellious" states and Great Britain, and entries in their composite cyphering book were replete with defiant illustrations and anti-British slogans. Although this patriotism had a strong impact on the *appearance* of pages, from a purely arithmetic content point of view the sisters merely studied the time-honored *abbaco* sequence of elementary and middle-level arithmetical topics.

What did seem to have a large effect on the implemented curriculum experienced by Martha and Elisabeth was the fact that they lived on a North Carolina plantation. Although their father, James Ryan, died in 1777—when Martha was preparing her section of the cyphering book—it seems that throughout the periods Martha and Elisabeth were preparing their sections of the cyphering book they were assisted by a knowledgeable tutor. This person often indicated that he—it is presumed the tutor was male—had been involved with the entries on a page by placing the initial "H" at the foot of the page. Careful examination of entries indicated that, undoubtedly, much of the writing in the composite manuscript, and even some of the calligraphic headings and decorations, had been done by "H." This constant, close checking by a tutor raises serious questions about the quality of the attained curriculum for Martha and Elisabeth.

Many of the examples solved in the composite book dealt with situations which would have been relevant to plantation life, and there is the distinct possibility that the problems were often carefully chosen so that they would be obviously relevant to the situation in which the two girls found themselves. Through topics like the four operations on whole numbers, and the application of the rules of three to fellowship, practice, simple and compound interest, and discount, much attention was given to plantation-relevant exercises on buying and selling, and investing money. Be that as it may, the arithmetic in this composite book reached a much higher level than would be found in other cyphering books prepared by girls in the eighteenth century. Indeed, most girls of the same period did not prepare a cyphering book at all.

Mary Walters' mathematical education from an ethnomathematical perspective. In Chapter 5 we identified the "Mary Walters" who prepared the cyphering book which we studied in that chapter as having been born in Kilkenny, Ireland, in about 1806. Mary migrated to North America sometime between 1806 and 1818, and she began preparing the cyphering book in 1820 in Baltimore, Maryland. She would marry James Henry O'Connor (also from Kilkenny) in 1826, and die in Taberg, Oneida County, New York, late in the nineteenth century.

As we were studying Mary Walters' cyphering book we found ourselves being transported, unexpectedly, into an educational world in which arithmetic was to be learned through ballads, poetry and song. In her cyphering book, Mary did not go beyond the rule of three, but the arithmetic she studied was framed within an overwhelmingly Irish literary tradition. Julie Henigan (2012) has drawn attention to the predominantly oral poetic culture of the Irish in the eighteenth century, tracing developments of form, genre, style, song and transmission in the traditional poetry of Ireland, in both Irish (Gaelic) and English. However, according to Henigan, it was neither literacy nor orality, but performance within community that most truly defined the tradition developed in education in Ireland in the eighteenth century. Reading Mary Walters' cyphering book made it clear to us that she saw arithmetic as contributing to her developing world-view, and not as something to be learned simply because it could be useful.

The poems and songs in Mary's cyphering book which incorporated arithmetical concepts were intended to educate Mary's Irish heart (Duffy, 1845) as well as help her learn to solve arithmetical problems. Her book featured poems which offered realistic and open approaches to all areas of life, including romance, death, and the consequences of promiscuous misbehaviour. The poems featured interesting scenarios, and defined problems which demanded solutions.

Mary's attained curriculum in mathematics was quite different from Charles Page's. We venture to suggest that Charles would probably have done much better than Mary on a written test of mathematics, but Mary's beautiful celebration of reaching the rule of three—by showing an attractively painted rose—suggested that somehow she had done a better job of integrating her mathematics education with her emotions and with other aspects of her everyday humanity within her community.

Abraham Lincoln's early mathematical education from an ethnomathematical perspective. In Chapter 6 we analysed the 22 extant pages of the cyphering book that young Abraham Lincoln prepared between about 1819 and 1826, when growing up in rural Indiana. The circumstances in which Abraham reached towards manhood in and around Pigeon Creek, and especially the kind of schooling he received at that time, have become the stuff of legend—but the investigation, in Chapter 6, of his cyphering book is the first major analysis of what is easily the oldest surviving handwritten manuscript of the future President.

As we studied Abraham's efforts at solving arithmetic word problems it became obvious to us that his mathematical development was heavily constrained by the circumstances in which he lived and also by the attitude that he developed that he had to follow standard procedures. In our analysis there emerged a person with huge determination, someone who would be not be satisfied unless he believed he understood what he was doing. Yet, somewhat ironically, his methods became very mechanical, because he simply did not have the opportunity to learn a range of methods for tackling the problems that he set out to solve. Thus, for example, because he had never had the opportunity to study vulgar or decimal fractions, or percentages, his attempts at solving simple and compound interest problems, and problems concerned with discount, were clumsy and inefficient. Undoubtedly, Abraham's problem-solving efforts in arithmetic would have been more impressive if he had studied decimal fractions fully, but how could he have known that? It was probably the case that no-one else living in or around Pigeon Creek knew how to make use of decimal fractions when solving such problems.

Nevertheless, the very determined and intelligent young Abraham did learn important mathematical concepts as a result of growing up within his community. Abraham made some cabinets (Bartelt, 2008; Donald, 1995), and while making them he would have reflected on ideas associated with measurement of length, area, volume, the four operations on whole numbers, and ratio and proportion. These ideas would probably have been reinforced by what he learned about measurement and space as he tilled the soil, harvested crops, and walked long distances—to school, to libraries, and to neighbours' homes. Abraham also carried water to his family's cabin, and this would have reinforced ideas of weight and capacity. Concepts associated with the measurement of time would have been acquired as a result of participation in family and community activities, and an intuitive understanding of fractional parts, equality, and inequality, would have come from sharing many objects and quantities among members of a large family.

All the instances of "everyday mathematics" mentioned in the last paragraph might seem to have been too obvious to be worth mentioning. The point, however, is that Abraham learned his arithmetic within a fairly remote community which was very different from the communities in which Charles Page, Mary Walters, or for that matter, Martha and Elisabeth Ryan, and William Beattie, learned their mathematics. Later in his life, Abraham might not have thought highly of the education he received at the schools he attended, but it did whet his appetite for mathematics. Six or seven years after he left school he would take up surveying, and to do that he needed to read textbooks which dealt with logarithms and trigonometry. And, still later, he would immerse himself in the strict logical development of Euclid's *Elements* (Hirsch & Van Haften, 2010), perhaps the greatest book on geometry ever written.

Tensions between the intended and implemented curricula, and community culture. We chose to discuss the cyphering books by Charles Page, Martha and Elisabeth Ryan, Mary Walters, and Abraham Lincoln with respect to ethnomathematical considerations because we believed we knew enough about their peculiar situations to be able to present our comments with some degree of authority. What has been said has made clear that, inevitably, there would be tensions between the ethnomathematical situations in which all learners found themselves and traditional *abbaco* (or Sacrobosco, or navigation) curricular sequences.

The unknown person who prepared the 7-page circa 1667 cyphering book, probably somewhere in New England, would have been living under very different circumstances from any of the others whose cyphering books we have studied. What kinds of mathematics education were appropriate to the pioneering circumstances faced by the early settlers? We know very little of the context in which Thomas Dixson prepared his cyphering book around 1630, but almost certainly he too was surrounded by his own peculiar circumstances. The same could be said about the other persons whose cyphering books we have considered—Thomas Prust, George Bickham, the midshipman who prepared the beautiful circa 1791 manuscript, and William Beattie. It would appear to be obvious to us that any decision to force the same intended curriculum on those people would have been foolish, and the different implemented curricula we have outlined in Chapters 2 through 11 testified to the fact that the students focused on different aspects of mathematics—yet, the arithmetic cyphering books were all shaped by the *abbaco* or Sacrobosco curricular sequences. Similarly, the three navigation cyphering books all seemed to have been framed by the Christ's Hospital navigation curriculum expectations.

Because there was no uniform criteria for assessing attained curricula, the differences in the implemented curricula did not generate the kind of curriculum chaos which could have resulted if students, everywhere—in London, Bertie County (North Carolina), Baltimore (Maryland), and Spencer County (Indiana), and elsewhere—had been pushed into receiving more or less the same intended and implemented curricula, and then forced to take the same standardized tests as a check on what they had been taught and what they had learned. The lesson for modern mathematics education should be clear.

Lesson 6: British and North American Cyphering Books Became Different

Our collection of about 500 cyphering books includes 350 North American cyphering books, 100 British cyphering books, and about 50 manuscripts from other nations. The two largest collections of British cyphering books are held by the Mathematical Association of Great Britain, with about 200 manuscripts—that collection was originally created by a mathematics educator, John Hersee—and by John Denniss, with about 120 manuscripts (Denniss, 2012; Stedall, 2012). As far as we know, our collection is the only one with sizeable numbers of *both* North American and British cyphering books. We make that point only to indicate that we have had ready access to a large data set of relevant primary sources, and that that has enabled us to analyze similarities and differences. It is likely that 7 of the 11 cyphering books that we have studied in detail in this volume originated in Great Britain—assuming that the first part of Thomas Prust's cyphering book was prepared in Great Britain, before Prust migrated to North America and handed his cyphering book on to his step-son, John Collings. Table 12.1 presents summary data with respect to the 11 manuscripts.

Analyses of the extraordinary cyphering books indicated that there were persistent qualitative differences between British and North American cyphering books. In particular, we have noted the following two major differences:

Table 12.1
Summary of Manuscripts Analyzed in this Book

Name of Person Who Prepared the Manuscript (if known)	Year(s) When the Manuscript was Prepared	Likely Place(s) Where Manuscript was Prepared; and Present Location	Chapter in which Manuscript is Analyzed	Special Features
Unknown author	c. 1667	New England; Ellerton & Clements collection	2	Oldest-known extant North American cyphering book.
Thomas Prust and John Collings	1702–1703 (Prust), and 1720–1721 (Collings)	Great Britain, and then Portsmouth (New Hampshire); Ellerton & Clements collection	3	In 1720 Prust handed his cyphering book on to his step-son, John Collings.
Martha and Elisabeth Ryan	c. 1780	Bertie County, North Carolina; Southern Historical Collection, Wilson Library, University of North Carolina	4	Mathematical content proceeds well beyond the rule of three.
Mary Walters	c. 1820	Baltimore, Maryland, USA; Ellerton & Clements collection	5	Arithmetic is studied through poetry and song—links with Irish traditions.
Abraham Lincoln	c. 1819–1826	Spencer County, Indiana; the 11 leaves are held in 11 different places in the United States	6	A recently discovered eleventh leaf was included in the analysis.
Thomas Dixson	1630 –1632	England; Plimpton Collection, Butler Library, Columbia University (New York)	7	This is almost 50 years older than any manuscript to which Denniss (2012) referred.
George Bickham	c. 1740	London, England; Ellerton & Clements collection.	8	Commercially prepared—the aim was to show fine calligraphy and penmanship.
Unknown author	c. 1791	Great Britain; Ellerton & Clements collection	9	Possibly prepared by a midshipman during one or more long voyage(s).
Charles Page	c. 1820–1825	London, England; Ellerton & Clements collection	10	Prepared in the Royal Mathematical School at Christ's Hospital.
William Beattie	c. 1810	Dumfriesshire, Scotland; Ellerton & Clements collection	11	Mensuration manuscript, prepared at Clarencefield Academy.
William Beattie	c. 1835–1836	London, England; Ellerton & Clements collection	11	Navigation manuscript, possibly prepared for King William IV.

1. The *physical appearance* of nineteenth-century British cyphering books tended to differ from that of North American cyphering books. Although Denniss (2012) noted that most British cyphering books were "properly bound, often in leather or vellum and usually quarto, though some in folio" (p. 1), that description does not fit most of the British manuscripts in our collection. Of the 100 British cyphering books that we hold, 46 have leather spine covers, but only 4 of those have full-leather binding. Another 5 have vellum

covers. The others have soft covers, or hard covers that are made from stiffened card. They are of the exercise-book variety, with dimensions about 23 cm by 18 cm (i.e., 9" by 7.5"). By contrast, of the 350 North American cyphering books that we hold, 70 percent were folio (with dimensions about 30 cm by 20 cm, that is say about 12" by 8"), with most of the others having an exercise-book appearance and dimensions. In both countries, the smaller books were especially common after 1820, when they were sold commercially as especially suitable for cyphering.

2. A higher proportion of British students than North American students copied notes from textbooks into their cyphering books. Both Dennis (2012) and Stedall (2012) claimed that with most of the British manuscripts that they had examined, students had copied rules and exercises from textbooks, especially from the numerous editions of Francis Walkingame's *The Tutor's Assistant.* But we have found that relatively few North American students prepared cyphering books by copying directly from textbooks. This could have been largely because up to about 1830 most North American school students studying arithmetic did not have ready access to even one arithmetic textbook. It was common, however, for rules and examples to be copied from existing cyphering books, and for teachers to dictate notes to their students (Ellerton & Clements, 2012).

Of the eleven extraordinary cyphering books that we examined in this book we think only one of the four prepared in America (that by Mary Walters) reflected a lot of individuality so far as mathematical content was concerned. Mary Walters seemed to revel in the Irish-background approach to arithmetic, featuring ballads, poetry and song. Abraham Lincoln's manuscript lacked the flair of Mary's. Although he was very determined to solve exercises using his methods, and was not happy merely to copy someone else's work, his methods tended to be consistent with those set out in textbooks written by authors like Thomas Dilworth, Zachariah Jess, and Stephen Pike.

Of the British cyphering books, those by Thomas Dixson, George Bickham, Charles Page, and William Beattie displayed fine penmanship and calligraphy but seemed to lack creativity so far as their approaches to mathematics were concerned. The only British cyphering book that, mathematically speaking, seemed to reflect much of the individual student's own thought was the circa 1791 navigation cyphering book with its beautiful miniature paintings. But even in that book, the author seemed not to understand the calculations involving logarithms and trigonometric functions. As the nineteenth century progressed the British cyphering books tended to become very predictable in their format and content. Between 1820 and 1880 the typical British cyphering book was a commercially-sold, lightly-lined exercise book. The first interior page would have a stamp on which the name of the person preparing the book could be written, along with the year or the place where the owner was preparing the book. Figure 12.2 shows a typical example.

The reader will probably be impressed with the appearance of the title page of Master Rogers's cyphering book. The book comprised 134 pages and was prepared at Brynford House School in Holywell, Wales, in 1827 and 1828. The calligraphy and the illustration in which the name of the student and his school appeared in Figure 12.2, were attractive—but the illustration was based on a stamp, and we have seen many British cyphering books whose title pages feature exactly this same stamp. After the decorative title pages would come page upon page of very neat calculations featuring beautiful penmanship and high calligraphy—but not much individuality so far as the mathematics was concerned.

Figure 12.2. A typical title page from a British cyphering book during the period 1820–1880.

Of our 100 British cyphering books, 60 were very much like that, and most of the others also lacked evidence of mathematical creativity. None of the British cyphering books that were prepared in standard exercise books were regarded by us as "extraordinary." The calculations seemed to have been performed mechanically—and the statements of rules and cases, and even the examples, were likely to have been copied from textbooks.

One should not criticize the British school systems for having consciously linked writing and arithmetic, because learning to write well was important in 18th- and 19th-century Great Britain. But our impression was that too often the writing component of arithmetic took on a life of its own and became more important than the possibility that students could be taught to think mathematically. We gained a similar impression when he studied all of the extraordinary cyphering books which were prepared by British students. It seemed that form had become more important than substance. One might say the same thing about the beautiful composite cyphering book prepared by Martha and Elisabeth Ryan in North Carolina, but there was something very special about that book—for its very existence testified powerfully to the fact that not only could females do arithmetic beyond the rule of three, but also that they could do it with flair. In the United States of America the same level of standardization of cyphering books that occurred with British cyphering books did not take place. Furthermore, although in the United States the cyphering book era closed in the early 1860s, in Great Britain it lingered on until the 1880s, or perhaps even beyond that.

Of the extraordinary cyphering books that we examined, those prepared by British authors tended to be more decorative than the North American books—a notable exception being, of course, Martha and Elisabeth Ryan's hauntingly beautiful composite cyphering book, prepared on a North Carolina plantation. Thus, for example, although the level of mathematics in William Beattie's mensuration and navigation cyphering books was high—having been based on reputable texts by William Hawney and John Hamilton Moore—the extent of mathematical initiative displayed by Beattie was small. That same comment could be made with respect to the cyphering books by Thomas Dixson, George Bickham, and Charles Page.

Some Researchable Questions and Final Comments

Researchable Questions

From the outset it was anticipated that this book would put flesh on the bones that we unearthed in our study of the cyphering tradition as it operated in North America between 1607 and 1861 (Ellerton & Clements, 2012). This present book has offered qualitative analyses of a set of 11 extraordinary cyphering books. Some of the books were prepared by British students and some by North American students.

One of our conclusions has been that in the nineteenth century the cyphering tradition in Great Britain developed different characteristics from the cyphering tradition in North America. That raises the interesting, researchable question—did different cyphering traditions develop in other nations in Europe, Africa, South America, and Asia, and if so what were their characteristic features? That question has many subquestions, because it is likely that in different regions within those nations different cyphering traditions emerged. Such differences would need to be investigated separately, and then some form of synthesis attempted. The question of how long the cyphering tradition lingered in the different nations could also be of interest.

With respect to the 11 extraordinary cyphering books that we have studied in this volume we have some specific questions:

- Are there any other very early extant North American cyphering books? Note that by "very early" we mean cyphering books from the period 1607–1680. With respect to that question we comment that after recovering from the shock of being able to purchase the circa 1667 manuscript that we studied in Chapter 2, we quickly became disillusioned with its low level of mathematics, and with the transparent need its author felt to reach the rule of three at any cost.
- To what extent were the early cyphering books in North America and in Great Britain influenced by Sacrobosco's curricular sequence? With respect to that question, note that we were surprised to find evidence that the Sacrobosco sequence had had some influence with respect to the circa 1667 manuscript, and to the cyphering books prepared by Thomas Dixson and Thomas Prust.
- To what extent was the large attention given to penmanship, calligraphy, and drawing in the manuscripts by the Ryan sisters, by George Bickham, by Charles Page and by the person who prepared the unknown circa 1791 manuscript, beneficial in the later lives of those who prepared them?
- Did William Beattie ever show, or give, either or both of the mensuration and navigation cyphering books to King William IV?
- It is often stated that during the nineteenth century the British Navy "ruled the waves" (see, e.g., Mahan, 1898). To what extent did the preparation of navigation cyphering books help in the training of efficient navigators?
- With the Christ's Hospital approach, did the high level of mathematics that was part of the normal curriculum "rub off" on students even though they did not fully understand what they were doing when they were copying material into their cyphering books?
- How well did Mary Walters learn arithmetical concepts as she prepared her cyphering book which included so much emphasis on Irish-heritage ballads, poetry and song?
- What was the influence of existing textbooks on those who prepared the 11 extraordinary cyphering books? We recognize that the influence on William Beattie was large, because he copied much from the texts by William Hawney and John Hamilton Moore. But what about the others? How often did Abraham Lincoln have the relevant pages of a textbook open, and next to him, when he was preparing his cyphering book?

Final Comments

Much more research is needed on the cyphering tradition, especially as that tradition affected what went on in the name of "school mathematics." We confess to feeling frustrated when, from time to time, we mix with mathematicians who want to colonize the history of school mathematics so that it becomes little more than a study of contributions which great mathematicians have made to mathematics education. We feel equally frustrated with mathematics teachers and educators who think that the study of the history of school mathematics beyond, say, the "New Math(s)," or the "national curriculum," or the "NCTM *Standards*," is an extravagant use of time when so much needs to be done to help to improve the existing state of school mathematics. Many of the present practices in school mathematics—like, for example, the tendency of students to scribble down calculations without explaining their reasoning—can be identified as having strong historical antecedents. Historical perspectives can throw light on important questions, like how much should we allow our students to copy information from a textbook? Do they need to copy anything (because, after all, it is already in their textbook)?

We have identified what we believe to be six important lessons that can be learned from the cyphering tradition. One might ask, "Why six lessons, and not seven or eight? Or five? Etc." Our answer to that question would be that the six lessons that we have identified were those which emerged for us as we investigated the cyphering tradition in a grounded way—from the data, and from our knowledge of the pertinent literature. Our data set has been extensive: we have now examined over 1000 cyphering books altogether—in addition to the 500 in our own collection, we have examined cyphering books in major archival collections within the United States and Great Britain. We also hold a personal collection of about 450 mathematics textbooks published in Great Britain or the United States of America during the period from 1600 through 1860. In addition, we have diligently gained access to, and read, as many books and documents relating to the history of school mathematics during that same period as could be made available to us.

We believe that the lessons identified relate not only to the history of school mathematics, but also to intended, implemented and attained mathematics curricula of today. We hope that these "lessons from the past" will be scrutinized, for subsequent discussion of them might mean that there will be a greater chance that some of the lessons will be noted and, perhaps, heeded.

References

Allanson, J. T. (1968). Mathematics for the majority. *Journal of Biological Education, 2*(2), 87–93.

Bartelt, W. E. (2008). "*There I grew up*": *Remembering Abraham Lincoln's Indiana youth.* Indianapolis, IN: Indiana Historical Society Press.

Bell, E. T. (1945). *The development of mathematics* (2nd ed.). New York, NY: McGraw-Hill.

Boyer, C. B. (1991). *A history of mathematics* (2nd ed.). New York, NY: John Wiley & Sons.

Broome, E. C. (1903). *A historical and critical discussion of college-entrance admissions requirements.* New York, NY: Columbia University.

Cocker, E. (1695). *Cocker's decimal arithmetick, ...* London, United Kingdom: George Sawbridge.

D'Ambrosio, U. (1986). Socio-cultural bases for mathematical education. In M. Carss (Ed.), *Proceedings of the Fifth International Congress on Mathematical Education* (pp. 1–6). Boston, MA: Birkhauser.

Denniss, J. (2012). *Figuring it out: Children's arithmetical manuscripts 1680–1880.* Oxford, United Kingdom: Huxley Scientific Press.

Dilworth, T. (1806). *The schoolmasters assistant; being a compendium of arithmetic both practical and theoretical*. New York, NY: G. Jansen.

Donald, D. H. (1995). *Lincoln.* New York, NY: Simon & Schuster.

Duffy, C. G. (Ed.). (1845). *The ballad poetry of Ireland* (4th ed.). Dublin, Ireland: J. Duffy.

Ellerton, N. F., & Clements, M. A. (2011). Unique mathematics books from a lost tradition. *The Guild of Book Worker's Journal. 11,* 28–39.

Ellerton, N. F., & Clements, M. A. (2012). *Rewriting the history of school mathematics in North America 1607–1861.* New York, NY: Springer.

Ellerton, N. F., & Clements, M. A. (2013, September). *The mathematics of decimal fractions and their introduction into British and North America schools.* Paper presented to the Third International Conference on the History of Mathematics Education held at Uppsala University Sweden, September 25–28, 2013.

Emigh, R. J. (2002). Numeracy or enumeration? The uses of numbers by states and societies. *Social Science History, 26*(4), 653–698.

Gaydos, T., & Kampas, B. (2010). *American and Canadian cyphering books, n.d., 1727–1864* (2nd ed.). Salem, MA: Phillips Library at the Peabody Essex Museum.

Hawney, W. (1729). *The compleat measurer; or, the whole art of measuring* (3rd ed.). London, United Kingdom: J. and J. Knapton.

Henigan, J. (2012). *Literacy and orality in eighteenth-century Irish song*. London, United Kingdom: Pickering & Chatto Publishers.

Henry, J. (1843). *An address upon education and common schools.* Albany, NY: Steam Press.

Hirsch, D., & Van Haften, D. (2010). *Abraham Lincoln and the structure of reason.* New York, NY: Savas Beatie LLC.

Jackson, L. L. (1906). *The educational significance of sixteenth century arithmetic from the point of view of the present time.* New York, NY: Teachers College Press.

Karpinski, L. C. (1925). *The history of arithmetic.* Chicago, IL: Rand McNally & Company.

Lave, J., & Wenger, E. (1998). *Communities of practice: Learning, meaning, and identity.* Cambridge, United Kingdom: Cambridge University Press.

Mahan, A. T. (1898). *The influence of sea power upon history 1660–1783.* Boston, MA: Little, Brown, and Company.

Pike, S. (1822). *The teachers assistant or a system of practical arithmetic; wherein the several rules of that useful science, are illustrated by a variety of examples, a large proportion of which are in Federal money. The whole is designed to abridge the labour of teachers, and to facilitate the instruction of youth.* Philadelphia, PA: Benjamin Warner.

Rickey, F., Crackel, T., & Silverberg, J. (2013, August). *George Washington's use of trigonometry and logarithms.* Paper presented at the meeting of the Canadian Society for History and Philosophy of Mathematics at Hartford, CT.

Roach, J. (1971). *Public examinations in England 1850–1900.* Cambridge, United Kingdom: Cambridge University Press.

Rotherham, W. (1852). *The algebraical equation and problem papers, proposed in the examinations of St. John's College, Cambridge from the year 1794 to the present time.* Cambridge, United Kingdom: Henry Wallis.

Russell, B. (1926). *On education, especially in early childhood.* London, United Kingdom: George Allen and Unwin.

Santayana, G. (1905). *The life of reason; or, the phases of human progress.* New York, NY: Charles Scribner's Sons.

Sheehan, J. H. (2005). How do we learn from history? *Perspectives, 43*(1), 5–6.

Simons, L. G. (1936). Short stories in colonial geometry. *Osiris, 1*, 584–605.

Stedall, J. (2012). *The history of mathematics: A very short introduction.* Oxford, United Kingdom: Oxford University Press.

Thayer, V. T. (1928). *The passing of the recitation.* Boston, MA: D. C. Heath and Company.

Watson, F., & Kandel, I. L. (1911). Examinations. In P. Monroe (Ed.), *A cyclopaedia of education* (Vol. 2, pp. 532–538). New York, NY: The Macmillan Company.

Wilson, J. I. (1820). *The history of Christ's Hospital from its foundation by King Edward VI.* London, United Kingdom: Author.

Wingate, E. (1624). '*L'usage de la règle de proportion en arithmétique*. Paris, France: Author.

Wingate, E. (1630). *Arithmetique made easie.* London, United Kingdom: Stephens and Meredith.

Combined References

A Society of Gentlemen. (1773). Review of Robertson's *The Elements of Navigation. The Critical Review: Or Annals of Literature, 35*, 130–134.

A Teacher of Arithmetic. (1812). *A key to Dilworth's Arithmetic; wherein every question is worked out at full length, in a short and comprehensive manner.* New York, NY: Smith & Forman.

Adams, D. (1822). *The scholar's arithmetic—Or Federal accountant.* Keene, NH: John Prentiss.

Adams, D. (1848). *Adams's new arithmetic: Arithmetic in which the principles of operating by numbers are analytically explained and synthetically explained, thus combining the advantages to be derived both from the inductive and synthetic mode of instructing.* New York, NY: Collins & Brother.

Alexander, J. (1709). *A synopsis of algebra,* London, United Kingdom: J. Barber.

Allan, G. A. T. (1984). *Christ's Hospital.* London, United Kingdom: Town & Country Books.

Allanson, J. T. (1968). Mathematics for the majority. *Journal of Biological Education, 2*(2), 87–93.

Allen, H. D. (1970). The verse problems of early American arithmetics. *The Journal of Rutgers University Libraries, 33*(2), 49–62.

Allen, H. D. (1990). Classic "measurement" problems feature practical fanciful. *The Mathematical Log, 34*(1), 1, 4, 6.

Allen, J. B. L. (1970). *The English mathematical schools, 1670–1720.* PhD thesis, University of Reading.

Arnold, I. N. (1891). *The life of Abraham Lincoln.* Chicago, IL: A. C. McClurg & Company.

Austin, P. B. (1885). William Beattie (1793–1875). In L. Stephen (Ed.), *Dictionary of national biography* (Vol. 4, pp. 25–27). London, United Kingdom: Smith, Elder & Co.

Ayres, J. (1682). *Arithmetick and writing.* London, United Kingdom: Author.

Azel W. Dorsey, Lincoln Pedagogue (1930, July 7). *Lincoln Lore, 65*, 1–2.

Babcock, T. H. (1829). *The practical arithmetic; in which the principles of operating by numbers are analytically explained and synthetically applied ... Adapted to the use of schools and academies in the United States.* New York, NY: G & C & H. Carvill.

Bache, A. D. (1839). *Report on education in Europe, to the trustees of the Girard College for Orphans.* Philadelphia, PA: Bailey.

Baker, H. (1568/1591). *The well-spring of sciences, which teacheth the perfect worke and practice of arithmeticke, both in whole numbers and in fractions.* London, United Kingdom: Thomas Purfoot.

Baker, H. (1687). *Baker's arithmetick teaching the perfect worke and practice of arithmeticke, both in whole numbers and in fractions, whereunto are added many rules and tables of interest, rebate and purchases, &c, also the art of decimal fractions intermixed with common fractions for the better understanding thereof, newly corrected and contracted and made more plain and ease by Henry Phillippes.* London, United Kingdom: J. Richardson and William Thackery.

Banta, D. D. (1906). The early schools of Indiana: Third installment from the papers of D. D. Banta first published in the *Indianapolis News* in 1892. *The Indiana Quarterly Magazine of History, 2*(1), 131–138.

Barrett, J. H. (1860). *Life of Abraham Lincoln, of Illinois.* Cincinnati, OH: Moore, Wilstach, Keys & Co.

Bartelt, W. E. (2008). *"There I grew up": Remembering Abraham Lincoln's Indiana youth.* Indianapolis, IN: Indiana Historical Society Press.

DOI 10.1007/978-3-319-02502-5, © Springer International Publishing Switzerland 2014

Basler, R. P., Pratt, M. D., & Dunlap, L. A. (Eds.). (1953). *The collected works of Abraham Lincoln.* New Brunswick, NJ: Rutgers University Press.

Beattie, W. (1831). *Journal of residence in Germany, written during a professional attendance on their Royal Highnesses the Duke and Duchess of Clarence ... during their visits ... 1822, 1825, and 1826.* London, United Kingdom: British Library.

Beattie, W. (Ed.). (1855). *Life and letters of Thomas Campbell.* New York, NY: Harper & Brothers, Publishers.

Beaujouan, G. (1988). The place of Nicolas Chuquet in a typology of 15th-century French arithmetics. In C. Hay (Ed.), *Mathematics from manuscript to print 1300–1600* (pp. 73–88). Oxford, United Kingdom: Clarendon Press.

Bell, E. T. (1945). *The development of mathematics* (2nd ed.). New York, NY: McGraw-Hill.

Bennett, T. (1815). *New system of practical arithmetic peculiarly calculated for the use of schools in the United States* (5th ed.). Philadelphia, PA: Bennett & Walton.

Benoit, P. (1988). The commercial arithmetic of Nicolas Chuquet. In C. Hay (Ed.), *Mathematics from manuscript to print 1300–1600* (pp. 96–116). Oxford, United Kingdom: Clarendon Press.

Berkin, C. (2005). *Revolutionary mothers: Women in the struggle for America's independence.* New York, NY: Vintage Books.

Bethlehem Digital History Project. (2013). *Susan Shimer cyphering book.* http://bdhp.moravian. edu/ bethlehem/bethlehem.html, retrieved November 15, 2013.

Beveridge, A. J. (1958). *Abe Lincoln in Indiana.* Fort Wayne, IN: Public Board for Allen County.

Bickham, G. (c. 1740). *Youth's instructor in the art of numbers, a new cyphering book in which is shewn variety of penmanship & command of hand: Engrav'd for the use of schools.* London, United Kingdom: William & Cluer Dicey.

Bickham, G. (Ed.). (1743). *The universal penman.* London, United Kingdom: Henry Overton.

Bidwell, J. (2013). *American paper mills: A directory of the paper trade with notes on products, watermarks, distribution methods and manufacturing techniques.* Hanover, NH: University Press of New England.

Bjarnadóttir, K. (2009, October 15). *Arithmetica—Arithmetic art, mathematics for all.* Paper presented at the Nordic Research Network on Special Needs Education in Mathematics (NORSMA) Conference, held at the University of Iceland.

"Blues" and the Royal Navy. (1900, November 24). *The Navy and Army Illustrated,* 245–248.

Bowditch, N. (1853). *The new American practical navigator.* New York, NY: E. G. W. Blunt.

Boyer, C. B. (1991). *A history of mathematics* (2nd ed.). New York, NY: John Wiley & Sons.

Brenan, M. (1935). *Schools of Kildare and Leighlin, A.D.1775–1835.* Dublin, Ireland: Gill and Son.

Broome, E. C. (1903). *A historical and critical discussion of college-entrance admissions requirements.* New York, NY: Columbia University.

Brown, J. C., & Coffman, L. D. (1914). *How to teach arithmetic: A manual for teachers and a text-book for normal schools.* Chicago, IL: Row Peterson and Company.

Burlingame, M. (2008). *Abraham Lincoln: A life* (2 vols). Baltimore, MD: The Johns Hopkins University Press.

Burton, W. (1833). *The district school as it was, by one who went to it.* Boston, MA: Carter, Hendee and Company.

Butler, L. S. (1976). *North Carolina and the coming of the Revolution, 1763–1776.* Zebulon, NC: Theo. Davis Sons, Inc.

Butler, W. (1819). *An introduction to arithmetic designed for the use of young ladies* (3rd ed.). London, United Kingdom: Simpkin, Marshall & Co.

Cajori, F. (1890). *The teaching and history of mathematics in the United States* (Circular of Information No. 3, 1890). Washington, DC: Bureau of Education.

Cajori, F. (1907). *A history of elementary mathematics with hints on methods of teaching.* New York, NY: The Macmillan Company.

Carlo, P. W. (2005). *Huguenot refugees in colonial New York: Becoming American in the Hudson Valley.* Brighton, United Kingdom: Sussex Academic Press.

Carpenter, A. (1998). *Verse in English from eighteenth-century Ireland.* Cork, Ireland: Cork University Press.

Carpenter, C. (1963). *History of American schoolbooks.* Philadelphia, PA: University of Pennsylvania Press.

Cavell, S. A. (2012). *Midshipmen and quarterdeck boys on the British Navy 1793–1815.* Woodbridge, United Kingdom: Boydell Press.

Champion, R. J. (1747). *The tutor's assistant in teaching arithmetic.* London, United Kingdom: Author.

Chapman, T. J. (1895). Some early schools of Pennsylvania. *Pennsylvania School Journal, 44*, 205–209.

Christ's Hospital (1595). *Dame Mary Ramsey's gift: Deeds relating to the maintenance by the Governors of Christ's Hospital of a grammar school in Halstead, Essex.* London Metropolitan Archives Reference CLC/210/G/BRB/041/MS13583.

Christ's Hospital (1823–1830). *General Committee minutes*, Volume 17. Handwritten manuscript, London Metropolitan Archives Reference CLC/210/B/005/MS12811/017.

Christ's Hospital (1836–1840). *General Committee minutes*, Volume 19. Handwritten manuscript, London Metropolitan Archives Reference CLC/210/B/005/MS12811/019.

Christ's Hospital (1857). *Register of Christ's Hospital Royal Mathematical School boys apprenticed to ships' masters 1816–1857.* Handwritten manuscript, London Metropolitan Archives Reference CLC/526/MS30338.

Christ's Hospital (1953). *The Christ's Hospital book.* London, United Kingdom: Hamish Hamilton.

Clark, K. M. (2013). "In these numbers we use no fractions": A classroom module on Stevin's decimal fractions. MathDL (The MAA Mathematical Sciences Digital Library).

Clark, P., & Murfin, L. (1995). *The history of Maidstone—The making of a modern county town.* Stroud, United Kingdom: Alan Sutton Publishing.

Clarke, P. (1996). A glimpse at Irish accounting history. *Irish Accounting Review, 3(*2), 23–40.

Clemens, S. (Mark Twain). (1885). *Adventures of Huckleberry Finn.* New York, NY: Charles L. Webster.

Clements, M. A. (1980). Analyzing children's errors on written mathematical tasks. *Educational Studies in Mathematics, 11*(1), 1–21.

Clements, M. A., & Ellerton, N. F. (2012, October 27). *The curious case of duodecimals in North American cyphering books and early textbooks.* Paper presented to the History and Pedagogy of Mathematics (HPM) Conference, The University of California at Berkeley, California.

Clements, M. A., Keitel, C., Bishop, A. J., Kilpatrick, J., & Leung, F. (2013). From the few to the many: Historical perspectives on who should learn mathematics. In M. A. Clements, A. J. Bishop, C. Keitel, J. Kilpatrick, & F. Leung (Eds.), *Third international handbook of mathematics education* (pp. 7–40). New York, NY: Springer.

Cobb, L. (1835). *Cobb's ciphering book, No. 1, containing all the sums and questions for theoretical and practical exercises in Cobb's Explanatory Arithmetic No 1.* Elmira, NY: Birdsall & Huntley.

Cocker, E. (1664). *The tutor to writing and arithmetic.* London, United Kingdom. H. Tracy.

Cocker, E. (1678/1725). *Cocker's arithmetick: A plain and familiar method ...* (42nd ed.). London, United Kingdom: H. Tracy.

Cocker, E. (1684/1720). *Decimal arithmetic, wherein is shewed the nature and use of decimal fractions in the usual rules of arithmetic, ...* (5th ed.). London, United Kingdom: J. Darby for M. Wellington.

Cocker, E. (1695). *Cocker's decimal arithmetick, ...* London, United Kingdom: George Sawbridge.

Cohen, P. C. (1982). *A calculating people: The spread of numeracy in early America.* Chicago, IL: University of Chicago Press.

Cohen, P. C. (2003). Numeracy in 19th-century America. In G. M. A. Stanic & J. Kilpatrick (Eds.), *A history of school mathematics* (Vol. 1, pp. 43–76). Reston, VA: National Council of Teachers of Mathematics.

Colburn, W. (1822). *Arithmetic upon the inductive method of instruction being a sequel to intellectual arithmetic*. Boston: MA: Cummings and Hilliard.

Colburn, W. (1824) *Arithmetic; being a sequel to first lessons in arithmetic*. Boston, MA: Cummings, Hilliard, & Co.

Coldham, P. W. (1990). *Child apprentices in America from Christ's Hospital, London, 1617–1778.* Baltimore, MD: Genealogical Publishers Co.

Colson, J. (1736). *An arithmetical copy-book.* London, United Kingdom: Author.

Courtenay, C. M., with illustrations by A. M. Turner & H. Kaucher (1917). *Abraham Lincoln.* New York, NY: Sam'l Gabriel Sons & Company.

Cracraft, J. (1971). *The church reform of Peter the Great.* London, United Kingdom: Macmillan.

Cremin, L. A. (1970). *American education: The colonial experience 1607–1783*. New York, NY: Harper & Row.

Cross, A. (2007). *By the banks of the Neva.* Cambridge, United Kingdom: Cambridge University Press.

Cubberley, E. P. (1920). *The history of education.* Boston, MA: Houghton Mifflin Company.

Cullen, L. (1988*). The hidden Ireland: Reassessment of a concept.* Dublin, Ireland: Lilliput.

Daboll, N. (1820). *Daboll's schoolmaster's assistant; improved and enlarged; being a plain practical system of arithmetic adapted to the United States*. Auburn, NY: Thomas M. Skinner.

D'Ambrosio, U. (1986). Socio-cultural bases for mathematical education. In M. Carss (Ed.), *Proceedings of the Fifth International Congress on Mathematical Education* (pp. 1–6). Boston, MA: Birkhauser.

Dash, J. (2000). *The longitude prize.* New York, NY: Frances Foster Books.

Davis, N. Z. (1960). Sixteenth-century French arithmetics on the business life. *Journal of History of Ideas*, *21*(1), 18–48..

Davis, V. J. (1890). *Jefferson Davis: Ex-President of the Confederate States of America* (Vol. 1). New York, NY: Belford Company.

De La Hunt, T. J. (1916). *Perry County: A history*. Indianapolis, IN: The W. K. Stewart Co.

De Morgan, A. (1847). *Arithmetical books from the invention of printing to the present time.* London, United Kingdom: Taylor and Walton.

De Morgan, A. (1860). Drawing Society of Dublin. *Notes and Queries, 2*(9), 444.

Denniss, J. (2012). *Figuring it out: Children's arithmetical manuscripts 1680–1880.* Oxford, United Kingdom: Huxley Scientific Press.

Dickens, C. (1850). *David Copperfield.* London, United Kingdom: Bradbury & Evans.

Dickinson, H. W. (2007). *Educating the Royal Navy.* London, United Kingdom: Routledge.

Dilworth, T. (1773). *The schoolmasters assistant: Being a compendium of arithmetic, both practical and theoretical* (17th ed.). Philadelphia, PA: Joseph Crukshank.

Dilworth, T. (1784). *The young book-keeper's assistant; shewing him, in the most plain and easy manner, the Italian way of stating debtor and creditor* (9th ed.). London, United Kingdom: Richard and Henry Causton.

Dilworth, T. (1792). *The schoolmasters assistant; being a compendium of arithmetic both practical and theoretical*. New York, NY: Hugh Gaine.

Dilworth, T. (1806). *The schoolmasters assistant; being a compendium of arithmetic both practical and theoretical*. New York, NY: G. Jansen.

Ditton, H. (1709). *An appendix to the algebra of John Alexander*. London, United Kingdom: J. Barber.

Doar, A. K. (2006). *Cipher books in the Southern Historical Collection*. Master of Science thesis, Wilson Library, University of North Carolina at Chapel Hill.

Donald, D. H. (1995). *Lincoln*. New York, NY: Simon & Schuster.

Dorfman, M. (1966a). Lincoln's arithmetic education: Influence on his life. *Lincoln Herald, 68*(2), 61–80.

Dorfman, M. (1966b). Lincoln's arithmetic education: Influence on his life: Part II. *Lincoln Herald, 68*(3), 108–120.

Doyle, D. N. (1980). *Ireland, Irishmen and revolutionary America*. Cork, Ireland: Mercier.

Duffy, C. G. (Ed.). (1845). *The ballad poetry of Ireland* (4th ed.). Dublin, Ireland: J. Duffy.

Duhem, P. (1985). *Medieval cosmology, theories of infinity, place, time, void, and the plurality of worlds*. Chicago, IL: Chicago University Press.

Dunlap, L. A. (1959). Lincoln's sum book. *Lincoln Herald, 61*(1), 6–10.

Dunleavy, J., & Dunleavy, G. (1991). *Douglas Hyde—A maker of modern Ireland*. Berkeley, CA: University of California Press.

Durkin, J. J. (1942). Journal of the Revd. Adam Marshall, schoolmaster, *U.S.S. North Carolina*, 1824–1825. *Records of the American Catholic Historical Society of Philadelphia, 53*(4), 152–168.

Ellerton, N. F., & Clements, M. A. (2008). An opportunity lost in the history of school mathematics: Noah Webster and Nicolas Pike. In O. Figueras, J. L. Cortina, S. Alatorre, & A. Mepúrlveda (Eds.), *Proceedings of the Joint Meeting of PME 32 and PME-NA XXX* (Vol. 2, pp. 447–454). Morelia, Mexico: Cinvestav-UMSWH.

Ellerton, N. F., & Clements, M. A. (2011). Unique mathematics books from a lost tradition. *The Guild of Book Worker's Journal. 11*, 28–39.

Ellerton, N. F., & Clements, M. A. (2012). *Rewriting the history of school mathematics in North America 1607–1861*. New York, NY: Springer.

Ellerton, N. F., & Clements, M. A. (2013, September). *The mathematics of decimal fractions and their introduction into British and North America schools*. Paper presented to the Third International Conference on the History of Mathematics Education held at Uppsala University Sweden, September 25–28, 2013.

Emerson, F. (1838). *Key to the North American arithmetic, Part second and third. For the use of teachers*. Boston, MA: G. W. Palmer & Co.

Emigh, R. J. (2002). Numeracy or enumeration? The uses of numbers by states and societies. *Social Science History, 26*(4), 653–698.

Erekson, K. A. (2012). *Everybody's history: Indiana's Lincoln Inquiry and the quest to reclaim a President's past*. Amherst, MA: University of Massachusetts Press.

Fenn, J. (1769). *First volume of the instructions given in the drawing school established by the Dublin-Society, pursuant to their resolution of the fourth of February, 1768; ... Under the direction of Joseph Fenn*. Dublin, Ireland: Alex M'Culloh.

Fisher, G. (1734). *Arithmetick in the plainest and most concise methods hitherto extant. With new improvements for dispatch of business in all the several rules. As also fractions, vulgar and decimals, wrought together after a new method, which renders both easy to be understood in their nature and use. ...* London, United Kingdom: Bettesworth and C. Hitch.

Franci, R. (2009). Mathematics in the manuscript of Michael of Rhodes. In P. O. Long, D. McGee, & A. M. Stahl (Eds.), *The book of Michael of Rhodes: A 15th century maritime manuscript* (Vol. 3, pp. 115–146). Cambridge, MA: MIT Press.

Freeman, F. N. (1913). Writing. In P. Monroe (Ed.). *A cyclopedia of education* (Vol. 5, pp. 819–827). New York, NY: The Macmillan Company.

Funk, A. L. (1969). *A sketchbook of Indiana history*. Rochester, IN: Christian Book Press.

Gaydos, T., & Kampas, B. (2010). *American and Canadian ciphering books, n.d., 1727–1864.* Salem, MA: Phillips Library at the Peabody Essex Museum.

Gibson, R., & Craig, J. D. (1822). *A treatise on practical surveying* (3rd ed.). Baltimore, MD: F. Lucas Jr., and Cushing & Jewett.

Goodspeed Bros. & Co. (1885). *History of Warwick, Spencer and Perry Counties, Indiana*. Chicago, IL: Author.

Griffis, W. E. (1909). *The story of New Netherland.* Boston, MA: The Riverside Press.

Grigsby, N. (1865, June 13). Interview with Nathaniel Grigsby. Reproduced in D. L. Wilson, R. O. Davis, & T. Wilson (Eds.). (1998). *Herndon's informants: Letters, interviews and statements about Abraham Lincoln* (pp. 111–115). Urbana, IL: University of Illinois Press.

Grimes, J. B. (1910). *Abstract of North Carolina wills.* Raleigh, NC: E. M. Uzzell and Co.

Hanks, D. (1865, June 13). Interviews with Dennis Hanks. Reproduced in D. L. Wilson, R. O. Davis, & T. Wilson (Eds.). (1998). *Herndon's informants: Letters, interviews and statements about Abraham Lincoln* (pp. 38–42, and pp. 103–106). Urbana, IL: University of Illinois Press.

Hans, N. A. (1951a). The Moscow School of Mathematics and Navigation. *The Slavonic and East European Review, 29*(73), 532–536.

Hans, N. A. (1951b). *New trends in education in the 18th century.* London, United Kingdom: Routledge.

Hathaway, J. R. B. (Ed.). (1900). *The North Carolina historical and genealogical register.* Edenton, NC: Clearfield.

Hawney, W. (1729). *The compleat measurer; or, the whole art of measuring* (3rd ed.). London, United Kingdom: J. and J. Knapton.

Heal, A. (1931). *The English writing-masters and their copy-books 1570–1800.* Cambridge, United Kingdom: Cambridge University Press.

Henigan, J. (2012). *Literacy and orality in eighteenth-century Irish song.* London, United Kingdom: Pickering & Chatto Publishers.

Henry, J. (1843). *An address upon education and common schools.* Albany, NY: Steam Press.

Henry, W., & Kingston, W. (1878). *The two supercargoes.* London, United Kingdom: Sampson, Low, Marston, Searle & Rivington.

Herndon, W. H., & Weik, J. W. (1936). *Herndon's life of Lincoln: The history and personal recollections of Abraham Lincoln.* New York, NY: Albert & Charles Boni.

Hersee, J. (1997). Multiplication is vexation. *Paradigm, 24,* 2–10.

Hirsch, D., & Van Haften, D. (2010). *Abraham Lincoln and the structure of reason.* New York, NY: Savas Beatie LLC.

Hodder, J. (1714). *Arithmetick, or that necessary art made most easie* (520th ed.). London, United Kingdom: N. & M. Baddington.

Hodgson, J. (1723). *A system of mathematics containing the Euclidean geometry, plane and spherical trigonometry, the projection of the sphere, both orthographic and stereographic, astronomy, the use of globes and navigation.* London, United Kingdom: Thomas Page, William and Fisher Mount.

Hopkins, W. (1833). *Elements of trigonometry.* London, United Kingdom: Baldwin and Craddock.

Howson, G. (1982). *A history of mathematics education in England.* Cambridge, United Kingdom: Cambridge University Press.

Howson, G., Keitel, C., & Kilpatrick, J. (1981). *Curriculum development in mathematics.* Cambridge, United Kingdom: Cambridge University Press.

Houser, M. L. (1943). *Young Abraham Lincoln mathematician*. Peoria, IL: L. O. Schriver.

Howson, G. (1982). *A history of mathematics education in England.* Cambridge, United Kingdom: Cambridge University Press.

Høyrup, J. (2005). Leonardo Fibonacci and *abbaco* culture: A proposal to invert the roles. *Revue d'Histoire des Mathématiques, 11*, 23–56.

Høyrup, J. (2008). The tortuous ways toward a new understanding of algebra in the Italian *Abbacus* School (14th–16th centuries). In O. Figueras, J. L Cortina, A. Alatorre, T. Rojano & S. Sepulveda (Eds.), *Proceedings of the joint meeting of PME 32 and PME-NA XXX* (Vol. 1, pp. 1–20). Morelia, Mexico. International Group for the Psychology of Mathematics.

Hunter, J. (1793). *An historical journal of the transactions at Port Jackson and Norfolk Island with the discoveries that have been made in New South Wales and in the Southern Ocean since the publication of Phillip's voyage, compiled from the official papers, including the journals of Governors Phillip and King, and of Lieut Ball.* London, United Kingdom: John Stockdale.

Hutton, C. (1764). *Schoolmaster's guide: Or a complete system of practical arithmetic adapted to the use of schools.* Newcastle Upon Tyne, United Kingdom: J. White and T. Saint.

Ifrah, G. (2000). *The universal history of numbers from prehistory to the invention of the computer.* New York, NY: John Wiley & Sons. Inc.

Indiana Department of Public Instruction. (1917). *Twenty-eighth biennial report of the State Superintendent of Public Instruction for the school years ending July 31, 1915 and July 31, 1916*. Fort Wayne, IN: Fort Wayne Printing Co.

Jackson, L. L. (1906). *The educational significance of sixteenth century arithmetic from the point of view of the present time.* New York, NY: Columbia Teachers College.

Jefferson, T. (1784). *Notes on the establishment of a money unit and of a coinage for the United States.* Handwritten manuscript held in the Library of Congress, Washington, DC.

Jefferson, T. (1785). *Notes on the establishment of a money unit and of a coinage for the United States.* Paris, France: Author.

Jenkins, J. (1813). *The art of writing.* Cambridge, United Kingdom: Author.

Jess, Z. (1821). *The American tutor's assistant, improved; or, a compendious system of decimal, practical arithmetic.* Philadelphia, PA: M'Carty & Davis.

Johnson, H. (1719). *New treatise of practical* arithmetic (4th ed.). London, United Kingdom: Thomas Wood.

Johnson, W. R. (1994). "Chanting choristers": Simultaneous recitation in Baltimore's nineteenth century primary schools. *History of Education Quarterly, 34*(1), 1–23.

Johnston, T. L. (1866). Interviews with Thomas L. D. Johnston. Reproduced in D. L. Wilson, R. O. Davis, & T. Wilson (Eds.). (1998). *Herndon's informants: Letters, interviews and statements about Abraham Lincoln* (pp. 531–533). Urbana, IL: University of Illinois Press.

Jones, H. S. (1954). Foreword by the Astronomer Royal. In E. G. R. Taylor, *The mathematical practitioners of Tudor & Stuart England 1485–1714* (pp. *ix–x*). Cambridge, United Kingdom: Cambridge University Press.

Karpinski, L. C. (1925). *The history of arithmetic.* Chicago, IL: Rand McNally & Company.

Karpinski, L. C. (1980). *Bibliography of mathematical works printed in America through 1850* (2nd ed.). New York, NY: Arno Press.

Kersey, J. (1683). *Mr Wingate's arithmetick, ...* (8th ed.). London, United Kingdom: J. Williams.

Kilpatrick, J. (2012). Foreword. In N. F. Ellerton & M. A. Clements, *Rewriting the history of school mathematics in North America 1607–1861* (pp. v–vi). New York, NY: Springer.

Kilpatrick, J., & Izsák, A. (2008). A history of algebra in the school curriculum. In C. E. Greenes & R. Rubenstein (Eds.), *Algebra and algebraic thinking in school mathematics: Seventieth yearbook* (pp. 3–18). Reston, VA: National Council of Teachers of Mathematics.

Kilpatrick, W. H. (1912) *The Dutch schools of New Netherland and colonial New York.* Washington, DC: US Bureau of Education.

Knox, V. (1788). *Liberal education: Or a practical treatise on the methods of acquiring useful and polite learning.* London, United Kingdom: Charles Dilly.

Kool, M. (1999). *Die conste vanden getale. Een studie van Nederlandstalige rekenboeken uit de vijftiende en zestiende eeuw, met een glossarium van rekenkundige termen* [*The art of numbers. A study of Dutch arithmetic books of the 15th and 16th century, with a glossary of arithmetical terms*]. Hilversum, The Netherlands: University College.

Koos, G. (2001). The Irish hedge schoolmaster in the American backcountry. *New Hibernia Review, 5*(2), 9–26.

Lam, L-Y (1966). On the Chinese origin of the galley method of arithmetical division. *The British Journal for the History of Science, 3*(1), 66–69.

Lave, J., & Wenger, E. (1998). *Communities of practice: Learning, meaning, and identity.* Cambridge, United Kingdom: Cambridge University Press.

Linehan, J. C., & Murray, T. H. (1898). *Irish Schoolmasters in the American colonies, 1640–1775, with a continuation of the subject during and after the War of Revolution.* Washington, DC: American-Irish Historical Society.

Linklater, A. (2002). *Measuring America: How the United States was shaped by the greatest land sale in history.* New York, NY: Plume.

Long, P. O., McGee, D., & Stahl, A. M. (Eds.). (2009). *The book of Michael of Rhodes: A 15th century maritime manuscript.* Cambridge, MA: MIT Press.

Loomis, E. (1868). *Elements of plane and spherical trigonometry, with their applications to mensuration, surveying, and navigation.* New York, NY: Harper & Brothers.

Lounsbury, R. G. (1930). Yankee trade at Newfoundland. *The New England Quarterly, 3*(4), 607–626.

Love, J. (1688). *Geodaesia, or the art of surveying and measuring of land, made easie, showing by plain and practical rules how to survey, protract, cast up, reduce or divide any piece of land whatsoever; with new tables for the ease of the surveyor in reducing the measures of land. Moreover, a more facile and sure way of surveying by the chain than has hitherto been taught. As also, how to lay out new lands in America, or elsewhere: and how to make a perfect map of a river's mouth or harbour, with several other things never yet published in our language.* London, United Kingdom: John Taylor.

Mahan, A. T. (1898). *The influence of sea power upon history 1660–1783.* Boston, MA: Little, Brown, and Company.

Mann, M. (1865). *Life of Horace Mann.* Boston, MA: Walker, Fuller, and Company.

Mansell, K. (2011). *Christ's Hospital in the Victorian era.* Whitton, United Kingdom: Ashwater Press.

Manwaring, E. (1925). *Italian landscape in eighteenth century England.* New York, NY: Oxford University Press.

Martin, G. H. (1897). *The evolution of the Massachusetts public school system: A historical sketch.* New York, NY: D. Appleton & Co.

Martin, N. (1800). *A new system of arithmetic, to which is added, a form of book-keeping by single entry.* Air, Scotland: J. & P. Wilson.

Massey, W. (1763). *The origin and progress of letters.* London, United Kingdom: Johnson.

McCusker, J. J., & Menard, R. R. (1985). *The economy of British America, 1607–1789.* Chapel Hill, NC: University of North Carolina Press.

McKenney, F. (1825). *A complete key to the Teacher's Assistant, or System of Practical Arithmetic, compiled by Stephen Pike.* Philadelphia, PA: McCarty & Davis.

McLachlan, H. (1931). *English education under the Test Acts: Being the history of the nonconformist academies, 1662–1820*; Manchester, United Kingdom: Manchester University Press.

McMillin, M. M. (2011a). *Gather the shadowmen: The lords of the ocean.* Richardson, TX: Hephaestus.

McMillin, M. M. (2011b). *Prince of the Atlantic: The war.* Richardson, TX: Hephaestus.

McMurry, C. A., & McMurry, F. M. (1903). *The method of recitation.* New York, NY: The Macmillan Company.

Michael, I. (1993). The textbook as a commodity: Walkingame's *The Tutor's Assistant. Paradigm, 12*, 2–10.

Middlekauff, R. (1963). *Ancients and axioms.* New Haven, CT: Yale University Press.

Miller, K. A. (1985). *Immigrants and exiles: Ireland and the Irish exodus to North America.* New York, NY: Oxford University Press.

Miller, R. L. (2006). *Lincoln and his world: The early years.* Mechanicsburg, PA: Stackpole Books.

Monaghan, E. J. (2007). *Learning to read and write in colonial America.* Amherst, MA: University of Massachusetts Press.

Money, J. (1993). Teaching in the market place, or "*Caesar adsum jam forte: Pompey aderat*": The retaining of knowledge in provincial England during the 18th century. In J. Brewer & R. Porter (Eds.), *Consumption and the world of goods* (pp. 335–377). London, United Kingdom: Routledge.

Monroe, W. S. (1917). *Development of arithmetic as a school subject.* Washington, DC: Government Printing Office.

Moore, J. H. (1779). *The practical navigator and seaman's new daily assistant, being a complete system of practical navigation, improved and rendered easy to any common capacity* (4th ed.). London, United Kingdom: B. Law.

Moore, J. H. (1791). *The practical navigator and seaman's new daily assistant, being an epitome of navigation, including the different methods of working the lunar observations, with every particular requisite for keeping a complete journal at sea.* London, United Kingdom: B. Law.

Moore, J. H. (1796). *The new practical navigator, being an epitome of navigation, explaining the different methods of working the lunar observations, and all the requisite tables used with the nautical almanac, in determining the latitude and longitude, and keeping a complete reckoning at sea; illustrated by proper rules and examples; the whole exemplified in a journal kept from*

England to the island of Teneriffe: also, the substance of the examination, every candidate for a commission in the Royal Navy, and Officer in the Honourable East India Company's service, must pass through previous to their being appointed; this with the sea terms, are particularly recommended to the attention of all young gentlemen designed for, or belonging to the sea. (12th ed.). Tower Hill, United Kingdom: E. Law.

Moore, J., & Perkins, P. (1681). *A new systeme of the mathematicks designed for the use of the Royal Foundation of the Mathematical School in Christ's Hospital.* London, United Kingdom: Robert Scott.

Moravian Publication Office. (1876). *Historical sketch of the Moravian Seminary for Young Ladies.* Bethlehem, PA: Author.

Morley, V. (2002). *Irish opinion and the American Revolution 1760–1783.* Cambridge, United Kingdom: Cambridge University Press.

Morrice, D. (1801). *The young midshipman's instructor (designed to be a companion to Hamilton Moore's Navigation): With useful hints to parents of sea youth, and to captains and schoolmasters of the Royal Navy.* London, United Kingdom: Knight and Compton.

National Park Service. (1998). *National Registrar of Historic Places Registration Form: Jonathan Hill Jacocks House.* Washington, DC: United States Department of the Interior.

Newby, W. G. (1958). *Town of Hertford bi-centennial, 1758–1958: And historic data of Perquimans County, North Carolina.* New Bern, NC: Owen G. Dunn.

Nicolay, J. G. (1904). *A short life of Abraham Lincoln.* New York, NY: The Century Co.

Northend, M. H. (1917). *Memories of old Salem: Drawn from letters of a great-grandmother.* New York, NY: Moffatt, Yard and Company.

Novaresio, P. (1996). *The explorers.* New York, NY: Stewart, Tabori & Chang.

O'Brien, M. J. (1920). *A hidden phase of American history: Ireland's part in America's struggle for liberty.* New York, NY: Dodd, Mead and Company.

Okenfuss, M. J. (1973). Technical training in Russia under Peter the Great. *History of Education Quarterly, 13*(4), 325–342.

Padfield, P. (2005). *Maritime power and the struggle for freedom: Naval campaigns that shaped the modern world.* Woodstock, NY: The Overlook Press.

Pearce, E. H. (1901). *Annals of Christ's Hospital.* London, United Kingdom: Hugh Reed.

Pedersen, O. (1985). In quest of Sacrobosco. *Journal for the History of Astronomy, 16,* 175–221.

Phillip, A. (1789). *The voyage of Governor Phillip to Botany Bay, with an account of the establishment of the colonies of Port Jackson & Norfolk Island, to which are added the journals of Lieuts. Shortland, Watts, Ball & Capt. Marshall.* London, United Kingdom: John Stockdale.

Pike, N. (1788). *A new and complete system of arithmetic, composed for the use of citizens of the United States.* Newbury-Port, MA: John Mycall.

Pike, S. (1811). *The teachers assistant; or a system of practical arithmetic; wherein the several rules of that useful science, are illustrated by a variety of examples, a large proportion of which are in Federal money. The whole is designed to abridge the labour of teachers, and to facilitate the instruction of youth.* Philadelphia, PA: Johnstonn (sic.) and Warner.

Pike, S. (1822). *The teachers assistant or a system of practical arithmetic; wherein the several rules of that useful science, are illustrated by a variety of examples, a large proportion of which are in Federal money. The whole is designed to abridge the labour of teachers, and to facilitate the instruction of youth.* Philadelphia, PA: Benjamin Warner.

Plowden, F. (1805). *An historical review of the State of Ireland from the invasion of that country under Henry II to Its union with Great Britain on the First of January 1801.* Philadelphia, PA: W. F. McLaughlin and Bartholomew Graves.

Plumley, N. (1976). The Royal Mathematical School within Christ's Hospital. *Vistas in Astronomy, 20*, 51–59.

Plummer, M. A. (2001). *Lincoln's rail-splitter: Governor Richard J. Oglesby.* Urbana, IL: University of Illinois Press.

Porter, T. (1995). *Trust in numbers: The pursuit of objectivity in science and public life.* Princeton, NJ: Princeton University Press.

Rawley, J. A. (1981). *The trans-Atlantic slave trade*. New York, NY: W. W. Norton.

Recorde, R. (1543/1549). *Grounde of the arts: Teaching the worke and practise, of arithmeticke.* London, United Kingdom: R. Wolff.

Reynolds, B. E. (1993). The algorists vs the abacists: An ancient controversy on the use of calculators. *The College Mathematics Journal, 24*(3), 218–223.

Reynolds, J. (2010, April 14). Reunited set of math problems gives clues to how Lincoln learned. *The State Journal-Register* (Springfield, IL), 1–2.

Rickey, V. F. (1987). Isaac Newton: Man, myth, and mathematics. *College Mathematics Journal, 18*, 362–389.

Rickey, F., Crackel, T., & Silverberg, J. (2013, August). *George Washington's use of trigonometry and logarithms.* Paper presented at the meeting of the Canadian Society for History and Philosophy of Mathematics at Hartford, CT.

Ring, B. (1993). *Girlhood embroidery: American samplers and pictorial needlework, 1650–1850.* New York, NY: Knopf Publishers.

Roach, J. (1971). *Public examinations in England 1850–1900.* Cambridge, United Kingdom: Cambridge University Press.

Robertson, J. (1767). *A general treatise of mensuration* (3rd ed.). London, United Kingdom: J. Nourse and J. Wilcox.

Robertson, J., & Wales, W. (1780). *The elements of navigation containing the theory and practice, with the necessary tables and compendiums for finding the latitude and longitude at sea, to which is added a treatise on marine fortification* (4th ed.). London, United Kingdom: Christ's Hospital.

Rotherham, W. (1852). *The algebraical equation and problem papers, proposed in the examinations of St. John's College, Cambridge from the year 1794 to the present time.* Cambridge, United Kingdon: Henry Wallis.

Russell, B. (1926). *On education, especially in early childhood.* London, United Kingdom: George Allen and Unwin.

Santayana, G. (1905). *The life of reason; or, the phases of human progress.* New York, NY: Charles Scribner's Sons.

Schwarz, C. J. (1920). Lincoln's schooling and pioneer schools. *The Fortnightly Review, 27*(8), 50–51.

Sheehan, J. J. (2005). How do we learn from history? *Perspectives, 43*(1), 5–6.

Simons, L. G. (1924). *Introduction of algebra into American schools in the 18th century.* Washington, DC: Department of the Interior Bureau of Education.

Simons, L. G. (1936). Short stories in colonial geometry. *Osiris, 1*, 584–605.

Sinclair, N. (2008). *The history of the geometry curriculum in the United States.* Charlotte, NC: Information Age Publishing, Inc.

Skemp, R. (1976). Instrumental understanding and relational understanding. *Mathematics Teaching, 77*, 20–26.

Sloan, K. M. (1986). *The teaching of non-professional artists in eighteenth-century England.* PhD thesis, Westfield College, University of London.

Sloan, K. (2000). *"A noble art": Amateur artists and drawing masters c.1600–1800*. London, United Kingdom: British Museum Press.

Small, W. H. (1914). *Early New England schools*. Boston, MA: Ginn.

Smallwood, M. B. (1964). *Some colonial and revolutionary families of North Carolina*. Washington, NC: Author.

Smith, D. E. (1908). *Rara arithmetica.* Boston, MA: Ginn and Company.

Smith, D. E. (1925). *History of mathematics: Volume II, Special topics of elementary mathematics*. Boston, MA: Ginn and Company.

Smith, E. B. (2005). "Let them compleately learn": Manuscript clues about early women's educational practices. In *A manuscript miscellany*, Summer 2005 Institute. Manchester, United Kingdom: Manchester University Press.

Smith, W. C. (1867). *Indiana miscellany: Consisting of sketches of Indian life, the early settlement, customs and hardships of the people.* Cincinnati, OH: R. P. Thompson.

Sobel, D. (1995). *Longitude: The true story of a lone genius who solved the greatest scientific problem of his time*. New York, NY: Walker and Company.

Somerset, A. (1980). *The life and times of William IV.* London, United Kingdom: Weidenfeld Nicholson.

Stedall, J. (2012). *The history of mathematics: A very short introduction.* Oxford, United Kingdom: Oxford University Press.

Steinfeld, R. J. (1991). *The invention of free labor: The employment relation in English and American law and culture 1350–1870.* Chapel Hill, NC: University of North Carolina.

Sterry, C., & Sterry, J. (1795). *A complete exercise book, in arithmetic: Designed for the use of schools in the United States*. Norwich, CT: John Sterry & Co.

Stewart, L. (1999). Other centres of calculation, or where the Royal Society didn't count: Commerce, coffee-houses and natural philosophy in early modern London. *British Journal of the History of Science, 32*, 132–153.

Struik, D. J. (1987). *A concise history of mathematics* (4th ed.). New York, NY: Dover Publications.

Sturtz, L. (2002). *Within her power: Propertied women in colonial Virginia.* New York, NY: Routledge.

Swett, L. (1889). Mr Lincoln's story of his own life. In A. T. Rice (Ed.), *Reminiscences of Abraham Lincoln by distinguished men of his* time (8th ed.) (pp. 455–468). New York, NY: North American Review.

Swetz, F. (1987). *Capitalism and arithmetic: The new math of the 15th century.* La Salle, IL: Open Court.

Tarbell, I. M., & Davis, J. M. (1896). *The early life of Abraham Lincoln containing many unpublished documents and unpublished reminiscences of Lincoln's early friends*. New York, NY: S. S. McClure.

Taylor, E. G. R. (1954). *The mathematical practitioners of Tudor & Stuart England 1485–1714.* Cambridge, United Kingdom: Cambridge University Press.

Taylor, E. G. R. (1956). All plain sailing. *Journal of Navigation, 9*(2), 230–232.

Taylor, E. G. R. (1966). *The mathematical practitioners of Hanoverian England 1714–1840.* Cambridge, United Kingdom: Cambridge University Press.

Thayer, V. T. (1928). *The passing of the recitation.* Boston, MA: D. C. Heath and Company.

Thornton, T. P. (1996). *Handwriting in America: A cultural history.* New Haven, CT: Yale University Press.

Travers, K. J., & Westbury, I. (Eds.). (1989). *The IEA Study of Mathematics I: Analysis of mathematics curricula.* Oxford, United Kingdom: Pergamon.

Trollope, W. (1834). *A history of the royal foundation of Christ's Hospital.* London, United Kingdom: Pickering.

Turnbull, H. W. (Ed.). (2008). *The correspondence of Isaac Newton 1676–1687* (Vol. II). Cambridge, United Kingdom: The Royal Society and Cambridge University Press.

Turnham, D. (1865). Interview with David Turnham. Reproduced in D. L. Wilson, R. O. Davis, & T. Wilson (Eds.), *Herndon's informants: Letters, interviews and statements about Abraham Lincoln* (1998, pp. 120–122). Urbana, IL: University of Illinois Press.

Underhill, D. C. (1825). *Tables of arithmetic made easy.* New York, NY: Mahlon Day.

Unger, F. (1888). *Die methodik der praktischen arithmetick.* Leipsic, Germany: Teubner.

University of Pennsylvania (1749–1768). *Trustees of the University of Pennsylvania 1740–1900, Vol. 1 (1749–1768).* Philadelphia, PA: University of Pennsylvania Archives.

Van Berkel, K. (1988). A note on Rudolf Snellius and the early history of mathematics in Leiden. In C. Hay (Ed.), *Mathematics from manuscript to print 1300–1600* (pp. 156–161). Oxford, United Kingdom: Clarendon Press.

Van Egmond, W. (1976). *The commercial revolution and the beginnings of Western mathematics in Renaissance Florence, 1300–1500.* PhD dissertation, Indiana University.

Van Egmond, W. (1980). *Practical mathematics in the Italian Renaissance: A catalog of Italian abbacus manuscripts and printed books to 1600.* Firenze, Italy: Istituto E Museo di Storia Della Scienza.

Van Natter, F. M. (1963). *Lincoln's boyhood: A chronicle of his Indiana years.* Washington, DC: Public Affairs Press.

Van Sickle, J. (2011). *A history of trigonometry education in the United States: 1776–1900.* PhD dissertation, Columbia University.

Venema, P. (1730). *Arithmetica of Cyffer-Konst, volgens de Munten Maten en Gewigten, te Nieu-York, gebruykelyk als mede een kort Ontwerp van de Algebra.* New York, NY: Jacob Goelet.

Vince, S. (1810). *A treatise on plane and spherical trigonometry; with an introduction, explaining the use of logarithms adapted to the use of students in philosophy.* Cambridge, United Kingdom: J. Smith.

Walkingame, F. (1785). *The tutor's assistant being a compendium of arithmetic and a complete question book* (21st ed.). London, United Kingdom: J. Scratcherd & I. Whitaker.

Walkingame, F. (1802). *The tutor's assistant, being a compendium of arithmetic, and a complete question-book.* London, United Kingdom: C. Whittingham.

Walsh, M. (1820). *A new system of mercantile arithmetic: Adapted to the commerce of the United States, in its domestic and foreign relations; with forms of accounts, and other writings usually occurring in trade* (4th ed.). Salem, MA: John D. Cushing.

Wallis, P. J. (1963). An early best seller: Francis Walkingame's "The Tutor's Assistant." *The Mathematical Gazette, 47*(361), 199–208.

Wardhaugh, B. (2012). *Poor Robin's prophecies: A curious almanac, and the everyday mathematics of Georgian Britain.* Oxford, United Kingdom: Oxford University Press.

Warren, L. A. (1931). Lincoln's Hoosier schoolmasters. *Indiana Magazine of History, 27*(2), 104–118.

Warren, L. A. (1959). *Lincoln's youth: Indiana years seven to twenty-one 1816–1830.* New York, NY: Appleton.

Waters, D. W. (1958). *The art of navigation in England in Elizabethan and early Stuart times*. New Haven, CT: Yale University Press.

Watson, F. (1902). The curriculum and text-books of English schools in the first half of the 17th century. *Transactions of the Bibliographical Society VI* (Oct 1900–March 1902) (pp. 159–179). London, United Kingdom: Blades, East & Blades.

Watson, F., & Kandel, I. L. (1911). Examinations. In P. Monroe (Ed.), *A cyclopaedia of education* (Vol. 2, pp. 532–538). New York, NY: The Macmillan Company.

Westbury, I. (1980). Change and stability in the curriculum: An overview of the questions. In H. G. Steiner (Ed.), *Comparative studies of mathematics curricula: Change and stability 1960–1980* (pp. 12–36). Bielefeld, Germany: Institut für Didaktik der Mathematik-Universität Bielefeld.

Wickersham, J. P. (1886). *A history of education in Pennsylvania.* Lancaster, PA: Inquirer Publishing Company.

Willetts, J. (1822). *The scholar's arithmetic designed for the use of schools in the United States.* Poughkeepsie, NY: Paraclete Potter.

Wilson, D. K. (1935). *The history of mathematical teaching in Scotland to the end of the eighteenth century*. London, United Kingdom: University of London Press.

Wilson, D. L., Davis, R. O., & Wilson, T. (Eds.). (1998). *Herndon's informants: Letters, interviews and statements about Abraham Lincoln.* Urbana, IL: University of Illinois Press.

Wilson, J. I. (1820). *The history of Christ's Hospital from its foundation by King Edward VI.* London, United Kingdom: Author.

Wilson, W. E. (1940). *The Wabash.* New York, NY: Farrar & Rinehart.

Wingate, E. (1624). '*L'usage de la règle de proportion en arithmétique*. Paris, France: Author.

Wingate, E. (1630). *Arithmetique made easie.* London, United Kingdom: Stephens and Meredith.

Wingate, E. (1671). *The clark's tutor for arithmetick and writing, or, a plain and easie way of arithmetick.* London, United Kingdom: Stephens and Meredith.

Workman, B. (1788). *A treatise on arithmetic in theory and practice.* Philadelphia, PA: J. McCulloch.

Yeats, W. B. (1928). *Coinage of Saorstát Éireann.* Dublin, Ireland. The Stationery Office.

Yeldham, F. A. (1926). *The story of reckoning in the Middle Ages.* London, United Kingdom: George A. Harrap.

Yeldham, F. A. (1936). *The teaching of arithmetic through four hundred years (1535–1935).* London, United Kingdom: Harrap.

Ziegler, P. (1973). *King William IV.* London, United Kingdom: Harper & Row.

Author Biographies

Nerida F. Ellerton has been professor within the Mathematics Department at Illinois State University since 2002. She holds two doctoral degrees—one in Physical Chemistry and the other in Mathematics Education. Between 1997 and 2002 she was Dean of Education at the University of Southern Queensland, Australia. She has taught in schools and at four universities, and has also served as consultant in numerous countries, including Australia, Bangladesh, Brunei Darussalam, China, Malaysia, the Philippines, Thailand, the United States of America, and Vietnam. She has written or edited 14 books and has had more than 150 articles published in refereed journals or edited collections. Between 1993 and 1997 she was editor of the *Mathematics Education Research Journal*, and since 2011 she has been Associate Educator for the *Journal for Research in Mathematics Education*. In 2012 Springer published the 223-page *Rewriting the History of School Mathematics in North America 1607–1861*, which she jointly authored with M. A. (Ken) Clements, and although this present volume is a follow-up to that book, it also reflects her growing interest in Lincoln studies.

M. A. (Ken) Clements's masters and doctoral degrees were from the University of Melbourne, and at various times in his career he has taught, full-time, in primary and secondary schools, for a total of 15 years. He has taught in six universities, located in three nations, and is currently professor within the Mathematics Department at Illinois State University. He has served as a consultant and as a researcher in Australia, Brunei Darussalam, India, Malaysia, Papua New Guinea, South Africa, Thailand, the United Kingdom, the United States of America, and Vietnam. He served as co-editor of the three *International Handbooks of Mathematics Education*—published by Springer in 1996, 2003 and 2013—and, with Nerida Ellerton, co-authored a UNESCO book on mathematics education research. He has authored or edited 25 books and over 200 articles on mathematics education, and is honorary life member of both the Mathematical Association of Victoria and the Mathematics Education Research Group of Australasia.

Nerida F. Ellerton and M. A. (Ken) Clements, *Abraham Lincoln's Cyphering Book and Ten other Extraordinary Cyphering Books*,
DOI 10.1007/978-3-319-02502-5, © Springer International Publishing Switzerland 2014

Valeria Aguirre Holguín is a doctoral student in Mathematics Education at New Mexico State University. She holds Master's and Bachelor's degrees in Mathematics from the University of Texas at El Paso and the University of Juarez respectively. She taught undergraduate courses in mathematics and mathematics education for five years at three universities in Mexico and in the United States of America, and she has also taught middle-school Mathematics, Physics and Chemistry for two years in Mexico. She served as a research assistant for MathSnacks, a National Science Foundation grant that emphasizes the improvement of middle-school mathematics teaching, and for Mathematically Connected Communities, a project funded by the New Mexico Public Education Department. In Summer 2013 she was an invited speaker at the XVI Bolivian Mathematical Society Conference. Her research interests include undergraduate students' understandings of mathematical definitions. Her curiosity about Lincoln's cyphering book was kindled by a course project in the history of American mathematics education.

Author Index

Nerida F. Ellerton and M. A. (Ken) Clements, *Abraham Lincoln's Cyphering Book and Ten other Extraordinary Cyphering Books*,
DOI 10.1007/978-3-319-02502-5, © Springer International Publishing Switzerland 2014

Subject Index

DOI 10.1007/978-3-319-02502-5, © Springer International Publishing Switzerland 2014

Printed by Publishers' Graphics LLC